EXCITATION-CONTRACTION COUPLING IN SMOOTH MUSCLE

EXCITATION-CONTRACTION COUPLING IN SMOOTH MUSCLE

Proceedings of the International Symposium on Excitation-Contraction Coupling in Smooth Muscle held in Louvain, Belgium, 12-14 July, 1977 and the Erwin-Riesch Symposium in Heidelberg, Germany, 15-16 July, 1977.

Editors:

R. CASTEELS
T. GODFRAIND
J.C. RÜEGG

1977

ELSEVIER/NORTH-HOLLAND
AMSTERDAM · NEW YORK · OXFORD

ISBN 0-444-80017-4

Published by:
Elsevier/North-Holland Biomedical Press
335 Jan van Galenstraat, P.O. Box 211
Amsterdam, The Netherlands.

Sole distributors for the U.S.A. and Canada:
Elsevier/North-Holland Inc.
52 Vanderbilt Avenue
New York, N.Y. 10017.

PRINTED IN THE NETHERLANDS

PREFACE

The idea of organizing an international symposium on excitation-contraction coupling in smooth muscle was conceived independently in both Leuven and Heidelberg. Because the scientific approach in the two areas is different and complementary, it was decided that we should join in organizing a symposium consisting of two parts, a first dealing with the role of the cell membrane in excitation-contraction coupling and a second dealing with biochemical aspects of this phenomenon. This symposium has had the major advantage of bringing together scientists working in different disciplines who otherwise would not easily meet. By this common meeting the different groups have become aware of the fact that excitation-contraction coupling is not regulated solely either at the membrane or at the contractile proteins. The conference has provided not only the occasion for colleagues with opposing ideas to present and discuss their work, but also the opportunity for all participants to view areas of controversy in the field of smooth muscle. Discussions have led to the undeniable conclusion that the term "smooth muscle" describes a group of muscles which are extremely varied and complex in function. Hence we have become aware of the dangers inherent in a too enthusiastic interpretation of fragmentary data or in extrapolating findings from one animal or one tissue to another. Better insight into fine regulatory systems of these muscles is going to require new approaches with different techniques. It is felt that progress will be enhanced by regular contact and a degree of cooperation among investigators in the different disciplines of smooth muscle physiology, pharmacology and biochemistry. The editors hope that the freedom left to the authors in expressing their own views will allow the reader to share the above statements.

The papers presented in the first part of the symposium held in Leuven concern the transfer of the excitation from the cell membrane to the contractile apparatus. All topics actually lead to a discussion of factors which can modify the concentration of intracellular ionized calcium.
In the morphological section there is an agreement about the ultrastructural organization of the contractile proteins. However, the morphological localization of calcium inside the smooth muscle cells remains controversial. Here we can expect important contributions from using the electron-probe technique. The diversity of smooth muscles was clearly demonstrated by the differences of their electrical response and by the varying contribution of external and cellular calcium to excitation-contraction coupling. There is as yet no unifying classification of the slow potential changes and this issue is further complicated by a varying dependence on external ions. The use of voltage clamp and flux studies in smooth muscle still presents many practical and theoretical difficulties. This is not

unexpected in view of the morphological heterogeneity of smooth muscle tissues. Perhaps the use of single isolated smooth muscle cells might help. The interaction of Na and Ca at the cell membrane and the existence and physiological role of a Na-Ca extrusion mechanism or an ATP-dependent Ca pump remain obscure and controversial. It has been brought forward that in studying subcellular structures we have to be aware of the fact that the origin of the microsomes has not yet been elucidated. Most contributors seem to agree that Ca-uptake by the microsomal fraction could represent a physiologically important mechanism, while the uptake by mitochondria would only intervene during extreme or unphysiological conditions. No data on regulation of the release of Ca by any of these structures has been brought forward. Also the study of Ca-exchange in smooth muscle cells remains controversial. The use of lanthanum has been introduced in order to delineate different Ca-fractions in smooth muscle. The advantages of this so called lanthanum method are, however, limited by the fact that although La largely blocks the Ca-uptake, it is less efficient in reducing the Ca efflux. In addition the effect of La on Ca bound to the extracellular matrix seems to vary according to the smooth muscle tissue. In spite of these deficiencies La can be a useful tool in estimating changes of biologically important calcium fractions. However, this method can give no information on changes of the cytoplasmic ionized calcium, nor on the kinetics of the transfer of cellular Ca across the membrane. The study of these two important parameters remains a challenge.

The second part of this book is an account of papers given at the ERWIN-RIESCH SYMPOSIUM on Biochemical Aspects of Excitation-Contraction Coupling. This meeting was held in Heidelberg. These papers are concerned with the action of Ca^{2+} ions and other intracellular messengers as well as the regulation of actomyosin ATPase, energy turnover and mechanical output of the smooth muscle cell. It is now well established that smooth muscle contraction entails a sliding of actin filaments and myosin filaments which is triggered by calcium ions. Structural considerations presented and biochemical evidence make it clear that the troponin-tropomyosin-system may play an important role in the mechanism which senses and responds to the intracellular calcium ion concentration. On the other hand, one of the light peptide chains attached to myosin will also be shown to have the properties of a "calcium switch." As yet it is still undecided whether calcium acts on the regulatory light chain directly, or whether it does so more indirectly by inducing the phosphorylation of the regulatory light chain (via activation of a specific light chain kinase). In view of this controversy, a synthesis incorporating the various mechanisms proposed in a complex scheme will be attempted. The possibility of a modulation of the response to calcium ions by phosphorylation of "regulatory proteins" must be borne in mind. However, at present, the evidence for involvement of cyclic nucleotides in these phosphorylation processes is still fragmentary and the role of cyclic nucleotides in regulation of smooth muscle tonus seems quite uncertain. In any case when a muscle contracts as a result of activation, myosin cross bridges interact with actin and this leads to increased energy turnover and mechanical output, i.e. force, power and stiffness. It is still a matter of debate whether mechanical stiffness reflects cross bridge properties and hence the extent of actin-myosin interaction in the case of smooth muscle. Further experimental work is clearly required to find an explanation for the high mechanical output obtained with a low myosin content and at a low rate of energy expenditure.

In summary, we hope that this book will throw light on areas of agreement and disagreement and that in so doing it will stimulate critical experiments to clarify our ideas on excitation-contraction coupling in smooth muscle.

R. Casteels (K.U. Leuven)
T. Godfraind (U.C. Louvain)
J.C. Rüegg (Universität Heidelberg)

ACKNOWLEDGEMENTS

The organizers acknowledge for the Leuven meeting (part I) the generous financial support by the NFWO-FNRS, by the Derde Cyclus Programma and the Faculty of Medicine of K.U. Leuven, and by the Belgian Ministry of National Education.

As for the Heidelberg meeting (part II), the ERWIN-RIESCH-STIFTUNG, Lorch/Württemberg, has made a very generous grant which is gratefully acknowledged.

We also acknowledge the substantial support of Dr. H. Schroeder and N. Frommann of the company Dr. KARL THOMAE, Biberach/Riss, in organizing the conference as well as for their generous hospitality.

In addition we thank the Kultusministerium of Baden-Württemberg, the Deutsche Forschungsgemeinschaft, the Medizinisch Pharmazeutische Studiengesellschaft e.V and the following companies : CIBA-GEIGY AG, FARBWERKE HOECHST AG, HOFFMANN-LA ROCHE AG, PHARMA SCHWARZ GMBH, SANDOZ AG, Dr. THIEMANN GMBH, also CARL ZEISS for contributions towards the costs of the Symposium.

CONTENTS

smooth muscle tonus seems quite uncertain. In any case when a muscle contracts as a result of activation, myosin cross bridges interact with actin and this leads to increased energy turnover and mechanical output, i.e. force, power and stiffness. It is still a matter of debate whether mechanical stiffness reflects cross bridge properties and hence the extent of actin-myosin interaction in the case of smooth muscle. Further experimental work is clearly required to find an explanation for the high mechanical output obtained with a low myosin content and at a low rate of energy expenditure.

In summary, we hope that this book will throw light on areas of agreement and disagreement and that in so doing it will stimulate critical experiments to clarify our ideas on excitation-contraction coupling in smooth muscle.

R. Casteels (K.U. Leuven)
T. Godfraind (U.C. Louvain)
J.C. Rüegg (Universität Heidelberg)

ACKNOWLEDGEMENTS

The organizers acknowledge for the Leuven meeting (part I) the generous financial support by the NFWO-FNRS, by the Derde Cyclus Programma and the Faculty of Medicine of K.U. Leuven, and by the Belgian Ministry of National Education.

As for the Heidelberg meeting (part II), the ERWIN-RIESCH-STIFTUNG, Lorch/Württemberg, has made a very generous grant which is gratefully acknowledged.

We also acknowledge the substantial support of Dr. H. Schroeder and N. Frommann of the company Dr. KARL THOMAE, Biberach/Riss, in organizing the conference as well as for their generous hospitality.

In addition we thank the Kultusministerium of Baden-Württemberg, the Deutsche Forschungsgemeinschaft, the Medizinisch Pharmazeutische Studiengesellschaft e.V., and the following companies : CIBA-GEIGY AG, FARBWERKE HOECHST AG, HOFFMANN-LA ROCHE AG, PHARMA SCHWARZ GMBH, SANDOZ AG, Dr. THIEMANN GMBH, also CARL ZEISS for contributions towards the costs of the Symposium.

Part I : Louvain, 12-14 July, 1977

MEMBRANE PHENOMENA AND EXCITATION-CONTRACTION COUPLING

Excitation-Contraction Coupling in Smooth Muscle, R. Casteels et al. eds.

A MORPHOLOGICAL STUDY OF THE MECHANICAL COUPLING BETWEEN SMOOTH MUSCLE CELLS

Giorgio Gabella
Department of Anatomy, University College London

Introduction. Force is generated in smooth muscle by the so-called contractile machinery, i.e. the sets of myofilaments. In this communication I address myself to the following problem: How are the forces generated by the individual and minute smooth muscle cells added up to each other to produce the "macroscopic" pull of the muscle? This problem has been investigated by light and electron microscopy, and by serial section analysis, on the taenia coli of the guinea-pig. The results are reported *in extenso* elsewhere[1]. The taenia coli was chosen because, being a discrete and conspicuous strip of longitudinal musculature, tension development along its length and shortening are the most important aspects of its mechanical behaviour.

In view of the minute size of smooth muscle cells (just over 1/300th of a cubic millimetre in volume), junctions between cells (mechanical coupling) and connections between cells and extracellular material (collagen fibrils in particular) may be expected to play a major role in the development of contraction. During isotonic contraction the transverse sectional area of the taenia increases in proportion to the amount of shortening and the approximative packing density of the muscle cells is reduced from 94,000 mm^{-2} (in taeniae fixed during a mild stretch) to 18,000 mm^{-2} (in taeniae fixed during isotonic contraction against 1 g load). However, the total number of muscle cells present in one complete transverse section of taenia changes only very little, suggesting that shortening and elongation of this muscle do not involve to any great extent sliding of the muscle cells past each other.

Size of muscle cells. The smooth muscle cells of the taenia coli of adult guinea-pigs have a volume of about 3,500 μm^3; when fixed in condition of mild stretch they measure about 500 µm in length. The cell surface of one cell is about 5,300 μm^2, and it is increased by about 73% by the presence of some 170,000 caveolae (ref. 2).

Dense bands. Between the caveolae, which are arranged in rows parallel to the cell length[3,4], there are regions where the inner aspect of the cell membrane is incrustated by electron-dense material (Fig. 9); these areas are known as dense patches, but I will refer to them as dense bands since they are stripes whose length can exceed 2 microns and which in mildly stretched muscles are approximately parallel to the cell length. In transversely sectioned muscle cells the dense bands appear as patches, sometimes regularly alternating along the cell profile with areas rich in caveolae and sarcoplasmic reticulum. The amount of the cell profile occupied by dense bands is about 50% (range 40-60%) at the cell's levels containing the nucleus; in the tapering parts of the cell, towards either end, the percentage of the surface occupied by dense bands increases and may reach 100% in some of the smallest profiles. Since the dense bands are attachment sites for the thin filaments [5,6,7], it appears that a substantial number of these are anchored to the cell membrane (or, so to speak, to the lateral walls of the cell) along its entire length. As already suggested by Pease & Molinari[5] much of the force generated by the myofilaments is directly applied to the cell membrane. Therefore, the number of contractile units (whatever is indicated with this term) working in parallel is greater than if they were arranged so as to develop tension mainly between the two ends of a cell (as is the case in most skeletal muscle fibres).

Basal lamina. A basal lamina is present over the entire surface of smooth muscle cells. In tangential sections the basal lamina shows an ill-defined meshwork of microfibrillar material, which is connected to other microfibrils in the intercellular space and to collagen fibrils. The basal lamina often appears more electron-dense in correspondence of the dense bands, and in the cell-to-cell anchoring apparatuses (see below) it becomes considerably thicker.

Collagen fibrils. Intramuscular collagen fibrils measure 30-35 nm in diameter, and show the usual cross-striation with a 64 nm period. In mildly stretched muscles most of the collagen fibrils run close to the direction of the muscle cells. On the other hand, in isotonically contracted muscles the collagen fibrils run almost orthogonal to the cell's longitudinal axis and appear wound in spirals of small

Fig. 1. Terminal part of a muscle cell with a convoluted profile. A prominent basal lamina, merging with that of the neighbouring muscle cells, is visible. Marker: 1 micron.
Fig. 2. Terminal part of a muscle cell which splits into three processes. Marker: 1 micron.
Fig. 3. End part of a muscle cell with longitudinal invaginations of the cell membrane and a conspicuous basal lamina. Marker: 1 micron.
Fig. 4. A complex muscle cell profile which serial sections revealed to be formed by two cells anchored to each other end-to-end by means of interdigitating processes. The basal lamina has a characteristic appearance. Marker: 1 micron.

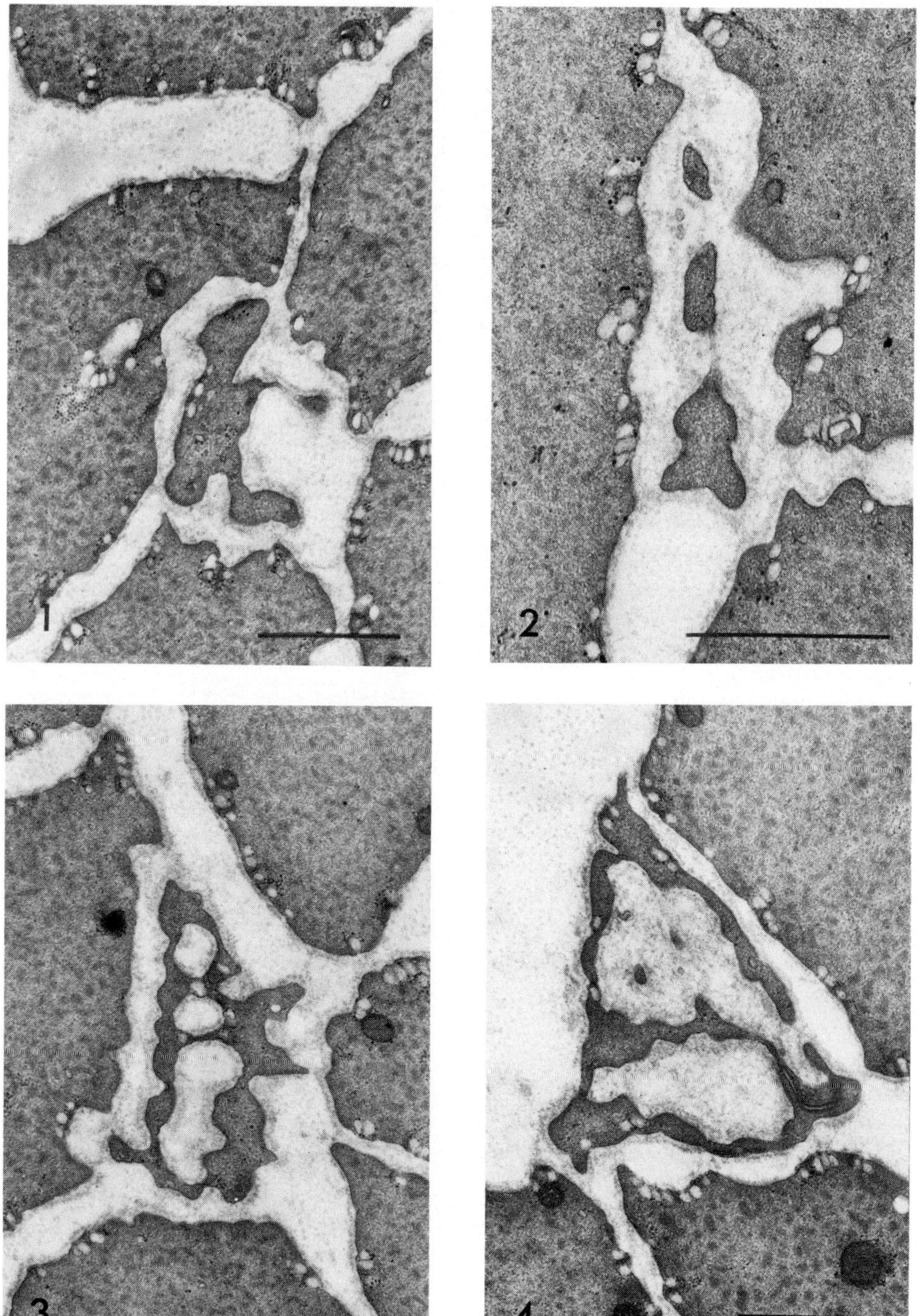
1
2
3
4

pitch around the muscle cells. How this extensive re-arrangement of the collagen fibrils takes place during contraction is not yet clear. The oblique arrangement of collagen fibrils and dense bands in isotonically contracted muscles suggests that a component of torsion operates during cell shortening. The orientation of myofilaments, on the other hand, remains close to the cell's longitudinal axis in contracted and stretched cells (Fig. 11).

Junctions of muscle cells to collagen fibrils. Collagen fibrils approach the basal lamina but it is not clear whether junctions of mechanical significance are formed; that they may be formed is suggested by the observation that during isotonic contraction the surface of the muscle cells becomes extremely convoluted with deep grooves and high folds, but collagen fibrils remain closely associated with the areas of the cell surface bearing the dense bands. The structures covered under the vague term of sarcolemma (plasma membrane, basal lamina, and the nearest collagen fibrils and microfibrils) are likely to play a major role in the transmission of tension and shortening. At present, however, the mechanical properties of the smooth muscle sarcolemma and its chemical composition are poorly understood.

Muscle cells' ends. Some muscle cells show near their ends finger-like invaginations of the cell membrane, which are parallel to the cell's longitudinal axis and are lined by a basal lamina (Fig. 3). The invaginations may contain collagen fibrils but more often they show amorphous material of medium electron-density. These structures are similar to those found at the muscle-tendon junctions in skeletal fibres[8,9], and they may serve a similar function, i.e. to anchor the skeletal fibre's end to the tendon or the smooth muscle cell's end to the intramuscular collagen framework.

Other muscle cells acquire near their ends a convoluted profile, with laminar and finger-like projections (Figs. 1, 2). This produces a relative increase of the cell surface. The cell membrane is covered by a thickened basal lamina, or material of similar electron-density to the basal lamina forms a thick layer around these profiles and appears in continuity with the basal lamina of neighbouring cells, towards which some of the cell processes extend.

Fig. 5. Representative micrographs from serial sections illustrating an end-to-side anchoring apparatus between two muscle cells. The level, expressed as the distance in microns from the beginning of the series, is: A: 11 µm; B: 13 µm; C: 16 µm; D: 25 µm. Marker: 1 micron.

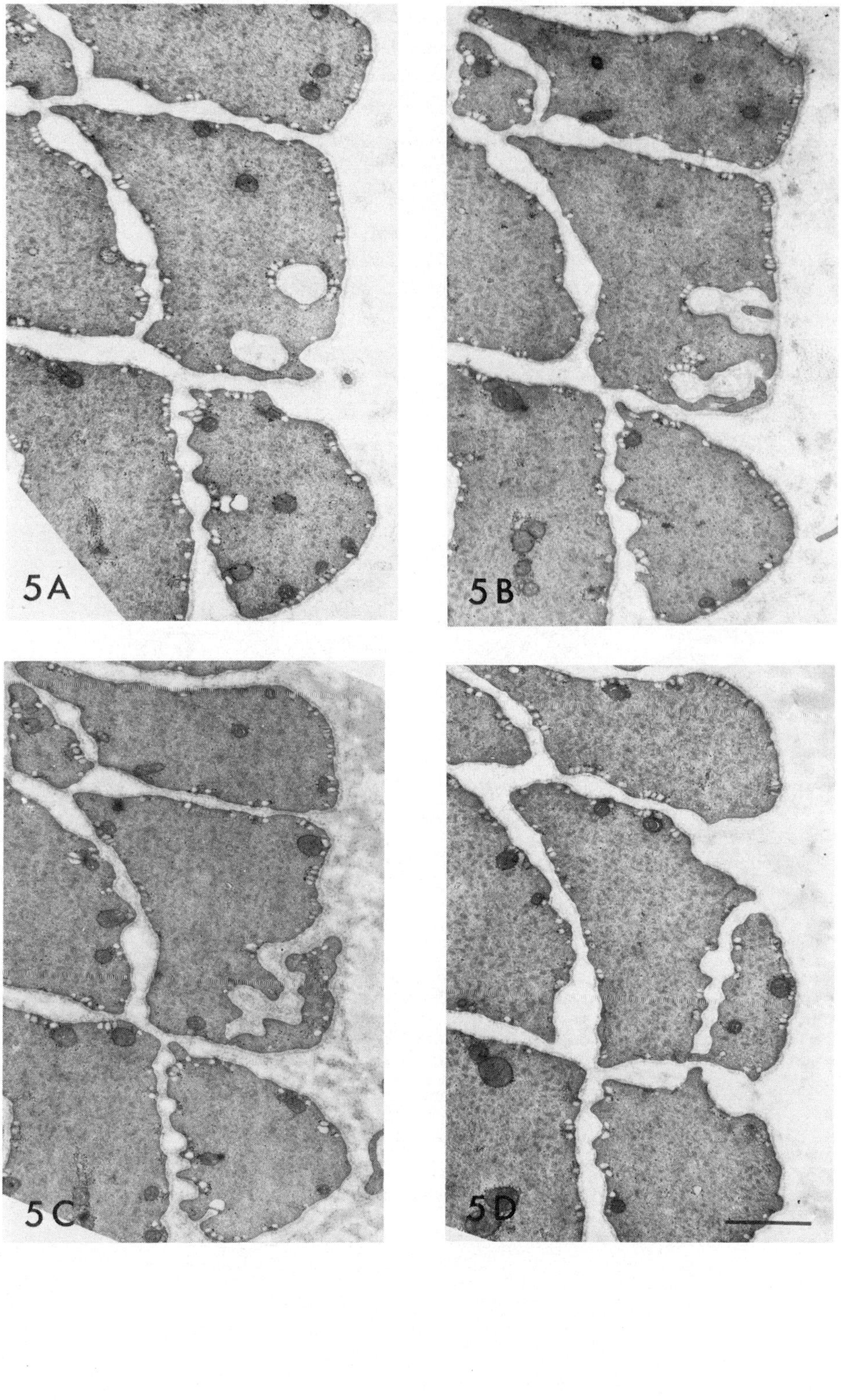
5A
5B
5C
5D

Cell-to-cell junctions. Well known cell-to-cell junctions are the intermediate contacts or desmosome-like junctions[10] (Fig. 9). In addition there are more elaborate structural specializations which link two cells together either end-to-end or side-to-side or end-to-side; they are best studied on serial sections through several microns of the cell's length (Fig. 5). They usually involve two muscle cells, but junctions of three or four cells were also found. These anchoring apparatuses are characterized by an increase of the surface of both cells by means of folds, invaginations and projections. Particularly elaborate are the end-to-end anchoring apparatuses, where laminar and finger-like processes from two cells interdigitate producing sectional profiles of great complexity (Figs. 4, 6). At

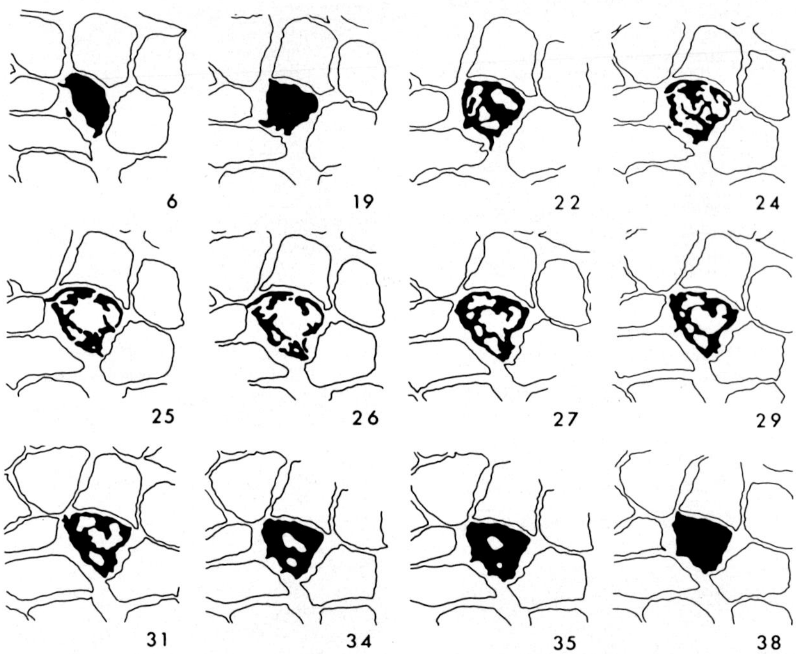

Fig. 6. Drawings from electron-micrographs of serial transverse sections. The number below each section is the distance in microns from the beginning of the series. Two muscle cells (shadowed) - which appear as plain profiles at 6 and 38, respectively - join each other end-to-end by means of complex laminar processes.

Fig. 7. A side-to-side anchoring apparatus between two muscle cells. A third muscle cell (bottom left) is also partly involved in this junction. Marker: 1 micron.
Fig. 8. An anchoring apparatus between two muscle cells as viewed in longitudinal section. Marker: 1 micron.

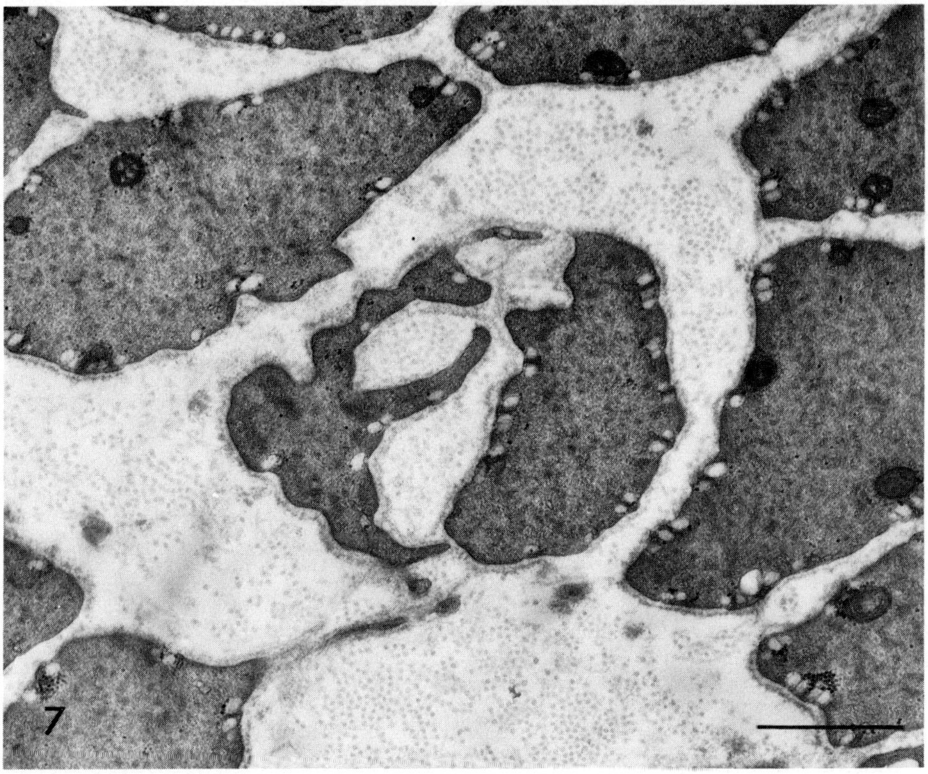
7

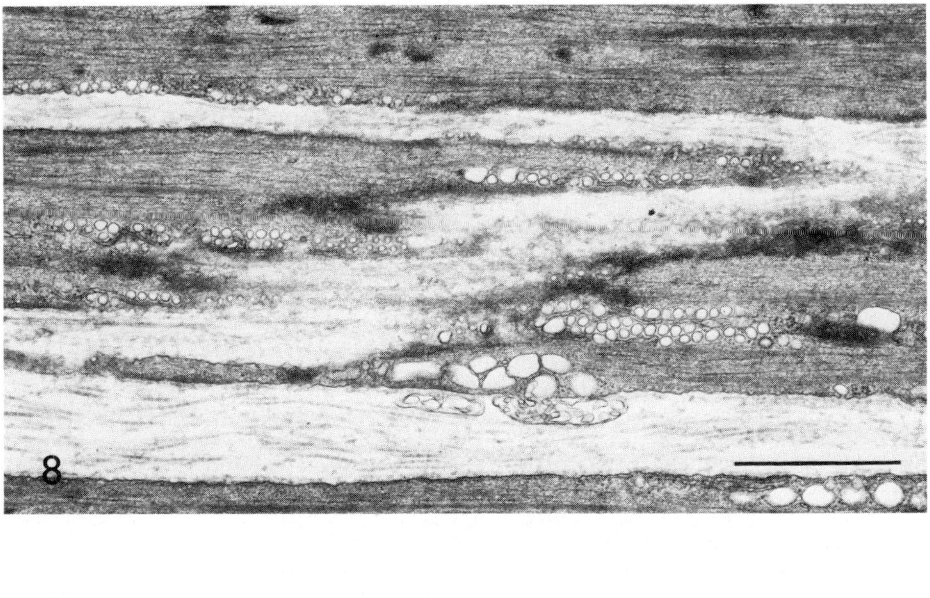
8

the level of these apparatuses the basal lamina is thicker than over the rest of the cell (Fig. 7); amorphous material of similar electron-density often fills gaps of hundreds of nanometres between the two cells. With the help of reconstructions based on serial transverse sections, anchoring apparatuses can also be recognized in longitudinal sections (Fig. 8). These various junctions are found in stretched and in contracted muscles, but how and to what extent they change their shapes during contraction is not yet clear.

Connective tissue septa. Collagen septa, clearly visible between groups of muscle cells in transversely sectioned taeniae, have been regarded as structures subdividing the muscle into anatomical bundles which would correspond to the bundles identified electrophysiologically. This notion is not supported by the analysis of serial sections, which shows that the groups of cells merge or divide into further groups within hundreds or tens of microns. Each septum extends for only a few hundred microns along the length of the taenia. At all muscle lengths the septa are approximately parallel to each other. In mildly stretched taeniae the axes of the muscle cells form an angle of few degrees with the longitudinal axis of the septa. On the other hand, in isotonically contracted taeniae the angle can be as large as 20-25°; in longitudinal sections cut tangential to the serosal surface it is apparent that the orientation of the muscle cells varies in adjacent groups of cells producing a kind of herring-bone pattern, with the muscle cells directed from one septum to another (Fig. 10). This arrangement is reminiscent of a multipennate skeletal muscle. It is suggested that the connective tissue septa should be regarded as a sort of diffuse, intramuscular tendon, and that they have the important function of bearing and transmitting the force generated by the muscle cells. This idea is supported by the observation that the collagen content (measured as hydroxyproline concentration in micrograms per milligrams of wet weight) of the taenia coli is 4.24 $\pm$ 0.19 (and in the guinea-pig myometrium is 4.81 $\pm$ 0.75), and that of sartorius is 1.25 $\pm$ 0.02. The present tentative interpretation is that each collagen septum gives attachment

---

Fig. 9. Muscle cells in transverse section showing thin, thick and 10 nm filaments, caveolae, dense bands and intermediate contacts. Marker: 0.5 microns.
Fig. 10. Tangential longitudinal section of an isotonically contracted taenia (against a load of 1 g). The angle between muscle cells and septa is over 20°. Hematoxilin and eosin. Marker: 100 microns.
Fig. 11. Longitudinal section of an isotonically contracted taenia. Myriad folds appear at the cell surface but the filaments run virtually parallel to the cell's length. Marker: 1 micron.

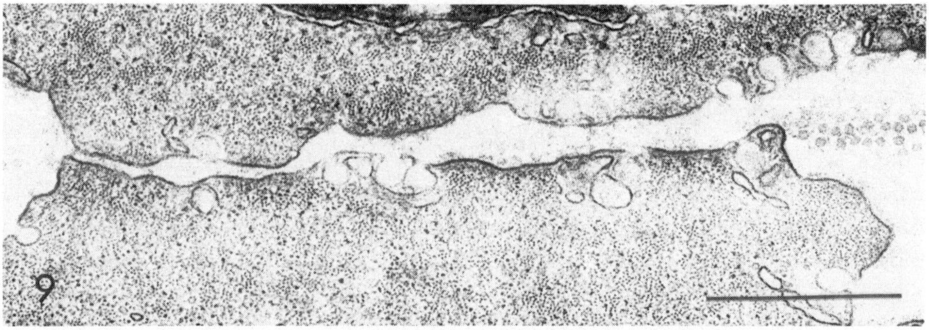

9

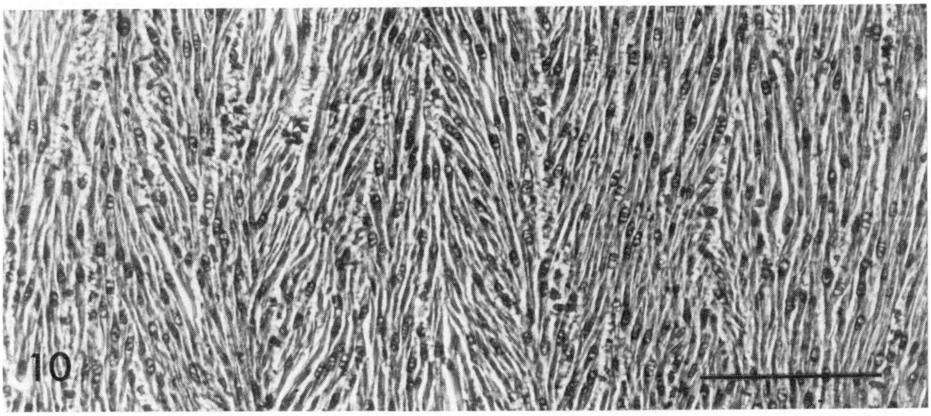

10

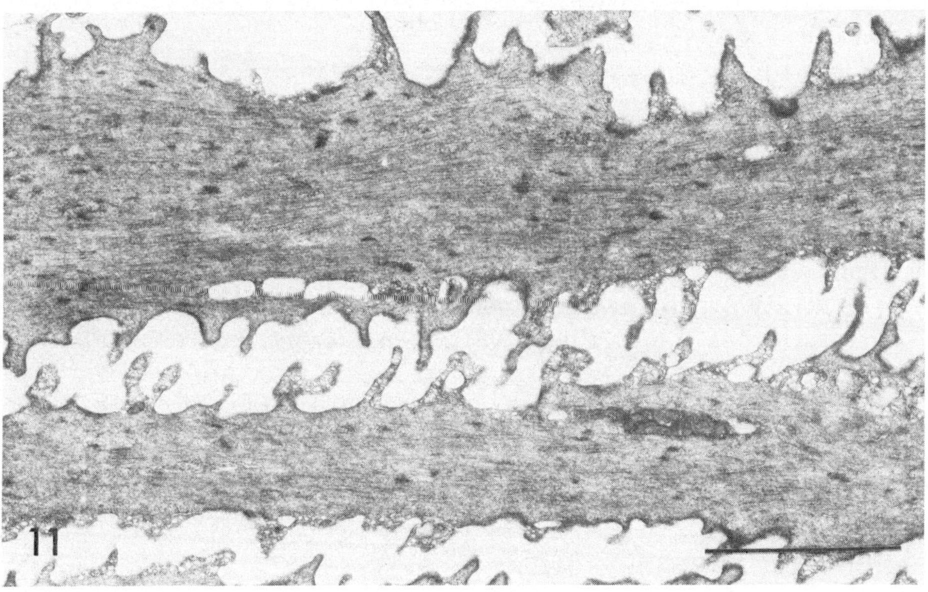

11

to several arrays of muscle cells pulling in opposite directions.

Arrangement of filaments. Shortening of a smooth muscle involves complex changes in shape and orientation of the muscle cells and in the arrangement of the intercellular material. However, the myofilaments maintain, in contracted and in relaxed muscles, an orientation close to the cell's longitudinal axis. They do not penetrate in most of the long folds which appear at the surface of the contracted muscle cells and they seem to remain as parallel to each other as they are in stretched muscles. On the other hand, a wide re-orientation of the intermediate, 10 nm filaments takes place during muscle shortening. Individual cells or small groups of cells in contracted muscles may display a spiral arrangement of filaments, as has been described in isolated muscle cells obtained from the toad stomach[11]. In taenia coli these cells are interpreted as having shortened by the active contraction of the neighbouring cells.

Acknowledgement. This work is supported by grants from the M.R.C. I thank Mr Bill Ogle for assistance with photography.

References.

1. Gabella, G. (1977) Cell Tiss. Res. (In the Press).
2. Gabella, G. (1976) Cell Tiss. Res. 170, 161-186.
3. Devine, C.E., Simpson, F.O. & Bertaud, W.S. (1971) J. Cell Sci. 8, 427-443.
4. Watanabe, H. & Yamamoto, T.Y. (1974) J. Anat. (Lond.) 117, 553-564.
5. Pease, D.C. & Molinari, S. (1960). J. Ultrastr. Res. 3, 447-468.
6. Prosser, C.L., Burnstock, G. & Kahn, J. (1960). Am. J. Physiol. 199, 545-552.
7. Rosenbluth, J. (1965) Science (N.Y.) 148, 1337-1339.
8. Gelber, D., Moore, D.H. & Ruska, H. (1960) Z. Zellforsch. 52, 396-400.
9. Ishikawa, H. (1965) Arch. histol. jap. 25, 275-296.
10. Henderson, R.M. (1975) In: Methods in Pharmacology, edited by E.E. Daniel & D.M. Paton, vol. 3, pp. 47-77. New York: Plenum Press.
11. Fisher, B.A. & Bagby, R.M. (1977) Am. J. Physiol. 232, C5-C14.

Excitation-Contraction Coupling in Smooth Muscle, R. Casteels et al. eds.

CYTOCHEMICAL STUDY OF THE INTRACELLULAR CALCIUM DISTRIBUTION IN SMOOTH MUSCLE

Surface Microvesicles and Cellular Calcium Homeostasis

L. M. Popescu
Department of Cytology and Histology
Faculty of Medicine, Bucharest
Rumania

INTRODUCTION ˣ

The essential role of Ca^{2+} in the regulation of the contraction-relaxation cycle of smooth muscle has been reviewed recently[1-4]. It has become clear from these studies that the *in situ* demonstration of the subcellular location of Ca^{2+} in smooth muscle is a subject of considerable interest. However, the direct electron microscopic visualization of Ca^{2+} is not possible since the size of this ion lies below the resolution limit of any electron microscope and, therefore, the use of cytochemical markers is required. In our laboratory, potassium oxalate has been used as specific cytochemical marker for Ca^{2+}. The principle of K-oxalate method[5] consists in the *in situ* Ca^{2+} immobilization, by precipitation with oxalate, as electron-opaque deposits of calcium oxalate. These deposits can be easily seen in the electron microscope, and their calcium content can be checked by X-ray analytical electron microscopy, which has the unique advantage to allow a direct chemical analysis of very small (< 200 Å) and well-defined areas from cells while they are being observed microscopically.

K-OXALATE METHOD : A TOOL FOR IN SITU CALCIUM LOCALIZATION

The validity of every ultracytochemical trapping reaction showing the location and the relative concentration of calcium *in situ* must be judged according to the following minimal criteria: (a) the possibility of the marker reagent to penetrate, before fixation, through the membranes of living cells; (b) its ability to precipitate Ca^{2+} promptly, before the ions can move in the tissue, and (c) the property of the precipitate not to diffuse during the subsequent preparative procedures for electron microscopy. As shown below, K-oxalate method fulfils these requirements.

Penetration of the oxalate anion through membranes: Since there is the following series of anion permeation through the biological membranes (Höber series) :

ˣ Abbreviations : SR, sarcoplasmic reticulum; SmV, surface microvesicles (synonyms: surface vesicles, micropinocytotic vesicles, subsarcolemmal vesicles, caveolae intracellulares); SDS, sodium dodecyl sulphate. All calibration lines are 0.1 μm unless stated otherwise.

$Cl^- > Br^- > I^- > HCOO^- >$ acetate $> NO_3^- > SO_4^{2-} > PO_4^{3-} >$ tartrate $>$ citrate $>$ oxalate $> F^-$, the polarized resting membrane of living smooth muscle cells is virtually impermeable to the oxalate anion. This could explain the repeated failures to localize Ca^{2+} by oxalate treatment of cells with intact membranes. In our laboratory[6,7], to introduce the oxalate anion into intact living cells, smooth muscle strips equilibrated in Krebs solution were incubated in a high-K medium containing oxalate, thus inducing depolarization, which minimized the role of the cell membrane as a a barrier to oxalate anion.

Chemical nature of the electron-opaque precipitates found in muscle cells after K-oxalate treatment: A series of chemical and physical tests (chelation with EGTA[8], electron probe microanalysis[9], X-ray and electron diffraction[10]) indicated that the electron-opaque deposits found in muscle cells after K-oxalate treatment consisted primarily of calcium oxalate. However, X-ray analytical electron microscopy[7, 11] showed that osmium was also present in the electron-opaque precipitates, thus raising the question whether osmium treatment is a prerequisite for the formation of precipitates (artefact of liquid-osmium fixation?) or whether osmium only "stains" the precipitates existing before fixation. Recent works in our laboratory has provided direct evidence for the formation of precipitates before osmium fixation. Less electron-dense precipitates, having the same localization, were found in specimens that were not post-fixed in osmium tetroxide[12].

SUBCELLULAR CALCIUM COMPARTMENTS REVEALED BY K-OXALATE METHOD

Four main Ca^{2+}-accumulating compartments are revealed by the K-oxalate method in smooth muscle cells of the guinea-pig taenia coli: SR, mitochondria, nucleus and SmV[6,7] (Figs. 1-9). Similar Ca^{2+} distribution was found in vascular smooth muscle with the same K-oxalate method[13]. A quantitative "map" of the intracellular Ca^{2+} distribution in one smooth muscle cell was established by correlating the cytochemical findings with morphometric estimation of the volume occupied by each subcellular component and with micro-chemical measurements of Ca^{2+} concentration in isolated subcellular fractions[6]. Assuming that the amount of Ca^{2+} translocated in one smooth muscle cell of the guinea-pig taenia coli during a contraction-relaxation cycle is about 10^{-5} picomoles[14], this amount would represent less than 2% of the nuclear Ca^{2+}, only 5% of the mitochondrial Ca^{2+} or about 20% of the Ca^{2+} stored in the SR. However, Ca^{2+} accumulation in sufficient abundance by various subcellular compartments does not by itself indicate their role in cellular Ca^{2+} homeostasis.

Sarcoplasmic reticulum: The conclusive evidence for Ca^{2+} accumulation inside the SR of intact smooth muscle cells in condition of relaxation, subsequent to a K^+-induced contraction[7] (Figs.1-4) shows that the smooth muscle SR participate in the regulation of intra-cellular free-Ca^{2+} levels during the contraction-relaxation cycle,

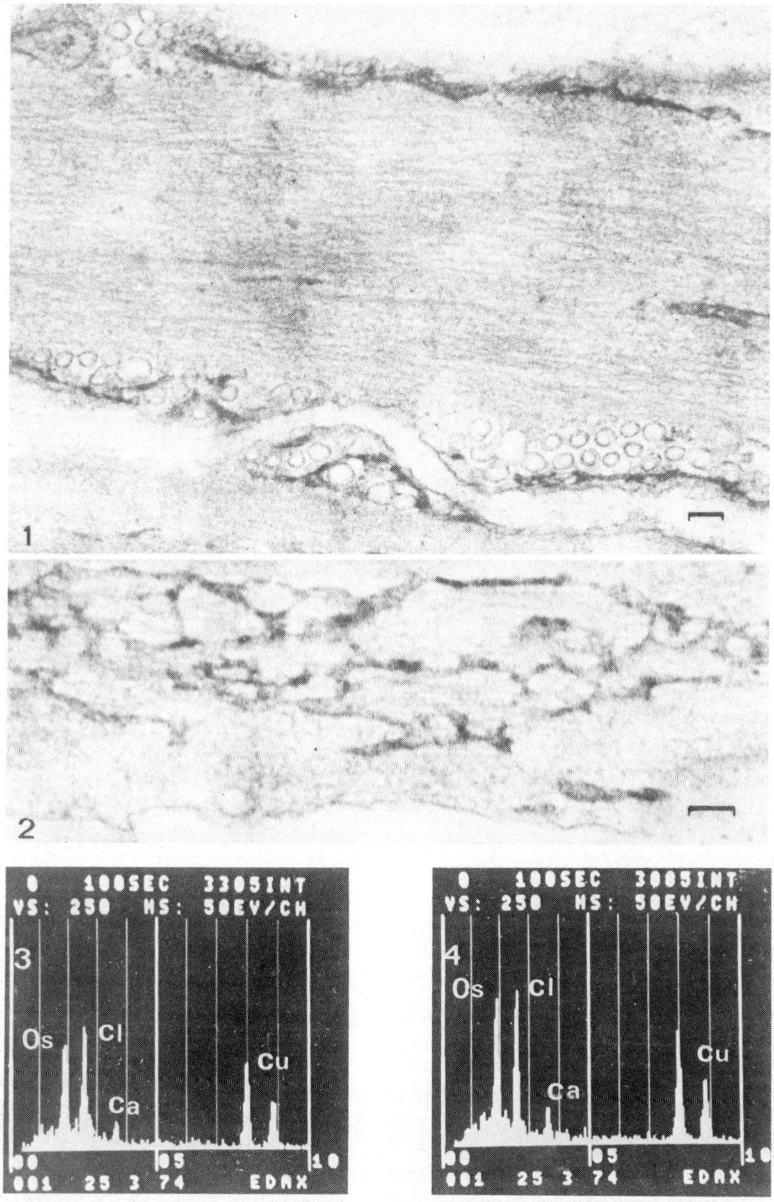

Figs. 1-4 : Ca^{2+} localization in SR. Guinea-pig taenia coli incubated in high-K oxalate medium containing sucrose. Muscle fixed in condition of relaxation after a K-contracture. Note the peripheral and deep SR filled with electron-opaque deposits of calcium oxalate. The X-ray spectra prove calcium accumulation in the peripheral (Fig. 3) and deep (Fig. 4) SR. The other peaks derive from fixative (Os), Epon (Cl) and grid (Cu). (From L.M. Popescu and I. Diculescu, J. Cell Biol., 67 (1975) 911).

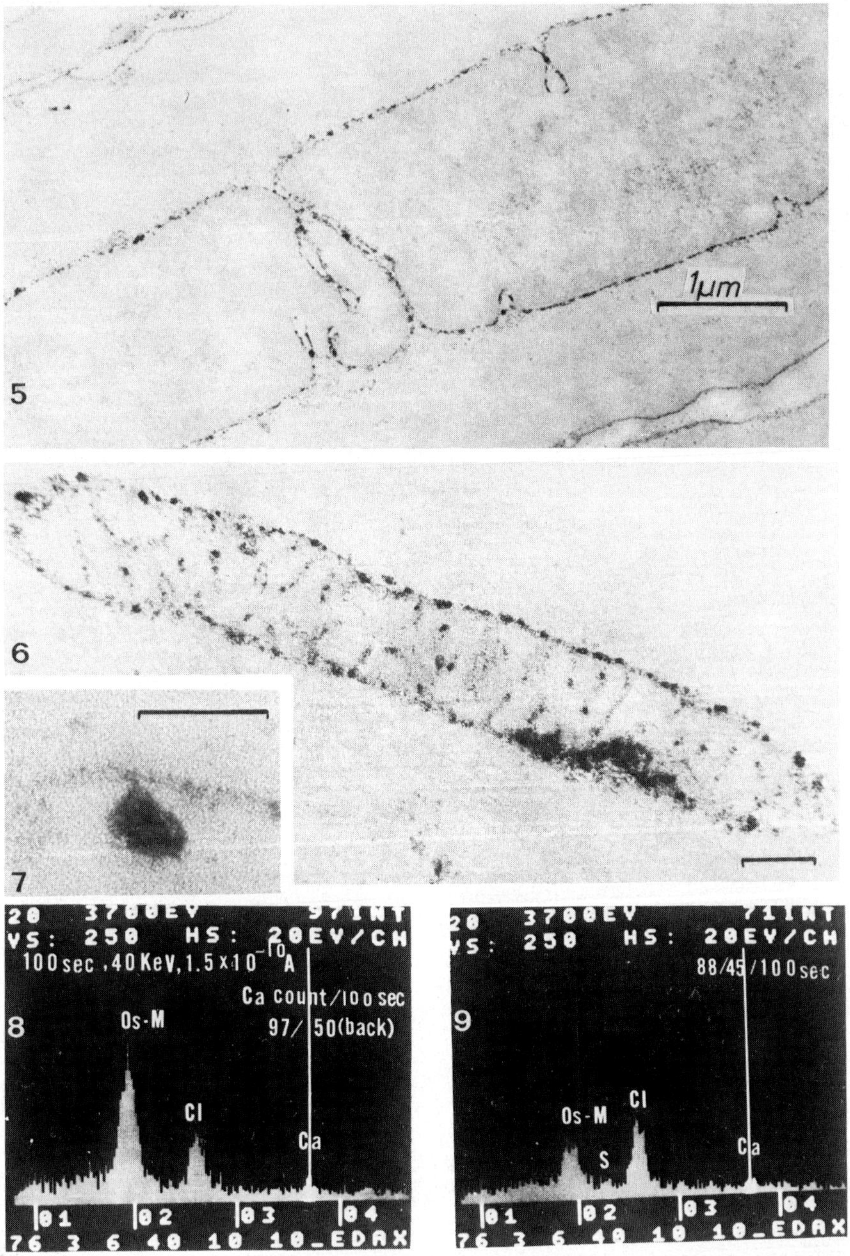

Figs. 5-9. Ca^{2+} localization in nucleus, mitochondria and SmV. Guinea-pig taenia coli incubated in high-K oxalate medium. Muscle fixed under relaxation after a K-contracture. Note calcium oxalate deposits in the nuclear envelope (Fig. 5), in a mitochondrion (Fig. 6) and SmV (Fig. 7). X-ray spectra show calcium accumulation in the nuclear envelope (Fig. 8) and mitochondria (Fig. 9). The peaks for Os and Cl appear due to the preparative technique for electron microscopy. (Figs. 8 and 9 from ref. 11)

as in skeletal muscle. Additional evidence is provided by the *in situ* cytochemical demonstration of a Ca^{2+}-ATP-ase (like in skeletal muscle SR) associated with the membrane of the smooth muscle SR (Fig. 13). Moreover, Ca^{2+} concentration of 22 mM found in smooth muscle SR[7] is quite similar with that (25 mM) reported for skeletal muscle SR[15]. The importance of the SR for the cellular Ca^{2+} homeostasis is emphasized by electron microscope images such as Fig. 2 showing that the network of SR tubules in smooth muscle cell is unexpectedly developed. In any case, the SR volume of about 3% of cell volume in guinea-pig taenia coli [6,16] seems to be an under-estimation.

Mitochondria : *In situ*, mitochondria are the main calcium-storage sites of the smooth muscle cytoplasm[6]. This finding together with the demonstration of an energy-dependent Ca^{2+} uptake in isolated mitochondria-enriched fractions (for review see refs. 17, 18) suggest that mitochondria can play a role in controlling the intracellular Ca^{2+} concentration. However, considering the intracellular topography of mitochondria and their *in vivo* movements, as well as new comparative Ca^{2+}-uptake experiments[19], these organelles seem to represent a "buffer" compartment for free cytoplasmic Ca^{2+} rather than a source for the release of activator Ca^{2+}.

Nucleus: This subscellular compartment has a Ca^{2+} content significantly higher than SR or mitochondria[6]. The cytochemical approach[6,13] revelaed the perinuclear cisterna as a sub-compartment of high-Ca^{2+} accumulation (Fig. 5). This fact is not surprising since the nuclear envelope is directly related to the SR. Fig. 14 shows the existence of a Ca^{2+}-ATP-ase (like in SR) associated with the external membrane of the nuclear envelope. This enzyme could perform Ca^{2+} accumulation from cytoplasm against a large concentration gradient. Therefore, further study of ATP-dependent Ca^{2+} uptake in isolated nuclear fraction is required. In any case, although the nucleus cannot participate directly in Ca^{2+} regulation during the contraction, any tentative model of the intracellular Ca^{2+} homeostasis in smooth muscle must include the contribution of the nucleus.

Surface microvesicles: The SmV occupy large areas of the smooth muscle cell membrane[14, 20, 21] and their total volume is comparable with the SR volume[6]. They can be considered as specific to smooth muscle ultrastructure, since such SmV are only occasionally seen in other types of muscle. A priori, taking into account the correlation between the fine structure and function, the SmV could be responsible for some specific functional features of smooth muscle. Therefore, the SmV role in the intracellular calcium homeostasis deserves a detailed discussion.

SURFACE MICROVESICLES - SITES OF CALCIUM EXTRUSION

In the last few years, several investigators suggested the existence of an efficient Ca^{2+}-extrusion pump in the periphery of smooth muscle cells[22-26], which opposes the continuous leak of Ca^{2+} through the membrane of resting cells. Since the smooth muscle cell has a high surface/volume ratio, which is considerably further increased by the SmV[27,28], it follows that the SmV role is, first of all, to enlarge the exchange area of the cell interior with the extracellular space, and they are therefore, the most likely candidate for the Ca^{2+} extrusion. To test the validity of this hypothesis a simple and informative approach would be to destroy chemically the SmV and to study concomitantly the outflow of Ca^{2+} from cells.

Calcium outflow from smooth muscle cell after SmV modification : *In vitro* treatment of smooth muscle with 100 mM SDS resulted in the disappearance of about 20% of the SmV and most of the persistent SmV were less well defined (Fig. 10). The only detectable modification of the ultrastructure was that of the SmV.

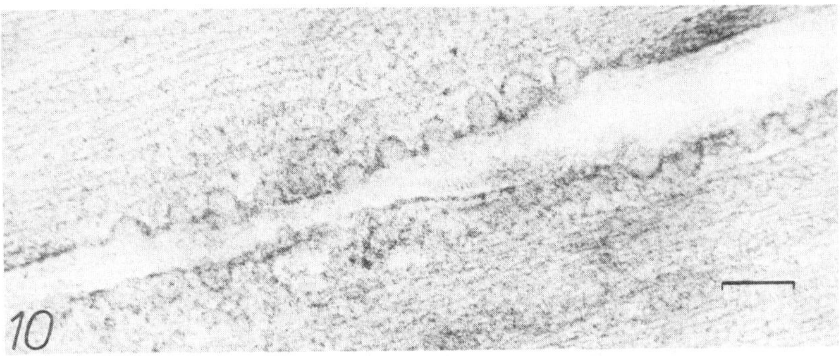

Fig. 10 Guinea-pig taenia coli. Tissue equilibrated in Krebs for 30 min. and then treated with 100 mM SDS. A modification of the normal aspect of the persistent SmV is evident. The surface membrane has an undulating aspect.

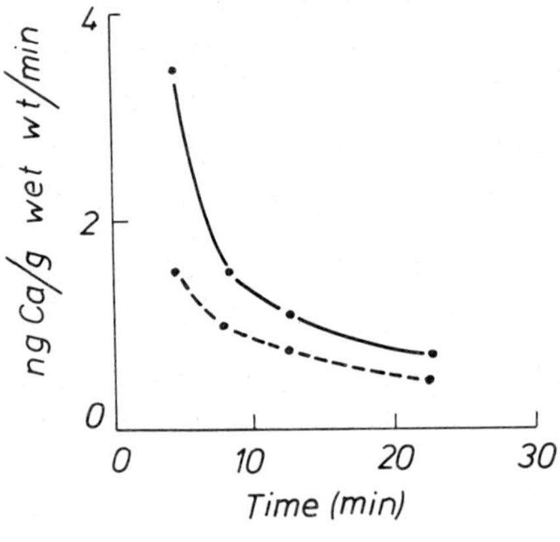

Fig. 11. Calcium outflow derived from measurements of calcium concentrations in bathing solution by atomic absorption spectrometry. •——• control muscles; •- - -• muscles treated with SDS.

Comparative estimations of Ca^{2+} outflow from intact cells and cells with SmV "wiped out" by SDS treatment (Fig. 11) support Ca^{2+} extrusion through SmV. Since the SmV have to transport Ca^{2+} from the cell interior against a large concentration gradient, the presence of a Ca^{2+}-transport ATP-ase in their membrane would be expected.

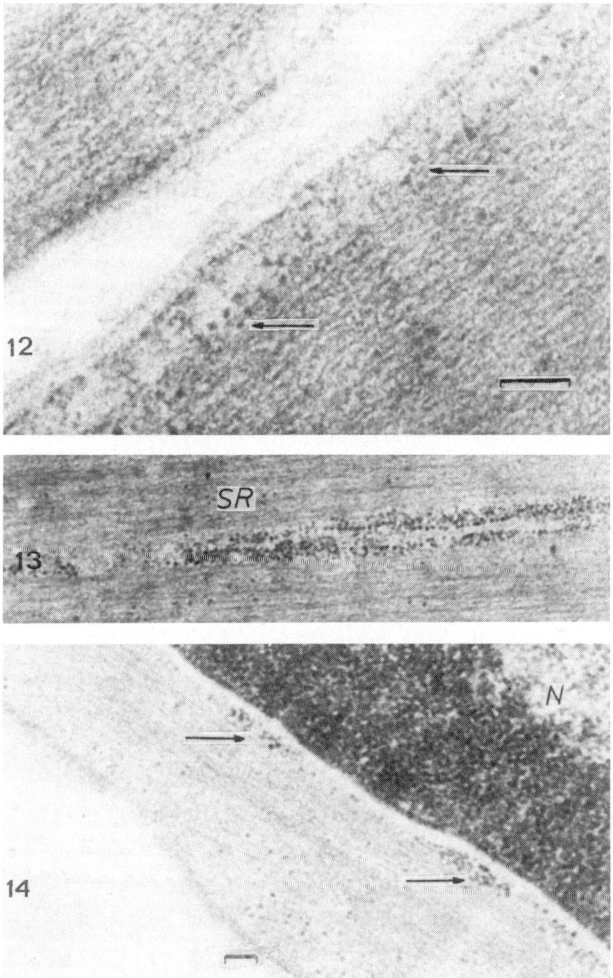

Figs. 12-14. Localization of Ca^{2+}-activated ATP-ase in smooth muscle cells from guinea-pig taenia coli, revealed by calcium-citro-phosphate technique. Brief fixation of tissue is required to prevent the enzyme inactivation. ATP splitting by Ca^{2+}-ATP-ase produces phosphate ions, which then react with calcium and citrate, added in the incubation medium, and form an insoluble reaction product - electron-dense granules showing the sites of enzyme activity. Such granules appear associated with SmV, SR and external membrane of the nuclear envelope.

Cytochemical evidence for Ca^{2+}-ATP-ase activity of SmV *in situ:* Calcium-citrophosphate technique, recently recommended for the *in situ* specific demonstration of Ca^{2+}-activated ATP-ase in the SR from skeletal muscle[29], was used (Figs. 12-14). Electron-dense granules showing the existence of Ca^{2}-ATP-ase were found at the level of surface membrane. More granules are associated with the SmV membrane than located on the non-vesicular membrane (Fig. 12). In addition, electron-dense granules are associated with SR membranes (Fig. 13) and with the external membrane of the nuclear envelope (Fig. 14). No such electron-dense granules appear in mitochondria. A correlation between the cytochemical findings and the freeze-etch observations of smooth muscle membrane systems [30, 31] is tempting. In skeletal muscle, the so called 9 nm particles on the SR membranes have been related to the Ca^{2+}-ATP-ase[32,33]. In smooth muscle, it was reported that the 9 nm particles were distributed more densely near the SmV than in the intervesicular areas of the surface membrane [30,31]. Moreover, 9 nm particles were seen on the SR and nuclear membranes, but they were not detected on the mitochondrial membranes[31]. Since a striking similarity of distribution between the cytochemically located Ca^{2+}-ATP-ase and the 9 nm particles is evident, it seems justified to suggest that also in smooth muscle at least a part of the 9 nm particles correspond to Ca^{2+}-ATP-ase.

To establish the functional significance of Ca^{2+}-ATP-ase shown cytochemically in SmV the required next step was the isolation of this enzyme. The starting premise has been that Ca^{2+}-ATP-ase belongs to the so called intrinsic or integral proteins which can be extracted from membranes with detergents. Indeed, in the proteic fraction extracted with 100 mM SDS from the surface membrane of the smooth muscle cells a Ca^{2+}-activated ATP-ase was identified.

Isolation and properties of the surface-Ca^{2+}-ATP-ase: In detergent-soluble form, the ATP-ase extracted from the surface membrane of smooth muscle cells (guinea-pig taenia coli and human pregnant myometrium) requires Mg^{2+} and is stimulated by Ca^{2+}. The optimal Mg^{2+} concentration is 5 mM, and the optimal free Ca^{2+} concentration is in the range 10^{-7} to 10^{-6} M. The low levels of Ca^{2+} that activate the ATP-ase indicate the extremely high affinity for Ca^{2+}. The enzyme requires the presence of phospholipids for activity and is clearly distinct from Na^{+}-K^{+}-dependent ATP-ase. Ouabain (0.1 mM) had no demonstrable effect. The ATP-ase activity, in the presence of 5 mM Mg^{2+}, is strongly inhibited by the addition of 0.5 mM EGTA. This confirms the existence of a site for Ca^{2+} on the enzyme itself. All the essential characteristics of the surface-Ca^{2+}-ATP-ase extracted from smooth muscle cells are similar to those reported for the membrane-bound Ca^{2+}-ATP-ase of the skeletal muscle SR[34,35]. In addition, the specific ATP-ase activity of 1-2.5 IU is comparable with that of Ca^{2+}-transport ATP-ase isolated from skeletal muscle SR: 3-5 IU for partially purified enzyme or 15-30 IU for the purified enzyme (for review see ref. 35).

CONCLUDING REMARKS

Bülbring and Kuriyama suggested in 1973[22] that cytoplasmic Ca^{2+} concentration in smooth muscle depends on two processes: internal redistribution of Ca^{2+} and Ca^{2+} pumping out of the cell. At present, it can be stated that two systems maintain the the intracellular Ca^{2+} homeostasis (Fig. 15):

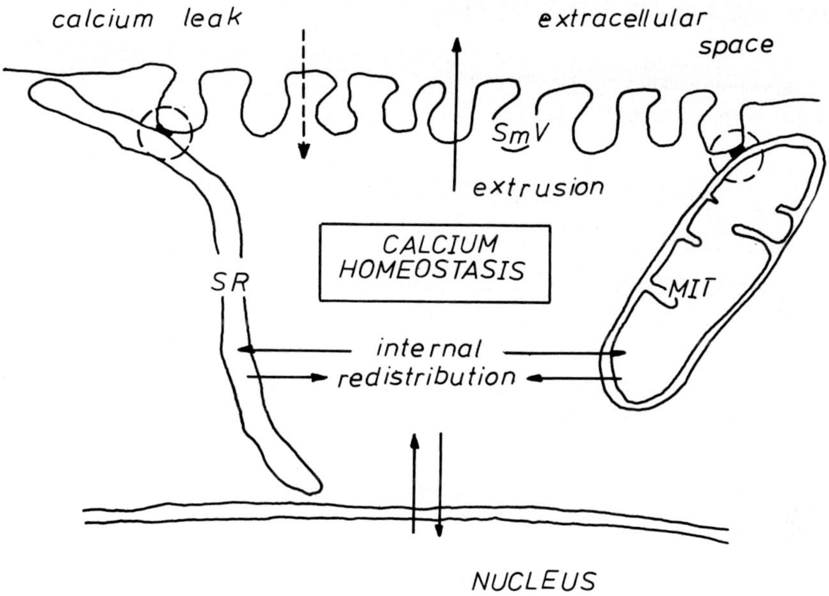

Fig. 15. A schematic model illustrating the role of sub-cellular components in the intracellular calcium homeostasis. Couplings between the two systems are encircled. Not to scale.

(1) the system of intracellular redistribution of Ca^{2+}, represented by the SR, mitochondria and nucleus, and

(2) the system of Ca^{2+} extrusion, represented by surface membrane with SmV, which pumps Ca^{2+} out of the cell.

The close spatial contacts ("couplings") of the SmV with peripheral SR or superficially located mitochondria might be the morphological basis for the functional coupling of the two systems.

REFERENCES

1. H. Kuriyama, T. Osa, Y. Ito and H. Suzuki, Adv. Biophys., 8 (1975) 115.
2. H. Huddart and N.R. Price, Comp. Biochem. Physiol., 54A (1976) 375
3. A.P. Somlyo and A.V. Somlyo, Fed. Proc., 35 (1976) 1288.

4. L. Hurwitz, in N.L. Stephens, The Biochemistry of Smooth Muscle, Univ. Park Press, Baltimore, 1977, p. 559.

5. I. Diculescu and L.M. Popescu, Proc. 5th Internat. Congr. Histo-chem.Cytochem., Bucharest, 1976, p. 82.

6. L.M. Popescu, I. Diculescu, U. Zelck and N. Ionescu, Cell Tiss. Res., 154 (1974) 357.

7. L.M. Popescu and I. Diculescu, J. Cell Biol., 67 (1975) 911.

8. I. Diculescu, L.M. Popescu and N. Ionescu, Exp. Cell Res., 68 (1971) 210.

9. I. Diculescu and L.M. Popescu, Exp. Cell Res., 82 (1973) 152.

10. I. Diculescu, L.M. Popescu, N. Ionescu and N. Butucescu, Z. Zellforsch., 121 (1971) 181.

11. I. Diculescu, L.M. Popescu and V. Mizuhira, unpublished.

12. I. Diculescu and L.M. Popescu, unpublished.

13. L. Jonas and U. Zelck, Exp. Cell Res., 89 (1974) 352

14. L.M. Popescu, Studia Biophys., 44 (1974) 141.

15. S. Ebashi and M. Endo, Prog. Biophys. Mol. Biol., 18 (1968) 123.

16. C.E. Devine, A.V. Somlyo and A.P. Somlyo, J. Cell Biol., 52 (1972) 690

17. S. Batra, in E. Carafoli et al., Calcium Transport in Contraction and Secretion, North Holland, Amsterdam, 1975, p. 87

18. G.D. Ford, Fed. Proc., 35 (1976) 1298

19. L. Raeymaekers, F. Wuytack, S. Batra and R. Casteels, Pflügers Arch., 368 (1977) 217.

20. G. Burnstock, in E. Bülbring et al., Smooth Muscle, Arnold, London, 1970, p.1

21. G. Gabella, in E. Bülbring and M.F. Shuba, Physiology of Smooth Muscle, Raven Press, New York, 1976, p.265

22. E. Bulbring and H. Kuriyama, Phil. Trans. R. Soc. Lond.B., 265 (1973) 223.

23. R. Casteels and C. van Breemen, Pflügers Arch., 359 (1975) 197.

24. H. Kuriyama, T. Osa, Y. Ito, H. Suzuki and K. Mishima, in E. Bülbring and M.F. Shuba, Physiology of Smooth Muscle, Raven Press, New York, 1976, p.185

25. P.K. Rangachari, M.G. Pernollet and M. Worcel, Europ. J. Pharmacol., 40 (1976) 291.

26. R.A. Janis and E.E. Daniel, in N.L. Stephens, The Biochemistry of Smooth Muscle, Univ. Park Press, Baltimore, 1977, p.653.

27. G. S. Wootton and P.J. Goodford, Cell Tiss. Res., 162 (1975) 75.

28. G. Gabella, Cell Tiss. Res., 170 (1976) 161.

29. M.A. Khan, J.M. Papadimitriou and B. Kakulas, Histochemistry, 43 (1975) 101.

30. G. S. Wells and M.W. Wolowyk, J. Physiol., 218 (1971) 11p.

31. C. E. Devine and D.G. Ryans, J. Ultrastruct. Res., 51 (1975) 293.

32. D. W. Deamer and R.S. Baskin, J. Cell Biol., 42 (1969) 296.

33. D. H. MacLennan, P. Seeman, G. H. Iles and C. C. Yip, J. Biol. Chem. 246 (1971) 2702

34. G. Inesi, Annu. Rev. Biophys. Bioeng., 1 (1972) 191.
35. D.H. MacLennan and P.C. Holland, in A. Martonosi, The Enzymes of Biological Membranes, Plenum, New York, vol. 3 1976, p.221

Excitation-Contraction Coupling in Smooth Muscle, R. Casteels et al. eds.

EFFECTS OF MEMBRANE POTENTIAL ON ACTIVATION OF CONTRACTION IN VARIOUS SMOOTH MUSCLES

Hirosi KURIYAMA, Yushi ITO and Hikaru SUZUKI

Department of Pharmacology, Faculty of Medicine, Kyushu University 60, Fukuoka, 812 JAPAN

INTRODUCTION

The term 'excitation-contraction coupling; E-C coupling', i.e. how changes in electrical potential of the sarcoplasmic membrane of the skeletal muscle induce contraction, was introduced by Sandow in 1952 (1). In the past two decades great progress has been made in this field. An essential step of E-C coupling in skeletal muscle is that the excitation or depolarization of the surface membrane induces a regenerative depolarization of the tubular system, (T-system), and thereby a release of stored Ca from the terminal cisternae of the sarcoplasmic reticulum (SR) (2, 3).

The term 'E-C coupling' is also applied to smooth muscle assuming an analogous explanation for these contraction-relaxation phenomena. However, in smooth muscle, the distribution of the T-system is not apparent, and a characteristic feature of the membrane is the presence of numerous 'inpocketings', flask-shaped vesicles with bottle-shaped microinvaginations on the membrane surface. A morphometric analysis made by Popescu (4) indicated that only 9% of all microvesicles (caveolae) form a close spatial relationship (coupling) with the peripherally located SR elements and 1% with the mitochondria.

Forsmann and Girardier (5) observed in the rat heart that cells with a well developed T-system had little caveolae. A large number of caveolae were present only in cells with a poorly developed T-system. This observation suggests a relationship between the T-system and caveolae.

Ishikawa (6) reported that the T-system of cultured striated muscle fibres is composed of chains of partially fused vesicles or caveolae. Since ontogenesis is a repetition of phylogenesis, it is conceivable that during evolution the primitive muscle cells are equipped with surface caveolae, and that in skeletal muscle chains of vesicles or caveolae are formed resulting in the T-system. While in the skeletal muscle, the Ca storage sites are mainly localized in the SR and not in the T-system, we could speculate that in smooth muscle microvesicles are an important Ca-storage site (4, 7, 8, 9). This assumption could imply that E-C coupling in skeletal and smooth muscle follows a different pattern.

It has been established that changes in free Ca^{++} concentration in the myoplasm of skeletal muscle regulate contraction-relaxation cycle, and that the key reaction of the contraction is triggered by the binding of Ca^{++} to the troponin-tropomyosin complex (2, 3). Kendrick-Jones et al (10) concluded from their studies in molluscan muscles that Ca^{++} regulates contraction by acting on myosin, because troponin and related components were absent, and because tropomyosin, although present on the thin filaments, was not required for regulation. These observations were confirmed in mammalian smooth muscles by Bremel (11). It was therefore postulated that the regulation of contraction in smooth muscle is myosin-linked. On the other hand, Ebashi and colleagues (personal communication) suggested that regulation is fundamentally the same as in skeletal muscle, even though the properties of the troponin-tropomyosin complex differ from those in the skeletal muscles.

Based on evidence obtained from physiological studies performed on various visceral smooth muscles, there is currently a consensus of opinion that the Ca^{++} which activates contractile protein can originate from two different sources. One source is the Ca^{++} present in the extracellular fluid including the Ca which is loosely bound to superficial sites in the muscle fibre, and the second source consists of a tightly bound pool of Ca^{++} that is sequestered in intracellular sites (9, 12, 13, 14, 15, 16).

Although, it is now possible to record the activity of a single cell, it is not yet possible to determine the relationship between the electrical activity and mechanical response recorded simultaneously from single cells, even though the single cell mechanogram was recorded by Fay and Delise (17).

This article is an attempt to describe the inadequacy of the present experimental methods and the difficulties encountered in trying to explain the experimental data of E-C coupling in smooth muscles. We will try to analyze the changes of the membrane potential which produce mechanical responses in relation to changes of Ca^{++} during the E-C coupling process.

CONTRACTION OF SMOOTH MUSCLE INDUCED BY ACTIVATION OF THE MEMBRANE

The membrane potential of smooth muscle fibres varies from one tissue to another. The smooth muscle membrane of the portal vein and the urinary bladder of the guinea-pig maintain a resting potential of only -40 mV, whereas this potential in the vas deferens is about -63 mV. These values are much lower than those recorded from the fast twitch muscle of the frog (-85 mV), and these low membrane potentials do not seem to be essential for the spontaneous activity of the membrane. For example in guinea-pig stomach, the cells of the longitudinal muscle of the fundus remain quiescent, although they present the same membrane potential as the antral cells which show spike activity. The quiescent smooth muscle cells of the aorta, trachea, and pulmonary artery consistently present a high membrane potential (-58 to -62 mV) (18, 19).

The ionic mechanism involved in spike generation has not been completely elucidated. The action potentials in many muscles are explained by a rapid influx of Ca^{++} because of the following reasons. (a) In low Na solution, the amplitude of the overshoot potential increases despite a reduction in E_{Na}. (b) In a low Ca-solution, the generation of a transient inward current is suppressed and the reversal potential level shifts as one could expect for a Ca-inward current. The reduction of Na in the medium does not affect the reversal potential level (16). Tetrodotoxin does not suppress spike activity in smooth muscle (20, 21). However, this cannot be taken as unequivocal evidence for a Ca-spike. In the denervated muscle preparation, the spike generated by an inward movement of Na^{+} is not suppressed by tetrodotoxin (22). (d) The spikes generated in smooth muscles are blocked by treatment with Mn^{++}. In the guinea-pig stomach, taenia coli, jejunum, ileum, rectum, urinary bladder, vas deferens and portal vein, the spikes are regarded as Ca-spikes. However, in the guinea-pig ureter, the membrane activity is composed of a rapid Ca-current followed by a Na-current during the plateau formation (23, 24, 25).

It is not yet certain whether the action potential in rat myometrium is a Ca-spike, a Na-spike or a mixed Na-, Ca- spike (26, 27, 28). The slow action potential evoked by noradrenaline or procaine in sheep carotid arteries is considered to depend on an Na-current (29), while in the pulmonary artery the chemically evoked spike seems to depend on Ca because it is blocked by treatment with Mn^{++} or EGTA.

The electrical parameters of the smooth muscle action potential vary from one tissue to another, and these parameters can be summarized for the guinea pig taenia coli as follows : membrane potential, -55 mV; spike amplitude, 60 mV; the maximum rate of rise of the spike, 20 V/sec; spike rise time, 10 msec; specific membrane resistance, 30 $k\Omega cm^2$; membrane capacitance, (C_m) 5 $\mu F/cm^2$; length constant of the tissue (λ), 1.5 mm; membrane time constant (τ_m), 100 msec; conduction velocity of excitation, 6 cm/sec (18, 30).

In the frog skeletal muscle the free Ca^{++} concentration in the resting state is estimated at 3×10^{-7} M. This value is far below the threshold concentration of Ca^{++} in frog myoplasm, viz, 2×10^{-6} M (50). In the glycerinated smooth muscle tissue of guinea-pig taenia coli, the minimum concentration of Ca^{++} required to activate the contraction is of the order of 5×10^{-7} M, and in the vascular muscles this value is slightly lower than that in taenia coli ($5 - 10 \times 10^{-8}$ M) (M. Hirata unpublished observations, M. Endo personal communication). Thus the influx of Ca^{++} during the active state of the membrane may explain the activation of the contractile proteins.

However, contraction induced by spikes seems to be a complex phenomenon, e.g. i) when Na^{+} in the medium is replaced with Li^{+} the membrane activity is somewhat enhanced but this activity is accompanied by a decrease in the tension development. This phenomenon was observed by Axelsson (31) who proposed that this dissociation was caused by a metabolic disturbance of the cells. ii) in hypertonic solution (prepared by adding sucrose), the electrical activity can be evoked by electrical stimulation after the cessation of the spontaneous activity, but these action potentials are not accompanied by a mechanical response (51). It has been demonstrated that in skeletal muscle (2) the effect produced by hypertonic solutions can be largely attributed to the osmotic effect and the concomitant increase of the concentration of electrolytes in the sarcoplasmic reticulum. In skinned muscle fibres, additions of 40 mM sucrose to the bathing solution suppresses the release of Ca^{++} induced by depolarization (32, 50). iii) a low K^{+} medium hyperpolarizes the membrane of taenia coli and enhances the spike amplitude but reduces the mechanical response (33). In the mesenteric artery, however, the contraction induced by acetylcholine is enhanced by solutions with lowered $[K]_o$. Therefore, the results obtained in low K^{+} media are not consistent. iv) a substitution of Ba^{++} for Ca^{++} in Krebs solution depolarizes the membrane and the action potentials present a prolonged falling phase (plateau formation) (34). However, the mechanical response is suppressed. It has been observed in glycerinated muscle that the affinity constant of Ba^{++} to the contractile protein is five times lower than that of Ca^{++} (M. Hirata, unpublished observation).

On the other hand, a simulation of the spike by application of outward current in Krebs solution under a current clamp condition, induces a much smaller tension response than that recorded during the normal spike generation (H. Kuriyama, Y. Ito and H. Suzuki, unpublished observation). Another argument against a close relation between spike and tension development is that although the longitudinal muscles of the fundus and of the antrum of the guinea-pig stomach have almost the same membrane potential, a clamping of the membrane of both muscles to the same depolarization level will generate an early Ca-inward current in the antrum but not in the fundus. The contraction evoked by the antrum is much larger than that evoked in the fundus (35).

These different observations suggest that not all the Ca^{++} penetrating into the cell during influx is utilized for contraction, and that part of it could be sequestered in intracellular stores. Depolarization induced by spike generation could in some smooth muscles also mobilize Ca from intracellular sites.

CONTRACTION OF SMOOTH MUSCLES INDUCED BY DEPOLARIZATION OF THE MEMBRANE

K-induced contracture

When excess K^{+} is applied to smooth muscle tissues, the induced contracture is composed of a phasic and a tonic response. The initial, rapidly developed tension (phasic) gradually declines to a level which is sustained for more than 10 min (tonic). Urakawa and Holland (36) proposed that for the phasic contraction of guinea-pig taenia coli, sufficient Ca^{++} is released from a cellular site to initiate contraction, whereas for the tonic contraction an adequate amount of Ca penetrates the membrane to initiate contraction. On the other hand, Imai and Takeda (37) postulated that the tonic response of the contracture was due to release of bound Ca^{++} from the smooth muscle. Selective suppression of the phasic and the tonic response of the contracture has been described by Golenhofen (38) and Boev et al. (39).

The ratio of the amplitude of phasic and tonic response differs from tissue to tissue. In general, tissues generating spikes produce a large phasic and a small tonic contracture while less excitable tissues as trachea, pulmonary artery and stomach fundus produce a small phasic and a large tonic contracture (40, 41, 42). In chicken stomach muscle an increase of the K^+ concentration from 5 times to 20 times the normal concentration reduces the amplitude of the tonic contracture (K. Tajima and H. Kuriyama unpublished observations). This could mean that the amplitude of the tonic contracture is not only due to an influx of Ca^{++} or to a release of Ca^{++} during depolarization, but also that the inactivation of the Ca-releasing mechanism and reabsorption of Ca^{++} at the sequestered sites plays a role in its maintenance.

In the SR of skeletal muscle, there are at least two different mechanisms for inducing a release of sequestered Ca^{++}, i.e. Ca-release induced by depolarization and by Ca itself. The former mechanism is not affected by caffeine or procaine but the second one is appreciably modified by these compounds (3, 43). In the smooth muscle, depolarization of the membrane by excess K^+, releases sequestered Ca and increases the influx of Ca^{++}. However, in several smooth muscles, the external Ca^{++} concentration, $(Ca)_o$, determines the amplitude of a K-induced contracture. For example, in the rabbit pulmonary artery, removal of $(Ca)_o$ in excess K^+ solution results in a rapid cessation of K-induced contracture (with a t 1/2 of 3 min) without reducing the depolarization of the membrane by K. This finding does not indicate a lack of stored Ca^{++} because noradrenaline can still produce a contracture even when the K-induced contracture has ceased (44).

Isotonic K-Krebs solution produces a contracture of the circular muscle of the antrum and longitudinal muscle of the fundus : in the antrum a large phasic and small tonic contracture and in the fundus a small phasic and large tonic contracture. During the generation of tonic contracture in circular muscle, application of inward current which repolarizes the membrane to its resting level produces a relaxation. A cessation of the inward current produces a rebound contracture. In the longitudinal muscle this relaxation and rebound contraction remain much smaller. These findings indicate that in some tissues the tonic contracture depends on the membrane potential, while in others this tonic contraction seems to be largely independent of membrane polarization.

Ito et al (44) have recently studied the effects of excess K^+ on the smooth muscle of pulmonary artery at different membrane potential levels. Because smooth muscle cells of this tissue do not generate spikes during electrical stimulation or during excess K, the interference with the Ca-spike is eliminated. Application of isotonic K^+ solution depolarizes the membrane from -58 mV to -5 mV and produces a marked contracture. When the membrane potential was fixed at the resting level, before, during and after application of excess K^+, contracture was still produced (80 % the control height). However, these findings immediatly raised the question of whether during application of excess K^+ solution the length constant of the tissue is reduced and whether therefore, the clamping conditions are inadequate. In fact, at 59 mM $(K)_o$, the length constant decreased from 1.45 mm to 0.68 mm, a value which is roughly twice the gap distance of the central pool. The amplitude of the contracture evoked by 59 mM K^+ was 0.5 times that produced by 118 mM K^+. When the depolarization induced by 118 mM K^+ was stimulated in Krebs solution by outward current the amplitude of the contraction was found to be 0.6 times the contracture induced by 118 mM K^+.

Here again, it is obvious that the properties of different smooth muscle cells vary. In guinea-pig taenia coli and stomach antrum, K^+ induced contracture is probably the result of both Ca-influx and release of sequestered Ca^{++}. In rabbit pulmonary artery and guinea-pig stomach and trachea, K^+ induced contracture is the direct result of the influx of Ca^{++}.

Chemically induced contracture

Acetylcholine (ACh), Carbachol (Carb), histamine, noradrenaline (NA) and various other compounds induce contraction with or without depolarization of the membrane. In many smooth muscle tissues ACh and Carb depolarize the membrane, increase the ionic conductance and produce contraction. These tension developments appear in the presence of $(Ca)_o$ or in K-rich and Ca-free solution, indicating that these compounds can besides allowing an influx of Ca^{++}, release sequestered Ca^{++}. We give here two examples of characteristics of chemically induced contracture : 1) When the longitudinal muscle of the stomach fundus and circular muscle of the stomach are immersed in Ca-free isotonic K-solution, containing EGTA the tissues completely relax. Application of Carb induces a tension development which on repeated application gradually lowers and finally ceases in both tissues. Addition of 2.5 mM Ca^{++} to the above mentioned solution elicits a contracture in the longitudinal muscle but not in the circular muscle. When the tissue was rinsed again in Ca-free EGTA K-solution, complete relaxation occurred. Addition of Carb now produces a much larger contraction in circular muscle than in the longitudinal one (35). This means that influx of Ca^{++} is directly utilized for contraction and that only part of the Ca^{++} is stored in the longitudinal muscle. In the circular muscle the influx of Ca^{++} is mainly sequestered in the storage sites. These differences might explain the tonic response during K-depolarization and the roles of storage sites. Moreover, compounds are capable of releasing the stored Ca directly, with or without changes in the surface membrane. 2) In the pulmonary artery, when the tissue was immersed in Ca-free EGTA solution, the NA induced contracture gradually decreased in amplitude. When the decay times of the contracture were plotted against the amplitude of the contracture on a logarithmic scale, the curve could be split up into three components, with half times of 3 min, 18 min and over 100 min respectively. The first exponential has the same decay time as the one obtained by K-rich solution in Ca-free medium. The 2nd and 3rd components probably depend on stored Ca. In fact, the K-induced contracture in pulmonary artery is mainly due to influx of Ca, while noradrenaline induced contracture also depends on sequestered Ca. Loss of contractile response in Ca-free solution is not correlated with a diminished electrical excitation by the drugs, as the amplitudes of depolarizations induced by these stimuli remain the same (44).

In the main pulmonary artery, noradrenaline at concentrations below 10^{-8} M, while eliciting a contraction, does not change the membrane potential or the membrane resistance. Increasing the concentration to 2 x 10^{-8} M induces depolarization and reduction of the membrane resistance. Also the contraction is markedly enhanced (41). At low concentrations of NA, the contraction depends on an influx of Ca^{++}, while at higher concentrations NA causes an increased Ca influx and a release of stored Ca. Such a release of Ca^{++} by chemical stimulation could imply that smooth muscle cells containing SR or other storage sites have selective receptors for these substances, and that their activation could induce a release of stored Ca independent of a change in the membrane potential. These conclusions are in agreement with the findings of Evans, Schild and Thesleff (45) on the depolarized myometrium.

EFFECTS OF SOME DRUGS ON Ca-RELEASING MECHANISM

In skeletal muscle fibres, caffeine labilizes the Ca^{++}-induced Ca^{++} release mechanism. In the skinned muscle fibre, a high concentration of caffeine accelerates the accumulation of Ca^{++} into the SR (2, 3, 46). Procaine has an antagonistic action against caffeine and this action is probably related to the stabilization or inhibition of Ca^{++}-induced Ca^{++} release from SR (32, 47). In smooth muscles of the guinea-pig alimentary canal, caffeine and procaine act in a different manner from that observed in skeletal muscle, i.e. caffeine acts not only on the Ca-releasing mechanism but also on the surface membrane (48).

In skeletal muscle thymol is known to be the more potent stimulator of Ca^{++} release from SR. Application of this compound consistently inhibits spike generation in smooth muscles of the stomach, ileum, taenia coli and rectum of guinea-pig. It is not known whether these effects are due to immobilization of Ca^{++} or to complete removal of Ca^{++} from the membrane.

The effects of caffeine and thymol vary according to the section of the alimentary canal of the guinea-pig. a) The effect of thymol (> 0.5 mM) in stomach consists of a depolarization of the membrane and an accompanying decrease in membrane resistance. Caffeine does not change the membrane potential or membrane conductance, although both thymol and caffeine completely block the spike and slow potentials. b) In the ileum, thymol hyperpolarizes the membrane, produces a reduction in the membrane resistance and completely inhibits spike generation. Caffeine depolarizes the membrane of the ileum, causes a slight increase of the membrane resistance and enhances the spike amplitude and frequency. c) In the taenia coli, thymol (0.5-1 mM) does not change the membrane potential, but markedly reduces the membrane resistance. It also blocks spike generation and suppresses the mechanical response. Caffeine depolarizes the membrane, reduces the membrane resistance, and blocks the membrane responses. d) In the rectum, thymol hyperpolarizes the membrane, decreases the membrane resistance and completely inhibits the spike generating mechanism. However, caffeine induces membrane depolarization, decreases the membrane resistance, and enhances the frequency of spikes (48).

The differences in actions of caffeine and thymol, and also the differences in tissue reaction should be further investigated. On the other hand, procaine and local anesthetics which block the excitability in the skeletal muscle act as excitatory agents on the smooth muscle tissues, i.e. they slightly depolarize the membrane of the spontaneously active tissues, increase the membrane resistance, enlarge the spike amplitude and increase the spike frequency, yet the mechanical response is suppressed. In electrically quiescent tissues, as trachea, pulmonary artery and carotid artery, procaine depolarizes the membrane, increases the membrane resistance, suppresses the rectifying property of the membrane and allows an outward current pulse to evoke a Ca-spike. However, the mechanical responses evoked by NA or excess K^{+} solution are suppressed by procaine in a dose dependent manner. When the relation between depolarization and contraction is studied by the voltage clamp method and during application of 1 mM procaine, we do not observe changes in the mechanical responses because spike generation somehow compensates for the suppression of mechanical response. If the procaine concentration is increased to 5 mM the suppression of mechanical response dominates the effects of spike generation (42, 44).

CONCLUSION

The effects of membrane potential changes on smooth muscle contraction have been outlined mainly from a phenomenological point of view. The subcellular regulation of E-C coupling mechanism (5) requires a great deal more study and the following areas remain to be elucidated.

i) The minimum concentration of Ca^{++} to activate the contractile protein in glycerinated smooth muscles varies from tissue to tissue, and the values range from 10^{-7} to 10^{-6} M. It should be clarified whether the contractile proteins of some smooth muscle tissues possess a stronger affinity for Ca^{++} than others or whether mechanisms different from those existing in skeletal muscle are involved.
ii) The influx of Ca^{++} during the active state of the membrane is due in some tissues to a Ca-spike. At least a part of this Ca^{++} induces an increase in $(Ca)_i$ and another part may directly be sequestered into the storage sites. During such a Ca-spike, Ca-release may also be induced by the depolarization. iii) Microvesicles and caveolae, which exist at the surface of the cell membrane, can represent accumulation sites of Ca^{++} as suggested by electron microscopic and histochemical studies. The role of these small organelles at the surface membrane should be one of the investigations given top priority.
iv) In the SR, mitochondria and nuclei Ca can be detected by histochemical methods (49). However, only a part of the SR is connected to either surface membrane or caveolae or both. It is not clear in which way the excitation is conducted from the surface membrane to the SR, how the remaining free SR receives information from the surface. While a role of the mitochondria as a Ca-modulator has been ruled out for skeletal and cardiac muscles (3), a final conclusion as to whether the mitochondria contribute to Ca-regulation in smooth muscle has not been reached.
v) It has not been resolved whether NA and ACh act only on the surface membrane or also on the Ca-storage sites. It also remains to be determined whether Ca is released from a depolarization and/or chemically-sensitive site or whether there exist discrete populations of depolarization-sensitive and chemically-sensitive Ca-releasing sites.

These questions and many more which include the phenomenon of relaxation and its relation to the changes in membrane potential will be subjects of intensive research in the next decade.

REFERENCES

1. Sandow, A. (1965) Pharmac. Rev. 17, 265-320.
2. Ebashi, S. (1976) Ann. Rev. Physiol. 38, 293-313.
3. Endo, M. (1977) Physiol. Rev. 57, 71-108.
4. Popescu, L.M. (1974) Studia Biophysica, Berlin Band 44, Heft 2. S 141-153.
5. Forsmann, W.G. and Girardier, L. (1970) J. Cell. Biol. 44, 1-19.
6. Ishikawa, H.J. (1968) J. Cell. Biol. 38, 51-66.
7. Gabella, G. (1973) Phil. Trans. Roy. Soc. Lond. B. 265, 7-16.
8. Devine, C.E., et al. (1973) Phil. Trans. R. Soc. Lond. B. 265, 17-23.
9. Somlyo, A.P. (1975) Cellular Pharmacology of Excitable Tissues, ed. Narahashi,

T. 360-407, Charles C. Thomas, Springfield.
10. Kendrick-Jones, J., et al. (1970) J. Mol. Biol. 54, 313-326.
11. Bremel, R.D. (1974) Nature 252, 405-406.
12. Lüllman, H. (1970) Smooth Muscle, ed. E. Bülbring, A.F. Brading, A.W. Jones & T. Tomita 151-165, Edward Arnold, London.
13. Hurwitz, L. and Suria, A. (1971) Ann. Rev. Pharmacol. 11, 303-326.
14. Prosser, C.L. (1974) Ann. Rev. Physiol. 34, 503-535.
15. Hurwitz, L. (1976) Smooth Muscle Pharmacology and Physiology, ed. M. Worcel & G. Vassort, INSERM,Paris.
16. Kuriyama, H., et al. (1975) Adva. Biophys. 8, 115-190.
17. Fay, F.S. and Delise, C.M. (1973) Proc. Nat.Acad. Sci. U.S.A. 70, 641-645.
18. Kuriyama, H. (1970) Smooth Muscle, ed. E. Bülbring, A.F. Brading, A.W. Jones & T. Tomita 366-395, Edward Arnold, London.
19. Bülbring, E. and Shuba, M.F. (1976) Physiology of Smooth Muscle, Raven Press New York.
20. Hotta, Y. and Tsukui, R. (1968) Nature 217, 867.
21. Kuriyama, H., et al. (1966) Br. J. Pharmac. Chemother. 27,366-376.
22. Redfern, P. and Thesleff, S. (1971) Acta. Physiol. Scand. 81, 557-564.
23. Kuriyama, H. and Tomita, T. (1970) J. Gen. Physiol. 55, 147-162.
24. Shuba, M.F. (1977 a) J. Physiol. 264, 837-851.
25. Shuba, M.F. (1977 b) J. Physiol. 264, 853-864.
26. Abe, Y. (1971) J. Physiol. 214, 173-190.
27. Osa, T. (1973) Jap. J. Physiol. 23, 199-216.
28. Kao, C.Y. and Mc Cullough, J.R. (1975) J. Physiol. 246, 1-36.
29. Keating, W.R. (1976) Smooth Muscle Pharmacology and Physiology, ed. M. Worcel & G. Vassort 177-182, INSERM, Paris.
30. Tomita, T. (1970) Smooth Muscle, ed. E. Bülbring, A.F. Brading, A.W. Jones & T. Tomita 197-243, Edward Arnold, London.
31. Axelsson, J. (1961) J. Physiol. 158, 381-398.
32. Thorens, S. and Endo, M. (1975) Proc. Jap. Acad. 51, 473-478.
33. Bülbring, E. and Kuriyama, H. (1963) J. Physiol. 166, 29-58.
34. Sakamoto, Y. and Kuriyama, H. (1970) Jap. J. Physiol. 20, 640-656.
35. Kuriyama, H., et al (1975) J. Physiol. 251, 317-331.
36. Urakawa, N. and Holland, W.C. (1964) Am. J. Physiol. 207, 873-876.
37. Imai, S. and Takeda, K. (1967) J. Physiol. 190, 155-170.
38. Golenhofen, K. (1975) Physiology of Smooth Muscle, ed. E. Bülbring, & M.F. Shuba 197-202, Raven Press, New York.
39. Boev, K., et al. (1975) Physiology of Smooth Muscle, ed. E. Bülbring & M.F. Shuba 203-208, Raven Press, New York.
40. Suzuki, H., et al. (1976) Jap. J. Physiol. 26, 303-320.

41. Casteels, R., et al (1977 a) J. Physiol. in press.
42. Casteels, R., et al (1977 b) J. Physiol. in press.
43. Endo, M. (1975) Proc. Jap. Acad. 51, 467-472.
44. Ito, Y., et al. (1977) J. Physiol. submitted.
45. Evans, D.H.L., et al., (1958) J. Physiol. 143, 474-485.
46. Weber, A. and Herz, R. (1968) J. Gen. Physiol. 52, 750-759.
47. Ford, L.E. and Podolsky, R.J. (1972) J. Physiol. 223, 21-33.
48. Ito, Y., et al. (1973) Jap. J. Physiol. 24, 217-232.
49. Somlyo, A.V. and Somlyo, A.P. (1976) Smooth Muscle Pharmacology and Physiology, ed. M. Worcel & G. Vassort 369-380, INSERM, Paris.
50. Ebashi, S. and Endo, M. (1968) Prog. Biophysic. Molec. Biol. 5, 125-183.
51. Tomita, T. (1966) J. Physiol. 183, 450-468.

This work was supported by research grant No.148094 of the ministry of Education of Japan.

Excitation-Contraction Coupling in Smooth Muscle, R. Casteels et al. eds.

ELECTRICAL AND MECHANICAL ACTIVITIES IN THE GUINEA-PIG STOMACH MUSCLE

Tadao Tomita and Yasuzi Sakamoto
Department of Physiology, School of Medicine,
Fukuoka University, Fukuoka 814, Japan

SUMMARY

Effects of removal of the external Na on electrical and mechanical properties were studied in the circular muscle of the guinea-pig stomach, which had a prominent slow wave generated by a potential-independent process. The membrane was depolarized and the muscle tone was increased in Na-free solution. An addition of 5 - 10 mM Na recovered these effects. Ouabain did not essentially affect the effects of Na, but lowering the temperature to 10°C blocked them. It has been speculated that electrogenic Na-Ca exchange processes are responsible both for depolarization and for repolarization of the slow wave and that Na affects the membrane potential by modifying these processes.

INTRODUCTION

A slow rhythmic depolarization of the membrane is characteristic of some of the gastrointestinal muscles. The depolarization lasting several seconds appears at a fairly constant intrinsic frequency ranging from 3 to 18/min in different muscles, under the conditions where a nervous contribution is thought to be blocked. The slow depolarization has various terms such as the slow wave (1,2), the basic electrical rhythm (3), the basic organ specific rhythm (4), the pace-setter potential or the synchronizer potential (5) and the control potential (6). The slow wave usually triggers a repetitive spike activity on the top of the depolarization and the strength of mechanical response is closely related with the spike activity. However, the slow wave without the spike activity may appear accompanied with no or weak contraction. The slow wave is believed to be generated in the longitudinal muscle layer of the small intestine (7,8,9) and also of the stomach of the cat (1). However, in the colon of the cat (10) and in the stomach of the guinea-pig (11,12,13,14), it has been concluded that the circular muscle generates the slow wave.

Several hypotheses have been presented for the mechanism of the slow wave. The slow wave in longitudinal muscles of the cat duodenum and jejunum is explained by a rhythmic electrogenic Na-pump, since the wave is blocked by removing Na or K, or by applying ouabain, and also since rhythmic inward currents corresponding to the slow wave can be recorded when the membrane potential is clamped at a steady level (15,16,17). On the other hand, the slow wave in longitudinal muscles of the rabbit jejunum is explained by conductance increase initially to Na and later to Cl (18). In this preparation, inhibition of a Na-pump by various procedures abolishes the slow wave but does not depolarize the membrane to the peak depolarization of the slow wave. This is taken as evidence against the electrogenic pump hypothesis. In the circular muscle of the guinea-pig stomach, it is assumed that the slow wave consists of two components; one is generated by a potential-independent, active transport process (the first component) and the other by a potential-dependent process (the second component) triggered by the first component, as shown in Fig. 1 (13,14,19).

Since the electrical properties of smooth muscles seem different in longitudinal and circular layers, in the stomach and the intestine, and also in different species, it is possible that the slow wave in different muscles originates from different mechanism. However, none of the above hypotheses is based on sufficient analyses. The present communication is concerned with further studies on the circular muscle of the guinea-pig stomach along the line of the last hypothesis.

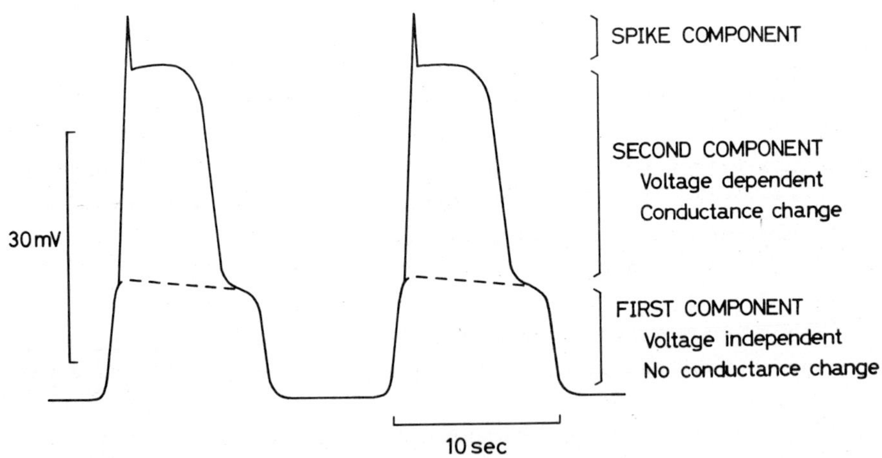

Fig. 1. Schematic diagram of slow waves in circular muscle of guinea-pig stomach, showing three components.

METHODS

The methods were essentially the same as previously described (13,19). After guinea-pigs were stunned and bled, a preparation was dissected in a circular direction from the antral region of the stomach free of mucosa. The single or double sucrose-gap method was mainly used for recording the electrical and mechanical activities. Occasionally the micro-electrode method was also employed to record transmembrane potentials from a piece of preparation exposed to the test solution, in order to confirm the results obtained with the sucrose-gap method. When only the mechanical responses were recorded, the preparation was suspended in a small organ bath (0.3 ml in capacity), and a strain gauge transducer and a potentiometric pen-recorder were used for recording.

RESULTS

Depolarization in Na-free solution: The membrane is depolarized by 10 - 20 mV when Na is completely replaced with choline, dimethyl diethanolamine, or tris-hydroxymethyl aminomethane (19). Substitution with Ca, Mg or sucrose also produced depolarization. During the onset of depolarization, prolongation of the slow wave was usually observed (Fig. 2a). Thus, it is possible that the depolarization in Na-free solution is partly due to a sustained generation of the slow wave or due to a blockade of a mechanism responsible for repolarization of the slow wave. The mechanism of the depolarization is difficult to explain. The depolarization may result from an increase in the Ca or Cl conductance, a decrease in the K conductance, a suppression of a hyperpolarizing electrogenic pump, a potentiation of a depolarizing electrogenic pump, or a combination of some of them.

Possible involvement of conductance changes in the depolarization: It is known that the uterine smooth muscle of the pregnant mouse is also depolarized by a removal of Na (20,21). In this preparation, the membrane resistance is decreased during the depolarization and Mn (0.6 - 1.2 mM) added to Na-free solution repolarizes the membrane. These results suggest that an increase in the Ca conductance may be the cause of depolarization for the uterine muscle. In the stomach muscle, Mn (0.2 - 1 mM) depolarized the membrane slightly and increased the frequency of slow wave. The depolarization started as if it was due to a prolongation of the first component, and subsided gradually, so that the slow wave did not show a clear first component after Mn application. At a high concentration (more than 1 mM), the slow wave was blocked within 10 min, and a sustained depolarization (5 - 10 mV) was observed. When Na was replaced with choline in the presence of Mn (0.3 mM), the depolarization developed gradually and the maximum depolarization was slightly smaller than in the absence of Mn. This depolarization was still observed after a complete abolition of the slow wave by 2 mM Mn. When Mn (0.5 mM) was applied about 5 min after exposing to Na-free solution, the muscle tone was decreased, but the membrane potential was not much affected (Fig. 2a). A removal of Ca showed more or less the same tendency to the effect of Mn. These results suggest that an increase in Ca conductance contributes to the early phase of depolarization, but this factor is probably small in the sustained depolarization. This is also indicated by the fact that the membrane resistance, measured after 5 min in Na-free solution, was increased during the depolarization. Depolarization to the same degree by applying outward current in Krebs solution or by increasing the external K concentration reduced the membrane resistance, suggesting that the increase in membrane resistance in Na-free solution is not simply due to depolarization.

<u>Possible involvement of active transport in the depolarization</u>: The changes in membrane potential caused by an alteration of the external Na concentration are opposite to those expected from the Goldman's equation. Furthermore, there is a critical concentration

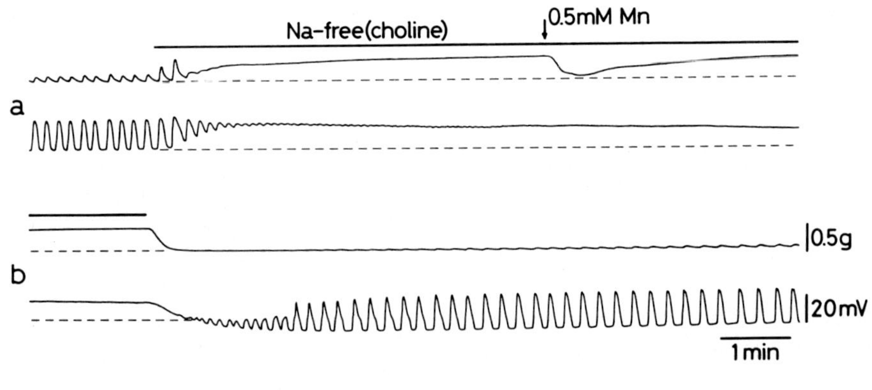

Fig. 2. Mechanical (top) and electrical (bottom) recordings with single sucrose-gap method. Mn (0.5 mM) was transiently applied 5 min after replacement of Na with choline (a). A rapid relaxation and recovery of slow wave on readmission of Krebs solution (b). Note a decrease in tension, without potential change by Mn.

between 5 and 10 mM which produces the depolarization or the repolarization. It is therefore likely that the effect of Na appears through some process, such as a carrier-mediated transport, rather than through a passive permeability. The depolarization in Na-free solution was not observed at a temperature below 10°C, while depolarization by excess K was not much affected by lowering the temperature. This also suggests that some active transport process is involved in the depolarization. In the Purkinje fibre of the dog heart depolarized in Na-free solution, a large hyperpolarization can be produced by adding 8 - 24 mM Na (22). This Na-induced hyperpolarization is explained to be due to an activation of a Na-pump because of a blockade by ouabain, a low temperature (9°C), or K removal, although other possibility such as a Na-dependent electrogenic Ca extrusion can not be disregarded. In the stomach muscle, the depolarization in Na-free solution also recovers by adding more than 5 mM Na, but this is not prevented by ouabain (2×10^{-6} M) (19) (Fig. 3). Ouabain produced prolongation of the first component at a low concentration (10^{-6} M) and depolarized the membrane by about 5 mV at a high concentration (10^{-5} M), but did not

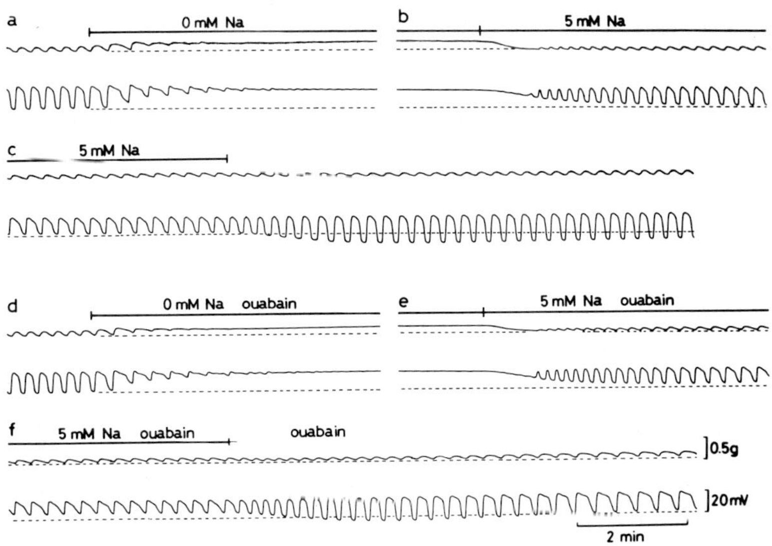

Fig. 3. Exposure to Na-free (choline) solution (a, b), addition of 5 mM Na (b, e) and readmission of Krebs solution (c, f). Ouabain (2×10^{-6} M) was added to Na-free solution in (d) and maintained in (e) and (f). Exposure to Na-free and 5 mM Na solution were for 10 min. Ouabain did not affect the recovery by adding Na, except prolongation of slow wave in Krebs solution.

usually abolish the slow wave. At a concentration of 5 x 10^{-7} M, ouabain completely blocked a K-induced hyperpolarization following a K-free treatment, which was supposed to be due to an activation of a Na-pump. In the presence of ouabain, a removal of Na still depolarized the membrane. Thus, the depolarization in Na-free solution does not seem to be the result of a suppression of the ouabain-sensitive Na-pump. Furthermore, developments of depolarization on removal of Na and repolarization of readmission of Na seem too quick to be explained by changes in intracellular Na concentrations affecting the Na-pump activity.

In many preparations, the membrane is depolarized by about 5 mV in K-free solution and the first component becomes difficult to distinguish from the second component (19). However, in some preparations with a prominent first component, the membrane potential was not much affected in K-free solution and the first component was also rather resistant. When Na was removed from K-free solution, there was a transient hyperpolarization followed by depolarization of various degrees in different preparations. When K was removed after depolarization had been fully developed in Na-free solution, the membrane was repolarized by 5 - 10 mV and a rhythmic oscillation of the membrane potential appeared. Therefore, in the absence of K or in the absence of Na, the membrane potential behaves, at least qualitatively, as if it were determined by ionic concentration gradients. Although the electrogenic Na-pump is partially involved in determining the membrane potential, some other processes seem to be contributed to the depolarization in Na-free solution or in the generation of the first component of slow wave.

Mechanical responses in Na-free solution: Each slow wave produces a phasic contraction. When Na was removed, there was a transient increase in phasic contractions, related to an increase in duration of the slow wave and often in spike activity, and the muscle tone, a level of the maximal relaxation between phasic contractions, was increased probably as a result of depolarization (Fig. 4). Addition of more than 5 mM Na to Na-free solution produced the complete relaxation of muscle tone and this process was much faster than repolarization of the membrane (Figs. 2a, 3b and 4). The relaxation by readmission of Na was not affected by ouabain (Fig. 3e). A rapid decrease in the tone was also observed by application of a chelating agent, such as EGTA (5 mM), or of an agent which suppresses the Ca conductance, such as Mn (0.5 mM)

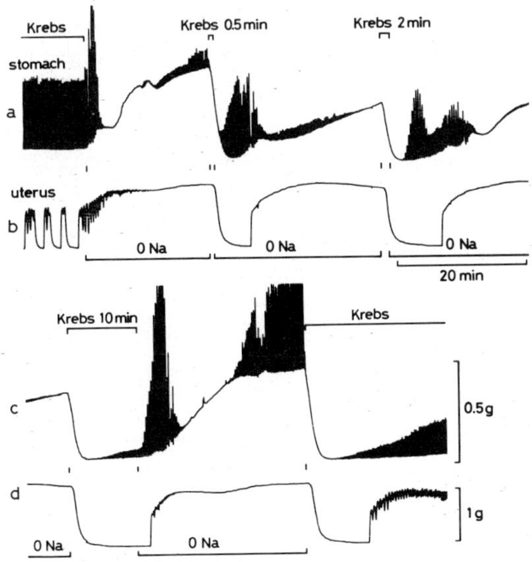

Fig. 4. Mechanical records of muscle strips of stomach (a, c) and uterus (b, d) in the same organ bath. During exposure to Na-free (choline) solution, perfusion was switched to Krebs solution for 30 sec, 2 min and 10 min as indicated. (c) and (d) are continuous records of (a) and (b).

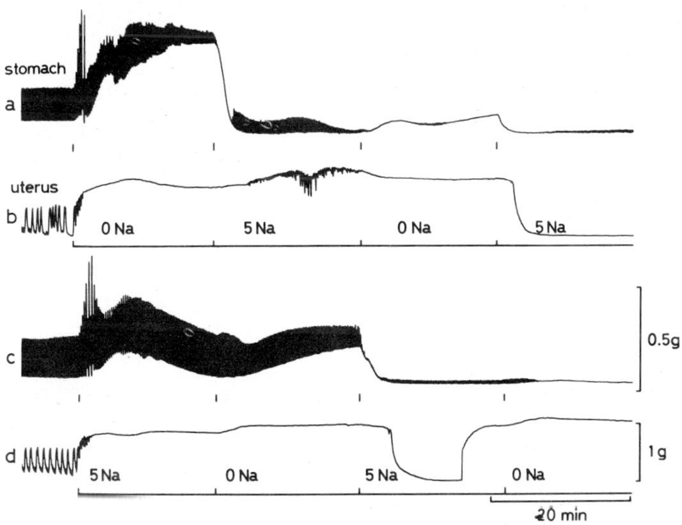

Fig. 5. Mechanical response of stomach (a, c) and uterine (b, d) muscles to exposure to Na-free (choline) and 5 mM Na solution as indicated. Before starting records of (c) and (d), the preparations were recovered in Krebs solution for 60 min. Note that second exposure to the same solution produced very different responses.

(Fig. 2) or Ca (10 mM), suggesting an increase in Ca conductance during depolarization in Na-free solution. Essentially similar results were obtained in the organ bath experiments, observing mechanical responses. After a transient increase in phasic contractions, there was usually a period of suppression of the contractions before the phasic contraction and the muscle tone increased again. The guinea-pig uterine muscle dissected in a longitudinal direction also produced an increase in muscle tone in Na-free solution. An increase in muscle tone, which had a close relationship with depolarization, produced by a second exposure to Na-free solution was dependent on the interval during which the normal Krebs solution was applied (Fig. 4). When the exposure to Krebs solution was short, the rate of development of muscle tone was slower. The maximum increase in tone was obtained at least 10 min after exposure to Krebs solution. Although 3 - 5 mM Na added to Na-free solution completely reduced the muscle tone, a second removal of Na following this treatment produced only very small contracture (Fig. 5a). Therefore, it is likely that a proper intracellular Na concentration appears to be prerequisite for a large increase in the tone in Na-free solution, and that once the tone is increased, Na is not necessary for maintaining. The first exposure to 5 mM Na solution produced a transient increase in the tone lasting about 10 min, and a complete removal of Na increased the tone again (Fig. 5c). After an exposure to Na-free solution, an addition of 5 mM Na produced a quick relaxation. Similar results were obtained by observing electrical responses. Thus, a treatment with Na-free solution increased the sensitivity to the inhibitory action of Na, probably by increasing a concentration gradient due to a loss of intracellular Na. The uterine muscle was less dependent on changes in Na concentration, compared with the stomach muscle.

DISCUSSION

In order to explain the generation of the first component of slow wave and the changes in membrane potential caused by an alteration of the external Na concentration, the following mechanism is tentatively hypothesized (Fig. 6). Inward movements of Ca and outward movements of Na are coupled and the external K is necessary for proper functioning of this process. The system carries a positive charge inward, causing depolarization, i.e. the first component. Under normal conditions, the Ca concentration gradient may drive the process. However, when the external Na is removed, the concentration gradient for Na is reversed and the process is facilitated to depolarize the membrane. After reduction of the external Ca or K, or after depletion of the intracellular Na, the depolarization becomes small and slow. Another mechanism contributing an early rapid depolarization may be an increase in Ca permeability, i.e. a potentiation and prolongation of the second component of slow wave. Ca binding would be increased by reducing the external Na and this leads to an increase in Ca permeability.

Repolarization of the first component is due to a similar mechanism to that for depolarization but working in the reversed direction, and the intracellular Na is necessary for proper functioning. This system carries a positive charge outward, repolarizing the membrane. A contribution of the intracellular Na is thought because of rather slow development of repolarization on readmission of Na. The Na concentration gradient may drive the process for repolarization. Therefore, a reduction of the intracellular Na concentration following a treatment with Na-free solution increases the sensitivity to readmission of Na in causing repolarization and relaxation.

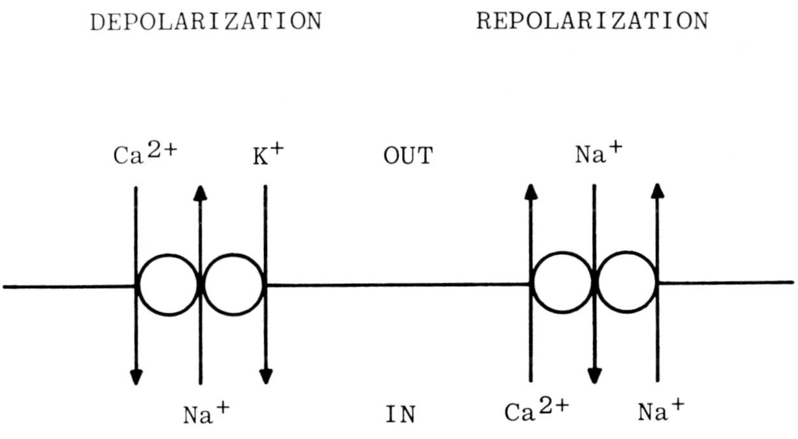

Fig. 6. A model of transporting system responsible for depolarizing and repolarizing phases of the first component of slow wave. Ca concentration gradient drives depolarizing system and Na concentration gradient repolarizing system. Movement of Ca also affects mechanical response.

REFERENCES

1. Papasova, M.P., Nagai, T. and Prosser, C.L., 1968, Am. J. Physiol. 214, 695-702
2. Prosser, C.L. and Bortoff, A., 1968, In: Handbook of Physiology, Alimentary Canal, IV, Am. Physiol. Soc. pp. 2025-2050
3. Bass, P., Code, C.F. and Lambert, E.H., 1961, Am. J. Physiol. 201, 287-291
4. Golenhofen, K. and Lammel, E., 1972, Pflügers Arch. 331, 233-243
5. Code, C.F. and Carlson, H.C., 1968, In: Handbook of Physiology, Alimentary Canal, IV, Am Physiol. Soc. pp. 1903-1916
6. El-Sharkawy, T.Y. and Daniel, E.E., 1975, Am. J. Physiol. 229, 1268-1276
7. Kobayashi, M., Nagai, T. and Prosser, C.L., 1966, Am. J. Physiol. 211, 1281-1291
8. Bortoff, A., 1965, Am. J. Physiol. 209, 1254-1260
9. Bortoff, A. and Sachs, F., 1970, Am. J. Physiol. 218, 576-581
10. Christensen, J., Caprilli, R. and Lund, G.F., 1969, Am. J. Physiol. 217, 771-776
11. Kuriyama, H., Osa, T. and Tasaki, H., 1970, J. gen. Physiol. 55, 48-62
12. Magaribuchi, T., Ohbu, T., Sakamoto, Y. and Yamamoto, Y., 1972, Jap. J. Physiol. 22, 333-352
13. Ohba, M., Sakamoto, Y. and Tomita, T., 1975, J. Physiol. 253, 505-516
14. Ohba, M., Sakamoto, Y. and Tomita, T., 1976, In: Smooth Muscle Pharmacology and Physiology, M. Worcel and G. Vassort (Eds), pp. 301-316, INSERM, Paris
15. Job, D.D., 1969, Am. J. Physiol. 217, 1534-1541
16. Liu, J., Prosser, C.L. and Job, D.D., 1969, Am. J. Physiol. 217, 1542-1547
17. Connor, J.A., Prosser, C.L. and Weems, W.A., 1974, J. Physiol. 240, 671-701
18. El-Sharkawy, T.Y. and Daniel, E.E., 1975, Am. J. Physiol. 229, 1287-1298
19. Ohba, M., Sakamoto, Y. and Tomita, T., 1977, J. Physiol. 267, 167-180
20. Osa, T., 1971, Jap. J. Physiol. 21, 607-625
21. Osa, T., 1973, Jap. J. Physiol. 23, 113-133
22. Wiggins, J.R. and Cranefield, P.F., 1974, J. gen. Physiol. 64, 473-493

Excitation-Contraction Coupling in Smooth Muscle, R. Casteels et al. eds
© 1977 Elsevier/North-Holland Biomedical Press

SENSITIVITY OF SPIKES AND SLOW WAVES IN SHEEP CAROTID ARTERIES TO BLOCKING AGENTS

W.R. Keatinge

Physiology Department
London Hospital Medical College
Turner Street, London, E.1. 2AD.
Great Britain

INTRODUCTION

Procaine and tetrodotoxin classically block voltage-dependent Na channels. They accordingly block Na-based action potentials of nerve and muscle, and slow the Na-based upstroke of the cardiac action potential (1,2,3). Verapamil and Mn are generally effective blocking agents for voltage-dependent Ca channels, blocking the Ca influx in nerve and cardiac muscle as well as the largely Ca-based action potentials of some smooth muscles (4).

However the spike channels in smooth muscle present anomalies. Na-based spikes are obtained easily in sheep carotid arteries (5,6) and with more difficulty in intestinal muscle (7,8). These are not blocked by tetrodotoxin. Slow waves in intestinal smooth muscle generally require Na to some extent (9,10) and since they are sometimes blocked by ouabain they have sometimes been attributed to a rhythmically active electrogenic Na pump.

Slow waves as well as spikes could be elicited from sheep carotid arteries in Na and Ca containing solutions (11) particularly when moderate concentrations of procaine were present (12), but it was found that slow waves could also be induced by K depolarization when Mn was present even in the absence of Na and Ca (13). The present experiments were designed to show whether there was any difference between the sensitivity of spikes and slow waves in these arteries to tetrodotoxin, procaine, and verapamil, and whether slow waves in the arteries were prevented by ouabain.

METHODS

The sucrose-gap method used for the present study has been described previously (11.) Helical strips of sheep carotid artery were placed in this apparatus and were subjected to K depolarization, with progressive replacement by K of Na in solution containing (mM) NaCl 133; $NaHCO_3$ 15, KCl 4.7 glucose 7.5 (Ca-free saline); or by replacement by K of Tris in solution containing Tris (Trisma base, Sigma, with HCl to bring pH to 7.35) 148; $KHCO_3$ 4.7; glucose 7.5 (Tris + Mn solution). Solutions were at 36±1°C and were gassed with 95% O_2 + 5% CO_2.Na EDTA (BDH), ouabain (BDH), verapamil (Pfizer) and tetraethylammonium (TEA, Schuchardt) were prepared as concentrated solutions and added to perfusion solutions as required. Since procaine was required in high final concentration, perfusion solutions with procaine (BDH) were made up specially, with the main ions (Na or Tris, and K) reduced by the same molar concentrations as the procaine added, and with pH adjusted by HCl.

RESULTS

Fig 1a shows spike discharges that developed spontaneously in an artery that had been in Ca-free saline for 30 min. It also shows acceleration of spike frequency when EDTA 1 mM was added. Spike discharges were recorded in seven out of eight similar experiments, either in the Ca-free saline alone or after addition of EDTA. Similar experiments again produced spike discharges in the presence of tetrodotoxin 10μM for 30 min (two experiments). Similar spikes were also obtained in

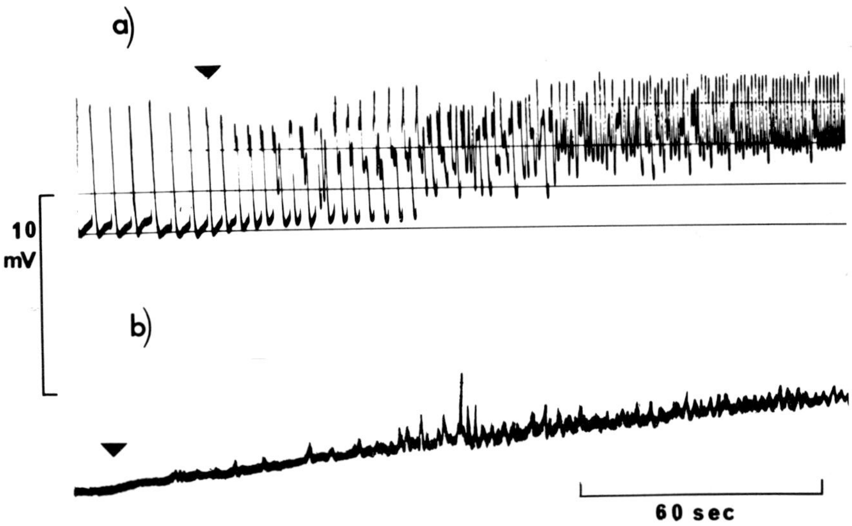

Fig 1. Spike discharges in Ca-free saline. EDTA 1mM added at arrow.

(a) Without blocking agent
(b) With verapamil 40μM present, and TEA 10 mM

similar experiments after 30 min in the presence of procaine 2 - 10 mM (13 out of 15 experiments) but not with procaine 20 mM or 40 mM (12 experiments). In similar experiments with verapamil instead of procaine, spike discharges were recorded with verapamil 2 - 10 µM (three out of five experiments). No spikes were obtained with verapamil 20µM (eight experiments), but Fig 1b shows that after addition of TEA 10 mM to reduce K permeability (14) spikes of an irregular pattern were obtained in the presence of verapamil 40µM (two out of five experiments). Even with TEA present no spikes could be obtained in the presence of verapamil 80 µM in any of eight experiments.

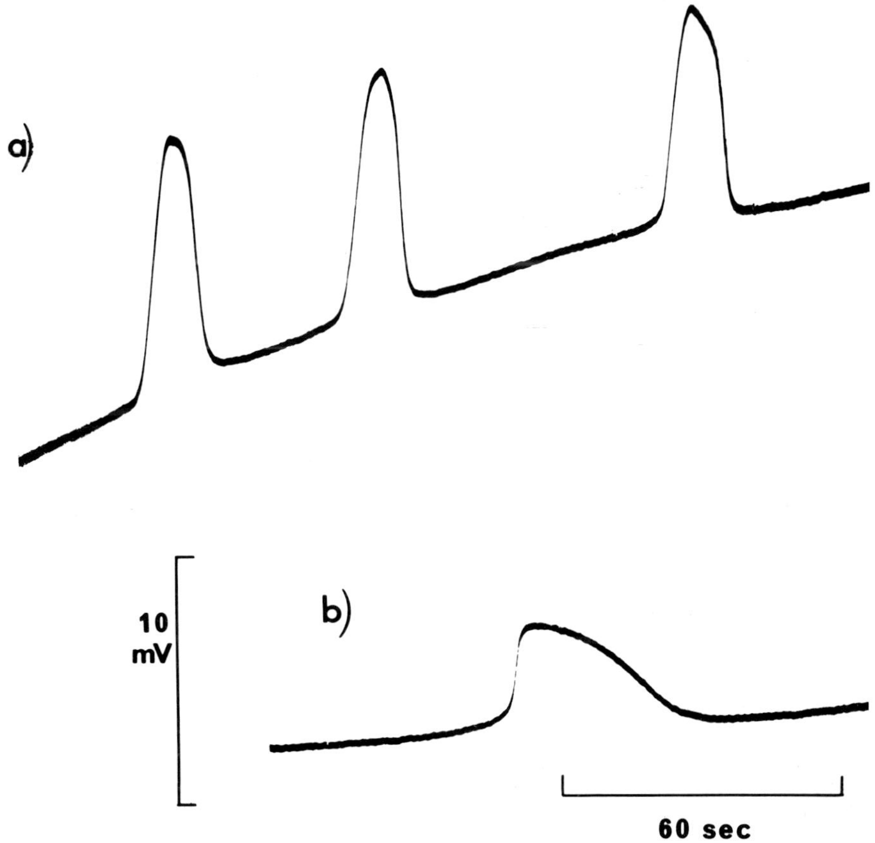

Fig 2. Slow discharges induced by K depolarization in Ca-free and Na-free Tris solution containing Mn.

K concentration increasing from 4.7 to 94mM.
Mn 5mM present throughout.
(a) Without blocking agent
(b) With verapamil 80µM present.

Fig 2a shows slow discharges that developed after 30 min in Na-free Tris solution containing Mn 5 mM when depolarization was produced by increasing K concentration to 94 mM. Slow discharges of this kind were obtained in two out of three such experiments. Similar experiments produced similar slow waves in the presence of tetrodotoxin 5μM (two out of four experiments). Procaine 40mM failed to prevent the slow waves (two experiments). Verapamil also failed to prevent slow waves, even in high concentration. Slow waves were recorded after 30 min in the presence of verapamil 80-160μM (two out of four experiments). Fig 2b shows one of these. Ouabain 1 mM did not prevent slow discharges from occurring during K depolarization in Tris solution with Mn (two out of three experiments).

DISCUSSION

It is clear that both Na-based spike discharges and Mn-based slow waves were markedly insensitive to conventional blockers of both Na channels and divalent cation channels, and that the insensitivity was particularly marked in the case of the slow waves. The results also show that the slow waves in this smooth muscle did not depend either on passive Na movement or the Na pump, since they could be obtained in the absence of Na and in the presence of ouabain. Although the apparent sensitivity of spikes to verapamil could be much reduced by the addition of TEA in order to reduce background K permeability, the spikes continued to be more easily suppressed than the slow waves. Since the spikes were also more easily suppressed than slow waves by procaine, which like TEA greatly reduces K permeability in these arteries, the findings are most easily explained by a spike channel that is readily permeable to Na and is blocked only by high concentrations of verapamil and procaine, and a slow channel that is permeable to Mn and is even less readily blocked by verapamil and procaine.

SUMMARY

Sheep carotid arteries produced spike discharges spontaneously in Ca-free saline, and produced slow waves when subjected to K depolarization with Mn present. The slow waves did not require either Na or Ca when Mn was present. The Na spikes and Mn waves were both unaffected by tetrodotoxin in a concentration of 5-10μM. The spikes were blocked by procaine 20 mM, while the slow waves could be obtained in the presence of procaine 40 mM. The spikes were blocked by verapamil 80μM, even when TEA had been added to reduce K permeability. Slow waves could still be obtained in the presence of verapamil 160μM, and also in the presence of ouabain 1 mM. The results indicate that the slow waves were not dependent on either passive Na movement or the Na pump, and probably involved a voltage-dependent divalent cation channel that was capable of admitting Mn and was markedly insensitive to conventional blocking agents for both Na and Ca channels.

REFERENCES

1. Kao. C.Y. (1966). Tetrodotoxin, saxitoxin and their significance in the study of excitation phenomena. Pharmac. Rev. 18 997-1049
2. Weidmann S. (1955). Effects of calcium ions and local anaesthetics on electrical properties of Purkinje fibres. J. Physiol. 129 568-582
3. Inoue F. & Frank G.B. (1962). Action of procaine on frog skeletal muscle. J. Pharmacol. exp. Ther. 136 190-196
4. Golenhofen K. & Lammel. E. (1972). Selective suppression of some components of spontaneous activity in various types of smooth muscle by iproveratril (Verapamil). Pfluegers Arch. 331 233-243
5. Keatinge W.R. (1968). Ionic requirements for arterial action potential. J. Physiol. 194 169-182
6. Keatinge W.R. (1968). Sodium flux and electrical activity of arterial smooth muscle. J. Physiol 194 183-200
7. Golenhofen K. & Petranyi P. (1969). Spikes of intestinal smooth muscle in calcium-free solution. J. Physiol. 200 136-137P
8. Bulbring E. & Tomita T. (1970). Effects of Ca removal on the smooth muscle of the guinea-pig taenia coli. J. Physiol. 210 217-232
9. Daniel E.E. (1965). Effects of intra-arterial perfusions on electrical activity and electrolyte contents of dog small intestine. Can.J. Physiol. Pharmacol 43 551-557
10. Connor J.A. Prosser. C.L. & Weems W.A. (1974). A study of pacemaker activity in intestinal smooth muscle. J. Physiol 240 671-701
11. Keatinge W.R. (1964). Mechanism of adrenergic stimulation of mammalian arteries and its failure at low temperatures. J. Physiol 174 184-205

12. Jacobs A. & Keatinge W.R. (1974). Effects of procaine and lignocaine on electrical and mechanical activity of smooth muscle of sheep carotid arteries. Brit. J. Pharmacol. 51 405-411

13. Keatinge W.R. (1976). Extracellular Ca and response of sheep carotid artery to depolarization. J. Physiol. 258 73-74P

14. Cooper W.D. GoodfordP.J. Hardy C.C. Herring J. Hind C.R.K. & Keatinge W.R. (1975). Effects of procaine and tetraethylammonium on ^{42}K efflux from smooth muscle of sheep carotid arteries. J. Physiol 246 70-71P.

Excitation-Contraction Coupling in Smooth Muscle, R. Casteels et al. eds.

ELECTRICAL PROPERTIES OF FRESHLY ISOLATED SINGLE SMOOTH MUSCLE CELLS

Joshua J. Singer and John V. Walsh, Jr.
Department of Physiology
University of Massachusetts Medical School
55 Lake Ave., N.
Worcester, Massachusetts 01605

INTRODUCTION

When intact tissue is employed for the experimental study of the electrical properties of the smooth muscle cell membrane, uncertainties may arise because of the presence of diffusional barriers and neural elements within the tissue and low resistance electrical coupling between the smooth muscle cells themselves. For example, interpretation of changes in potential due to current injection through a micropipette at or near the recording site must take into account the geometry of the smooth muscle cell syncytium[1]. To overcome difficulties such as these we have carried out electrophysiological experiments on freshly isolated single smooth muscle cells[2]. The experiments are performed on the same day as the isolation in order to avoid long-term changes which may occur upon culturing the cells. The cells appear to be basically unaltered by the isolation procedure[3] and have a number of characteristics which make them a promising preparation for studies of the electrical properties of smooth muscle: 1) they contract reversibly in response to both extracellular[4,5] and intracellular stimulation; contraction may be elicited many times; 2) they contract upon exposure to high concentrations of external K^+ [4,6]; 3) cholinergic receptors are intact since the cells contract in response to carbachol and acetylcholine and the K_I for atropine antagonism is of the same order of magnitude as in intact tissue[6]; 4) Beta adrenergic receptors are also intact since isoproterenol inhibits the contractile response to carbachol and this inhibition is antagonized by propranolol[7]; 5) as shown below, action potentials can be recorded.

METHODS

Suspensions of isolated smooth muscle cells were prepared by enzymatic digestion of the stomach muscularis from <u>Bufo marinus</u> by a modification[5,8] of the procedure originally described by Bagby et al.[4] Before each experiment, aliquots of the cell suspension in standard amphibian physiological saline were usually diluted 11 to 1 with a 2 mM Tris-HCl buffer (pH 7.2-7.4) containing 108 mM NaCl, 3 mM KCl, 1 mM $MgSO_4$, 10 mM $CaCl_2$ and 11.1 mM glucose. Changes in ionic concentrations in the Tris buffer were made isosmotically. The experiments were carried out at higher than physiological concentrations of Ca^{++} which improved the quality and stability of the recordings of membrane potential, increased the input resistance and made it easier to elicit action potentials (cf. Stefani and Steinbach[9]).

Recordings of membrane potential were made with high resistance (80-250 MΩ) 3 M KCl filled micropipettes. Cells were impaled under direct observation using phase contrast or differential interference contrast optics at approximately 500X magnification (Fig. 1). Upon penetration the cell contracts and then relaxes partially.

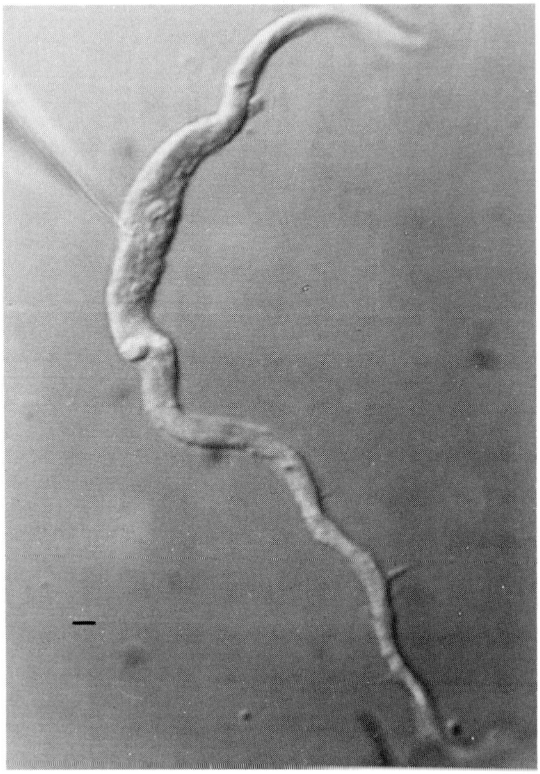

Fig. 1. Photomicrograph of an isolated single smooth muscle cell (Nomarski optics). Electrode can be seen above the cell. Calibration bar = 5 μ

Current steps were injected through the recording pipette using the current injection capability of the W-P Instruments M701 microprobe system; the membrane potential was monitored simultaneously. The potential drop across the electrode caused by the injected current was balanced out just before penetrating the cell by determining the range of current injection over which the electrode behaved in a linear manner. The M701 current monitor was used to measure the injected current and its accuracy was verified by using precision resistors and a W-P Instruments 180 virtual ground.

Cell membrane area was calculated (assuming a cylindrical profile) from the planar area obtained from an enlarged photomicrograph of each cell.

RESULTS

Action Potential: An important indication of the viability of these cells as well as a prerequisite for using them to study electrical activity is the ability to produce an action potential. A single action potential can be elicited from these cells with a depolarizing step of current (Fig. 2A) or at the offset of a hyperpolarizing step (Fig. 2B,C). Either sort of stimulus at times produced a brief train of action potentials (Fig. 2D). Action potentials were readily obtained in low sodium (7.4 mM) medium as demonstrated in Fig. 2E suggesting a Ca^{++} dependence of the action potential. (A more negative resting potential, also shown in Fig. 2E, was often seen in low Na^+ medium.) Tetraethylammonium chloride (TEA) (25 mM or greater) increased the duration of the action potential by a factor of 3 or more, suggesting that K^+ contributes to the repolarizing current.

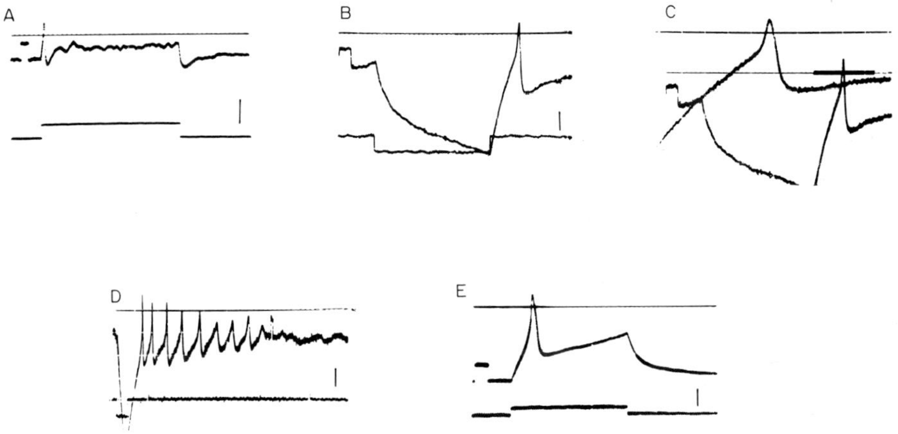

Fig. 2. Action potentials from isolated cells elicited by current passage through the recording pipette. In each case the uppermost trace indicates zero potential, the middle trace records the membrane potential and the lowermost trace monitors the current. Calibration pulses appear as upward deflection: +20 mV, 100 msec at the beginning of the membrane potential trace; current calibration bar = 100 pA at the end of the current trace.

A & B. Action potential elicited by depolarizing and hyperpolarizing steps of current.

C. Upper trace shows the same action potential as the lower trace but on an expanded (4X) time scale to give an idea of the time course of the action potential.

D. Multiple action potentials elicited at the break of a single hyperpolarizing pulse. (Calibration on membrane potential trace occurs just after the train of spikes.)

E. Action potential elicited in low sodium (7.4 mM) solution. In A-D, the sodium concentration is 110 mM. Thus, it appears that the action potential can be calcium mediated. This result is obtained even when the usual calcium (c.f. Methods) is lowered slightly, as it was in this case.

Under our recording conditions, these cells, unlike the cultured isolated taenia coli smooth muscle cells studied by Purves et al.[10], do not produce any spontaneous activity.

Evidence that the isolated cell behaves an an isopotential surface: Both theoretical considerations and experimental observations suggest that the length constant, λ, of the isolated cell is sufficiently greater than its actual length so that the cell behaves as an isopotential surface at least during subthreshold perturbations of the membrane potential.

Theoretically, the isopotential condition was demonstrated using the "short cable" equations given by Weidmann[11]. To do this input resistance was calculated from the steady-state plateau potential produced by a small hyperpolarizing current step passed through the balanced recording pipette; cell length and a calculated average diameter were obtained from photomicrographs; and internal resistance was assumed to be 200 Ω-cm. Using these values, it was shown that an alteration in the steady-state potential at one end of the cell was often in excess of 98% of that at the opposite end where current was assumed to be injected. Since the cell contracts upon impalement and relaxes only partially, it is then shorter and wider than in the original relaxed state, thus giving an even better approximation to the isopotential condition.

Experimentally, in response to a small step of hyperpolarizing current the membrane potential decays as a single exponential. Within the limits of the detection procedure, the exponential decay indicates that the isolated cell constitutes an isopotential surface, at least when "passive" properties are considered. Thus, the entire membrane may be modelled as a single parallel RC circuit.

Membrane Capacity: The membrane time constant in media containing different concentrations of Na^+ and Ca^{++} varied by over an order of magnitude in a series of 17 cells. This variation results primarily from large variations in input resistance from 100 MΩ (our minimal acceptable level for these studies), to over 3000 MΩ. Higher input resistances were obtained in higher Ca^{++}, lower Na^+ media. However, the specific membrane capacity calculated for each cell from the time constant, input resistance, and measures of membrane surface area varied by only a small amount; Cm = 1.51 ± .56 $\mu F/cm^2$ (mean ± S.D., n = 17).

Current-Voltage Relationship: The cells exhibited outward-going rectification when the current-voltage relation was obtained by using long steps of current. A typical example is illustrated in Figure 3. Many cells showed an even more marked rectification with a slope close to vertical in the region of depolarizing currents. As is evident from the traces in Figure 3A, the membrane potential did not always plateau when the hyperpolarization exceeded a certain level but continued to increase as if the membrane conductance were decreasing. In many cells

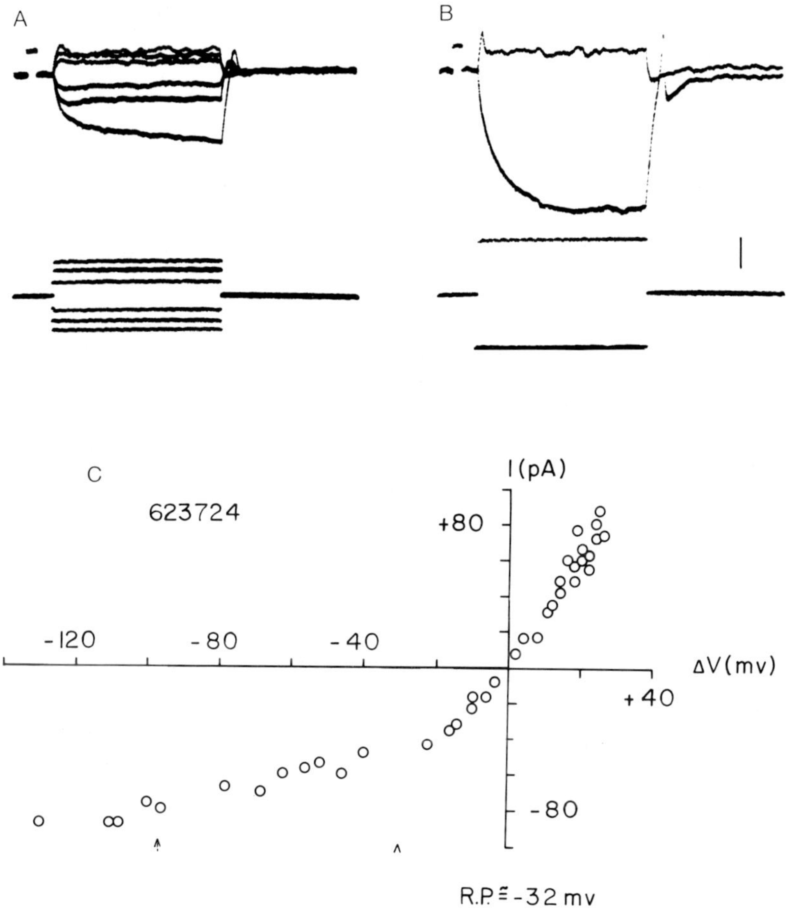

Figure 3.

A & B. Superimposed traces of deviations from resting potential (above) caused by the current steps (below). Calibration pulse at onset of voltage trace = + 20 mV, 100 msec; current bar = 50 pA, at end of trace in B.

C. Current-voltage relationship obtained from the series of records illustrated in A and B. The steady-state value of the membrane potential was plotted at each current level, except when a plateau was not attained; then the value at the end of the excursion was taken. At larger negative excursions (beginning at Λ in the graph and illustrated by the most negative going trace in A), the potential deviation did not plateau but continued downward. At still larger hyperpolarizations, the response often flattened out but with increased irregularity in the trace (illustrated in B and beginning at ↑ in the graph).

this continued hyperpolarization was less marked than in the cell illustrated. When the membrane was polarized to still more negative values, the potential plateaued again (arrow in Fig. 3C) but with more variation in the trace Fig. 3B. Furthermore, in some cases large hyperpolarizing currents cause the potential excursion to pass through a maximum and return to a less negative steady-state value. In summary, the current-voltage relation which we have observed may involve a series of complex conductance changes on the hyperpolarizing limb as well as in the depolarizing region where active membrane properties are of obvious importance.

No indication of inward going (anomalous) rectification has been seen in the large series of cells which have now been examined. Since anomalous rectification could be masked by an outward going rectification due to the external potassium concentration being lower than the internal concentration[12], experiments were carried out in which the external potassium was raised to 108 mM by replacing NaCl and KCl with K_2SO_4 and sucrose. In this sort of high potassium, low chloride medium the current-voltage relationship was either linear or displayed an outward going rectification which was less marked than at normal potassium levels. However, there was no indication of anomalous rectification.

In the presence of TEA (45.5 mM or greater), there was a marked decrease in outward going rectification. This consisted of a pronounced decrease in chord conductance when depolarizing current was injected, but with no apparent change in the conductance when a low level of hyperpolarizing current was injected. The decrease in outward going rectification is consistent with the broadening of the action potential caused by TEA.

Discussion: Although the results presented here are preliminary in nature, they have the advantage of being direct because the isolated cell removes many of the difficulties posed by a syncytium. For example, our mean value for specific membrane capacity is at the low end of the range suggested by Tomita[1] for taenia on the basis of an elegant and rigorous but indirect method which demands an accurate fiber radius, conduction velocity, and an assessment of the total internal resistance. With the isolated cell, only a measure of the cell surface area and the exponential decay of membrane potential in response to a step of current is required, both of which are readily made. Indeed our mean value for C_m may be falsely high so that the actual value may be close to that of many cells[12], including the value of the surface membrane capacity for skeletal muscle[13]. Similarly, in studies of the current-voltage relation, the syncytium prohibits the use of intracellular current injection as Tomita[14] has pointed out. However, the use of external polarization makes it difficult to estimate the actual current injected into the cell from which the recording is made. In contrast, when current-voltage relations are measured in the isolated cell, the exact amount of injected current can be monitored. Furthermore, when tissue is used, linearizing effects on rectifying properties which result from the spread of current through the syncytium complicate analysis of I-V relations, especially when ionic substitutions are made. These substitutions may result in an alteration of the membrane resistance and consequently the length constant. Such difficulties arising from the linearizing effects of ionic substitutions are encountered when potassium is raised to bring out possible anomalous rectification[15]. Because the isolated cell is essentially an isopotential surface, linearizing effects associated with a complex cable do not arise; thus, outward going rectification can be demonstrated readily as can the absence of anomalous rectification, even in high potassium solutions.

In summary, our initial results which have dealt primarily with passive properties of the membrane and provided measures of membrane capacity and rectification are in general agreement with the best measurements obtained in tissue preparations. It is pleasing to find this sort of agreement for it suggests that the isolated cell is in some ways representative of smooth muscle in tissue from a variety of sources. Finally, the preliminary results presented here provide evidence that the isolated cell preparation lends itself to quantitative measures of membrane properties.

ACKNOWLEGEMENTS

Supported by N.I.H. Grant HL 14523 and the grant #13-512-756 from the American Heart Association, Massachusetts Affiliate, Inc.

REFERENCES

1. Tomita, T. (1970) Electrical Properties of Mammalian Smooth Muscle. In: Smooth Muscle. E. Bülbring, A.F. Brading, A.W. Jones, and T. Tomita (Eds.) pp. 197-243, Williams and Wilkins Co., Baltimore.
2. Walsh, J.V., and Singer, J.J. (1977) XXVII International Congress of Physiological Science.
3. Fay, F.S., Cooke, P.H., and Canaday, P.G. (1976) Contractile response of isolated smooth muscle cells. In: Physiology of Smooth Muscle, E. Bulbring and M.F. Shuba (Eds.) pp. 249-264, Raven Press, New York.
4. Bagby, R.M., Young, A.M., Dotson, R.S., Fisher, B.A., and McKinnon, K. (1971) Nature 234:351-352.
5. Fay, F.S., and Delise, C.M. (1973) Proc. Natl. Acad. Sci. U.S.A. 70:641-645.
6. Fay, F.S., and Singer, J.J. (1977) Am. J. Physiol. 232:C144-C154.
7. Honeyman, T.H., and Fay, F.S. (1975) Fed. Proc. 34:361.
8. Singer, J.J., and Fay, F.S. (1977) Am. J. Physiol. 232:C138-C143.
9. Stefani, E., and Steinbach, A.B. (1969) J. Physiol. 203:383-401.
10. Purves, R.D., Mark, G.E., and Burnstock, G. (1973) Pflügers Arch. 341:325-330.
11. Weidmann, S. (1952) J. Physiol. 118:348-360.
12. Cole, K.S. (1972) Membranes Ions and Impulses, University of California

Press, Berkeley.

13. Hodgkin, A.L., and Nakajima S. (1972) J. Physiol. 221:121-136.
14. Tomita, T. (1966) J. Theoret. Biol. 12:216-227.
15. Mekata, F. (1976) J. Physiol. 258:269-278.

Excitation-Contraction Coupling in Smooth Muscle, R. Casteels et al. eds.
© 1977 Elsevier/North-Holland Biomedical Press

ACTION OF ANGIOTENSIN II ON THE MEMBRANE POTENTIAL AND CONTRACTION OF RAT MYOMETRIUM

G. HAMON and M. WORCEL

U 7 INSERM Physiologie et Pharmacologie
Hôpital Necker - 75015-PARIS

INTRODUCTION

Angiotensin II (AtII) produces a contraction of visceral and vascular smooth muscles of different species (1). OHASHI et al (2) and SOMLYO and SOMLYO (3) have shown that this peptide produces a depolarisation of guinea pig taenia coli and vascular muscle, respectively.

In order to examine further the mechanism of action of AtII we have attempted to study the effects of the peptide on rat myometrium membrane potential.

MATERIALS AND METHODS

Uterine strips were obtained from female Wistar rats weighing 220-250g pretreated the day before the experiment with diethylstilboestrol dipropionate 0.5mg/Kg. The membrane potential and the isometric contractions were measured simultaneously by the method of BÜLBRING and TOMITA (4). The solutions used have the following composition (mM): 1) physiological salt solution (PSS), NaCl 120.8, KCl 5.9, $CaCl_2$ 2.5, $MgCl_2$ 1.2, NaH_2PO_4 1.2 and glucose 11.5 gassed with 95% O_2 and 5% CO_2 or the same solution in which $NaHCO_3$ and NaH_2PO_4 are replaced by tris HCl and gassed with 100% O_2. 2) KCl isotonic solution and isotonic sucrose solution of resistivity higher than 1 MΩ/cm. The experiments were performed at 35°C.

RESULTS

AtII produces a depolarisation of the myometrium with a threshold at concentrations from 5×10^{-9}M to 10^{-8}M.

At this dose, the membrane potential ocillates slowly, with spikes arising during the depolarisation phase. At the higher concentration of 10^{-6}M, which produces a maximum contraction of rat myometrium, AtII induces a depolarisation which usually attains 20mV. This depolarisation is accompanied by an increase in the input conductance. Both effects fade soon, even during the continuous perfusion of the strip suggesting a rapid desensitisation of the preparation (Fig.1).

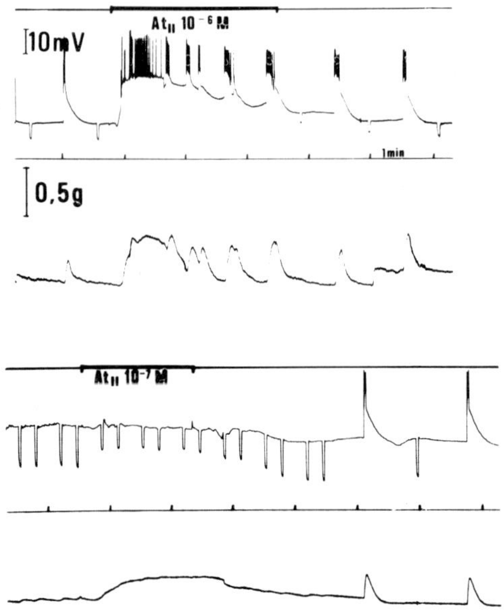

a.

b.

Figure 1. Effects of AtII in PSS. a) AtII 10^{-6}M produces a maximal depolarisation and contraction. b) During the perfusion of the muscle with AtII 10^{-7}M, the membrane potential is held at its base level, by an hyperpolarising current, adjusted manually. The input conductance increases independently from any variation in the membrane potential. AtII produces a tonic contraction.

Coincident with the depolarisation and spike production there is a contraction which does not seem to be exclusively dependent on a variation of the membrane potential. Indeed, when the membrane potential is held at baseline level by the injection of current, AtII 10^{-6}M is still capable of producing a tonic contraction coincident with an increase in membrane conductance.

In order to know which ions are involved in the increase in the input conductance produced by the peptide, we have examined the effects of AtII after different ionic substitutions.

Na^+ substitution. Myometrial strips were exposed during 10min to a solution in which all Na is replaced by Li. After this period, AtII 10^{-6}M produces a biphasic response. There is an initial, transitory hyperpolarisation of less than 5mV, lasting about 10sec, followed by a depolarisation of an amplitude varying between 5 and 10mV. Since these results might be explained by the partial permeation of Na^+ channels by Li^+ we have examined the action of the peptide after replacing Na^+ by other substituents. In the presence of a Mg^{2+} replaced solution, AtII 10^{-6}M again produces a transitory hyperpolarisation, the depolarisation being completely supressed after more than 15min of exposure (Fig.2).

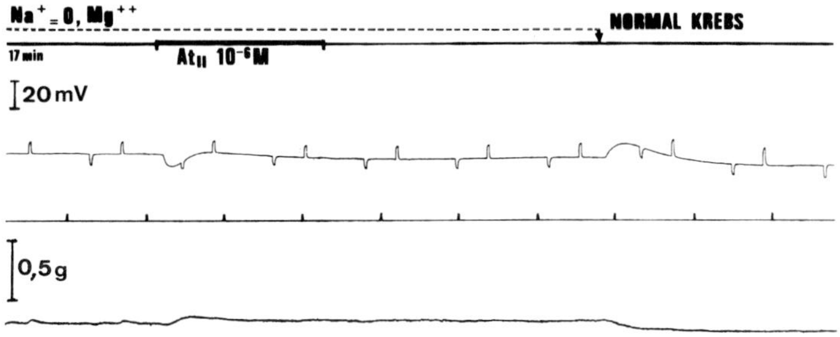

Figure 2. Effects of AtII 10^{-6}M, applied after 18min of exposure to a solution in which Na^+ has been substituted by Mg^{2+}. AtII produces a transitory hyperpolarisation followed by a brief depolarisation of 1 to 2mV;

This result suggests that the depolarisation induced by AtII is due to an increase in the membrane conductance to Na^+.It must be emphasized again that even if the depolarising action of AtII is supressed in the absence of extracellular Na^+ under these conditions, the exposure of the muscle to the peptide is accompanied by a weak tonic contraction. Even if the participation of the Na^+ ions in the depolarisation appears preponderant, an involvement of Cl^- ions cannot be excluded since E_{Cl^-} = -33mV in this preparation (7). Indeed the activation of another ionic conductance during depolarisation could explain why the hyperpolarisation observed in the absence of extracellular Na^+ is shortlasting.

Chloride substitution. Two different substituents have been used to replace extracellular Cl^-. In the first series of experiments Cl^- has been replaced by NO_3^-. Under these conditions there is a rapid depolarisation which does not seem to be due only to the inevitable change in the junction potential, since it is accompanied by a series of bursts of spikes associated with phasic contractions. AtII 10^{-6}M, administered more than 10min later still induces a depolarisation of aproximately the same size as observed in the presence of Cl^-. In order to confirm these findings, extracellular Cl^- was partially replaced in the second series of experiments by the nonpermeant anion cyclohexanesulfamate. The total or partial replacement of NaCl by Na cyclohexanesulfamate is rapidly followed by a depolarisation and spike discharge. Under these conditions after more than 7min following the total or partial replacement of Cl^- by cyclohexanesulfamate, the depolarising response to AtII is not changed significantly. The previous results do not confirm the participation of Cl^- in the increase in membrane conductance induced by AtII.

Potassium exclusion. As could be expected, the effects observed when all K^+ is eliminated from the bathing solution are rather complex. Immediately after the exposure to 0 K^+ solution the membrane depolarises rapidly. The membrane potential stabilises at a level of 10mV less negative than the control. This depolarising effect seems to be due to the inhibition of the Na^+-K^+ pump, which is electrogenic in this tissue (5), since there is a rapid hyperpolarisation which attains 20mV on readmitting K^+ in the bathing solution. Under exposure to a 0 K^+ solution, the size of the depolarisation obtained with AtII 10^{-6}M is not changed. This result, even if difficult to interpret in terms of the fate of the membrane permeability to K^+, excludes any effect of AtII on the Na^+-K^+ pump. This finding is in contradiction with the results presented by TÜRKER et al (6) who claimed that AtII acts through the stimulation of the Na^+-K^+ pump. Furthermore, we have tested the action of AtII 10^{-6}M after exposing the preparation to 10^{-3}M ouabain . Again, the polypeptide still has the same depolarising effect after blockade of the pump. Nevertheless, it must be noted that when the Na^+-K^+ pump is blocked either by ouabain or in 0 K^+ solution the rate of the desensitisation observed during the application of AtII 10^{-6}M is diminished indicating that stimulation of the electrogenic Na^+-K^+ pump may be at least partly responsible for the observed tachyphylaxis. The preceding experiments do not provide any information about a possible increase in K^+ permeability as was suggested by the short hyperpolarisation observed in the absence of extracellular Na^+. In order to clarify this point the action of AtII was examined in the presence of a solution in which Na^+ was replaced by Mg^{2+} and after addition of tetraethylammonium chloride (TEA). In most cases 10 or 20 minutes after exposure to TEA the

the hyperpolarisation observed in the absence of Na^+ is suppressed (Fig. 3).

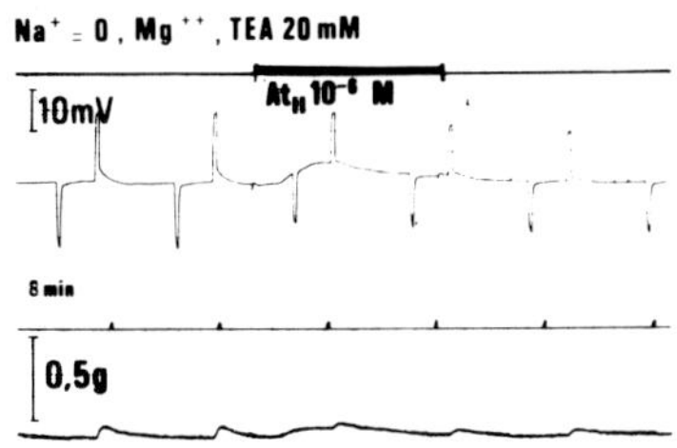

Figure 3. Action of AtII 10^{-6}M after 10min in a 0 Na^+, Mg^{2+} substituted solution, in the presence of 20mM TEA.

Nevertheless, in certain cases the hyperpolarisation is not completely eliminated. This occasional failure may be due to an insufficient diffusion of TEA to the internal surface of the cell membranes due to difficult diffusion in the central node of the sucrose gap. Another way to examine the problem is to reduce the E_{K^+} by increasing the external concentration of K^+ to 14.8mM. Under these conditions, and when the experiment is performed in the absence of Na^+, substituted by Mg^{2+}, the membrane potential does not change significantly. If the internal concentration of K^+ varies only slightly during a short exposure to this solution (15min), E_{K^+} should not be far from Em. (E_{K^+} = -64mV). As expected the initial hyperpolarisation induced previously by AtII 10^{-6}M is not suppressed.

The previous results suggest that AtII produces a simultaneous increase in G_{K^+} and G_{Na^+}. When exposed to a physiological solution, AtII would produce only a depolarisation due to the masking of the hyperpolarisation by the preponderance of the more important effect on G_{Na^+}.

Ca^{2+} substitution. In some experiments, $CaCl_2$ was replaced either by glucose or by Mg^{2+}. After 10min exposure to a medium in which all Ca^{2+} was substituted, AtII still produces a depolarization of the same size as in the control state, together with an enhancement of the input conductance. These effects are accompanied by a weak contraction. If a second application of AtII 10^{-6}M is repeated after a longer exposure to the "0 Ca^{2+}" solution (25min), the same depolarising effects are obtained but the contraction is this time either very weak or suppressed. In order to exclude further any possible primary effect on $G_{Ca^{2+}}$, we have tested the action of AtII in the presence of a solution containing Ca^{2+} 2.5mM and Mn^{2+} 5.0mM. Mn^{2+} blocks the diffusional passage of Ca^{2+} through the membrane and the Ca-Na exchange mechanism (8). Nonetheless, this solution does not affect the response to AtII, which produces as usual an increase in membrane conductance. As can be seen in Fig. 4, when the membrane potential is stabilized electrically, Mn^{2+} does not suppress the tonic contraction induced by AtII.

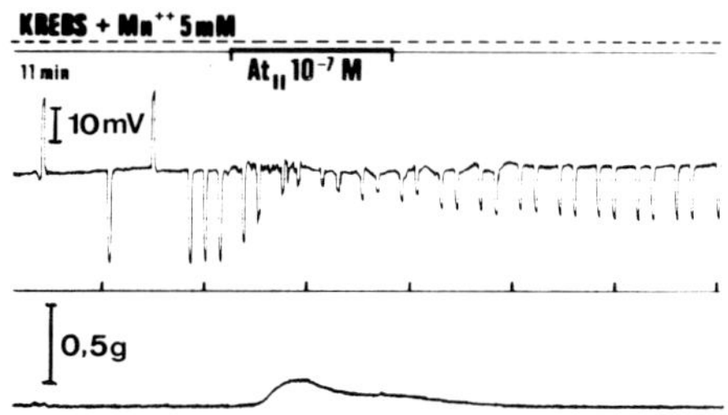

Figure 4. Effects of AtII 10^{-7}M in the presence of a PSS with 5mM Mn^{2+} added. The membrane potential is maintained at the base level, by an hyperpolarising current.

DISCUSSION

The peptide induces a rapid increase in the input conductance, immediately followed by a depolarisation, and bursts of spikes. A depolarisation in rat myometrium could be due to an increase in either G_{Na^+}, G_{Cl^-} or $G_{Ca^{2+}}$ given the ionic distribution in this tissue (7). Experiments with Na^+ substituted, Li^+ and Mg^{2+} replaced solutions indicate that the peptide induces a primary increase in G_{Na^+}, but in the absence of extracellular Na^+ an initial hyperpolarisation is disclosed which is supressed by TEA or a slight increase in $[K^+]_e$ to 14.8mM. These results indicate that the initial hyperpolarisation is the consequence of an enhancement in G_{K^+} which probably appears at the same time or before the activation of G_{Na^+}. The activation of G_{K^+} limits the depolarisating action of the peptide Studying the action of AtII on ^{24}Na, ^{42}K and ^{36}Cl fluxes we have previously found (9) that the peptide increases primarily Na^+ fluxes even in depolarised myometrial preparations. Nonetheless, there was no detectable primary change in K^+ fluxes. This discrepancy might be explained by the lack of precision of the isotope technique.

When exposed to a physiological salt solution, the sequence of electrical and mechanical phenomena induced by a high concentration of AtII in uterine smooth muscle is the following: 1) there is a rapid initial depolarisation which initiates, at a certain threshold, the discharge of bursts of spikes. 2) simultaneously with the depolarisation, and before any spike production, starts a tonic contraction 3) coincidently with the spikes, phasic contractions are superimposed on the tonic contraction. 4) once the desensitivation starts, and during the washout of the peptide, the relaxation of the muscle follows the fading of the membrane depolarisation, with occasional phasic contractions which are simultaneous with the spikes. Our results confirm previous findings (10 - 11) since it seems to be possible to distinguish in this tissue two kinds of contraction: phasic, dependent on the spike discharge and tonic, non-related to changes in membrane potential. The source of Ca^{2+} during the phasic contractions seems to be $[Ca^{2+}_e]$ since they disappear completely when the muscle is perfused with the "0 Ca^{2+}" solution. Indeed, it has been demonstrated previously (ABE, 12; MIRONNEAU, 13; and VASSORT, 14-15), that in uterine smooth muscle the ascending phase of the spike is the result of a potential dependent activation of $G_{Ca^{2+}}$, and that every spike is associated with a phasic contraction. Nonetheless, sometimes, it is possible to obtain a tonic contraction with AtII, even in the absence of any rapid or slow depolarising step. This effect does not seem to be dependent on a primary increase in G_{Ca^+}. The residual contraction could still be due to an increase in the Na^+_i through the Ca^{2+}– Na^+ exchange mechanism. But this possibility seems excluded since Mn^2 blocks not only the diffusion of Ca^{2+} through the membrane, but also Ca^{2+}-Na^+ exchange(8). There is an alternative explanation for the residual tonic contraction induced by the peptide under certain conditions: the cation could be released from intracellular stores. It is impossible at the present time to localise these stores, but they seem to be of a small size and closely dependent on extracellular Ca^{2+}, since in the presence of "0 Ca^{2+}", the first application of AtII induces only a weak tonic contraction. This response decreases markedly or disappears with the second stimulation with the peptide. In summary, the contraction of rat uterine smooth muscle induced by AT II presents two components: 1) a phasic contraction which is the result of the increase in intracellular Ca^{2+} given by the activation of $G_{Ca}2+$ during the spikes and 2) a tonic contraction due to the release of Ca^{2+} from intracellular stores. Our work confirms previous findings by VAN BREEMEN et al (16) who observed two components in the contraction induced by noradrenaline in rabbit aorta. Similar results were obtained by MIRONNEAU and GROSSET (17) and GROSSET and MIRONNEAU (18) who studied to mode of action of prostaglandins on rat uterus and frog atria. By the use of a different technique, BAUDOIN and MEYER (19) have previously shown in vascular smooth muscle, that AtII decreased the binding of Ca^{2+} to plasma membranes. Unfortunately, probably due to a failure in the preparation it has not been possible to show any effect of the peptide on membrane fractions of rat myometrium (20).

REFERENCES

1. WORCEL, M., PAPADIMITRIOU, A., and MEYER, P. (1971) J.Pharmacol. exp.Ther., 179, 42-46.
2. OHASHI, H., NONOMURA, Y. and OHGA, A. (1967) Jap.J.Pharmacol. 17, 247-257.
3. SOMLYO, A.V. and SOMLYO, A.P. (1968) J.Pharmacol.exp.Ther., 159, 129-145.
4. BULBRING, E. and TOMITA, T. (1969) Proc.R.Soc. B172, 89-102.
5. TAYLOR, G.S., PATON, D.M. and DANIEL, E.E. (1970) J.Gen.Physiol. 56, 360-375.
6. TÜRKER, R.K., PAGE, I.H., and KHAIRALLAH, P.A. (1967) Arch.Int. Pharmacodyn.Ther., 165, 394-404.
7. HAMON, G., PAPADIMITRIOU, A. and WORCEL, M. (1976) J.Physiol. (Lond) 254, 229-243.
8. RUSSEL, J.M. and BLAUSTEIN, M.P. (1974) J.Gen.Physiol., 63, 144-167.
9. HAMON, G. and WORCEL, M. (1973) J.Physiol.(Lond) 232, 99-100P.
10. BOEV, K., GOLENHOFEN, K. and LUKANOV, J. in "Physiology of smooth muscles", Bulbring E. and Shuba M.F. eds. Raven Press New-York (1975)
11. MANDREK, K and GOLENHOFEN, K. in "Smooth muscle Pharmacology and Physiology". Worcel and Vassort eds. Colloques INSERM Paris (1976).
12. ABE, Y. (1971) J.Physiol. (Lond), 214, 173-190.
13. MIRONNEAU, J. (1973) J.Physiol. (Lond), 233, 127-141.

14. VASSORT, G. (1973) J.Physiol. (Lond), 237, 50-51.

15. VASSORT, G. (1975) J.Physiol. (Lond), 252, 713-734.

16. VAN BREEMEN, C., FARINAS, B.R., CASTEELS, R., GERNA, P. WUYTACK, F. and DETH, R. (1973) Phil.Trans.R.Soc. (Lond), B265, 57-71.

17. MIRONNEAU, J. and GROSSET, A. (1976) Pflügers Arch., 366, 79-81.

18. GROSSET, A. and MIRONNEAU, J. (1976) J.Physiol. (Paris), 72, 9A.

19. BAUDOIN-LEGROS, M. and MEYER, P. (1973) Br.J.Pharmacol., 47, 377-385.

20. RANGACHARI, P.K., PERNOLLET, M.G. and WORCEL, M. (1976) Eur.J. Pharmacol., 40, 291-294.

Excitation-Contraction Coupling in Smooth Muscle, R. Casteels et al. eds.
© 1977 Elsevier/North-Holland Biomedical Press

MEMBRANE POTENTIAL AND CONTRACTION IN THE EAR ARTERY OF THE RABBIT

G. Droogmans and R. Casteels
Laboratorium voor Fysiologie
Gasthuisberg
B-3000 LEUVEN (BELGIUM)

INTRODUCTION

In vascular smooth muscle cells the contraction can be initiated by a change in membrane potential, i.e. a graded depolarization or action potentials. It has also been proposed (1,2) that the activation by a chemical stimulus may occur independently of any electrical process and such activation has been called pharmaco-mechanical coupling.

In the present study both modes of activation and their corresponding sources of activator Ca have been investigated in the isolated rabbit ear artery. In addition, experimental evidence is brought forward, indicating that Na-Ca exchange plays a minor role, if any at all, in the regulation of the free intracellular Ca-concentration and the vascular tone in this preparation.

MATERIALS AND METHODS

Experimental preparations and solutions : Rabbits (2-3 kg) were stunned and bled. The ear artery was isolated and transferred to a warmed and oxygenated Krebs solution and cleaned under a dissection microscope of its connective tissue.

The standard physiological solution was a Hepes buffered modified Krebs' solution containing (in mM) : Na 135.5; K 5.9; Ca 1.5; Mg 1.2; Cl 143.8; HEPES 11.6; glucose 11.5 (35°C, pH 7.3).

Solutions with increased $(K)_o$ were prepared by replacing Na by K, keeping the sum $(Na)_o + (K)_o$ constant. Isotonic K-Krebs' solution was obtained by replacing all the Na by K. K-free solutions were obtained by replacing KCl by NaCl. In order to prevent the effects of noradrenaline released by the adrenergic nerve fibres, phentolamine (10^{-5} M) was added to these solutions.

Na-free solutions were prepared by replacing NaCl by choline-Cl; atropine (100 mg/l) was added in order to prevent the cholinomimetic effect of choline. In the solutions containing TEA-Cl (tetraethylammonium-Cl) isotonicity was maintained by reducing NaCl.

Ca-free solutions always contained 2 mM EGTA.

Electrical recording : Intracellular recordings of membrane potential were made by means of glass microelectrodes filled with 3 M KCl and having a resistance of 60-100 MΩ (3). Constant current pulses with a duration from 100 ms to 1 s could be applied through two external platinum electrodes placed longitudinally along the preparation. A second extracellular microelectrode was mounted close to the intracellular electrode and the voltage between both electrodes was recorded differentially in order to reduce the stimulus artifact.

Mechanical recording: Helical strips of the ear artery were cut and mounted in an organ bath of 10 ml. Tension was recorded by means of an isometric force transducer.

Efflux studies: The preparations were loaded to saturation for several hours in Krebs' solutions containing the radioactive isotope. Washout was then performed in a nonradioactive solution (4,5).

RESULTS AND DISCUSSION

Vascular tone as a function of the membrane potential. Membrane potential and the accompanying changes in mechanical tension have been measured at different external K concentrations. The relationship between both quantities is represented in fig. 1. Tension is expressed as a percentage of the contraction induced by a

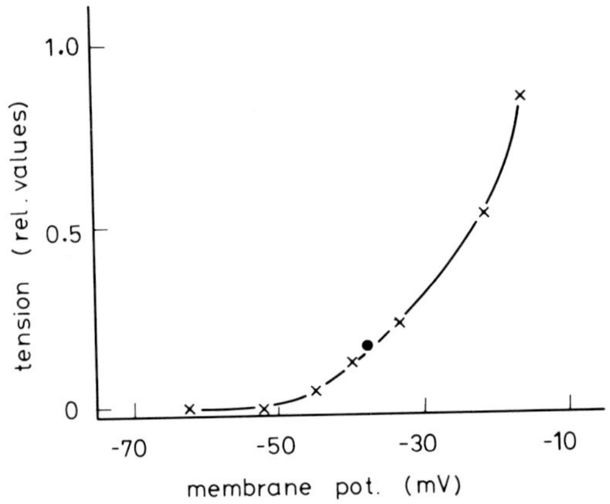

Fig. 1. Relation between membrane potential (mV) and mechanical tension, expressed as a fraction of a maximal noradrenaline contraction. The crosses correspond to data obtained at different $(K)_o$, the point represented by the dot is obtained in a solution containing 10 mM TEA-Cl.

supramaximal dose of noradrenaline. It can be observed that in order to induce a measurable change in vascular tone of the rabbit ear artery the membrane has to be depolarized by at least 10 mV from its value under normal physiological conditions. However, once the cells have been depolarized beyond this threshold value, small changes in resting potential cause pronounced modifications of the mechanical tension. Therefore it can be concluded that if the membrane depolarization is an essential link in the stimulus-contraction coupling by vaso-active substances, they should have a pronounced effect on the resting potential.

The contraction induced by K-depolarization depends on external Ca, because it does not occur in Ca-free Krebs' solutions. In addition, when the Ca-influx is inhibited by La (1 mM) no contraction occurs in isotonic K-Krebs' solutions. These findings suggest that the increase in vascular tone accompanying the membrane depolarization relies on an increased influx of Ca. The latter is probably due to an increase of the Ca-permeability of the cell membrane on depolarization.

The contraction induced by K-depolarization is also completely relaxed when La (1 mM) is added to the isotonic K-Krebs' solution. Because all the Na-ions have been replaced by K-ions, there is no inwardly directed electrochemical gradient for Na under these conditions. Therefore this relaxation cannot be explained by a Na-Ca exchange mechanism, powered by the Na-gradient. An accumulation of Ca into intracellular stores is also unlikely because no significant changes in the exchangeable Ca-content could be detected by exposing the tissues to isotonic K-Krebs' solution (6).

Action potentials and contractions induced by TEA. Tetraethylammonium (10-15mM) depolarizes the smooth muscle cells of the rabbit ear artery by 25 to 30 mV. This depolarization is accompanied by an increase in mechanical tension. As can be observed in fig. 1, this contraction fits the relation between vascular tone and resting potential.

When the membrane is depolarized by TEA to about -35 mV, spontaneous bursts of action potentials can occur (fig. 2) which are accompanied by phasic contractions.

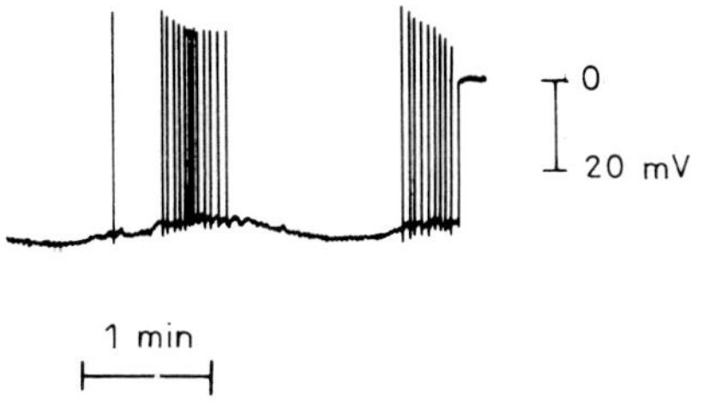

Fig. 2. Spontaneous action potentials observed in a Krebs' solution containing 10 mM TEA-Cl.

If this threshold potential is not reached these action potentials can be induced either by increasing $(K)_o$ or by applying depolarizing current pulses. It should be pointed out that these stimulation procedures do not induce action potentials in the absence of TEA. Similar findings have been observed in the rabbit pulmonary artery (7), and in guinea-pig stomach muscle (8).

The amplitude and the kinetic parameters of these action potentials strongly depend on external Ca. A reduction of $(Ca)_o$ to 0.5 mM interrupts the occurrence of action potentials. Increasing $(Ca)_o$ from 1.5 to 4.5 mM augments the amplitude of the action potentials from 43 ± 0.5 mM (n = 20) to 51 ± 0.5 mV (n = 20). Their maximum rate of rise is approximately doubled, while their maximum rate of repolarization is increased by a factor of three. As a consequence the duration of the action potential is markedly reduced.

Ca antagonists, such as D600 (.25 mg/l) and Mn (1 mM), inhibit both the spontaneous action potentials and the accompanying contractions, without affecting the resting potential significantly.

Also in Na-free solutions in which NaCl is replaced iso-osmotically by choline-Cl or $CaCl_2$, it is possible to induce action potentials by electrical stimulation.

It can be concluded from these observations that the inward current of these action potentials is probably mainly carried by Ca-ions. This would imply the existence of Ca- selective channels which can be activated by membrane depolarization.

Effect of noradrenaline on membrane potential and vascular tone.

The contraction induced by noradrenaline consists of a phasic and a tonic component (9,10). However in the concentration range 10^{-9} M to 10^{-5} M, the drug has neither a transient nor sustained effect on the membrane potential. A similar electrical quiescence during sympathomimetic stimulation has also been described in other vascular smooth muscle preparations, such as the rabbit pulmonary artery (1).

In contrast to the contraction induced by membrane depolarization, noradrenaline still induces a transient contraction in a Ca-free solution or in a solution containing 1 mM La. The amplitude of these contractions is dose-dependent and decreases as a function of the duration of the preceding exposure to Ca-free solution or to the La containing solution.

These findings suggest that the activator Ca is released from an intracellular store, which is gradually depleted in the absence of extracellular Ca or when the Ca-influx is inhibited. The finding that a second stimulation by noradrenaline does not evoke a second contractile response implies that the activator Ca does not remain available for a second activation.

In order to find out whether this Ca-fraction, once it has been released, remains stored inside the cell or is extruded through the plasma membrane, we have investigated the effect of noradrenaline stimulation on the ^{45}Ca-efflux from tissues exposed either to a Ca-free solution or to a Krebs' solution containing 1.5 mM Ca and 10 mM La. Under both experimental conditions noradrenaline induces a transient stimulation of the ^{45}Ca-efflux (fig. 3), suggesting that at least part of the activator Ca is removed from the cytoplasm through the cell membrane. Moreover, in the solution containing La and 1.5 mM Ca this extrusion is directed against the electrochemical gradient of Ca, and therefore must be active.

These findings also indicate that La exerts a different effect on the inward and outward movement of Ca. If both the influx and efflux of Ca occur by means of the Na-Ca exchange carrier one should expect that La affects both fluxes.
It is therefore difficult to reconcile this finding with a transmembrane Ca-transport that occurs mainly via a Na-Ca exchange mechanism.

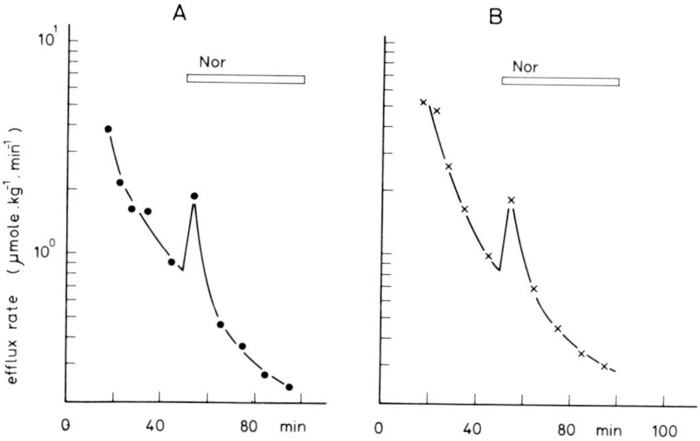

Fig. 3. Stimulation of the ^{45}Ca-efflux by 10^{-6} M noradrenaline in a Ca-free solution (A) or in a Ca-containing solution with 10 mM La (B).

Role of Na-ions in activation and relaxation.
The antagonism between $(Na)_o$ and $(Ca)_o$ on mechanical tension in rabbit aorta, as well as the contraction induced by K-free solution or ouabain fit the Na-Ca exchange hypothesis proposed by Reuter et al (11) and Blaustein (12). In order to find out whether the same mechanism also applies to the rabbit ear artery, we have investigated the effect of reducing $(Na)_o$ or increasing $(Na)_i$ on the resting potential and on the contractility of this tissue.

Effect of Na-free solutions : Exposure of the rabbit ear artery to a Na-free solution induces, as in the rabbit aorta, a contraction with an amplitude of about 25 to 30 per cent of a maximal noradrenaline contraction. Although this finding is consistent with Na-Ca exchange the following observations do not fit this hypothesis.

After a pretreatment during 90 min in a Na-free, Ca-free solution the tissue is relaxed, and $(Na)_i$ is vanishingly small. Readmission of Ca induces a contraction which is not significantly different from the Na-free contraction without this pretreatment (fig. 4). This finding suggests that the entrance of activator Ca can occur independently of an exchange for internal Na and consequently through a pathway that is different from the hypothetical Na-Ca carrier.

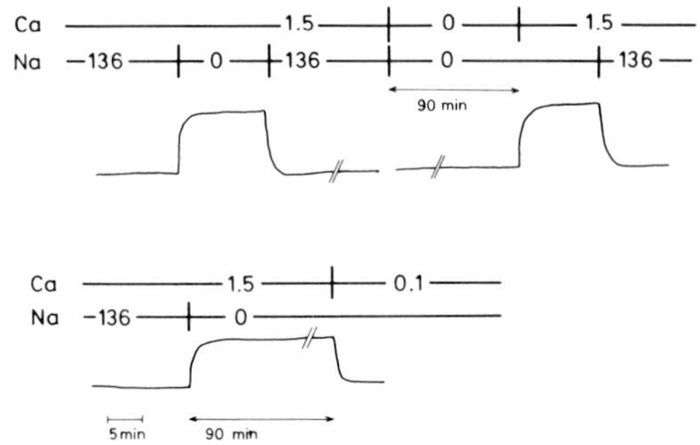

Fig. 4. Contraction induced by adding Ca in a tissue pretreated in Na-free, Ca-free solution (upper record); relaxation of the Na-free contraction by reducing $(Ca)_o$ to 0.1 mM (lower record).

Also after a similar pretreatment in a Na-free but Ca containing solution it is possible to induce complete relaxation, and therefore a decrease in $(Ca)_i$, by reducing $(Ca)_o$ to 0.1 mM (fig. 4). This reduction of $(Ca)_i$ cannot be powered by the Na-gradient, suggesting that Ca-ions can leave the cytoplasm by a mechanism different from Na-Ca exchange. A sequestration of Ca into intracellular Ca-compartments is also unlikely, because no significant changes in the exchangeable Ca content of the tissue could be detected under Na-free conditions (unpublished observation).

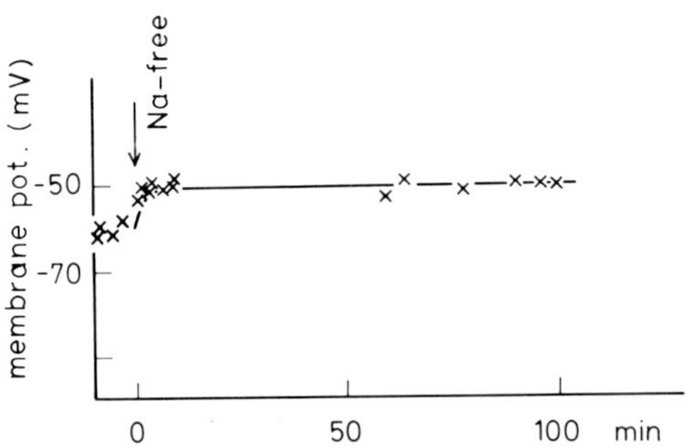

Fig. 5. Effect of Na-free solution on the membrane potential of the rabbit ear artery.

The Na-substitution is accompanied by a depolarization to about -50 mV (fig. 5). This value corresponds to the threshold potential for contraction induced by $(K)_o$. It is therefore unlikely that this depolarization is the only trigger for the Na-free contraction, unless the threshold potential for the activation of the Ca-influx is shifted to more negative values.

Effect of K-free solution: Exposure to a K-free solution induces in the rabbit ear artery, after a delay of 50-60 min, a contraction which reaches its maximal value after about another 60 min. The amplitude of this contraction amounts to about 30 percent of a maximal noradrenaline contraction. External Na and Ca have opposite effects on this amplitude. On readmission of external K, this contraction is relaxed within a few minutes.

During a 4 hr exposure to K-free medium no significant changes in the resting potential could be observed. However on readmission of K the cell membrane hyperpolarizes promptly to about -100 mV (fig. 6).

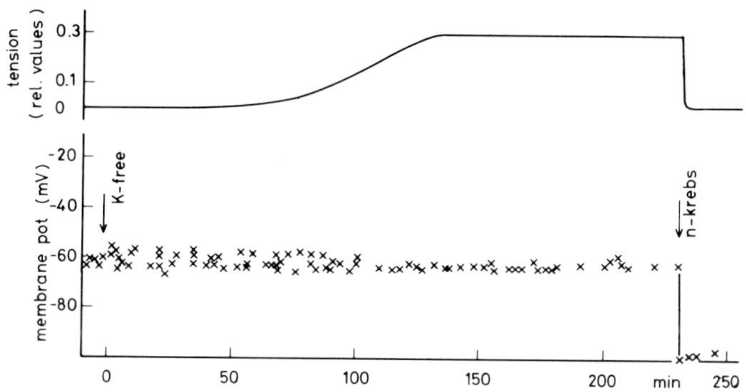

Fig. 6. Effect of K-free solution on membrane potential and mechanical tension of the rabbit ear artery.

The gain in intracellular Na during the K-free treatment can be estimated roughly if one assumes that it compensates for the loss in K, as has been observed in the guinea-pig's taenia coli (13). The rate of K-loss can be estimated from the rate of ^{42}K-efflux in K-free solutions. This K-depletion can be approximated by an exponential with a rate constant k = .0053 ± 0.0005 min^{-1} (n = 10). If one further choses steady-state values for $(Na)_i$ and $(K)_i$ similar to those found in the guinea-pig's taenia coli (4), the time course of $(Na)_i$ during the K-depletion can be described roughly by the following equation (t = time of exposure to K-free solution):

$$(Na)_i = 13 + 164 \ (1 - \exp(-kt))$$

For t = 60 min, $(Na)_i$ = 58 mM, i.e. an increase by a factor of 4.4. According to the model proposed by Blaustein (14) a small increase in $(Na)_i$ should produce a marked increase in vascular tone. Yet there is no measurable change in the tension development of the ear artery in spite of the pronounced increase in $(Na)_i$. Also the relaxation is completed before significant changes in $(Na)_i$ could occur.

It can be concluded from these experiments with altered Na-gradients that the correlation between contraction, as a measure of $(Ca)_i$, and the Na-gradient is not good and cannot fully explain the observed phenomena. A model which predicts a loose coupling between Na and Ca movements seems therefore more appropriate than the Na-Ca exchange hypothesis with its very tight coupling to describe the experimental observations. A possible alternative is that Na and Ca-ions compete for the same channel, and that a reduction in the Na-gradient favours the influx of Ca-ions. This model fits the observed contractile responses, and also explains the shift of the threshold potential for contraction to more negative values.

CONCLUDING REMARKS

1. The results discussed above show that contraction in the rabbit ear artery may be mediated either by electrical or by non-electrical means. The contraction induced by membrane depolarization (graded or all-or-none) is caused by an increased influx of Ca. On the other hand, the activation by noradrenaline is independent of extracellular Ca: this contraction can be induced by a release of Ca from intracellular pools. However, the refilling of these stores requires extracellular Ca.

2. We have also reported some observations which do not fit the Na-Ca exchange hypothesis. These results do not fully disprove the existence of Na-Ca exchange. Nevertheless they show that its role in the regulation of vascular tone should not be overstressed.

REFERENCES

1. Su, C., Bevan, J. and Ursillo, R. (1964) Circ. Res. 15, 20-27.
2. Somlyo, A.V. and Somlyo, A.P. (1968) J. Pharmacol. Exp. Ther. 159, 129-145.
3. Hendrickx, H. and Casteels, R. (1974) Pflügers Arch. Eur. J. Physiol. 346, 299-306.
4. Casteels, R. (1969) J. Physiol. (Lond.) 205, 193-208.
5. Casteels, R. and van Breemen, C. (1975) Pflügers Arch. Eur. J. Physiol. 359, 197-207.
6. Droogmans, G., Raeymaekers, L. and Casteels, R. (1977) J. Gen. Physiol. 70 (in the press).
7. Mekata, F. (1971) J. Gen. Physiol. 57, 738-751.
8. Ito, Y., Kuriyama, H. and Sakamoto, Y. (1970) J. Physiol. (Lond.) 211, 445-460.
9. Bevan, J. and Waterson, J. (1971) Circ. Res. 28, 655-661.
10. Steinsland, O., Furchgott, R. and Kirpekar, S. (1973) Circ. Res. 32, 49-58.
11. Reuter, H., Blaustein, M. and Haeusler, G. (1973) Phil. Trans. R. Soc. Lond. B. 256, 87-94.
12. Blaustein, M. (1974) Rev. Physiol. Biochem. Pharmacol. 70, 33-82.
13. Casteels, R., Droogmans, G. and Hendrickx, H. (1971) J. Physiol. (Lond.) 217, 281-295.
14. Blaustein, M. (1976) The Physiologist 15, 525-540.

Excitation-Contraction Coupling in Smooth Muscle, R. Casteels et al. eds.
© 1977 Elsevier/North-Holland Biomedical Press

SUMMARY OF DISCUSSION ON "MEMBRANE POTENTIAL FLUCTUATIONS AND THE ACTIVATION OF CONTRACTION"

BY C.L. PROSSER (University of Illinois)

The papers of this session introduced the diversity of electrical activity and of modes of activation of contraction in smooth muscles. It was agreed initially that contractile activity requires an increase in intracellular free Ca^{++} and that these ions can come either (1) from membrane or extra-membrane sources (via action potentials) or (2) from intracellular stores. Agents such as acetylcholine (guinea pig stomach) or norepinephrine (arterial muscle) can stimulate contractions (1) by causing membrane depolarization or (2) by initiating intracellular release of Ca^{++} without depolarization and in the absence of extracellular calcium. The intracellular action of acetylcholine is antagonized by atropine and arterial contraction by NE is potentiated by slight depolarization.

Endogenous electrical activity consists of rhythmic slow potentials and spikes. Some smooth muscles are electrically active only when depolarized by pulses (isolated stomach fibers of toad) or by ionic alterations such as reduced Na_o or Ca_o or increased K_o or when treated with transmitter substances or hormones. Spike depolarization is generally caused by increase in conductance of Ca^{++} and is blocked by agents specific for Ca-channels. However, Na^+ interacts with Ca^{++} in some muscles and Na^+ can, in the absence of Ca_o use channels normally used by calcium (carotid artery, many other smooth muscles after treatment with EGTA). Isolated fibers resemble many other excitable cells in their electrical parameters, such as outward (delayed) rectification.

There appear to be several kinds of rhythmic spontaneous slow potentials and no unifying classification based on ionic mechanisms has been developed. Those in guinea-pig stomach consist of an initial voltage-independent and a second voltage sensitive component (in addition to spikes). The first component is lost in a Na-free medium when depolarization accompanies an increase in membrane resistance ; recovery of membrane potential and of lost slow waves due to added Na_o following depolarization in Na-free medium is unaffected by ouabain. It is argued that, while the initial phase of slow potential in guinea-pig stomach is metabolically controlled, it may not be due to a Na-K pump. More direct evidence for rhythmicity of an electrogenic Na-K pump has been presented for longitudinal muscle of cat intestine. The pace-setting process is not in the ion pump and not in a redox oscillation ; it is, however, sensitive to Ca_o. The second component of slow potential of guinea-pig stomach continues rhythmically in absence of Na_o, is reduced in amplitude in low Ca_o and is correlated with contractions. It more closely resembles rhythmic spikes of other smooth muscles. A different sort of slow wave appears in carotid artery when depolarized by K in the absence of Na or Ca, provided Mn is present. These non-physiological waves are not affected by agents which block Na and Ca channels. Insufficient attention has been given to the differences underlying control of frequency and of amplitude of various types of slow wave.

Contractions which are triggered by membrane depolarization can be modulated by various physiological agents. In guinea-pig stomach, gastrointestinal hormones at concentrations lower than needed to stimulate secretion can enhance motility and prolong the plateau (second slow potential component) or increase spikes. In uterus, prostaglandin (PGF_2) causes depolarization, enhances slow waves and increases spiking ; the spikes are Ca-sensitive, the slow potentials Na-sensitive. Another kind of physiological modulation was indicated for cat intestine where rhythmic slow potentials originating in longitudinal muscle are amplified by Ca-sensitive events in the circular muscle ; passive electrical spread between the two layers appears to occur via interstitial cells.

Species differences (guinea-pig vs cat), organ differences (stomach, intestine, uterus) and muscle layer differences preclude generalizations. Activity changes caused by physiological agents may be due to depolarization, to changes in membrane resistance and to actions on intracellular organelles. The membrane of a given cell may have parallel paths of electrogenesis, and cells which are interconnected in a three-dimensional network may differ from well insulated cells. It is concluded that smooth muscles are more complex than the more specialized striated muscle.

Excitation-Contraction Coupling in Smooth Muscle, R. Casteels et al. eds.

LIMITATIONS AND POSSIBILITIES IN SMOOTH MUSCLE VOLTAGE-CLAMP

Nels C. Anderson Jr.
Departments of Physiology and Ob-Gyn
Duke University Medical Center
Durham, N.C. 27710

The history of voltage-clamp studies on smooth and cardiac muscle is brief and filled with ususually intense controversy. This course of events may well have been predicted from our earlier understanding of the structure and geometry of these tissues. Clearly it was well known prior to the late sixties when voltage-clamp experiments began, that although the functional electrical syncytium model seemed to account for the apparent cable properties of smooth muscles,[1,2] the equivalent circuit for a unit length of this cable was definitely not a simple parallel RC element.

Nonetheless, from this work the basic rationale for voltage-clamp experiments had been defined, i.e. that conditions of space-clamp or current clamp, a prerequisite for voltage-clamp, could be satisfied by working with a short length of a small diameter multicellular strand of smooth muscle.

In March, 1967, after two seasons of squid axon voltage-clamp work at Woods Hole with the double sucrose gap, we were prepared to adapt this procedure to smooth muscle. It was possible to obtain a resting membrane potential or "gap potential" and to inject depolarizing and hyperpolarizing current pulses which clearly revealed the non-linear behavior of the "node" membrane. (fig 1). The membrane potential instability at threshold in the face of a constant current was again, a clear demonstration of excitable cell behavior. At this point it was virtually impossible not to attempt voltage-clamp even in face of the apparent differences between a strip of muscle and a giant nerve fiber. The experiment could be tried and so it was. The results of that first voltage-clamp experiment are shown in figure 2. Although not totally unexpected, the clarity of the current records and their striking similarity to our squid data was, at the time, assurance enough that voltage-clamp analysis of smooth muscle membrane ionic conductance properties was possible. Indeed these early data provided the basis for certain fundamental conclusions on the excitation process in uterine smooth muscle. First it was clear that the transient inward current was a smooth, graded function of membrane potential with no trace of threshold. The inward current at first

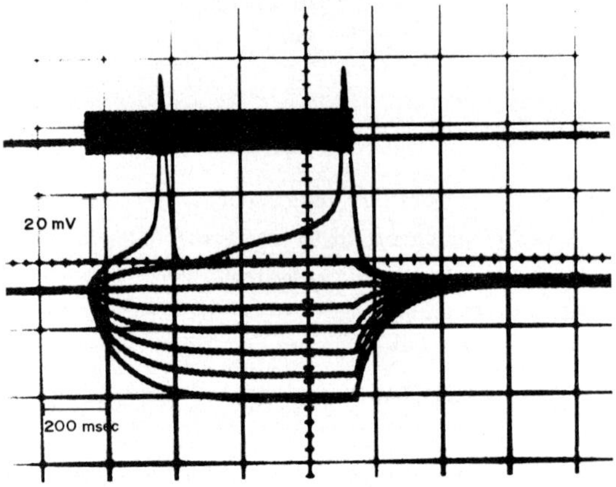

Fig. 1. Uterine smooth muscle membrane potential responses to depolarizing and hyperpolarizing constant current pulses in the double sucrose gap. Reproduced with the permission of the J.Gen.Physiology.

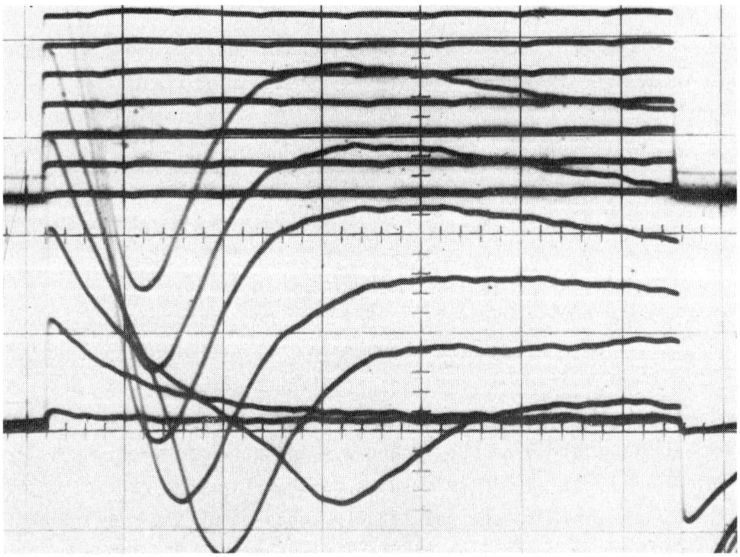

Fig. 2. Voltage-clamp record from uterine smooth muscle in response to step voltage commands.

increased in amplitude and speed with increasing potential steps, at larger potential steps the rate of onset further increased and its amplitude decreased. The late current gradually developed into a sustained outward current increasing in amplitude at larger depolarizations. These data alone gave a new level of understanding of such basic processes as threshold, and action potential generation and repolarization.

It was shortly thereafter, that many of us began a more rigorous analysis of the effects of practical concessions upon the experimental results. This analysis took several forms and the results were and remain equivocal with respect to evaluating the quality of voltage-clamp data. For example, it was relatively easy to establish that voltage clamp of a cable with Hodgkin-Huxley parameters showed voltage inhomogenity during a step voltage command, [3] but this did not evaluate the deviation of net current measurements under experimental conditions from those expected under ideal voltage-clamp conditions. Clearly with the double sucrose-gap technique, one is inevitability working with a short cable i.e., the axoplasmic or myoplasmic resistance remains unchanged as compared with the axial wire technique. For axons this appeared to present no significant problem, in other preparations such as cardiac and smooth muscle the effect remained unclear. Furthermore, the multicellular preparations were also compromised by a potentially large resistance in series with much of the "nodal" membrane.

An experimental approach to this problem which has been more helpful involves the use of an independent voltage sensing microelectrode to monitor "nodal" membrane potential during step voltage commands. Tarr and Trank [4] were among the first to use this technique and showed in cardiac muscle that with large depolarizing commands the potential control of many cells within the "node" is lost. Based on this evidence of nonisopotential control of membrane potential, Tarr and other cardiac physiologists have terminated their voltage-clamp experiments.

Similar experiments with an independent voltage sensing microelectrode have now been done with smooth muscle. [5] While this work "substantially supports the theoretical doubts which (have been) raised," the ingenious "short cable" preparations also described in this paper gives a direct measure of maximum cable length over which voltage control might be expected, 150μ or less. These data are interesting to compare with results from Inomata and Kao (1976) and our own observations. Kao reports that when the "node" width exceeded approximately 100μ, evidence for loss of voltage control appeared in the voltage and current records. This has consistently been our experience as well with uterine smooth muscle and the rabbit mesotubarium. Bolton [5] has also evaluated the homogenity of voltage control for slow (25mV/sec) ramp commands. Under these experimental conditions, large depolarizing potentials can be attained without loss of voltage control. This result supports the position that the voltage-clamp system can be used for the study of slow potential events in smooth muscle.[6,7,8]

Undoubtedly, the most intensive and general indictment of multicellular voltage-clamp was presented by Johnson and Lieberman.[9] I do not propose to review here again this evidence for nonisopotential control of membrane potential. Rather I would like to comment on two specialized properties of smooth muscle which have been studied under voltage-clamp conditions. One concerns the steady-state inactivation curve, h_∞, and secondly, the response of spontaneously active smooth muscle to voltage-clamp.

It is expected that the voltage-clamp technique should reveal unique and specialized membrane properties of smooth muscle. The steady state inactivation curve [10] represents a striking departure from the familiar relationship in squid axon both in terms of the relative steepness of the curve and the apparent hyperpolerization inactivation. The unsettled question is whether this represents a true characteristic of smooth muscle membrane and may therefore be of major functional significance [10] or whether this is yet another contribution of membrane series resistance. This dilemma demonstrates the nature of the problem in interpreting smooth muscle voltage-clamp data. That is, while it is precisely voltage-clamp results which deviate from the familiar squid model in which we hope to find new understanding, it is inevitably here that we also find the challenge of experimental limitations the greatest.

Subthreshold membrane potential oscillatory behavior has also been studied under voltage-clamp conditions.[6,7,8] As part of our effort to study this behavior in reproductive smooth muscles and to look for relationships to the spike mechanism, we initiated voltage-clamp track and hold experiments on the rabbit mesotubarium superious.[11] With this preparation we were unable to obtain records of large slow pacemaking events as occasionally observed in uterine smooth muscle.[10] However, spontaneous action potential activity was observed with clear prepotential type pacemaker activity (fig 3). If the voltage-clamp was applied during a spontaneous burst of action potentials repetitive spontaneous transients frequently (fig 4) but not always (fig 5) appeared in the current records. If under these conditions the membrane was further polarized by increasing the holding potential, it was possible to abolish these spontaneous current transients (fig 6). My interpretation of these results is that the node membrane equivalent circuit must be considered as at least a two element model and that one element is often remote and relatively inaccessable to the voltage command. However, when the holding potential is further increased this eventually brings the entire node area below threshold and the spontaneous action currents stop.

The published voltage-clamp work by a number of laboratories [12,13,5,14,6,8], either satisfy certain operational criteria or experimental conditions (e.g. slow ramps) for voltage-clamp. Clearly all of these investigators are sensitized to the many problems. It seems reasonable therefore to conclude, in the absence of any obvious alternative, that these experiments be accepted in the qualified con-

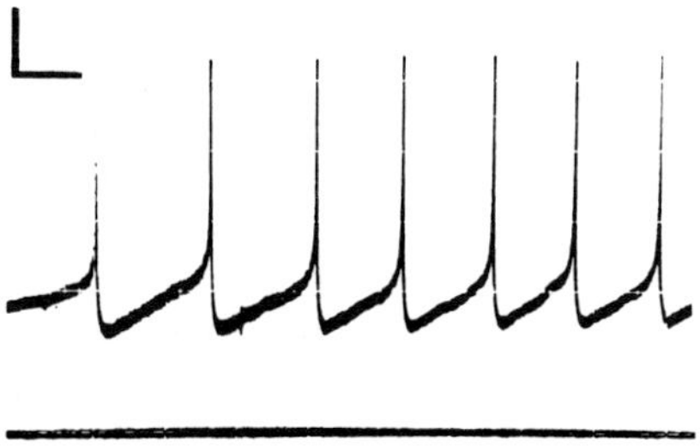

Fig. 3. Spontaneous action potentials from the mesotubarium superious. 2 sec; 10 mV.

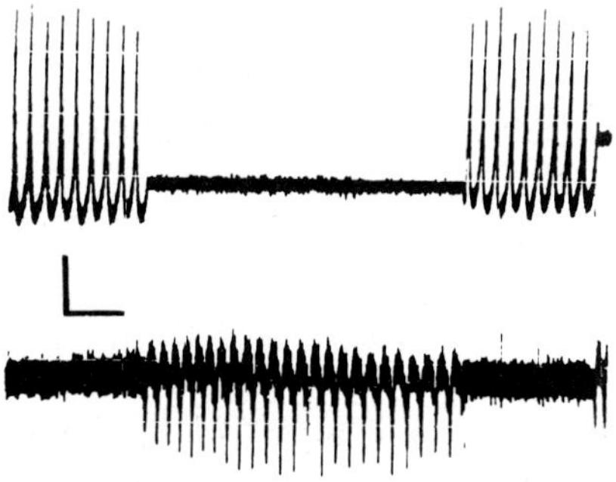

Fig. 4. Voltage-clamp during a spontaneous burst of action potentials. Spontaneous transients appear in the current record. 5 mV; 20 nA; 5 sec.

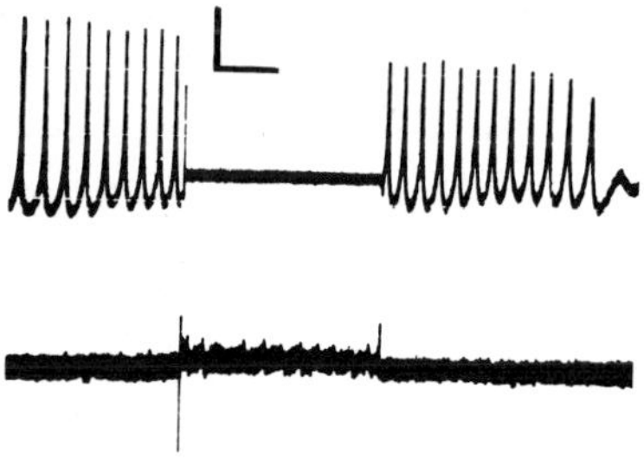

Fig. 5. Voltage-clamp during a spontaneous burst of action potentials in which there was little evidence of current instability. 5 mV; 20 nA; 5 sec.

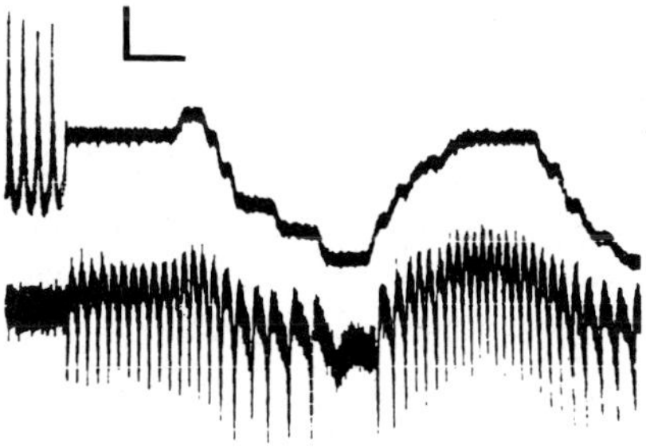

Fig. 6. Voltage-clamp during a spontaneous burst of action potentials in which the current transients were suppressed by hyperpolarizing the membrane. 5 mV; 20 nA; 5 sec.

text in which they are presented. To the extent that these data are consistent with and are corroborated by other nonvoltage-clamp data, will allow increased confidence in their interpretation.

Possibilities

Opportunities for inovation in experimental approaches to smooth muscle voltage clamp seem promising if for no other reason that the uncertainty surrounding presently used systems.

New approaches must be subject to detailed analysis and their assumptions must be strongly supported experimentally and theoretically. This is not to suggest that we restrict our thinking and work to the familiar. On the contrary, it seems an appropriate time for exploring new ideas and techniques. However, I do not presume to be able to direct how this feat should be performed. Nonetheless, there are a number of ideas which have been suggested at various times as well as examples of voltage-clamp work in other systems which may be useful.

Choice of experimental preparation is obviously the first and in all probability the most critical if not fateful step. However, it is at this level that our options seems to be significantly expanding. Possible multicellular preparations include not only the standard preparations of the past but also cultured synthetic strands[15] or small clusters of cells.[16]

The possibility of direct measurements on single isolated smooth muscle cells has seemed remote and until now unnecessary. We know of no giant smooth muscle cells, however, there is a considerable range in size. In reviewing this literature Burnstock (1968) concluded that the largest cells are found in the pregnant uterus (10μ X 600μ) and in urodel amphibians (12μ X 1200μ). Our recent work with smooth muscle from Amphiuma stomach has shown that these cells, in the living relaxed state, range from 15-25μ in diameter at the nucleus and approximately 1000μ long (unpublished observations).

Other isolated cell strategies include the possibility of forming a larger "cell" through cell fusion techniques. However, the problems with this approach are obvious and many.

Voltage-clamp Techniques

All voltage-clamp techniques operate on the assumption that the voltage sensing system gives an accurate measure of the membrane potential during the time and over the membrane area in which current flows. This assumption, as we now know, is never ideally true and at times may be a poor reflection of the membrane potential. In the system most widely used for smooth muscle voltage clamp, the double sucrose-gap, both current injection and voltage sensing are extracellular. A hybrid system involving a microelectrode for voltage sensing and a single sucrose gap for control of current injection, has been a widely used technique in cardiac muscle.

A strategy recently suggested by Attwell and Cohen[18] for evaluation of multicellular voltage clamp data is to compare results from two different experimental approaches. The best alternative to the double sucrose-gap technique for smooth muscle is unclear. Norma and Irisawa [19] have recently adapted the double microelectrode method [20] to strands of sinoatrial node tissue which were shortened by ligation. With this procedure preparations 0.2 - 0.3mm wide and 0.2 - 0.3mm long were obtained. Spacial voltage homogenity was evaluated with a third microelectrode not in the feedback circuit. Their results indicate that the preparation was isopotential during step depolarizations. This procedure seems adaptable to smooth muscle preparations and could provide comparative data on such phenomena as the anomalous h_∞ curve. These authors also point out that there were no repetitive spontaneous currents during the voltage-clamp suggesting that the underlying pacemaker mechanisms were primarily voltage and time dependent currents. This observation could also be used as a criterion for their ability to voltage-clamp.

The double microelectrode voltage-clamp technique has also been applied recently to cultured neuroblastoma cells with spectacular results.[21] These cells are 60μ - 110μ in diameter. As with the small sinoatrial node preparation, when the current passing electrode was in a bridge circuit, the recorded voltage response was the same at both electrodes, thus demonstrating spacial voltage homogeneity. Another relevant point for our consideration was that when projections from the cell exceeded 200μ, evidence for poor voltage control appeared in the current record.

Kostyuk et.al.[22] have presented a technique for voltage-clamp of single isolated snail neurons using large external electrodes. A more recent paper however, includes a microelectrode to redcue the access/series resistance problem.[23] The basic rationale of this procedure is to electrically isolate a patch of membrane by inserting the cell into a small hole manufactured ina thin sheet of plastic. Under these conditions it has been possible to not only voltage-clamp the preparation but to change the internal ionic composition by dialysis.[22]

The question which remains is how to take advantage of these, or other developments, in the voltage clamp of smooth muscle. We are presently in the process of isolating single smooth muscle cells from <u>Amphiuma</u> stomach. These relatively giant fibers suggest the possibility of single cell voltage-clamp. However, the long narrow geometry of these cells at rest is not preferred for either the microelectrode or pore[22] clamp methods. Fortunately, during isolation procedures many cells will contract to an elipsoid or round shape. These cells range in diameter up to 90μ! Judging from vital dye exclusion technique many of these large round cells are viable.

Our present experimental technique is similar to the procedure used for snail neurons. The chamber consists of two closed compartments separated by a thin (10μ) plastic sheet in which a small (15μ - 20μ) hole has been manufactured.

This apparatus is placed on an inverted phase microscope where the pore and lower compartment are viewed at high magnification. Isolated cells are perfused into the lower chamber and captured in the pore by applying a small, variable negative pressure to the upper chamber. This procedure has not yielded voltage clamp data as of this date, however, it is presented in the context of hopeful possibilities. The double microelectrode technique also appears to be a definite possibility with these cells.

In Cole's recounting of the era in which the squid giant axon finally yielded to the persistent efforts of a number of investigators, he refers to the development of the current clamp and voltage-clamp concepts as the events which tamed the squid axon.[24] I think we must agree that smooth muscle has not yet been completely tamed. However, we now understand the nature of its rebellion to control and hopefully will be able to bring to bear new and more powerful techniques to further restrain its behavior.

References

1. Tomita, T., J. Physiol. (London) 183,450 (1966).
2. Abe, Y. and T. Tomita., J. Physiol. 196,87 (1968).
3. Anderson, N.C. Jr., Fidel Ramon and J.W. Moore. Computer Simulation of the Double Sucrose-Gap Voltage-Clamp. The Physiologist 14, No. 3 (1971). (Abstract)
4. Tarr, M. and J.W. Trank., Biophysical J. 14,627 (1974).
5. Bolton, T.B., J. Physiol. 250,175 (1975).
6. Connor, J.A., C.L. Prosser and W.A. Weems., J. Physiol. 240,671 (1974).
7. Ohba, M., Y. Kakamoto and T. Tomita., Spontaneous Rhythmic Activity of the Smooth Muscle of the Guinea-Pig Stomach and Effects of Ionic Environment. In Smooth Muscle Pharmacology and Physiology. ed., M. Worcel and G. Vassart. INSERM, Paris (1975).
8. Ohba, M., Y. Sakamoto and T. Tomita., J. Physiol. 267,167 (1977).
9. Johnson, E.A. and M. Lieberman., Physiology 33,479-532 (1971).
10. Anderson, N.C. and Fidel Ramon., Interaction Between Pacemaker Electrical Behavior and Action Potential Mechanism in Uterine Smooth Muscle. In Physiology of Smooth Muscle. ed., E. Bulbring and M.F. Shuba.pp. 53. Raven Press, New York (1976).
11. Anderson, N.C., Membrane Potential Oscillations. In Smooth Muscle Pharmacology and Physiology. ed., M. Worcel and G. Vassort. INSERM, Paris (1965).
12. Anderson, N.C. Jr., J. Gen. Physiol. 54,145 (1969).
13. Kao, C.Y. and J.R. McCullough., J. Physiol. 246,1 (1975).
14. Mironneau, J., J. Physiol. 233,127 (1973).

15. Lieberman, M., T. Sawanobori, N. Shigeto and E.A. Johnson. Physiological Implications of Heart Muscle in Tissue Culture. In Developmental & Physiological Correlates of Cardiac Muscle. ed., M. Lieberman and T. Sano. Raven Press, New York (1976).
16. DeHaan, R.L. and H.A. Fozzord., J. Gen. Physiol. 65,207 (1975).
17. Burnstock, G., Structure of Smooth Muscle and Its Innervation in Smooth Muscle ed. Edith Bulbrings, A.F. Brading, A.W. Jones and T. Tomita. Williams and Wilkins Co. (1970).
18. Attwell, D. and I. Cohn., Prog. Biophys. Molec. Biol. 31,201 (1977).
19. Noma, A. and H. Irisawa., Pflugers Arch. 364,45 (1976).
20. Deck, K.A., R. Kern and W. Troutwein., Pflugers Arch. Gen. Physiol. 280,50 (1964).
21. Moolenaar, W.H. and Ilan Spector., Science 196,331 (1977).
22. Kostyuk, P.G., O.A. Krishtal and V.I. Pidoplichpo., Nature 257,691 (1975).
23. Kostyuk, P.G., O.A. Krishtal and V.I. Pidoplichpo., Nature 267,70 (1977).
24. Cole, K.S., Membranes, Ions and Impulses. Univ. of California Press (1968).

This work supported in part by NIH-NICHD grants HD02742 and HD08807.

Excitation-Contraction Coupling in Smooth Muscle, R. Casteels et al. eds.

RECENT EXPERIMENTS IN VOLTAGE-CLAMP STUDIES ON SMOOTH MUSCLES

C.Y. Kao
Department of Pharmacology
State University of New York
Downstate Medical Center
Brooklyn, N.Y. 11203 U.S.A.

Since the first successful adaptation of a double sucrose-gap voltage-clamp method to a mammalian smooth muscle (1), progress in the area has been relatively limited when viewed against somewhat greater expectations. Although a major reason for the limited progress is undoubtedly the technical difficulties of the method, there is also among students of smooth muscles a state of pessimism and cynicism questioning possible values of such studies. By the nature of the multicellular preparations used, and some inherent properties of such preparations, there can be no question that there are some serious limitations on the accuracy of the voltage control possible. However, certain problems in smooth muscle physiology cannot be properly studied without using some method which would control, within reasonable limits of error, the driving forces on charge-carriers across the surface membrane. From this point of view then, in spite of some imperfections in the present voltage-clamp technique, useful information has been obtained from some mammalian smooth muscles.

Basically, all present voltage-clamp studies are based on small bundles of multicellular preparations on which a double sucrose-gap technique is applied. In this technique, two cuffs of high resistance sucrose are applied to the bundle. By replacing extracellular electrolytes, these sucrose cuffs, in effect convert the bundle into 3 segments which are interconnected to one another primarily through intracellular current paths. The center segment, often called the "node," is the active portion from which membrane voltage is monitored, and to which current is applied.

Different variations are used for securing a stable "node" for study. Some investigators rely on rubber diaphragms to separate the node from the sucrose cuffs; others use vaseline seals (2,3) and still others rely on liquid interfaces between flowing sucrose and saline solutions (1,4,5). In most studies, a primary objective was to make the length of the node relatively short in comparison with the space constant of the preparation in an attempt to reduce any non-uniformity in voltage profile from one edge of the node to the other. In one instance (6), the length of the node was deliberately increased to lessen the degree of shunting by leakage current. A key difference between the latter study and the others is that the sucrose used in that study contained 10^{-5} M of $CaCl_2$, following a practice of many cardiac physiologists to maintain the viability of the preparation. My own preference is to use very high resistance sucrose without any $CaCl_2$ and to work relatively fast to avoid prolonged exposure to sucrose. The reasons for my preference are: 1) Within 20-30 minutes, there is no detectable benefit of incorporating $CaCl_2$; 2) The shunt resistance through the sucrose-gap is about 10 times lower in the presence of $CaCl_2$ than in the absence of the salt; and 3) Prolonged exposure to sucrose have other effects, such as an increase in the longitudinal access resistance, which increase technical problems of voltage control or produce deleterious effects on the preparation.

In addition to the length of the node, the total amount of tissue in the node, and hence the thickness of the preparation, is also an important consideration in any attempt at adequate voltage control. Indeed, the small amount of extracellular fluid separating the different layers of individual cells in a bundle is the major contributor to the series resistance and consequent problems of voltage control. The thinner the preparation, the more uniform is the voltage control from the surface to the deep layers. Considering the limitations imposed by the series resistance, and the futility of eliminating or compensating for the largest portion of it, a realistic approach to voltage-clamp studies on mammalian smooth muscles is to reduce it to the lowest possible level that is consistent with survival of the preparation, and then to maintain the series resistance as constant as possible under different experimental conditions. Similarly, although the length and thickness of the nodal tissue are important considerations, it is equally important to maintain the amount of nodal tissue constant under different experimental conditions.

A convenient test for constancy of series resistance and nodal tissue is to impose a small voltage step (ca. 5 mV) at a fast time scale (e.g., 2 msec/cm) and to check the initial current displacement and the discharge time-constant. The former leads to an estimate of the series resistance, and the latter, total membrane capacitance. An additional advantage of estimating the membrane capacitance is that the measure is an indication of the amount of membrane under study. In the absence of any measurement of the membrane area, total electrode currents expressed on the basis of unit capacitance (e.g., μA/μF) can be useful procedure for comparing the membrane currents in different conditions. Considerable progress in voltage-clamp studies on smooth muscles would be realized by the wider adoption of procedures to estimate and maintain constant series resistances and total membrane capacitances, and to express observed currents on the basis of unit capacitance.

Some specific examples of accomplishments in voltage-clamp studies on smooth muscles might be listed:

(1) <u>Clarification of the nature of ionic currents in various smooth muscles:</u>

One controversial problem in this area has been the nature of the charge-carrier responsible for the action potential. Although there are many observations on the effects, or absence of effects, of altering the concentration of one or another ion, on the spontaneous, drug-induced, or electrically elicited action potential, a common flaw in all these observations is that the role of an ion as a charge-carrier cannot be clearly distinguished from its role in regulating membrane conductance.

By the voltage-clamp technique, it has been possible to distinguish between these roles, and to show that in the pregnant-rat myometrium, the early current (responsible for the rising phase of the action potential) is carried primarily by Na^+ (4), whereas in the guinea-pig taenia coli, it is carried primarily by Ca^{2+} (5). In each tissue, the evidence was accumulated not only from altering the concentrations of these ions in the solution, but also in showing that the effects of these alterations on the driving force can be distinguished from any possible effects on membrane conductance. In the myometrium, it was also possible to reverse the driving force and to show that the direction of the early current was reversed.

Other investigators (1,2,7) have also concluded that Na^+ is a charge-carrier in the myometrium, but they further attributed a somewhat greater role to Ca^{2+} as a possible charge-carrier than we have. The basis of their conclusion is the blocking effects on the early current of di- and tri-valent cations, Mn^{2+}, Co^{2+}, and La^{3+}. We have not relied on this line of evidence as heavily, because of our belief that the actions of these multivalent cations are not specific.

For the late current (responsible for the falling phase of the action potential), voltage-clamp experiments have provided direct electrophysiological evidence that the major charge-carrier is K^+. The evidence depends on the ability to control the driving force on K^+ in such a manner that the direction of the late current can be reversed (4,5). Without using a voltage-clamp approach, this evidence cannot be obtained.

(2) Estimation of some ionic equilibrium potentials:
The establishment of ionic equilibrium potentials is of importance not only for electrophysiological studies, but also for studies of ionic movements in smooth muscles. In the pregnant-rat myometrium E_{Na} has been estimated to be about 25 mV (1,4), a value very close to that obtained from chemical analyses and partitioning (see 8). Because of technical problems and some overlap in the early and late currents in the guinea-pig taenia coli, the equilibrium potential for the early current in this tissue is less certain (5).

In both the pregnant-rat myometrium and the guinea-pig taenia coli, the E_K has been determined as being about 15 to 20 mV more negative than the natural resting potential. These values are lower than those estimated from chemical data, but they may represent more realistic values.

It is to be hoped that the estimation of the ionic equilibrium potentials will help to close some fruitless controversies over the extracellular space in different smooth muscles. The ultimate aim of the various attempts to estimate the extracellular space is related to the partitioning of some ions inside and outside the smooth muscle cell, and to establish some ionic equilibrium potentials for estimating the limits of driving forces. The voltage-clamp technique, through determinations of the reversal potential, can provide a relatively direct estimate of the most likely intracellular activity of an ion.

(3) Studies on actions of neurotransmitters and pharmacological agents: On the pregnant-rat myometrium, Mironneau (9) has found that oxytoxin can increase the early inward current. On the guinea-pig taenia coli, Bolton (6) has found an inward current due to application of carbamylcholine. I have not had great successes in studying the actions of depolarizing drugs, and have worked primarily with hyperpolarizing compounds. In this regard, my colleagues and I have found that the hyperpolarizing actions of some catecholamines on the rat myometrium and the guinea-pig taenia coli is not caused by any effects on membrane conductances, but by an effect on the electromotive force of the potassium cell and a selectivity change in favor of K^+ (10). These mechanisms of action are novel and point to the importance of considering biochemical and metabolic actions of some neurotransmitters and drugs on smooth muscles as a means by which important electrophysiological and contractile consequences can be produced.

These are some of the achievements of using a voltage-clamp approach to studying some problems in smooth muscles. It is a narrow and rather personally-biased account with which not everyone is in agreement. Regardless of possible controversies and imperfections in the technique, they do show that meaningful progress can be attained even at the present stage of technical development. With continued refinement in technique, further progress should be possible. For example, in the guinea-pig myometrium, Vassort (3) has proposed the existence of at least two different outward K^+ currents, the faster of the two overlapping with the early inward current to an appreciable degree. The main evidence upon which this new current is established is that there is a hump in this early outward current suggestive of some inactivation process (see also 11), and that tetraethylammonium ions would reduce this component and simultaneously increase the early inward current. My own outlook now is somewhat different, considering the hump and overlap to be consequences of technical limitations. Our evidence, obtained on the pregnant-rat myometrium, is that in a small node (total capacitance of about 0.1 μF), there is no hump suggestive of inactivation, and that TEA while reducing the outward current has very little effect on the inward

Figure 1. Effect of tetraethylammonium chloride on electrical properties of pregnant-rat myometrium. 19 day pregnant uterus. TEA 10 mM in Krebs' bicarbonate solution. 37° C. (a) Constant-current responses. Two negatives have been superimposed for printing. Action potential in Krebs solution was shorter, and became prolonged after 4 min. in TEA. (b and c) Membrane currents in voltage-clamped conditions. Bottom traces in each panel indicate steady-state holding potential and depolarizing (upward) and hyperpolarizing (downward) steps imposed. Holding potential was -48 mV, which was also natural resting potential. Depolarizing steps to 12.5 mV in (b) and 42.5 mV in (c). Top traces indicate membrane currents, inwards (downward) and outward (upward). Two negatives have been superimposed for each panel, and no correction for leakage currents has been made. Series resistance: 7.9 KΩ in Krebs' solution, 9.3 KΩ in TEA; membrane capacitance 0.21 μF in Krebs, and 0.18 μF in TEA. In each panel, inward currents in Krebs solution and after TEA are superimposed, indicating constancy of series resistances and membrane capacitances (see ref. 4). For outward currents, initial dips represent sodium currents uncorrected for leakage, and sustained portions represent potassium currents. Note absence of any humps on outward currents. Note also that TEA reduced late outward currents without appreciably affecting early currents. Calibrations in right lower corner: vertical bar represents 50 mV for both (b) and (c), 0.2 μA for (b) and 0.5 μA for (c); horizontal bar represents 20 msec. For (a), calibrations are 20 mV and 50 msec.

current (Fig. 1). Because of the lessons learned here, I am optimistic that with experience and continued refinement of technique, it would be entirely possible to clarify many more problems in smooth muscle physiology. No doubt there will be difficulties, but substantial progress can be attained if the expectations are realistic.

REFERENCES

1. Anderson, N.C., 1969, J. Gen Physiol. 54, 145-165.
2. Mironneau, J., 1974, Pflügers Arch. ges. Physiol. 352, 197-210.
3. Vassort, G., 1975, J. Physiol. (London), 252, 713-734.
4. Kao, C.Y. and McCullough, J.R., 1975, J. Physiol. (London), 246, 1-36.
5. Inomata, H. and Kao, C.Y., 1976, J. Physiol. (London), 255, 347-378.
6. Bolton, T.B., 1975, J. Physiol. (London), 250, 175-202.
7. Anderson, N.C., Ramon, F.C. and Snyder, A., 1971, J. Gen. Physiol., 58, 322-339.
8. Kao, C.Y., 1977, Electrophysiological properties of the uterine smooth muscle, in Biology of the Uterus, R.M. Wynn (Ed), pp. 423-496, Plenum, New York.
9. Mironneau, J., 1976, Pflügers Arch. ges Physiol., 363, 113-118.
10. Kao, C.Y., Inomata, H., McCullough, J.R. and Juan, J., 1976, Voltage-clamp studies of the actions of catecholamines and adrenergic blocking agents on mammalian smooth muscles, in Smooth Muscle Physiology and Pharmacology, M. Worcel and G. Vassort (Eds.), pp. 165-176. l'Inst. Nat. de la Santé et de la Recherche Médicale, Paris.
11. Kao, C.Y., 1973, Ionic currents in a pregnant myometrium. In Foetal and Neonatal Physiology, Proc. of the Sir Joseph Barcroft Centenary Symposium, pp. 584-588, Cambridge University Press, Cambridge.

Excitation-Contraction Coupling in Smooth Muscle, R. Casteels et al. eds.

Na, Ca AND CONTRACTION IN THE SMOOTH MUSCLE OF THE GUINEA-PIG TAENIA COLI

A. F. Brading
Department of Pharmacology,
South Parks Road,
Oxford,
England.

INTRODUCTION

The calcium required for contraction in the guinea-pig taenia coli may come from the extracellular solution, for example during the action potential, or through increased Ca permeability induced by K^+ or agonists such as acetylcholine. Internal calcium ion concentration may also be raised by release of calcium from internal stores, which could be membrane bound calcium and calcium in internal compartments such as the sarcoplasmic reticulum and mitochondria. Na ions may play several roles in regulating the levels of intracellular calcium. In this paper the role of Na has been investigated in experiments on tension responses, as the degree of contraction existing in the muscle at any time is likely to be a reflection of the free calcium ion concentration in the cells.

Na AND Ca EXTRUSION

With normal levels of $[Ca]_o$ the electrochemical gradient greatly favours calcium entry. As tracer experiments have shown that Ca is permeant, and exchanges readily between the extracellular and intracellular compartments, an efficient calcium extrusion mechanism must be postulated to account for the low Ca^{2+} in the resting tissue. In several tissues, including smooth muscle[1, 2] it has been postulated that a Na/Ca exchange mechanism could be involved, and that in some way the potential energy of the Na gradient might be used to extrude calcium ions against their electrochemical gradients. Casteels and his coworkers[3] have however suggested that Na ions may not be involved in Ca extrusion from the taenia coli, but that an active pumping mechanism may exist using ATP as an energy source.

Tension experiments show that pieces of taenia coli placed in Na free media, after an initial period of contraction depending on the tissue and substitute used, can maintain a relaxed condition. Similarly high-Na tissues, prepared by exposing them to K^+ free solution for several hours, also can remain relaxed. It is thus certain that these tissues can maintain a low intracellular calcium concentration in the absence of a Na gradient across the membrane. It is noticeable however that such tissues quite readily go into contracture if their calcium permeability is increased for instance by several doses of carbachol. High-Na tissues can also be made to contract by removing extracellular Na and thus reversing the Na gradient. Such contractures can in both cases be reversed by reintroducing Na into the external medium, suggesting that Na may be in some way involved in regulating intracellular Ca.

Na AND Ca ENTRY

There also is evidence that intracellular Na is necessary for efficient penetration of calcium in conditions where calcium permeability is raised. In solutions with normal extracellular calcium (2.5 mm) addition of high calcium for periods of up to several minutes, will suppress spontaneous rhythms, but will not induce contracture. This is also true of Na free tissues. In tissues exposed to zero calcium solutions, the membrane permeability to calcium seems to be high, and introduction of calcium will cause a transient contraction. This is often followed by a more prolonged period of contractions when the calcium is washed off. In Na free conditions however both the transient and prolonged contractions, become small, and are often abolished.

In calcium free Na solutions addition of carbachol (5×10^{-5}M) in the presence of calcium ions leads to contraction, graded according to the duration of application, and concentration of calcium present. In Na free solutions however these contractions are much less easy to elicit, although the tissue can be made to contract by long application of carbachol in high calcium solutions.

Other methods of increasing calcium entry such as depolarising the tissue with about 70 mm K also suggest that intracellular Na facilitates the Ca entry, since in HCO_3 buffered solutions, the threshold concentration of extracellular calcium necessary to produce contracture is lowered in tissues bathed in 50/50 K/Na solutions, than in 50/50 K/Na substitute. With higher levels of extracellular calcium (25 mm) depolarised tissues with Na reach a steady level of contracture at a tension less than the maximum possible, whilst in the absence of Na the contracture will reach the maximal tension of which the tissue is capable, suggesting that the presence of Na may be important in regulating Ca extrusion as well as Ca entry.

Na AND Ca STORES

Tissues exposed to Ca-free EGTA Krebs solution rapidly lose their ability to contract to a supramaximal dose of carbachol, or high-K. If such tissues are exposed to Ca containing solutions for a few seconds (which may produce only small transient contraction) subsequent stimulation by carbachol (in Ca-free solution) will initiate a contraction, the size of which depends on the concentration of Ca, and duration of exposure, but which may be large. One dose of carbachol (supramaximal) will normally totally exhaust this store of calcium, although exposure to isotonic KCl is not such an effective stimulus. If the same procedures are carried out in Na-free solution, the tissues lose their ability to refill the calcium store. It thus appears that Na ions not only can increase Ca inflow to the contractile machinary, but also can greatly facilitate the filling of Ca stores, even when the Ca application does not result in significant contraction of the tissue.

POSSIBLE EXPLANATION OF RESULTS

The results could be explained if Na ions are necessary for accumulation of calcium in the sarcoplasmic reticulum, which lies close to the plasma membrane in this tissue. It has previously been suggested[4] that the S.R. normally contains high Na, and it may be that this is necessary for the existence of a potential difference between the S.R. and the intracellular compartment. This potential could be essential for a low P_{Ca} of the S.R. membrane, as it appears to be for the plasma membrane. Accumulation of Ca in the S.R. could then be due to an active Ca pump from the cell to the S.R., combined with a low P_{Ca} of the membrane. Depolarisation of the S.R. would increase P_{Ca} and allow release of calcium to the contractile machinary. Ca may also enter the S.R. from the extracellular medium, possibly through a Ca/Na ion exchange mechanism, but in the absence of Na in the S.R. calcium would not be held in this compartment. Carbachol would then be postulated to both increase the calcium permeability of the plasma membrane, and also depolarise the S.R. whereas high K might depolarise the plasma membrane, but be less effective in depolarising the S.R.

Loss of calcium from the cell in normal conditions may also be facilitated by a Na/Ca ion exchange from the S.R., helping to reduce the Ca gradient against which the Ca pump would be working.

REFERENCES

1. Reuter, H., Blaustein, M.P. & Haeusler, G. (197). Na-Ca exchange and tension development in arterial smooth muscle. Phil. Trans. R. Soc. Lond. B. 265 87-94.
2. Katase, T. & Tomita, T. (1972). Influences of sodium and calcium on the recovery process from potassium contracture in the guinea-pig taenia coli. J. Physiol. 224, 489-500.
3. Raeymaekers, L., Wuytack, F. & Casteels, R. (1974). Na-Ca exchange in taenia coli of the guinea-pig. Pflugers Arch. 347, 329-340.
4. Brading, A.F. (1975). Sodium/sodium exchange in the smooth muscle of the guinea-pig taenia coli. J. Physiol. 251, 79-105.

Excitation-Contraction Coupling in Smooth Muscle, R. Casteels et al. eds.

THE ROLE OF Na-Ca EXCHANGE IN THE REGULATION OF TONE IN VASCULAR SMOOTH MUSCLE

Mordecai P. Blaustein
Department of Physiology and Biophysics
Washington University School of Medicine
St. Louis, Missouri 63110, U.S.A.

INTRODUCTION

The factors underlying tonic myogenic contraction ("tone") in vascular smooth muscle[1,2] are poorly understood, although it appears that a "catch" mechanism is probably not involved[3]. Because calcium ions induce contraction in vascular smooth muscle[4], as in most other types of muscle, one possibility is that the tone may be due to maintenance of free Ca^{2+} (in the sarcoplasm) above the contraction threshold. Then, since Ca^{2+} is known to cross the sarcolemma of vascular smooth muscle[5], we would expect: a) that there is a Ca extrusion pump in the sarcolemma which must play a critical role in regulating the ionized Ca^{2+} concentration in the sarcoplasm ($[Ca^{2+}]_i$) and, therefore, tone, and b) that small changes in Ca influx or efflux (so that $[Ca^{2+}]_i$ is altered slightly) may have a significant effect on the maintained tension. As will be shown below, experimental observations on vascular smooth muscle are consistent with these expectations. Available data indicate that Ca extrusion involves an exchange of Na^+ ions for Ca^{2+}, driven by energy from the Na electrochemical gradient; thus, small changes in the Na gradient may affect $[Ca^{2+}]_i$ and, therefore, tone. These interactions may also play a role in the well-documented association between dietary salt (Na) intake and hypertension[6].

Na-Ca EXCHANGE IN VASCULAR SMOOTH MUSCLE

Leonard[7] was the first to show that manipulations which may be expected to raise the intracellular Na concentration - either removal of external K, or introduction of cardiac glycosides - rapidly induce contraction in arterial strips. A prompt increase in tension is also observed when external Na is removed[8,9]. These effects are illustrated in Fig. 1. Fig. 1B shows that, after loading the tissue with (internal) Na, removal of external Na causes a more rapid and greater rise in tension than is seen before Na loading (Fig. 1A). The increase in tension, due to external Na removal, can be accounted for by a net gain of Ca[9] due to the increased Ca influx[5,9] and decreased efflux[9]. It has been more difficult to demonstrate a significant net gain of Ca under conditions of Na loading - perhaps because the contractures observed under these circumstances are more modest (compare Fig. 1C with 1A). Moreover, in view of the many difficulties encountered in attempting to determine accurately the (intracellular) Ca content of vascular smooth muscle, it seems most reasonable to rely on tension changes as a very sensitive and reliable measure of variation in the Ca concentration in the sarcoplasmic space (cf. ref. 4). With this in mind, the aforementioned experiments appear to support the conclusion that lowering external Na and/or raising internal Na tends to drive Ca into the cells, thereby causing tension to increase.

The corollary, of course, is that external Na also plays an important role in (net) Ca efflux; this idea is supported by the observations[9,10] that Ca efflux (from ^{45}Ca-loaded arterial strips) and relaxation (following contractures) are slowed when external Na is removed in the absence of external Ca.

In sum, these findings are all consistent with the view that in vascular smooth muscle, as in many other types of cells[11], a transport mechanism which derives energy directly from the sodium electrochemical gradient, and mediates the exchange of Na for Ca, plays a critical role in regulating the gradient of Ca across the plasmalemma (sarcolemma). If the energy for Ca transport is derived exclusively from the Na electrochemical gradient, the maximum Ca concen-

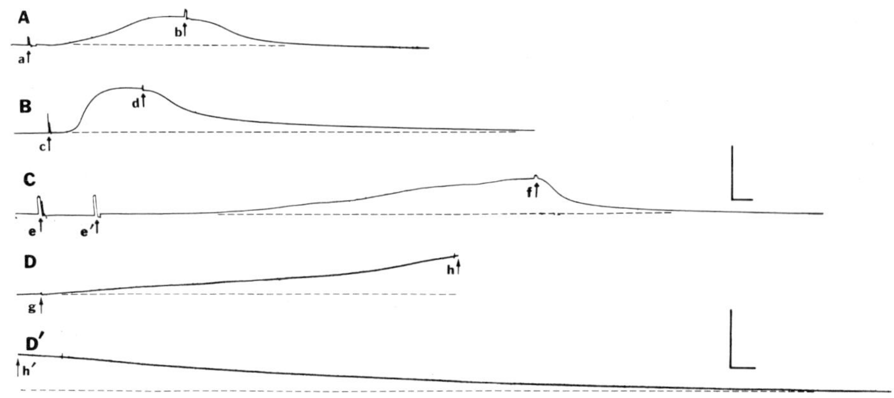

Fig. 1. Effect of cations and of ouabain on tension of rabbit aortic strips. Resting load was 0.5 g in all cases. A: increase in tension due to exposure (at a) to Na-free Li-Tyrode solution (all 137 mM NaCl replaced by LiCl). Tissue relaxed promptly when returned to standard (Na-containing) Tyrode (at b). B: between records A and B, aortic strip was incubated for 30 min in K-free Ca-free Na-containing Tyrode to load muscle with Na. Exposure to K-free Na-free Li-Tyrode (with 1.8 mM $CaCl_2$), at c, caused a more rapid and greater rise in tension than seen in A. Standard Tyrode was reintroduced at d. C: data from another aortic strip showing effect of external K removal (at e and e') on tension; solutions all contained 1.8 mM $CaCl_2$. Strip promptly relaxed when external K was added back (at f). Temperature, 37°C. Calibration (A-C): vertical (tension) = 2 g (A and B) or 1 g (C); horizontal (time) = 1 min. D and D': data from a third aortic strip illustrating effect of 10^{-6} M ouabain, introduced into bathing solution at g, on tension. Between h and h' (9.3 min), ouabain was washed out with four changes in fresh Tyrode solution. Calibration: vertical (tension) = 1 g; horizontal (time) = 1 min. (from ref. 10, with permission).

tration gradient across the sarcolemma, $[Ca^{2+}]_o/[Ca^{2+}]_i$, will be given by

$$1. \qquad \frac{[Ca^{2+}]_o}{[Ca^{2+}]_i} = \left(\frac{[Na^+]_o}{[Na^+]_i}\right)^n \exp\{(n-2)\, V_m F/RT\}.$$

$[Na^+]_o/[Na^+]_i$ is the concentration gradient of free Na^+ across the sarcolemma, V_m is the resting membrane potential, and F, R and T have their usual meanings; n indicates the number of Na^+ ions exchanged for each Ca^{2+} by the transport mechanism. Data from barnacle muscle[12], squid axons[13] and mammalian presynaptic nerve terminals[14] indicate that, in these preparations, n has a value of about 3. Unfortunately, there are no comparable data available for vascular smooth muscle, although there is some indication[9] that n may be at least 2; a value of 3 seems likely if $[Ca^{2+}]_i$ is maintained below 10^{-6} M (see below). Under any circumstances, as a consequence of the operation of this Na-Ca exchange mechanism, small changes in $[Na^+]_i$ should be reflected by changes in $[Ca^{2+}]_i$ in the same direction.

THE ROLE OF Na-Ca EXCHANGE IN REGULATING $[Ca^{2+}]_i$

Although the properties of the Na-Ca exchange mechanisms in giant barnacle muscle[12] and vascular smooth muscle are rather similar, there is one interesting and crucial difference which may provide an important clue about the regulation of tone in vascular smooth muscle. This difference concerns the responsiveness to changes in the Na gradient. Unlike the smooth muscle (see Fig. 1), normal barnacle muscle does not contract when either external Na is removed or internal Na is increased (by treating with cardiac glycosides or removing external K); however, Na-loaded barnacle fibers do contract when placed in Ca-containing Na-free media[15]. The free $[Ca^{2+}]_i$ in resting barnacle muscle is normally buffered at about $2\text{-}8 \times 10^{-7}$ M, below the contraction threshold[16]; consequently, the combined effects of removing external Na and raising internal Na are required to induce a sufficiently rapid gain of Ca so that the intracellular buffering mechanisms are overwhelmed, and $[Ca^{2+}]_i$ can reach the contraction threshold.

The much greater sensitivity of arterial muscle to changes in the Na gradient (see Fig. 1) implies that a somewhat different situation prevails here. The most straightforward explanation is that, in "resting" vascular smooth muscle, $[Ca^{2+}]_i$ is normally very close to the contraction threshold - and, in vessels with myogenic tone, is even above the contraction threshold (see $[Ca^{2+}]_i$ - tension diagram in Fig. 2). In this case, small changes in Ca influx and/or efflux and, therefore, in $[Ca^{2+}]_i$, as will occur when the Na gradient is altered, should be reflected by a change in tension.

The preceding discussion focuses exclusively on movements of Ca across the sarcolemma; however, it is well known that, even in vascular smooth muscle[17], Ca is sequestered by intracellular organelles such as the sarcoplasmic reticulum and mitochondria. We must, therefore, inquire about the interplay between these organelles and the sarcolemma, in the regulation of free intracellular $[Ca^{2+}]$. In this regard it will be helpful to refer to the diagram of Ca compartmentalization shown in Fig. 3. For convenience, all organelles which sequester Ca have been lumped into the compartment labelled "saracoplasmic reticulum". It will be assumed, for reasons given elsewhere[10], that there is no direct connection between these organelles and the extracellular space; in other words, Ca leaving these organelles cannot enter the extracellular space without traversing the sarcoplasmic space ("sarcoplasm"). According to this model, the Ca pumps in the sarcolemma and in the membranes lining the intracellular organelles (labelled "2" and "3", respectively, in the diagram), must operate independently so as to maintain $[Ca^{2+}]_i$ at the appropriate resting level. Because the volume of the intracellular organelles is limited, these organelles cannot serve as infinite sources or sinks for Ca.

Now, if the Ca "pump" in the sarcolemma ("2" in Fig. 3) involves an Na-Ca exchange mechanism (see above), with $[Na^+]_o$ and $[Ca^{2+}]_o$ maintained constant (by renal and other regulatory mechanisms), $[Ca^{2+}]_i$ will be altered if $[Na^+]_i$ is (equation 1). Under these circumstances, there will also be a movement of Ca^{2+}

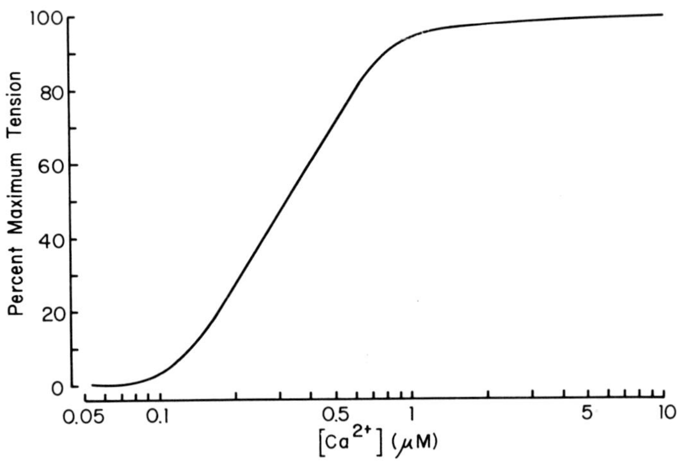

Fig. 2. Effect of concentration of free (ionized) Ca^{2+} on tension developed by glycerinated vascular smooth muscle fibers from hog carotid artery. Conditions: temperature, 20°C; 150 mM KCl, 20 mM histidine buffer (pH 6.6), 5 mM ATP, and 5 mM Mg, 4 mM $EGTA_{total}$. Free $[Ca^{2+}]$ was varied by changing ratio $Ca_{total}/EGTA_{total}$ appropriately. Curve was redrawn from Fig. 1 of Filo et al.[4] (from ref. 10, with permission).

from the sarcoplasmic space into the sarcoplasmic reticulum - but only until the ratio, $[Ca^{2+}]_{SR}/[Ca^{2+}]_i$ returns to its initial level (as governed by the ratio of the unidirectional rate coefficients, k_{-3}/k_3; see Fig. 3 and ref. 10). Note that, because the extracellular fluid is an enormous reservoir for Ca, this internal redistribution will not be sufficient to return $[Ca^{2+}]_i$ to its initial level (only the ratio is restored). In other words, a very small increase in $[Na^+]_i$ may cause a substantial increase in $[Ca^{2+}]_i$ because of the power law relationship (equation 1). And, if $[Ca^{2+}]_i$ is maintained above the contraction threshold, as may be the case in vascular smooth muscles with myogenic tone (see above), an increase in $[Ca^{2+}]_i$ should be directly reflected by an increase in tension (Fig. 2). The fact that tension in arterial strips is rapidly altered in response to changes in the Na gradient (Fig. 1), is certainly consistent with this view. Also consistent is the observation[18] that, even in normal man, pharmacologic doses of cardiac glycosides directly increase peripheral vascular resistance.

In sum, the foregoing considerations support two conclusions: 1. Myogenic tone in vascular smooth muscle is a consequence of the fact that $[Ca^{2+}]_i$ is maintained above the contraction threshold. 2. Under these circumstances, the long-term regulation of $[Ca^{2+}]_i$ is primarily under the control of the sarcolemmal Ca pump - which presumably involves an Na-Ca exchange mechanism.

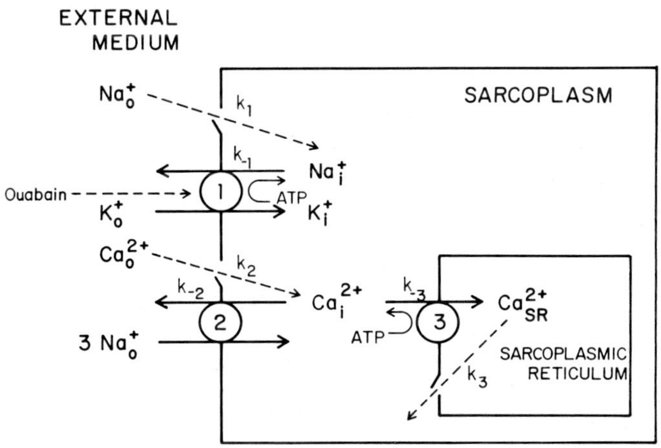

Fig. 3. Block diagram of compartmentalization of Ca in muscle fibers. Main pathways and transport systems for Na and Ca are shown: 1, ATP-dependent, ouabain-sensitive Na-K exchange pump; 2, sarcolemmal Na-Ca exchange mechanism (shown operating in the "forward", or Ca extrusion mode); 3, sarcoplasmic reticulum (SR) ATP-dependent Ca pump. Most of entering Na and Ca is assumed to come through voltage-sensitive conductance channels associated with the action potential. k's indicate rate coefficients for fluxes of Na and Ca across sarcolemma and SR membranes in the directions indicated (see text for further details; from ref. 10, with permission).

Na^+ IONS AND THE ETIOLOGY OF HYPERTENSION

As noted in the Introduction, there is a well-documented association between sodium intake and some forms of hypertension. There is increasing evidence that Na transport may be abnormal in the erythrocytes[19] and leucocytes[20] of humans with essential hypertension; the white cell data indicate that the cardiac glycoside-sensitive Na-K exchange pump is inhibited. Furthermore, Jones[21] observed that aortic smooth muscle from spontaneously hypertensive rats has a decreased ability to accumulate K and extrude Na, as compared to normal controls, and Overbeck et al.[22] found that a factor which inhibits the Na-K pump in vascular smooth muscle (the transport system labelled "1" in Fig. 3) is present in the blood plasma of hypertensive dogs. This factor may be the natriuretic hormone - the hormone which apparently reduces renal proximal tubular reabsorption of Na by inhibiting the Na-K exchange pump at the baso-lateral (peritubular) surface of the tubular epithelium[23]. If the plasma levels of natriuretic hormone are, indeed, elevated in patients with so-called plasma-expanded (increased blood plasma volume) hypertension[24], the hypertension could be readily accounted for by the following sequence of events: increased salt (Na) and plasma volume → increased circulating levels of natriuretic hormone (which should promote excretion of salt and water) → inhibition of the Na-K pump in vascular smooth muscle fibers → rise in $[Na^+]_i$ → rise in $[Ca^{2+}]_i$ (because of Na-Ca exchange) → increased myogenic tone → increased peripheral vascular resistance. Note that, because of the power law (perhaps cubic) relationship between $[Na^+]_i$ and $[Ca^{2+}]_i$ (equation 1), and the non-linear relationships between $[Ca^{2+}]_i$ and tension (Fig. 2) and between tension and peripheral resistance[25], a small change in $[Na^+]_i$ may have a profound effect on the peripheral resistance.

This hypothesis may also help to explain the efficacy of chronic natriuretic therapy in the treatment of mild hypertension. The diuretics should reduce plasma volume and total body Na stores, so that a significant reduction in the level of circulating natriuretic hormone would be expected; consequently all of the alterations mentioned in the preceding paragraph would be reversed.

CONCLUSIONS AND SUMMARY

The hypothesis presented above ties together several well-documented observations concerning vascular smooth muscle, viz.: 1. Some vascular smooth muscle has an inherent, or "myogenic" tone. 2. Tension in vascular smooth muscle rapidly increases when $[Na^+]_i$ is increased or $[Na^+]_o$ is decreased. 3. Ca uptake is increased, and Ca efflux decreased, by removing external Na. 4. There is an association between the dietary intake of sodium, and hypertension. The unifying hypothesis, which helps to explain the latter association, is based upon three fundamental notions: 1. In vascular smooth muscle with myogenic tone, $[Ca^{2+}]_i$

is normally maintained above the contraction threshold. 2. In this tissue $[Ca^{2+}]_i$ and, therefore, tension, is controlled by an Na-Ca exchange mechanism. Consequently, in intact animals, where $[Na^+]_o$ is normally maintained within narrow limits, changes in $[Na^+]_i$ may be a critical determinant of peripheral vascular resistance. 3. In so-called volume-expanded hypertension, which may include many patients with "essential hypertension," there may be elevated levels of natriuretic hormone in the blood plasma. This hormone may partially inhibit the Na-K pump in vascular smooth muscle, thereby causing an increase in $[Na^+]_i$ and, by the mechanisms outlined above, an elevation of blood pressure.

ACKNOWLEDGEMENTS

I thank Mrs. A. L. Guinn for secretarial assistance, and Mr. W. DiPalma and Mrs. S. McConnell for help with the illustrations. I gratefully acknowledge the support of the American Heart Association, the Muscular Dystrophy Association of America, and the National Institute of Neurological and Communicative Diseases and Stroke.

REFERENCES

1. Folkow, B. and Öberg, B. (1961) Acta Physiol. Scand. 53, 105.
2. Uchida, E. and Bohr, D.F. (1969) Circ. Res. 25, 549.
3. Rüegg, J.G. (1971) Physiol. Rev. 51, 201.
4. Filo, R.S., Bohr, D.F. and Rüegg, J.C. (1972) Science 147, 1581.
5. van Breemen, C., Farinas, B.R., Gerba, P. and McNaughton, E.D. (1972) Circ. Res. 30, 44.
6. Dahl, L.K. and Love, R.A. (1954) Arch. Int. Med. 94, 525.
7. Leonard, E. (1957) Amer. J. Physiol. 189, 185.
8. Bohr, D.R., Seidel, C. and Sobieski, J. (1969) Microvasc. Res. 1, 335.
9. Reuter, H., Blaustein, M.P. and Haeusler, G. (1973) Phil. Trans. R. Soc. Lond. B. 265, 87.
10. Blaustein, M.P. (1977) Amer. J. Physiol. 232, C165.
11. Blaustein, M.P. (1974) Rev. Physiol. Biochem. Pharmacol. 70, 33.
12. Russell, J.M. and Blaustein, M.P. (1974) J. Gen. Physiol. 63, 144.
13. Blaustein, M.P. and Russell, J.M. (1975) J. Membrane Biol. 22, 285.
14. Blaustein, M.P. and Ector, A.C. (1976) Biochim. Biophys. Acta 419, 295.
15. Blaustein, M.P. (1976) Physiologist 19, 525.
16. Hagiwara, S. and Nakajima S. (1966) J. Gen. Physiol. 49, 807.
17. Devine, C.E., Somlyo, A.V. and Somlyo, A.P. (1972) J. Cell Biol. 52, 690.
18. Mason, D.T. and Braunwald, E. (1964) J. Clin. Invest. 43, 532.
19. Wessels, F., Zumkley, H. and Losse, H. (1970) Z. Kreislaufforsch. 59, 415.

20. Edmonson, R.P.S., Thomas, R.D., Hilton, P.J., Patrick, J. and Jones, N.F. (1975) Lancet 1, 1003.

21. Jones, A.W. (1973) Circ. Res. 33, 563.

22. Overbeck, H.W. Pamnani, M.B., Akera, T., Brody, T.M. and Haddy, F.J. (1976) Circ. Res. 38, Suppl. II, II-48.

23. DeWardener, H.E. (1973) in Orloff, J. and Berliner, R.W., eds., Handbook of Physiology, Section 8, Renal Physiology. American Physiological Society, Washington, D.C. P. 677.

24. Haddy, F.J. and Overbeck, H. W. (1976) Life Sci. 19, 935.

25. Burton, A.C. (1954) Physiol. Rev. 34, 619.

Excitation-Contraction Coupling in Smooth Muscle, R. Casteels et al. eds.

IONTOPHORETIC APPLICATION OF ACETYLCHOLINE TO SMOOTH MUSCLE

T. B. Bolton
St. George's Hospital Medical School
London, S.W.17 ORE
United Kingdom.

Iontophoretic application of acetylcholine to smooth muscle has two main advantages over methods which involve the addition of the drug to the solution bathing the muscle, viz: exposure can be restricted to a small region and can be made very brief. Diffusion of the drug following its release at the tip of the pipette would be expected to be the main factor determining both the spatial and temporal distribution of the drug at later times. The better spatial and temporal resolution of the technique may enable more to be learnt about the kinetics of drug-receptor interaction and about the factors which affect the concentration of drug in the solution in the immediate vicinity of the receptor (the biophase).

The response of taenia smooth muscle to a brief pulse of acetylcholine (typically 50ms, 2.5nC) never had a latency less than 100ms and the time course of the response was often very similar to that observed by others upon stimulation of cholinergic nerves. The similar latency of responses to nerve-released and iontophoretically applied acetylcholine may indicate rate limiting processes either before acetylcholine binding to the receptor or between that event and the conductance change in the membrane. However, other explanations for the latency and time course of the iontophoretically evoked response should also be considered.

The method which was used for these experiments has been described elsewhere (1). It involved the introduction of a narrow strip (0.3 - 0.5mm wide) of longitudinal smooth muscle from the taenia of the guinea-pig caecum into a form of double sucrose-gap apparatus (2). The membrane potential of smooth muscle cells in the "nodal" region was recorded by microelectrode and the tip of the iontophoretic pipette was brought up to the surface of the muscle, close (10 - 25μm) to the point of recording. Acetylcholine, carbachol, or other agonist at the muscarinic receptors, was then released by passing outward current pulses through the pipette, after first hyperpolarizing the muscle by passing inward current to prevent the discharge of action potentials and contraction.

A response of the membrane potential obtained upon the iontophoretic release of acetylcholine is shown in Fig. 1. Certain features are notable. The sensitivity to acetylcholine was less than at the end-plate of skeletal muscle. In most experiments sensitivity was in the range of 5 - 50mV/nC which can be compared with the end-plate where sensitivities of 100 - 700mV/nC are usual (e.g. 3, 4, 5). The sensitivity of smooth muscle is, however, similar to that of denervated skeletal muscle (e.g. 6, 7). The time elapsing between the release of the transmitter from the pipette and the peak of the response was variable in smooth muscle but was within the range 300 - 2000ms, which is much greater than at the end-plate where 10 - 100ms is typical (e.g. 3, 5). Similar values to those seen at the end-plate are found in denervated muscle despite its lower sensitivity (6, 8). The latency between the release of acetylcholine and the appearance of the response is 0.1 - 10ms, depending how close the pipette is to the end-plate, or denervated muscle membrane (3, 5, 9) whereas in smooth muscle latency was never less than 100ms and could on occasions exceed 1s (Fig. 1 and ref. 1). Thus the response elicited by the iontophoretic application of acetylcholine to muscarinic receptors of smooth muscle had a much greater latency and time to peak than responses involving nicotinic receptors of skeletal muscle, even though the sensitivity of smooth muscle to acetylcholine is no less than that of denervated skeletal muscle.

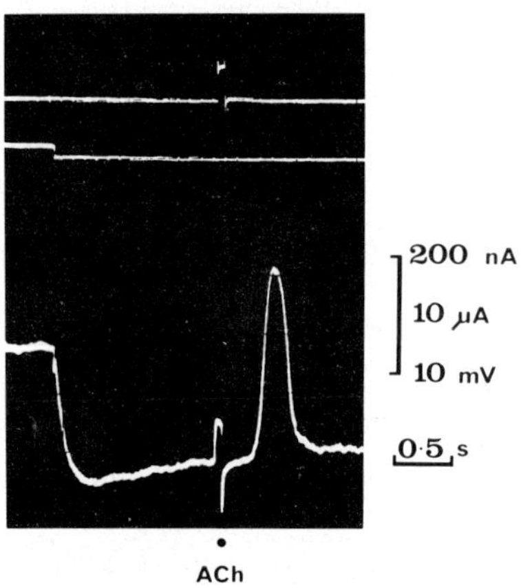

Fig. 1. Response of the membrane potential (lower trace and 10mV calibration) of taenia smooth muscle to acetylcholine (ACh). The membrane was first hyperpolarized by passing inward current as indicated in the middle trace (10μA calibration). Iontophoretic current (upper trace and 200nA calibration) caused an electrical artefact. Notice the latency and time course of the acetylcholine response.

The distance between the point of release of acetylcholine and the regions of smooth muscle membrane which are sensitive to the drug will be important in determining the latency and time-course of the response. The tip of the iontophoretic pipette was less than 1μm in diameter so that its position could not be seen during an experiment. However it was clear that on many occasions the tip of the iontophoretic pipette was touching the surface of the muscle, either because "dimpling" of this could be seen or because the resulting movement of the muscle dislodged the recording electrode. It must be concluded, therefore, that acetylcholine was released in close proximity to smooth muscle cells. Assuming that the most superficial smooth muscle cells are sensitive to acetylcholine, then the nearest acetylcholine receptors cannot have been much more than a cell diameter away, say 5µm (i.e. 5×10^{-4}cm).

It is possible to calculate the concentration of acetylcholine which would be expected in the vicinity of the receptors at various times after its release. If we suppose that the surface of the muscle is effectively infinite, that acetylcholine is deposited instantaneously at time zero, x cm above this surface, and that the stimulus, S, obtained via all the acetylcholine receptors is directly proportional to the integral of the concentration over the surface then (10)

$$S \propto \frac{M}{(\pi Dt)^{0.5}} \exp(-x^2/4Dt) \qquad \text{Eqn. 1}$$

M (moles) is the amount of acetylcholine deposited instantaneously at time zero, D (cm^2/sec) is the diffusion coefficient for acetylcholine in solution, t(sec) is the time elapsing since its release and π has its usual significance. If the peak of the response at t = T coincides with maximum stimulus to the receptors then (10)

$$T = x^2/2D \qquad \text{Eqn. 2}$$

If we insert $D = 1.5 \times 10^{-5}$ (11) and $x = 5 \times 10^{-4}$ in equation 2, then $T \approx 0.008$ or 8ms. We would expect therefore, that if the pipette was as close as 5µm to the nearest sensitive smooth muscle cells, the response would peak at about 8ms. In fact peak responses occurred at 0.3 - 2s, i.e. the responses were some 30 - 200 times slower than expected. It should be pointed out that the time-course of the much faster responses elicited by iontophoretic application of acetylcholine at the end-plate, are well described by diffusion equations (e.g. 3, 5). It would seem therefore that when acetylcholine is applied iontophoretically to the surface of smooth muscle, the situation is more complicated than envisaged in this simple model.

Other discrepancies were noted between the time course of responses and those expected on the basis of diffusion from a point source into an infinite homogeneous medium (the equation used in this case was that describing diffusion from a point source into an infinite sphere, 1). It was found that the latency of responses decreased as the amount of acetylcholine released was increased (Fig. 2). There was also a significant correlation between size of acetylcholine response and the time elapsing between release and its peak effect. Responses often showed more than one peak, and the effect on latency and rate of rise of the responses of moving the position of the pipette tip with respect to the muscle were also not easily fitted to a simple diffusional model (1). Thus the important features of the observed acetylcholine receptors which were difficult to fit to a simple diffusional situation were the latency, the slow time course, and the bi-or polyphasic shape.

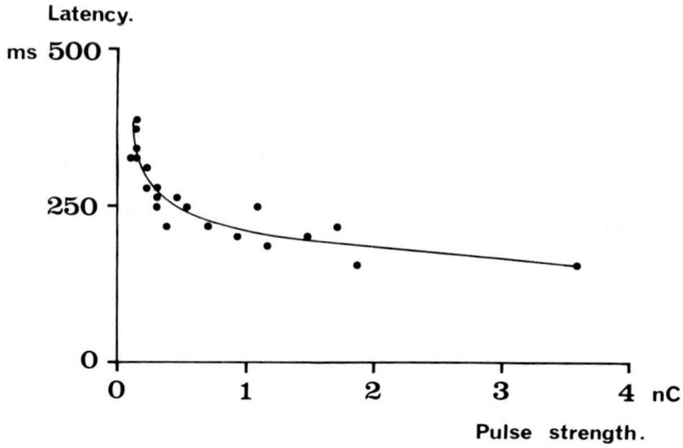

Fig. 2. The latency of acetylcholine responses decreased as the amount of acetylcholine released was increased. Points obtained by applying a number of differently sized acetylcholine pulses in the same experiment. Reproduced with permission from ref. 1.

Possible reasons for the slow time course

A number of possible reasons can be suggested for the slow time course of the response to iontophoretic application of acetylcholine to smooth muscle:-

1. Desensitization of receptors in the immediate vicinity of the tip of the iontophoretic pipette may occur. It may be that muscarinic receptors are more prone to desensitization than the nicotinic variety. If a ring of desensitized receptors was formed about the pipette tip, then acetylcholine would need to diffuse further to activate non-desensitized receptors and the time course of the response would be slowed. If T = 0.5s (Fig. 1) then this distance would be about 30 - 70μm depending on the exact form of the diffusional equations which is applicable. It would be expected, however, that if desensitization was the cause of the latency, that altering the retaining current on the pipette would change the latency of the responses. This was not found to be the case.

2. The sensitivity to acetylcholine may not be uniformly distributed over the smooth muscle surface. Sensitive regions may not be superficially located but may occur only in crevices between the smooth muscle cells of the bundles perhaps adjacent to nerves. This would not only increase the distance over which diffusion had to take place, but might also restrict the rate of diffusion if the crevices were narrow.

3. The rate of diffusion of acetylcholine after its release from the pipette may be slower than in free solution. Plausible causes could be that the receptors are covered by some material which retards diffusion or acetylcholine diffusion may be slowed by some component of the extracellular matrix.

4. If binding of acetylcholine occurs to an appreciable extent to constituents of the extracellular space (including acetylcholine receptors on the surfaces of cells), it could reduce the concentration of acetylcholine which ultimately reaches the receptors (12, 13).

5. Events following receptor activation and before the increase in conductance of the membrane may be rate-limiting. A possibility is that some time is needed to generate a messenger substance such as cyclic GMP.

Area or volume of muscle directly affected by acetylcholine

By use of diffusion equations, some idea of the area or volume of muscle directly affected by acetylcholine can be obtained. Suppose that the surface of the muscle is effectively impermeable to acetylcholine, (compared to the solution overlying it) and that the tip of the pipette is just touching the muscle surface then the concentration C(moles. cm^{-3}) at a distance r (cm) and time t (sec) after release will be given by the equation for diffusion into an infinite half sphere. If we take the concentration of acetylcholine which just produces an effect to be 5×10^{-8} molar, then we can by rearranging that equation obtain an expression for the radius, r, of the circle of muscle exposed to this concentration, or greater, which is

$$r = \sqrt{4Dt \ln\left[\frac{M}{4C(\pi Dt)^{1.5}}\right]} \qquad \text{Eqn. 3}$$

If $M = 7.5 \times 10^{-15}$ eq., $D = 1.5 \times 10^{-5}$ cm^2/sec, $C = 5 \times 10^{-11}$ moles cm^{-3}, and t = 0.5 sec, the radius of this circle is found to be about 130µm. Thus, when the peak of the acetylcholine response is observed about 0.5sec after release (Fig. 1) the diameter of the circle of muscle affected is about 0.25mm. In fact, because the solution over the muscle is flowing, and because the muscle surface is probably permeable to acetylcholine, the area will be somewhat less than this.

In the case of the tip of the pipette being lodged some way into the muscle strip, diffusion would be expected to be retarded to some extent. If it was not, the radius of the spherical volume of muscle exposed to a concentration of 5×10^{-8} molar or more would be about 120µm. If the diffusion coefficient was reduced, say to a $\frac{1}{7}$ (11) then the radius would be about 60µm. These calculations can only be regarded as giving a rough idea of the size of the affected volume, but they do indicate that affected region may not be insignificant compared to the size of the region of active muscle in the "node", about 0.5mm x 0.5mm x 0.2mm. When analogues of acetylcholine were used, which are not hydrolysed by cholinesterase, then bi- or polyphasic responses were much more common (Fig. 3). Presumably the volume or area of their influence is greater because their action is terminated mainly by diffusion out of the muscle, which increases the chances of their penetration to more than one

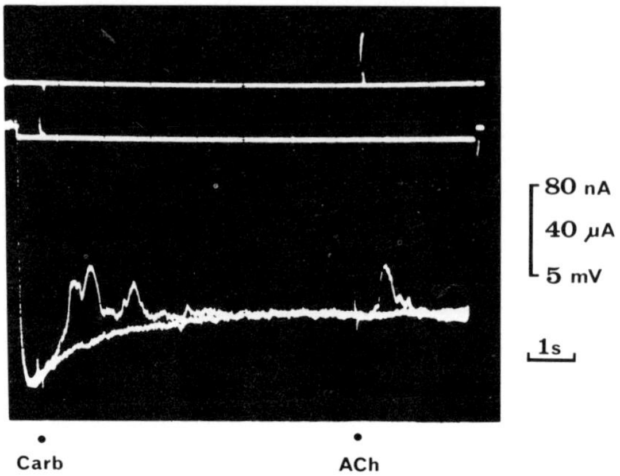

Fig. 3. Comparison of responses to carbachol (Carb) and acetylcholine. Explanation as Fig. 1 except that carbachol iontophoretic current monitor is inverted. Notice the multipeaked nature of the carbachol response especially. Two electrotonic potentials have been superimposed, one without drug responses.

sensitive region, thus giving rise to multiple peaks on the responses.

Conclusions

Some inferences can perhaps be drawn from the results described so far. In the first place it seems possible that some parts of smooth muscle are more sensitive to acetylcholine than others (alternatively, acetylcholine receptors may not all be equally accessible to iontophoretically applied drug). It is difficult to explain the latency of responses if the most superficial smooth muscle cells are sensitive, and rapidly responding, because the response ought to begin much sooner than was observed. We can say however, that by the time the peak of the acetylcholine response has developed, it would be expected that a spherical or circular region of smooth muscle of the order of 100 - 200μm in diameter will have become exposed to a suprathreshold concentration of drug. This has important implications for the electronic spread of depolarization, as it is known that point polarization decays very rapidly (space constant about 30μm) while polarization with external electrodes of an essentially disc-shaped area at right angles to the longitudinal axis of a smooth muscle strip, decays much more slowly (space constant about 1 - 2mm) (see 14).

The polarization produced by acetylcholine applied iontophoretically would be expected to show a spatial decay somewhere between these two extremes.

It may be that rate limiting processes operate either immediately prior to acetylcholine binding to the receptor or between that event and the conductance change in the membrane. If so, it would be expected that the actions of both nerve-released and iontophoretically applied acetylcholine would be affected, thus providing an attractive explanation for the similar times courses of responses to cholinergic nerve stimulation and to iontophoretically applied acetylcholine.

REFERENCES

1. Bolton, T. B. (1976) On the latency and form of the membrane responses of smooth muscle to the iontophoretic application of acetylcholine or carbachol. Proc. R. Soc. Lond. B. 194, 99-119.
2. Bolton, T. B. (1975) Effects of stimulating the acetylcholine receptor on the current-voltage relationships of the smooth muscle membrane studied by voltage clamp of potential recorded by micro-electrode. J. Physiol. Lond. 250, 175-202.
3. Del Castillo, J. and Katz, B. (1955) On the localization of acetylcholine receptors. J. Physiol. Lond. 128, 157-181.
4. Dreyer, F. and Peper, K. (1974) Iontophoretic application of acetylcholine: advantages of high resistance micropipettes in connection with an electronic current pump. Pflügers Arch. 348, 263-272.
5. Dreyer, F. and Peper, K. (1974) The acetylcholine sensitivity in the vicinity of the neuromuscular junction of the frog. Pflügers Arch. 348, 273-286.
6. Dreyer, F. and Peper, K. (1974) The spread of acetylcholine sensitivity after denervation of frog skeletal muscle fibres. Pflügers Arch. 348, 287-292.
7. Miledi, R. (1960) The acetylcholine sensitivity of frog muscle fibres after complete or partial denervation. J. Physiol. Lond. 151, 1-23.
8. Katz, B. and Miledi, R. (9964) The development of acetylcholine sensitivity in nerve-free segments of skeletal muscle. J. Physiol. Lond. 170, 389-396.
9. Katz, B. and Miledi, R. (1965) Tho measurement of synaptic delay, and the time course of acetylcholine release at the neuromuscular junction. Proc. R. Soc. B. Lond. 161, 483-495.

10. Purves, R. D. (1977) The time course of cellular responses to iontophoretically applied drugs. J. Theor. Biol. 65, 327-344.
11. Krnjevic, K. and Mitchell, J. F. (1960) Diffusion of acetylcholine in agar gels and in the isolated rat diaphragm. J. Physiol. Lond. 153, 562-572.
12. Thron. C. D. and Waud, D. R. (1968) The rate of action of atropine. J. Pharmac. exp. Ther. 160, 91-105.
13. Colquhoun, D., Henderson, D. and Ritchie, J. M. (1972) The binding of labelled tetrodotoxin to non-myelinated nerve fibres. J. Physiol. Lond. 227, 95-126.
14. Tomita, T. (1975) Electrophysiology of mammalian smooth muscle. Prog. Biophys. Molec. Biol. 30, 185-203.

Excitation-Contraction Coupling in Smooth Muscle, R. Casteels et al. eds.

EVIDENCE FOR A COMPONENT OF THE OUTWARD CURRENT MEDIATED BY CALCIUM INFLUX IN UTERINE SMOOTH MUSCLE

J. Mironneau, J.P. Savineau and A. Rahéty
Laboratoire de Physiologie Animale et de Physiologie Cellulaire des Structures Contractiles (CNRS : ERA 111), Université de Poitiers, F - 86022 Poitiers, France

SUMMARY

1. The outward current underlying delayed rectification has been studied in uterine smooth muscle using a double sucrose gap apparatus.

2. The outward current was reversibly reduced in calcium-free solution and in the presence of manganese ions or D600, or increased in high-calcium solution.

3. In reference solution, when depolarizing steps activated the outward currents to its maximal value, the tail currents measured at the end of the pulses were made up of two exponentially declining components. The slower of the two components was suppressed in calcium-free solution or increased in high-calcium solution.

4. The reversal potential of the tail currents was not affected by alterations of the external calcium concentration.

5. Tetraethylammonium ions blocked both outward currents.

6. It is concluded that the delayed rectification is composed of two components. One component is voltage dependent ; the other component is dependent on the calcium influx.

INTRODUCTION

In a variety of tissues an increase in the intracellular calcium concentration has been shown to increase the outward current (1,2). A similar association has been suggested for smooth muscle (3). In uterus, prostaglandin E1 acts essentially by increasing the intracellular calcium concentration (4). Simultaneously, the outward current is enhanced. This increase in the outward current intensity has been correlated to a shift of the activation curve towards less positive values of voltage indicating that a larger outward current can be obtained for a given depolarization in the presence of prostaglandin.

The present paper reports an investigation into the effects of raising or reducing the internal calcium on the outward current in uterine smooth muscle. Two indirect methods have been used : in the first method a decrease in cell calcium was obtained by blocking the calcium influx with different antagonists or by perfusing the preparation with a calcium-free solution ; in the second method an increase in calcium was obtained by increasing the calcium influx with a high-calcium solution. The voltage clamp experiments reported here show a component of the outward current which is dependent on the presence of calcium ions in the external solution. It is suggested that the activation of this outward current component is the result of a transient increase in calcium at the inner side of the membrane which sets the level of the outward current.

MATERIAL AND METHODS

A double sucrose gap technique has been used to polarize and voltage clamp small strips (70-120 μm in diameter ; 3-4 mm in length) of pregnant rat myometrium (18-20 days). The perfusion bath and the electronic apparatus have previously been described in detail (5).

Solutions : Physiological solutions had the following composition
1. Reference solution (mM) : NaCl 130 ; KCl 5.6 ; $CaCl_2$ 2 ; $MgCl_2$ 0.24 ;glucose 11. The solution was aerated with O_2 and was buffered by Tris-HCl (8.3 mM) at pH 7.4.
2. The following inhibitors of permeability were used : manganese chloride (5 mM), D600(5 μM) and tetraethylammonium chloride (20 mM).
3. Differences in calcium concentration were obtained by increasing (6 mM) or removing (with EGTA added as 1 mM) the calcium without alteration of the other constituents.

All solutions were maintained at 30 ∓ 1°C.

Uniformity of the voltage clamp : Direct evidence of reliability of voltage clamp can be obtained by measuring independently the voltage in the test gap, inside or just outside the strip, with a microelectrode. When outside, the microelectrode measured the external voltage which should be constant since, in this device, the feedback amplifier maintained the exterior of the strip at the ground potential. When impaled, the microelectrode measured the internal voltage which should be the command voltage. The record on Fig. 1A(a) illustrates the inward current elicited by a depolarization of +30 mV (V_c) and the controlled voltage (Ve) at the tip of the microelectrode just outside the membrane. The external voltage remained almost constant (+2.5 mV at the time to peak inward current)

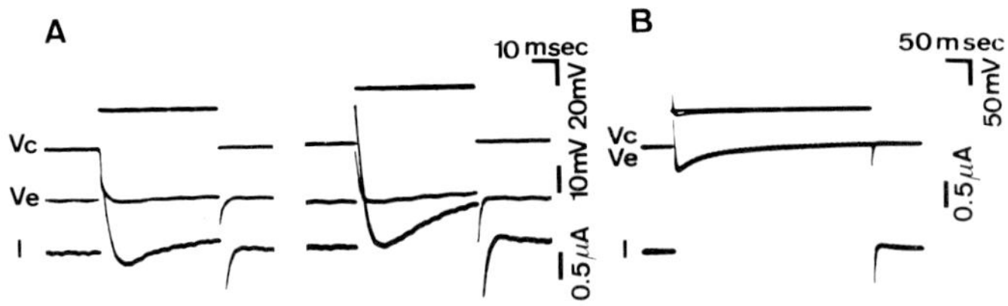

Fig. 1. Simultaneous records of membrane current (I) and controlled potential (Ve) induced by command depolarization (Vc). In A, the microelectrode is supposed to be just outside the membrane, the external surface of which is held at ground potential. In B, the microelectrode is inside the cell and the transmembrane potential can be measured (with the permission of the Journal of Physiology).

except for the fast transients corresponding to the capacitive current. The same accuracy of the external voltage control is obtained for a higher depolarization (+40 mV ; Fig. 1A,b). On impaling with a microelectrode (Fig. 1B), good agreement between the command potential (Vc) and the transmembrane potential (Ve) was found during the flow of a large outward current. These results suggest that cells in the test compartment did not escape voltage control when inward or outward currents developed. This microelectrode test was used routinely to select bundles suitable for voltage clamp analysis.

RESULTS

1. Effects of calcium antagonists and calcium-free solution on the outward current

In uterine smooth muscle, manganese, lanthanum and D600 all block the inward calcium current (6,7). However, the outward current was also reduced by calcium antagonists. Fig. 2A shows that the outward current (measured after 200 msec) was reduced by about 20% at +50 mV in the presence of 5 mM manganese chloride in the solution. The effect was almost fully reversible and was not the result of a shift of the current-voltage relationship along the voltage axis. D600 and calcium-free solutions had similar effects.

A depolarizing clamp pulse was followed by a tail of outward current at the resting potential whose amplitude indicated the amount of conductance turned on at the end of the pulse. Maximal tail amplitudes were plotted as a function of the pulse potential, giving a global activation curve of the outward current. Exposure to manganese, D600 or calcium-free solutions (Fig. 2B) diminished the tail current at all potentials. However, analysis had indicated that the calcium-free curve was simply a scaling factor of the reference curve and that, presumably, the activation process was not modified.

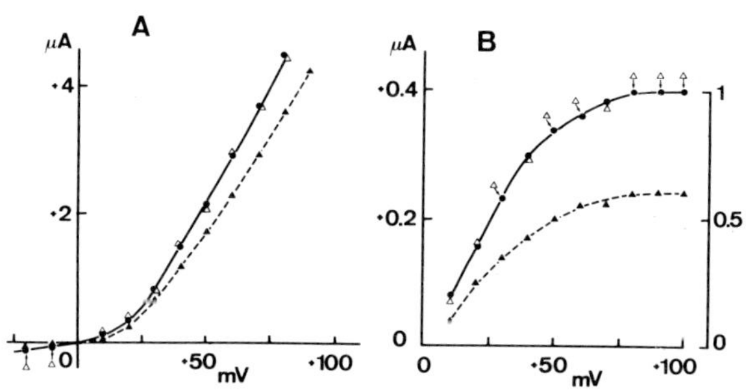

Fig. 2. (A) Effect of manganese on the outward membrane current as a function of the pulse potential. ● , current in reference solution, ▲ , current in 5 mM manganese solution (5 min), △ , current after return to reference solution (10 min).

(B) Effect of a calcium-free solution on the tail current amplitude plotted as a function of the potential. ● , tail current in reference solution, ▲ , tail current in calcium-free solution (8 min), △ , normalized tail current in calcium-free solution.

Fig. 3. shows a semi-logarithmic plot of the tail current produced when the bundle was repolarized to its resting potential following a depolarization of +45 mV for 300 msec. In reference solution the tail current can be graphically separated into two components, both of which decline exponentially, but with different time constants.

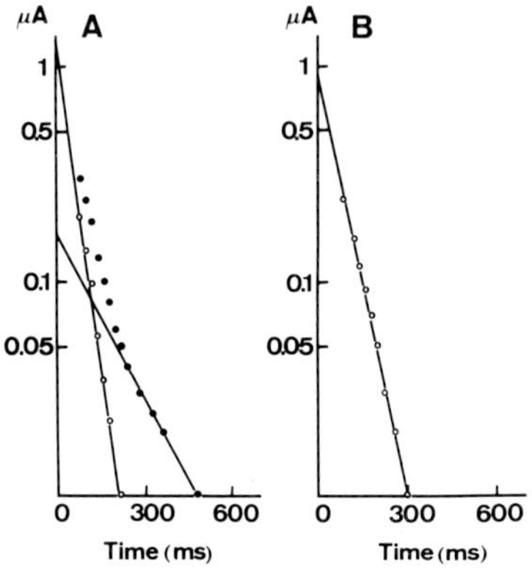

Fig. 3. Effects of calcium-free solution on tail currents recorded after a 45 mV pulse. Abscissa : time after repolarization. Ordinate : membrane current (logarithmic scale). ●, total current ; o, fast component obtained by subtraction of extrapolated slow component from total current. A, reference solution ; B, after 8 min in calcium-free solution.

It is the slower of the two components which is associated with calcium-mediated outward current because it is this component which is suppressed in calcium-free solution. Finally, the plot of the tail current as a function of the amplitude of the holding potential suggested that the reversal potential of the outward current was not modified in calcium-free solution or in the presence of calcium antagonists.

From these results it is suggested that, in uterus, a component of the outward current is mediated by calcium influx.

2. Effects of external high-calcium solution on the outward current

It has been previously shown that an increase in the external calcium concentration leads to an increase in the maximal intensity of the inward current (7). The current-voltage relationship was measured first in reference solution, then in 6mM-calcium solution. A typical result is shown in Fig. 4A, where the outward current in high-calcium solution is shifted upwards by about 30% at 50 mV, when compared with reference solution.

Fig. 4B shows that exposure to high-calcium solution increased the tail amplitude at all potentials. However, if the upper curve is scaled down by a constant factor, the normalized high-calcium curve runs roughly parallel to the control curve indicating minor modification of the activation process against voltage. Semi-logarithmic plots of the tail current produced when the bundle was repolarized to its resting potential after having been depolarized by 50 mV for 280 ms are analysed in Fig. 5(A,B). The time constant of the fast component was unaffected by high-calcium solution while the time constant of the slow component was increased from 143 to 177 msec.

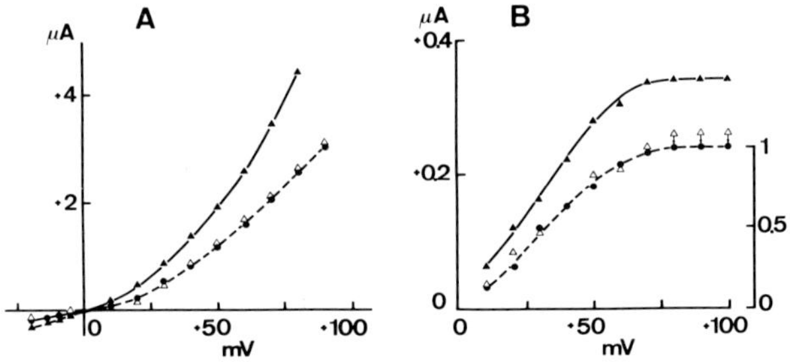

Fig. 4. (A) Effect of high-calcium solution (6 mM) on the outward current as a function of the pulse potential. ● , current in reference solution ; ▲ , current in high-calcium solution (10 min) ; △ , current after return to reference solution (8 min). (B) Effect of high-calcium solution on the tail current plotted as a function of voltage. ● , tail current in reference solution ; ▲ , tail current in high-calcium solution (10 min); △ , normalized tail current in high-calcium solution.

3. Tetraethylammonium (TEA) ions block both components of the outward current

Some contaminations from the calcium inward current in the results presented above may be suspected because of its slow inactivation. To eliminate this possibility, TEA ions were used in an attempt to determine the ionic nature of both components of the outward current. TEA ions are considered to be inhibitors of the potassium permeability in many preparations. In reference solution or in manganese solution TEA depressed the size of the outward current to the same (Fig. 5C) extent. This supports the conclusion that both components of the outward current are mainly carried by potassium ions.

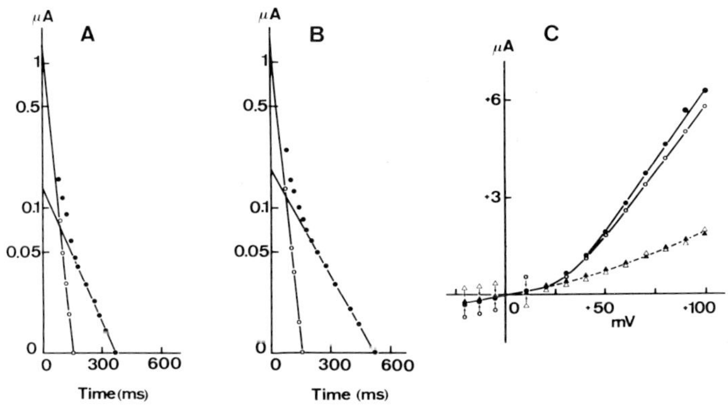

Fig. 5. (A,B) Effect of high-calcium solution on tail currents recorded after a 50 mV pulse. Abscissa : time after repolarization. Ordinate : membrane current (logarithmic scale). ● , total current ; o , fast component obtained as in Fig. 3. A, reference solution ; B, after 10 min in high-calcium solution. (C) Effect of TEA ions (20 mM) on the outward current ; ● , current in reference solution ; ▲ , current after adding TEA (5 min) ; o , current after return in reference solution (10 min) ; △ , current after adding manganese + TEA (5 min).

DISCUSSION

It has been shown in this report that the time and voltage-dependent outward current can be modified in a number of ways : (1) high-calcium solution increased the current ; (2) calcium-free solution and agents which block calcium inward current reduced the outward current ; (3) the tail current can be separated into two components, only one of which depends on the presence of external calcium ; (4) TEA blocks both components of the outward current.

These results suggest that a component of the outward current in uterine membrane is mediated by calcium influx. Following the suggestion of Meech (8) it is proposed that the influx of calcium through the inward channel would induce a large increase in the potassium permeability. It may be that calcium neutralizes negative charges on the inside of the membrane shifting the activation curve of the outward current in a hyperpolarizing direction. This is observed in the presence of prostaglandin E1 which increases the internal calcium concentration by facilitating the release of calcium from intracellular stores. However, prostaglandin E1 has no effect on the inward current intensity (4). Alternatively calcium could combine with a more specific site and have a direct effect in opening a potassium channel. There is evidence which supports this possibility : (1) the activation curve of the outward current is not shifted along the voltage axis in different external calcium concentrations indicating that the amplitude of the delayed conductance can be modified ; (2) there is no modification of the reversal potential of the outward current in different calcium solutions ; (3) a component of the tail current depends on the presence of extracellular calcium.

In conclusion it is suggested that, in uterus, the effects of calcium activation and potassium activation are balanced as in other tissues (1,2,9). It seems that the interaction between calcium influx and calcium mediated potassium activation may play an important role in the regulation of the action potential duration in smooth muscles.

This work was supported by grant from DGRST, France.

REFERENCES

1. Meech, R.W. and Standen, N.B. (1975) J. Physiol. 249, 211-239.
2. Bassingthwaighte, J.B., Fry, C.H. and McGuigan, J.A.S. (1976) J. Physiol. 262, 15-37.
3. Tomita, T. and Watanabe, H. (1973) Phil. Trans. R. Soc. B 265, 73-85.
4. Grosset, A. and Mironneau, J. (1977) J. Physiol. (in press).
5. Rougier, O., Vassort, G. and Stämpfli, R. (1968) Pflügers Arch. 301, 91-108.
6. Anderson, N.C., Ramon, F. and Snyder, A. (1971) J. Gen. Physiol. 58, 322-339.
7. Mironneau, J. (1973) J. Physiol. 233, 127-141.
8. Meech, R.W. (1974) J. Physiol. 237, 259-277.
9. Vassort, G. (1975) J. Physiol. 252, 713-734.

Excitation-Contraction Coupling in Smooth Muscle, R. Casteels et al. eds
© *1977 Elsevier/North-Holland Biomedical Press*

CALCIUM AND CAT INTESTINAL SMOOTH MUSCLE

B.A. CURTIS AND C.L. PROSSER
University of Illinois
Department of Basic Medical Sciences, Peoria 61605
and
Department of Physiology and Biophysics, Urbana 61801

Na-Ca interactions in spike production. Spikes in intestinal muscle (circular or longitudinal) appear to reflect increased calcium conductance. Spikes, either triggered or spontaneous, are reduced in amplitude in low Ca_o and are absent after treatment with Ca-free Krebs solution; they are blocked by Mn^{++} or Co^{++}, also by verapamil. These effects of reduced Ca_o are not due to change in resting potential since this is relatively insensitive to Ca_o in the range of 0-1 mM [1] and in taenia coli, spikes are abolished in Ca-free medium even when depolarization is prevented [2]. In guinea-pig vas deferens, spike amplitude decreased by 27 mV per 10-fold change in Ca_o [3].

Figure 1 shows the decrease in height of triggered spikes in longitudinal muscle of cat intestine at decreasing concentrations of Ca_o. Similar results were obtained with circular muscle fibers [4]. Spike height was relatively insensitive to Na_o, i.e., effects of Na and Ca were not additive. Figure 1 shows that when both Na_o and Ca_o were altered so that the ratio $Na/\sqrt{Ca}$ was kept constant, the spike amplitude remained relatively constant [4]. These observations, together with similar ones on taenia [2] indicate that, although Na cannot serve as the principal current carrier in spike electrogenesis, this ion modulates Ca conductance, possibly by competing for anionic binding sites in the membrane.

Evidence for Na^+ entry in Ca^{++} channels after treatment with chelators. Circular muscle of cat intestine can generate brief (100 msec) or prolonged (1 sec) spikes which probably result from increased Ca-conductance. Irrigation of these muscles with Ca-free physiological salt solution may not eliminate spikes, at least not until after a very long time. However, when millimolar amounts of EGTA or EDTA were added to the Ca-free solution, spiking ability was lost almost immediately. After several minutes in the Ca-free solution with a chelator, rhythmic prolonged potentials appeared (Figure 2) [5]. The membranes were depolarized by 25-30 mV from resting potentials of -65 mV and the square-topped waves were often of several seconds duration. Depolarization by the waves was nearly to zero and both rising and falling phases were very rapid. Rhythmicity persisted for many minutes. These waves differ in form and ionic dependence from plateau potentials such as are seen after barium.

The rhythmic EGTA-induced potentials disappeared when Na was replaced by Li, Tris, or choline; they were insensitive to TTX and were unaffected by variations in anions or Mg. However, the prolonged potentials were abolished by very low concentrations of Mn, Co, La or verapamil (Figure 2). After 10-15 minutes in 5mM

EGTA, voltage-current curves of cat circular intestinal muscle lost their normal inward (anomalous) rectification. The chelator-induced waves were seen in a variety of smooth muscle preparations. The nature of the rhythmicity is unknown, but appears that when membrane-bound Ca^{++} is removed by a chelator, the channels normally used by Ca^{++} becomes permeable to Na^{+}. The prolonged square-topped potentials are non-physiological, but they do indicate that under special conditions of treatment with a chelator, Na can pass in Ca channels.

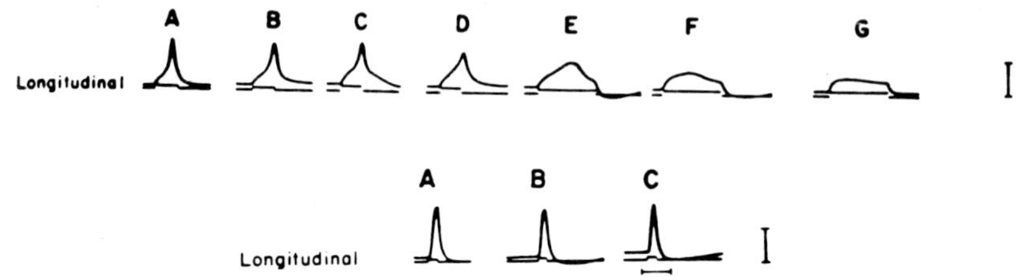

Figure 1. Effects of reduced calcium concentration on triggered spikes in strips of of longitudinal muscle from cat intestine, recorded by sucrose gap. Upper series with Na maintained at 136 mM but calcium varied; $CaCl_2$ in mM A, 5.0; B, 3.5; C, 2.5; D, 1.6; E, 0.8; F, 0.4; G, 0.0. Lower series both Na and Ca varied to keep relatively constant ratios of $Na/\sqrt{Ca}$; A, NaCl 58.7 mM, $CaCl_2$ 0.8 mM; B, NaCl 120 mM, $CaCl_2$ 2,5 mM; C, NaCl 181.4mM, $CaCl_2$ 4.8 mM. Calibration 10 mM, 200 msec. From reference 4.

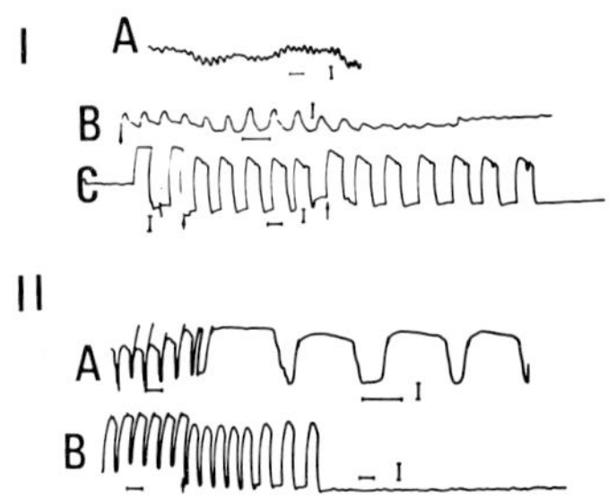

Figure 2. Rhythmic activity in circular strips.

I. Effects of blocking agents of Ca^{++} channels; cat intestine, effect of $CoCl_2$.

A. Normal slow waves.
B. Replacement of Krebs solution with Ca-free plus 5 mM EGTA at beginning of record, slow waves gradually disappear, 5 sec, 1 mV.
C. Prolonged potentials after 20 minutes in EGTA, gain reduced at first arrow, 2 mM $CoCl_2$ added at second arrow, prolong potentials stop.

II. Effect of D-600.

A. Prolonged potentials after many minutes in 0-Ca, EGTA containing Krebs solution. Recording at slow (10 sec), then faster (5 sec) speed.
B. D-600 10^{-6}M added at arrow, 10 sec. 0.5 mM; C, normal Krebs replaced 15 sec before record started, recovery of slow waves. 10 sec calibration.

Ca-Na interactions in rhythmic "slow waves". Longitudinal muscle of small intestine of several mammalian species generates rhythmic "slow waves" which normally set the frequency of bursts of spikes and resulting segmental contractions. Much evidence, particularly that from Na efflux and voltage clamping, indicates that the slow waves are generated by an oscillating Na-K pump [6]. The slow waves are markedly reduced in amplitude by agents which block Na transport (e.g., ouabain and K-free medium). Interference with metabolism reduces amplitude of the slow waves and concurrently shifts metabolic balance toward reduced intermediates [7]. However, frequency is relatively insensitive to blocking of metabolic or ion pump activities.

The only physiological treatment (other than by temperature) which we have found to alter the frequency of slow waves is variation in external calcium. In Ca-free medium the frequency is gradually reduced and this effect is enhanced by addition of Mg. Elevated Ca_o increases frequency slightly. The mechanism of action of the Ca^{++} on slow-wave frequency is unknown but models have been suggested which postulate interactions between Ca and Na transport systems [8].

Li, ^{45}Ca efflux and surface Ca. The efflux of ^{45}Ca from 800μ x 1 cm strips of circular smooth muscle from cat jejunum into Na Krebs does not decline as a single exponential, but as the sum of many exponentials. Even after 250 min. of efflux, the apparent time constant obtained from the data points did not agree with the final rate constant from the muscle. We concluded that compartmental analysis is of no great value in this preparation.

Figure 3 shows the ^{45}Ca efflux from a muscle bundle and the tension from that bundle. The continuously curving nature of the efflux can be observed even though a least squares straight line has been drawn through the points. Acetycholine activates muscle bundle tension and increases the ^{45}Ca efflux from the bundle. Replacement of 116 mM NaCl in the Krebs by LiCl results in a large increase in ^{45}Ca efflux, while leaving the muscle excitable. This example illustrates a particularly large increase in ^{45}Ca efflux with Li, a more typical response would show a 2 or 3 time increase in ^{45}Ca efflux lasting much longer. It is probable that the total ^{45}Ca lost is very similar in each case.

Calcium free and EGTA Krebs caused a similar increase in ^{45}Ca efflux. After calcium free or EGTA treatment, Li no longer releases additional ^{45}Ca. Since EDTA cannot penetrate intact cell membranes [9], we conclude that EDTA and Li are releasing a ^{45}Ca store on the surface membrane. Such a membrane bound Ca store has been described in taenia coli [10,11] and heart muscle [12]. Treatment of taenia coli with Li Krebs did not cause the increase in ^{45}Ca efflux seen with jejunal muscle.

Twenty minutes in Li Krebs reduces the total calcium in the bundles. There was a great deal of variation in Ca content between batches of strips but all batches showed a reduction of Ca content in Li. The average Ca content in Na

Krebs was 2.0mM Ca/kg wet wt. In one set of 6 muscle bundles, the 3 in Li Krebs contained 40% less Ca than the 3 strips in Na.

When sucrose and TEA were substituted for Na, there was no significant change in the efflux rate. Cesium caused both a transient and sustained increase in the rate of loss of ^{45}Ca. La^{+++} likewise caused a large increase in ^{45}Ca efflux. These results show that the surface bound Ca in cat jejunum is much more loosely bound than that of taenia coli.

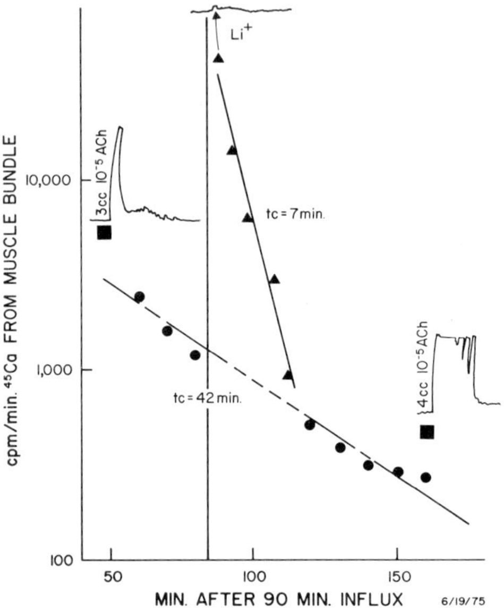

Figure 3. ACh induced contractions in circular smooth muscle of cat jenunum increased ^{45}Ca efflux. The replacement of 116 mM NaCl by LiCl in Krebs increased the ^{45}Ca efflux markedly, but did not induce contractions. We conclude that the ^{45}Ca efflux remaining after 60 min. in Li Krebs is a transmembrane efflux.

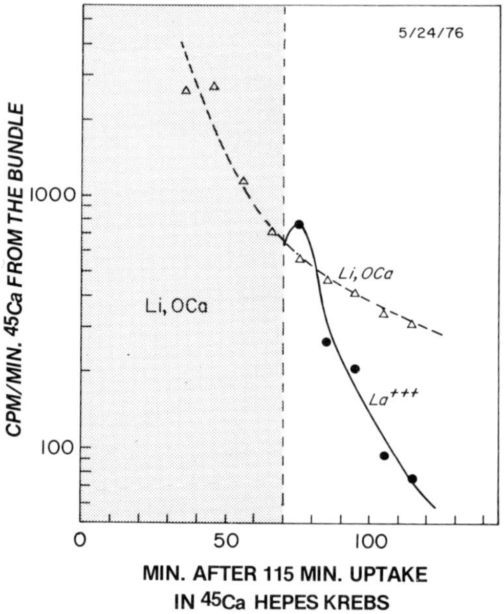

Figure 4. After 70 min. in Li, Ca-free Krebs, La inhibits the transmembrane ^{45}Ca efflux; Li has removed all of the surface ^{45}Ca. If surface ^{45}Ca had remained, La would have caused a large increase in efflux. Li also reduces the total bundle Ca by competing with Ca for surface anionic sites.

Ca binding to the membrane is probably at an anionic site as proposed by Goodford (1972) for Ca membrane binding in taenia coli. We have obtained the following evidence consistant with anionic binding of surface Ca: 1. Li displaces Ca from surface sites and occupies these sites with a higher affinity than Ca. 2. ^{45}Ca release by ions of a size similar to Ca ions such as La^{+++}, Cs^{+}, but no Ca release by large ions such as TEA. 3. Non-ionic substitutes for Na, such as sucrose do not release surface bound Ca.

Transmembrane Ca efflux. After 50-70 min. in Li Krebs, all of the surface ^{45}Ca has been lost. When the muscle bundle is treated with La^{+++} as shown in Figure 4, there was a profound fall in ^{45}Ca efflux. This experiment shows one of the generalized effects of La^{+++} on excitable tissue: blocking of Ca efflux from intracellular stores (van Breemen, et al). Figure 4 also confirms that there is no extracellular ^{45}Ca left after 60 min. in Li Krebs for the La^{+++} to release. We conclude that the efflux remaining after prolonged Li treatment is from deeper, intracellular stores; a transmembrane efflux.

The energy source for Ca efflux in smooth muscle is of considerable interest. Several authors have suggested a Na_o-Ca_i coupled exchange similar to that in Squid axon [13] and Barnacle muscle [14]. Other authors have postulated a Ca efflux directly linked to metabolic energy. The transmembrane ^{45}Ca efflux following efflux in Li is well suited for testing the sensitivity of Ca efflux to external sodium.

When the muscle is quiescent, returning the muscle bundle to Na Krebs after a long efflux into Li has no effect upon the rate of ^{45}efflux as shown in Figure 5.

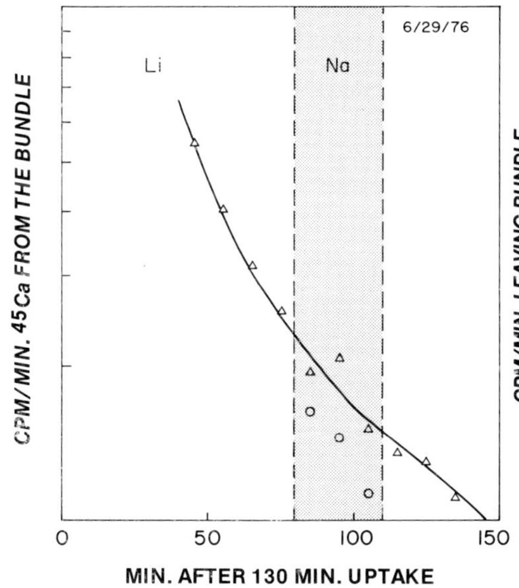

Figure 5. Returning the resting muscle bundle to Na, after a long exposure to Li does not increase the transmembrane ^{45}Ca efflux. Neither will placing the muscle in TRIS or arganine Krebs decrease the efflux. We conclude there is no Na dependent Ca efflux in resting circular smooth muscle from cat intestine.

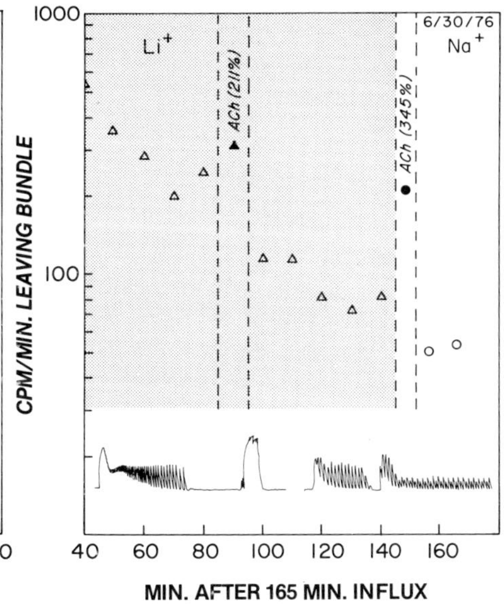

Figure 6. ACh induced contraction increases the ^{45}Ca efflux more in Na Krebs than in Li Krebs. Since the contraction pattern is similar in Li and Na, we conclude that the total Ca movement is the same in Na and Li Krebs; the extra ^{45}Ca efflux in Na Krebs must be the result of turnover of a previously untouched ^{45}store.

This lack of effect has been observed in 11 pairs of bundles. In addition, the same results were obtained in 9 pairs of bundles in Ca-free Li Krebs followed by Ca-free Na Krebs. Substitution of the organic cations Tris or arganine instead of Li did not alter the ^{45}Ca efflux. There is a 20% fall in efflux in sucrose Krebs but this fall is difficult to interpret because of the large change in ionic strength of the solution. We conclude that there is no Na sensitive Ca efflux in resting cat intestinal smooth muscle.

During cyclical mechanical activity, calcium must be moving into and out of the muscle cell much more rapidly than at rest. Hence, during cyclical activity, any effect of external sodium upon Ca efflux should be more pronounced. When the muscle bundles were stimulated electrically (100cps, 5 sec.), the ^{45}Ca efflux in either Li and Na Krebs doubled in the six bundles treated in both Na and Li Krebs.

The muscle bundle was also stimulated to contract by Acetyl Choline, as in Figure 6. While the muscle was in Li Krebs, the ^{45}Ca efflux increases, on the average, 5 times above resting level when the muscle was stimulated by ACh. These averages were obtained from the results in 4 bundles contracting successively in Li and Na and 4 bundles in Li and 3 bundles in Na after long Li.

Since the maximum tension and the rate of rise and fall of tension was the same when ACh stimulated circular smooth muscle in either Li and Na Krebs, and since no tonus developed in Li Krebs, we conclude that the total Ca movement in and out of the cell was the same in both Li and Na Krebs; there is no evidence for a Na stimulated Ca efflux.

The origin of the extra ^{45}Ca efflux must be from some previously untapped intracellular ^{45}Ca store such as the mitochondria. Smooth muscle mitochondria are quite capable of sequesting large amounts of ^{45}Ca [15]. It is possible that the ACh stimulates the mitochondria, either directly or through a messenger, to increase their metabolism and increase Ca turnover, releasing extra ^{45}Ca, in the process.

In conclusion, we have found no evidence for a sodium dependent calcium efflux either at rest or during activity. Since the Ca electrochemical gratient is certainly into the cell, we must conclude that Ca efflux is directly coupled to metabolism. We have poisoned the metabolic processes of muscle bundles with IAA, DNP, and N_2, but irreversible contractues forced us to abandon that type of experiment.

Supported by NIH Grants NS 12038 and AM 12786

REFERENCES;

1. Liu, J., C.L. Prosser and D.D. Job. (1969) Ionic Dependence of Slow Waves and Spikes in Intestinal Muscle. Amer. J. Physiol. 217; 1542-1547.
2. Brading, A.F., E. Bulbring and T. Tomita. (1969) Effect of Sodium and Calcium on Action Potentials of Smooth Muscle of Guinea-pig Taenia Coli. J. Physiol. 200: 637-654.
3. Bennett, M.R. (1967) Effects of Cations on Electrical Properties of Smooth Muscle Cells of Guinea-pig Vas Deferens. J. Physiol. 190: 465-479.
4. Connor, C. and C.L. Prosser. (1974) Comparison of ionic effects on longitudinal and Circular Muscle of Cat Jejunum. Amer. J. Phys. 226: 1212-1218.

5. Prosser, C.L., D.L. Kreulen, R.J. Weigel and W. Yau. (1977) Prolonged Potentials in Gastrointestinal Muscles Induced by Calcium Chelation. Amer. J. Physiol. In Press.

6. Connor, J.A., C.L. Prosser and W. Weems. (1974) A Study of Pacemaker Activity in Intestinal Smooth Muscle. J. Physiol. 240: 671-701.

7. Connor, J.A., D.L. Kreulen and C.L. Prosser. (1976) Relation Between Oxidative Metabolism and Slow Rhythmic Potentials in Mammalian Intestinal Muscle. Proc. Nat. Acad. Sci. 73: 4239-4243.

8. Prosser, C.L. (1977) Biolectric Rhythms in Gastrointestinal Muscles. Fed. Proc. In Press.

9. Baker, P.F. (1972) Transport and Metabolism of Ca Ions in Nerve. Prog. Biophysics. 24: 177-224.

10. van Breemen et al. (1973) Factors Controlling Cytoplasmic Ca Concentration. Phil. Trans. R. Soc. Lond. B 265: 57-71.

11. Goodford, P.J. and M.W. Wolowyk. (1972). Localization of Cation Interactions in the Smooth Muscle of Guinea-pig Taenia Coli. J. Physiol. 224: 521-535.

12. Langer, G.A., L.M. Mudd and M.V. Ricchiuti. (1976) The Effect of Sodium Deficient Perfusion on Calcium Exchange in Cardiac Tissue Culture. J. Mol. & Cell Cardiology. 8: 321-328.

13. Baker, P.F., M.P. Blaustein, A.L. Hodgkin and R.A. Stenhart. (1969) The Influence of Calcium on Sodium Efflux in Squid Axons. J. Physiol. 200: 431.

14. Russell, J.M. and M.P. Blaustein. (1974) Ca Efflux from Barnacle Muscle Fibers. Dependence on External Cations. J. Gen. Physiol. 63: 144-167.

15. Raeymaekers, L. et al. (1977) A Comparative Study of the Calcium Accumulation by Mitochondria and Microsomes Isolated from Smooth Muscle of Guinea-pig Taenia Coli. Pflugers Archiv. 368: 217-223.

Excitation-Contraction Coupling in Smooth Muscle, R. Casteels et al. eds.

CALCIUM SYSTEMS OF SMOOTH MUSCLE AND THEIR SELECTIVE INHIBITION

K. Golenhofen, B. Wagner, A.H. Weston

Department of Physiology, University of Marburg/Lahn,
Deutschhausstrasse 2, D-3550 Marburg/Lahn, West Germany

Based on studies with 'calcium-antagonistic' drugs, the concept that mammalian smooth muscle contains two different calcium activation systems- a P-system and a T-system - has been developed (1). The effects of the P-antagonists nifedipine (Bayer) and D600 (methoxyverapamil, Knoll) and the T-antagonist nitroprusside sodium (NP) (Merck) have now been studied in the smooth muscle of guinea-pig portal vein and stomach.

METHODS

Series 1 : Measurements in portal vein. Guinea-pigs (700 g mean body weight) were stunned and bled and a 15-20 mm in situ length of portal vein was excised. The preparation was tied to a glass tissue holder and placed in a thermostatically-controlled 50 ml glass tissue bath at 35°C. Isometric mechanical activity under an initial resting tension of about 10 mN was recorded. Noradrenaline (NA) dose-response experiments were carried out by increasing the NA concentration every 2 min from 10^{-7} to 10^{-6}, 10^{-5} and 10^{-4} mol/l. After two control tests, the NA application was repeated in the presence of an antagonist at the concentrations 10^{-8}, 10^{-7}....10^{-3} mol/l. Control experiments with untreated preparations were performed in parallel.

Series 2 : Measurements in stomach muscle. For the simultaneous recording of electrical and mechanical activity, the technique of partial fixation was used (2). One end of the preparation was fixed over a perspex plate of 4 mm diameter and the other, free, end was connected to a transducer for tension recording (total length of the preparation about 15 mm). The relatively-strongly stretched part over the perspex plate was used for intracellular recording of the electrical activity with glass microelectrodes. Continuous recordings in one and the same cell over several hours were possible under these conditions, even during maximum changes of the tension development. Microelectrodes of resistance 30-150 MOhm were used. These were filled with 3 mol/l KCl which was purified with a Millipore filter (0.1 µm pores) to reduce tip potentials. The preparation was mounted in a thermostatically-controlled, continuously-perfused perspex organ bath at 35°C.

In both series of experiments, the physiological salt solution was equilibrated with 95% O_2 and 5% CO_2 and had a pH of 7.4. The experiments with nifedipine were carried out under sodium light.

RESULTS AND CONCLUSIONS

Studies in portal vein : Nifedipine had effects on portal vein activity which were qualitatively similar to those previously described for verapamil and D600 (3,4). In its specific concentration range, a complete suppression of the spontaneous mechanical activity and a partial suppression of the NA-induced activation were produced. The inhibitory threshold was below 10^{-8} mol/l and 10^{-7} mol/l was sufficient for a full suppression of spontaneous activity, as seen in Fig. 1. It is further shown in Fig. 1 that an increase of the nifedipine

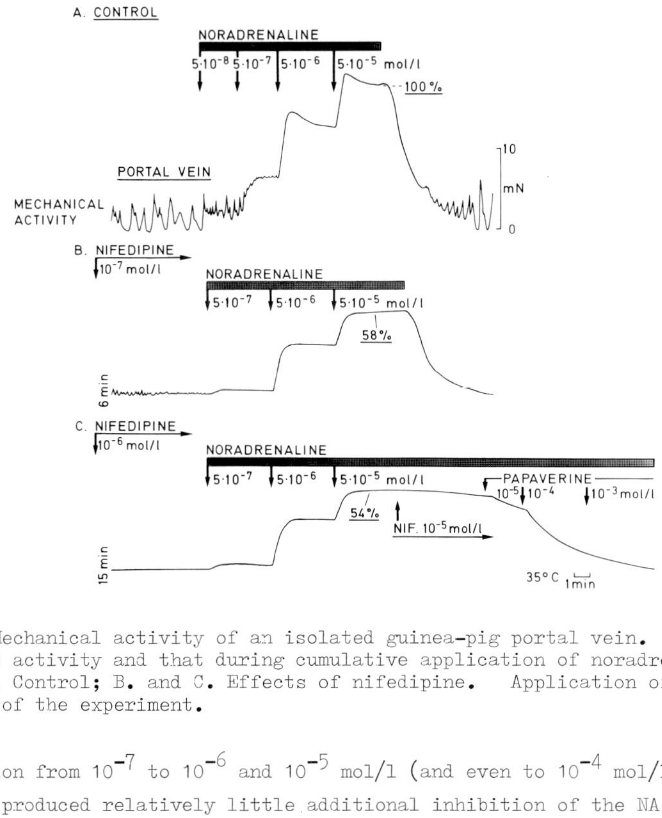

Fig. 1. Mechanical activity of an isolated guinea-pig portal vein. Spontaneous activity and that during cumulative application of noradrenaline are shown. A. Control; B. and C. Effects of nifedipine. Application of papaverine at the end of the experiment.

concentration from 10^{-7} to 10^{-6} and 10^{-5} mol/l (and even to 10^{-4} mol/l, not shown in Fig. 1) produced relatively little additional inhibition of the NA response. This led to a characteristic plateau in the dose-response curve of nifedipine when the effect on the maximum NA response was plotted as a function of the nifedipine concentration (Fig. 2).

The threshold for D600 was found to be higher and the plateau in the dose-response curve was less pronounced. Increasing the concentration of nitroprusside sodium (NP) from 10^{-8} to 10^{-3} mol/l produced relatively little reduction in either the spontaneous phasic activity or in the responses to NA. This small reduction increased continuously over the whole concentration range and was significantly different from the control experiments only at the highest NP concentrations (10^{-4} and 10^{-3} mol/l). Papaverine was able to suppress those parts of the activation which were resistant to nifedipine and NP (Fig. 1). More details of the dose-response experiments with nifedipine, D600 and NP will be reported elsewhere.

Definition of the P- and T-systems : The results show that in guinea-pig portal vein, nifedipine is a more potent and more specific antagonist of the P-component of the activity than D600 and verapamil. In Fig. 2, the ideal dose-response curve of a P-inhibitor is compared with the results obtained with nifedipine and it is clear that the results are close to the ideal situation.

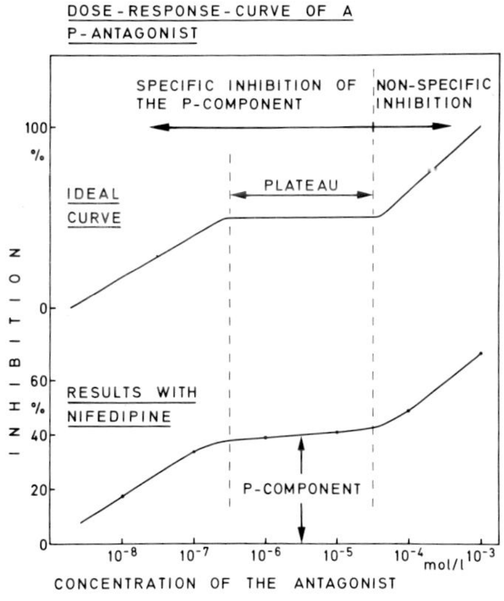

Fig. 2. Dose-response curve of a P-antagonist. Top : ideal curve; Bottom : results with nifedipine. The points in the lower curve are the means of 8 experiments uncorrected for the concurrent control measurements. The correction had no significant effect on the shape of the curve.

The plateau over a 100-fold concentration range is broad enough to allow selective inhibition of the P-component. This high specificity of nifedipine is at present the best foundation for a differentiation of the P- and T-systems. Consequently, the P-system can be defined as that part of the calcium activation process which can be blocked by nifedipine in its specific concentration range (at the middle of the plateau, 10^{-6} to 10^{-5} mol/l). An antagonist of comparable specificity for the other part of the activation process is not yet available. The T-system, therefore, is best characterised at the moment by its resistance to specific nifedipine blockade.

Studies in stomach muscle : D600 had effects on stomach smooth muscle similar to those described earlier for verapamil (5,6). Spike discharges and phasic mechanical activity were suppressed; the D600-resistant tonic components of stomach activation, which are particularly pronounced in fundus strips, were always spike-free. An example is given in Fig. 3. In the electrical activity, only the spike discharges were suppressed whereas slow fluctuations of the basic membrane potential remained unaltered. The maximum membrane potential (maximum value during the hyperpolarizing phase of the slow fluctuations) remained unchanged over 8 such experiments whereas slight depolarisations were observed during verapamil treatment (5). This indicates a greater specificity of D600 compared with verapamil. Acetylcholine (ACh) was not able to regenerate the spike discharges at D600 concentrations which produced a full, specific P-blockade (around 10^{-6} mol/l, Fig. 3).

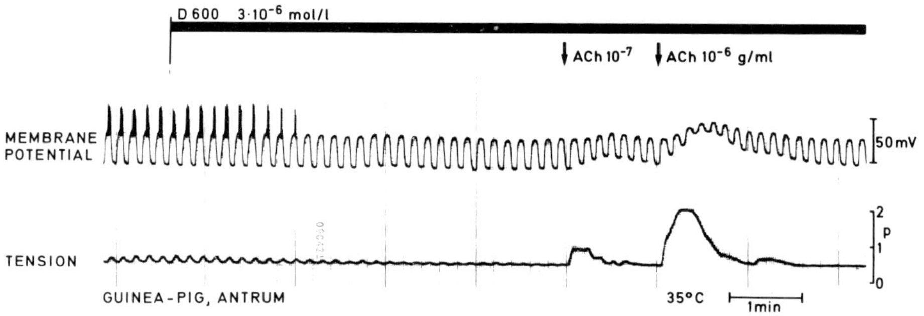

Fig. 3. The effect of D600 on a spontaneously active stomach muscle strip from the antrum region. Top : Membrane potential recorded intracellularly; Bottom : Tension development. Continuous exposure to D600 by adding the drug to the perfusing solution. Acetylcholine (ACh) was added to the organ bath at the arrow and washed out afterwards by continuous perfusion.

The main effect of NP was the suppression of tonic spike-free components of activity in the fundus and corpus regions of the stomach (7). It can, however, also inhibit spontaneous spike discharges of an antrum preparation as shown in Fig. 4. This effect was combined with a hyperpolarization of the membrane of 5-10 mV (6 experiments). The basic slow waves remained unaltered as during D600 treatment. In contrast to the effect of D600, the suppression of the spike discharges during NP application could easily be counteracted by ACh application. A spontaneous recovery of spike discharges and phasic mechanical activity could also be observed several minutes after NP application.

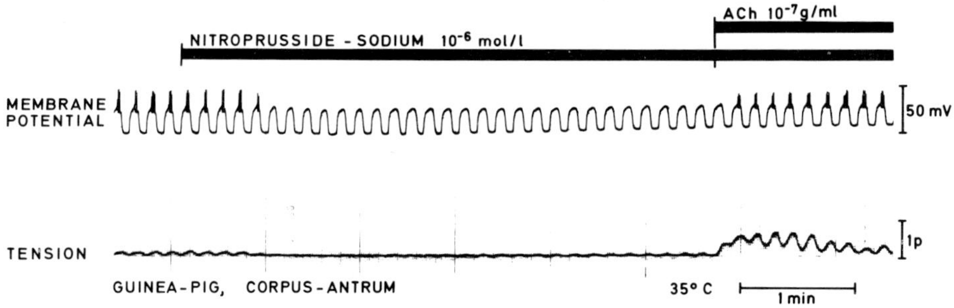

Fig. 4. Results of an experiment similar to that shown in Fig. 3. Exposure of the preparation to nitroprusside sodium.

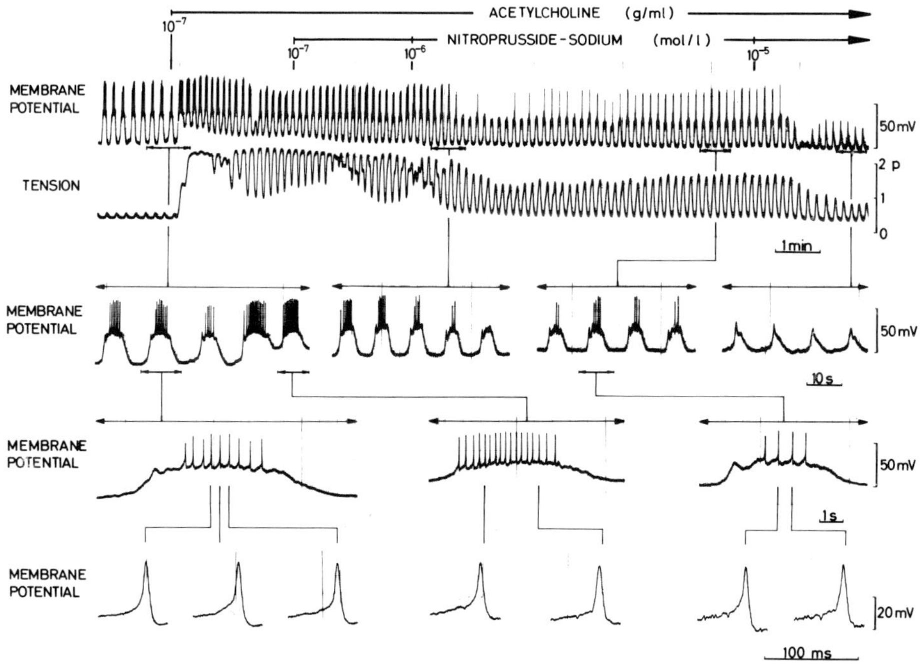

Fig. 5. Membrane potential recorded intracellularly and tension development of an antrum preparation. Upper part : slow tape replay. Lower traces : replay of the marked sections on expanded time scales showing details of the electrical activity. Exposure to nitroprusside sodium and acetylcholine as indicated.

More details of the effect of NP on electrical activity are shown in Fig. 5. An antrum preparation was first stimulated by ACh to near-maximum activity. The membrane became depolarised and the number of spikes per slow wave and the amplitude of the phasic contractions was increased. NP (10^{-7} mol/l) produced a partial repolarisation of the membrane with little effect on mechanical activity. At 10^{-6} mol/l, sufficient for an almost complete suppression of the tonic components of mechanical activation (7), NP produced a further repolarization of the membrane, a transient, complete suppression of the spike discharges in the impaled cell and a reduction of the contraction amplitude in the whole preparation. The fast recordings in the lowest trace in Fig. 5 show that the spike itself is unaltered in size and shape even at NP 10^{-6} mol/l. Only higher concentrations of NP (10^{-5} mol/l in Fig. 5) altered the electrical activity. Both the basic waves and the spikes were affected, in combination with a greater reduction of mechanical activity. It can be concluded that the typical T-inhibiting effect of NP in stomach muscle is independent of any effect on the spike mechanism. The effect of NP on the P-component seems to be produced by a polarising effect on the cell membrane which shifts the threshold for spike discharges (for more details about the results in stomach muscle, see 8).

REFERENCES

1. Golenhofen, K. In : E. Bülbring and M.F. Shuba (Eds.), Physiology of Smooth Muscle, Raven Press, New York 1976, pp. 197-202.
2. Golenhofen, K., Loh, D.v. and Milenov, K. (1970) Pflügers Arch., 315, 336-356.
3. Golenhofen, K., Hermstein, N. and Lammel, E. (1973) Microvasc. Res., 5, 73-80.
4. Golenhofen, K. and Hermstein, N. (1975) Blood Vessels, 12, 21-37.
5. Golenhofen, K. and Lammel, E. (1972) Pflügers Arch., 331, 233-243.
6. Golenhofen, K. and Wegner, H. (1975) Pflügers Arch., 354, 29-37.
7. Boev, K., Golenhofen, K. and Lukanow, J. In : E. Bülbring and M.F. Shuba (Eds.), Physiology of Smooth Muscle, Raven Press, New York 1976, pp. 203-208.
8. Wagner, B. (1976) Untersuchungen zur Differenzierung von P- und T-Aktivierungssystemen an der Magenmuskulatur von Meerschweinchen. Inaug.-Diss., Marburg/Lahn.

Excitation-Contraction Coupling in Smooth Muscle, R. Casteels et al. eds.

EXTRACELLULAR MAGNESIUM AND CONTRACTION OF VASCULAR SMOOTH MUSCLE

Burton M. Altura and Bella T. Altura
Department of Physiology, SUNY - Downstate Medical Center
Box 31, Brooklyn
New York

INTRODUCTION

Ever since the work of Hazard and Wurmser in 1932[1], it has been repeatedly demonstrated that elevations in serum Mg can produce profound and rapid vasodilatation and hypotension[2,3]. On the other hand, acute hypomagnesemia in animals and man is often associated with rises in blood pressure and elevations in peripheral vascular resistance[2,3].

We have been attempting to determine whether these acute vascular changes seen on lowering or elevating extracellular magnesium ions ($[Mg^{++}]_o$) are due to direct effects on the vascular smooth muscle cells or to indirect actions[3-12]. Among other findings, it has been reported that withdrawal of $[Mg^{++}]_o$ from the bath medium can differentially affect drug- and hormone-induced contractions and relaxations of isolated rat, rabbit and canine arteries[3-7,9-14]. In addition, it has been demonstrated that withdrawal of $[Mg^{++}]_o$ can induce potent contractile responses in some of these isolated arteries[3,5,6,8,11,12], as well as in rat portal veins[12,15-17]. The strength of these contractions appears to be dependent upon the amount of $[Ca^{++}]_o$[6,8,12,16]. We have tentatively hypothesized that these acute effects of changes in $[Mg^{++}]_o$ are in large measure a result of effects on calcium permeability, binding and translocation as well as membrane stability rather than on other cellular systems such as inhibition of Na^+, K^+-ATPase activity[3,5-9,11].

In this report, we: (a) present additional work in support of this hypothesis; (b) indicate the effect of $[Mg^{++}]_o$ in venous smooth muscle; and (c) present some new findings on isolated, perfused mesenteric arterioles 15-20μm in size.

EFFECTS OF EXTRACELLULAR Mg ON ARTERIAL AND VENOUS TONE

Besides modulating drug and hormone-induced responses in vascular muscle, extracellular Mg^{++} also appears to be able to directly alter baseline tension or tone[3,5,6,8,10,12]. Decrements in $[Mg^{++}]_o$ dose-dependently result in rapid elevations in tension development (Fig. 1). These elevations in mechanical activity in, for example, isolated rat aorta are markedly reduced and rapidly disappear as the $[Ca^{++}]_o$ is lowered or chelated (Fig. 2). These mechanical responses, produced by withdrawal of $[Mg^{++}]_o$, are potentiated by the addition of Ca EDTA to the bath fluid (Fig. 2), suggesting that influx of Ca^{++} is necessary for these contractile responses[6,8,10,12]. In contrast to the contractions seen on lowering $[Mg^{++}]_o$, an elevation of

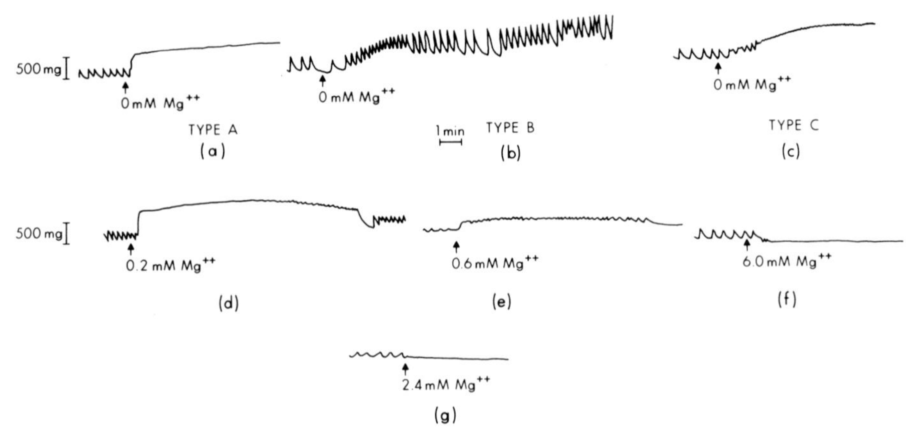

Fig. 1. Influence of external magnesium concentration on spontaneous mechanical activity and resting tension in isolated rat aortic strips. Panels a–c represent three different types of contractile responses observed in zero $[Mg^{++}]_o$.

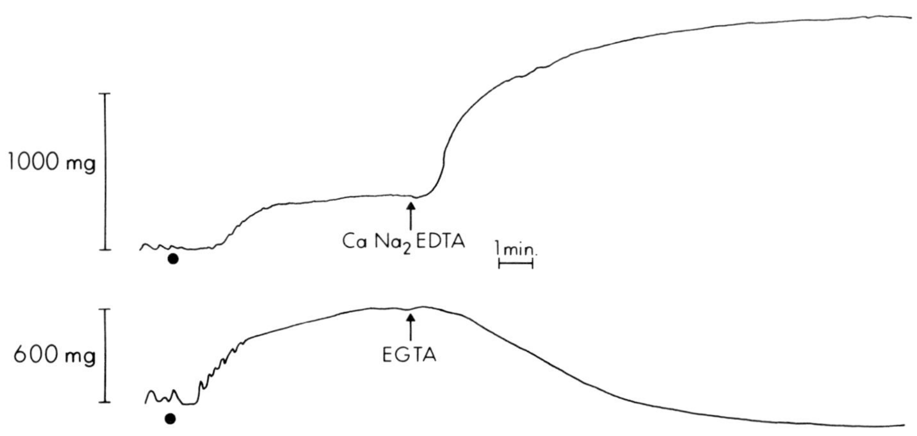

Fig. 2. Influence of 5mM $CaNa_2EDTA$ and EGTA (added at arrows, respectively) on contractions of rat aortic strips produced by exposing the tissues to Mg^{++}-free Krebs-Ringer solution (at dots).

the $[Mg^{++}]_o$ to as little as 2.4mM inhibits the spontaneous mechanical activity seen in isolated rat aorta (Fig. 1)[6, 10]. Raising the $[Mg^{++}]_o$ to 6mM not only completely inhibits all prior spontaneous mechanical events but, in addition, lowers baseline tension (Fig. 1); these results are not due to changes in osmolarity[6], but are probably due to a blocking of Ca^{++} influx.

In view of these findings, we turned our attention to other blood vessels in the rat. Table I shows the typical results one obtains when isolated rat portal veins are exposed to Mg^{++}-free Krebs-Ringer solution. Withdrawal of $[Mg^{++}]_o$ results in a rapid enhancement of the spontaneously-evoked mechanical responses and increases in rhythmic contractility (Table I), similar to that shown previously[12, 15-17]. Elevations of the $[Mg^{++}]_o$ above the normal 1.2mM dose-dependently lowers the frequency and the contractile tension (Table I). Ten mM $[Mg^{++}]_o$ completely obliterates development of all spontaneous mechanical events (Table I). Although not shown, addition of 30mM surcrose, instead of the 10mM $MgCl_2$, fails to either attenuate rhythmic contractility or contractile tension[12].

Other data, presented elsewhere, indicate that addition of 5mM EGTA to isolated rat portal veins, exposed to Mg^{++}-free solution, rapidly relaxes the spontaneous-induced enhancement of contractile tension to zero[12, 16]. In addition, EGTA completely inhibits development of rhythmic contractility[12, 16], as one might expect from the dependence of this tissue on $[Ca^{++}]_o$[18]. Addition of 5mM Ca EDTA fails, however, to exert any significant effects on the enhanced development of spontaneous mechanical events observed upon withdrawal of $[Mg^{++}]_o$[12, 16], thus quite unlike the potentiation of contraction noted for rat aortic smooth muscle (Fig. 2)[6]. We feel this may indicate that rat portal vein has a relatively small amount of external and/or loosely membrane-bound Mg when compared to the rat aorta.

Table 2 indicates that addition of $[Ca^{++}]_o$ or $[Sr^{++}]_o$ to the rat portal veins enhances the potentiated contractile tension observed in the absence of $[Mg^{++}]_o$. Although now shown, the magnitude of each mechanical spike is directly proportional to the concentration of added $[Ca^{++}]_o$ or $[Sr^{++}]_o$ under these conditions[12, 16]. In contrast to these effects, we have observed that addition of $[Ni^{++}]_o$ or $[Co^{++}]_o$ to portal veins exposed either to a Mg^{++}-free solution, or to a solution containing Mg^{++}, results in dose-dependent, and eventual complete, inhibition of all spontaneous mechanical events (Table 2), presumably as a result of blocking $[Ca^{++}]_o$ influx; i.e., divalent cations such as Ni^{++} and Co^{++} are known to inhibit uptake of Ca^{++} in excitable tissues[3, 8].

These studies show, rather nicely, that while $[Mg^{++}]_o$ may regulate membrane permeability to, and binding of, Ca in vascular muscle, $[Mg^{++}]_o$ and $[Ca^{++}]_o$ have separate roles and are not interchangeable in membrane stabilization and contraction events.

TABLE 1

INFLUENCE OF Mg^{++} ON SPONTANEOUS MECHANICAL ACTIVITY IN ISOLATED HEPATIC-PORTAL VEINS OF RATS

$[Mg^{++}]_o$ (mM)	N	Frequency (mechanical responses/min ± SE)	Strength of average spontaneous contraction (mg ± SE)
0	104	4.99 ± 0.12*	379.9 ± 23.5*
1.2	104	3.64 ± 0.16	239.5 ± 21.3
5.0	18	1.16 ± 0.12*	102.4 ± 16.6*
10.0	18	0*	0*

*Significantly different from values in 1.2mM Mg^{++} and all other responses ($P < 0.01$).

TABLE 2

INFLUENCE OF DIVALENT CATIONS ON RESPONSES OF RAT HEPATIC-PORTAL VEINS INDUCED BY WITHDRAWAL OF $[Mg^{++}]_o$ IN THE PRESENCE OF 2.5mM $[Ca^{++}]_o$

Cation added (1.25mM) in absence of $[Mg^{++}]_o$	N	Frequency (mechanical responses/min ± SE)	Strength of spontaneous contractions (Mg ± SE)
Mg^{++}-free	26	4.39 ± 0.24	489.5 ± 33.4
Mg	19	3.32 ± 0.19*	270.8 ± 46.5*
Ca	9	3.78 ± 0.46	600.0 ± 33.6*
Sr	7	3.42 ± 0.34	627.9 ± 44.6*
Ni	12	0**	0**
Co	10	0**	0**

*Significantly different from values in absence of $[Mg^{++}]_o$ ($P < 0.05$).

**Significantly different from all other paired values.

SPONTANEOUS CONTRACTION OF ARTERIOLAR SMOOTH MUSCLE UPON WITHDRAWAL OF $[Mg^{++}]_o$

Previous results clearly demonstrate that acute changes in extracellular Mg can alter responsiveness and tone of arteries and veins (vide supra). However, it is necessary to determine whether these results can be extrapolated to the sites of peripheral vascular resistance, namely the arterioles. Preliminary results obtained on eight, isolated rat mesenteries, perfused via the ileo-colic artery under constant pressure, and monitored with a closed-circuit television microscope recording system[19], indicate that both ileo-colic vein outflow and arteriolar lumen sizes are rapidly and progressively reduced when $[Mg^{++}]_o$ is simultaneously withdrawn from the perfusate and superfusate (Fig. 3). Although not shown, other preliminary findings utilizing isolated perfused rat mesenteries indicate that withdrawal of $[Mg^{++}]_o$ differentially affects drug- and hormone-induced constrictions, somewhat similar to that reported for isolated arteries[3-7, 9, 11-15]. Overall, these microcirculatory findings lend support to the roles for $[Mg^{++}]_o$ which have been hypothesized so far for the larger mammalian blood vessels (see above).

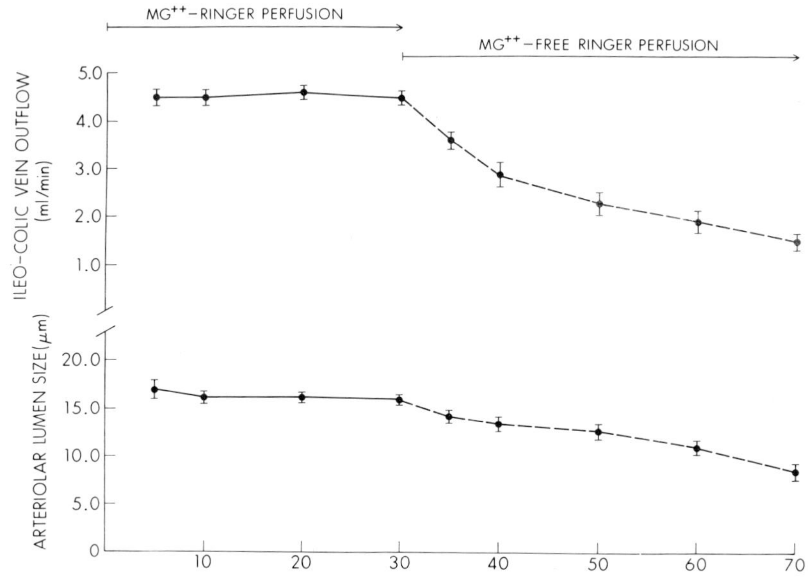

Fig. 3. Influence of withdrawal of $MgCl_2$ on rate of perfusion and arteriolar tone in isolated, perfused rat mesentery. N = 8. Values are given as mean ± SE. All of the mean values during perfusion with the Mg^{++}-free solution are significantly different from controls (paired *t* test, $P < 0.01$).

INHIBITORY EFFECTS OF ELEVATED $[Mg^{++}]_o$ ON VASOACTIVE DRUG-INDUCED CONTRACTIONS IN ISOLATED ARTERIAL MUSCLE

As mentioned before, elevations in $[Mg^{++}]_o$ can produce vasodilatation in intact or perfused vascular beds. Since the vasodilatation induced by changes in extracellular Mg^{++} is extremely rapid[1-3, 12], and since the intracellular Mg^{++} in vascular muscle is relatively insensitive to acute changes in $[Mg^{++}]_o$[20], it is likely that this hypermagnesemic-induced relaxation is due to some alteration at the cell surface[2, 3, 5, 6, 12]. Progressive elevations in $[Mg^{++}]_o$ above physiologic levels will dose-dependently produce progressive, increased inhibition of the contractile action of all vasoactive substances (with the possible exception of prostaglandins, see refs. 9, 15), irrespective of whether one studies intact or isolated vascular smooth muscle[2, 3]. The results presented in Table 3 confirm that the contractile actions of a variety of agonists, including agents like potassium ions and ouabain, are depressed by even slightly elevated levels of $[Mg^{++}]_o$. These results can not be attributed to hyperosmolarity; substitution of sucrose for $MgCl_2$ does not attenuate these agonist-induced contractile responses (Table 3). These inhibitory effects of elevated $[Mg^{++}]_o$ are probably a result of a block of Ca^{++} influx at the membrane and membrane stabilization[2, 3, 6, 17].

TABLE 3

INHIBITORY EFFECTS OF ELEVATED $[Mg^{++}]_o$ ON VASOACTIVE DRUG-INDUCED CONTRACTIONS IN RAT AORTA

Vasoactive agent	N	Developed tension (mg $\pm$ SE)		
		1.2mM $[Mg^{++}]_o$	4.0mM $[Mg^{++}]_o$*	12.0mM Sucrose*
Epinephrine	4	606.2 ± 63.9	303.7 ± 56.8**	700.0 ± 72.6
Angiotensin	4	450.0 ± 25.0	200.0 ± 25.0**	506.8 ± 42.8
KCl	6	861.2 ± 65.3	270.0 ± 76.4**	802.6 ± 76.4
Serotonin	4	475.0 ± 12.5	312.5 ± 43.7**	492.8 ± 38.8
Ouabain	6	295.8 ± 58.5	54.2 ± 23.8**	306.2 ± 64.2

Vasoactive drug concentrations used were: Epinephrine (0.5ng/ml); angiotensin (0.01µg/ml); KCl (20mM); serotonin (0.1µg/ml); and ouabain (3.4×10^{-5}M).

*All tissues preincubated with 4mM Mg^{++}-Krebs-Ringer or 12mM sucrose Krebs-Ringer for 10 min.

**Significantly different from control values in 1.2mM Mg^{++}-Krebs ($P < 0.03$).

EVIDENCE AGAINST MEDIATION BY INHIBITION OF Mg-ACTIVATED Na^+, K^+-ATPase

Active transport of sodium and potassium across mammalian membranes is dependent upon a splitting of ATP, which in turn is dependent upon membrane Na^+, K^+-ATPase activity, an enzyme system having an obligatory requirement for Mg^{++}[21]. Since this enzyme Na^+, K^+ pump is now thought to play an important role in regulation of vascular smooth muscle function[22], it has recently been suggested that the acute vascular effects of changes in $[Mg^{++}]_o$ may be brought about as a consequence of actions on the membrane Na^+, K^+-ATPase activity[2]. Numerous, recent experiments[3,5,11], employing, among other agencies, ouabain and K^+-free solutions (two situations known to inhibit membrane Na^+, K^+-ATPase systems, see refs. 21, 22), fail to confirm any role for the latter system in the $[Mg^{++}]_o$-dependent changes in tone and reactivity observed previously in arterial smooth muscle. Furthermore, it has been demonstrated that aortic smooth muscle does not exhibit any significant alteration in either Na^+, K^+ or water content when placed in a Mg^{++}-free environment for periods up to 60 min[5], despite the facts that: (a) reactivity to a variety of vasoactive stimulants is differentially altered[5]; (b) baseline tension is elevated in several of these preparations[3,5,6,8,11,12]; and (c) the Ca content is significantly elevated[5]. Although it is possible that Mg^{++}-dependent vascular muscle membrane Na^+, K^+-ATPase may be involved in some of the cardiovascular effects observed with chronic, long-term changes in $[Mg^{++}]_o$[2,3], it is unlikely that this enzyme system plays a dominant role in the acute $[Mg^{++}]_o$-dependent responses observed here or elsewhere (see above).

ACKNOWLEDGMENTS

This work was supported, in part, by Research Grants HL-18015 and MH-26236 from the U.S.P.H.S.

REFERENCES

1. Hazard, R. and Wurmser, L., 1932, Comp. Rend. Soc. de biol. 110, 525-528.
2. Haddy, F.J. and Seelig, M.S., 1977: Magnesium and the arteries. In: Second International Symposium on Magnesium. M. Cantin and M.S. Seelig (Eds), Spectrum Publications - Holliswood (in press).
3. Altura, B.M. and Altura, B.T., 1978: Magnesium in vascular muscle and its influence on reactivity. In: Prostaglandins in Cardiovascular Function. S. Greenberg, T.F. Burks and P.J. Kadowitz (Eds), Marcel Dekker Inc. - New York (in press).
4. Altura, B.M., 1970, Experientia 26, 1089-1090.
5. Altura, B.M. and Altura, B.T., 1971, Am. J. Physiol. 220, 938-944.
6. Altura, B.M. and Altura, B.T., 1974, Microvasc. Res. 7, 145-155.
7. Altura, B.M., 1975, Am. J. Physiol. 228, 1615-1620.
8. Altura, B.M. and Altura, B.T., 1976, Proc. Soc. Exp. Biol. Med. 151, 752-755.
9. Altura, B.M., Altura, B.T. and Waldemar, Y., 1976, Artery 2, 326-336.
10. Altura, B.M. and Altura, B.T., 1977: Magnesium-calcium interactions and contraction of isolated arterial smooth muscle. In: Second International Symposium on Magnesium. M. Cantin and M.S. Seelig (Eds.), Spectrum Publications - Holliswood (in press).

11. Altura, B.M. and Altura, B.T., 1977, Artery 3, 72-83.
12. Altura, B.M. and Altura, B.T., 1977, Blood Vessels (in press).
13. Jurevics, H.A. and Carrier, Jr., O., 1973, Am J. Physiol. 225, 1479-1485.
14. Turlapaty, P.D.M.V., Carrier, Jr., O. and Jurevics, H., 1975, J. Pharmacol. Exp. Ther. 192, 372-380.
15. Altura, B.M. and Altura, B.T., 1976, Fed. Proc. 35, 2360-2366.
16. Altura, B.M. and Altura, B.T., 1977: Influence of magnesium on contractile activity in isolated rat arterial and venous smooth muscle. In: Second International Symposium on Magnesium. M. Cantin and M.S. Seelig (Eds), Spectrum Publications, Inc. - Holliswood (in press).
17. Sigurdsson, S.B. and Uvelius, B., 1977, Acta Physiol. Scand. (in press).
18. Sigurdsson, S.B., Uvelius, B. and Johansson, B., 1975, Acta Physiol. Scand. 95, 263-269.
19. Altura, B.M., 1971, Microvasc. Res. 3, 361-384.
20. Palatý, V., 1974, J. Physiol. (London) 242, 555-569.
21. Skou, J.C., 1965, Physiol. Rev. 45, 596-617.
22. Anderson, D.K., 1976, Fed. Proc. 35, 1294-1297.

Excitation-Contraction Coupling in Smooth Muscle, R. Casteels et al. eds.

^{36}CHLORIDE EFFLUX FROM NORADRENALINE-STIMULATED RABBIT AORTA INHIBITED BY SODIUM NITROPRUSSIDE AND NITROGLYCERINE

V.A.W. Kreye, R. Kern, and I. Schleich
IInd Physiological Institute, University of Heidelberg
D-6900 Heidelberg, Im Neuenheimer Feld 326, Germany

INTRODUCTION

In the absence of extracellular calcium, sodium nitroprusside (NP) and nitroglycerine (GTN) retain their vasodilator properties (1,2,3, 4). Thus, mechanisms other than interference with transmembrane calcium fluxes must be responsible for their action. Recently, Häusler and Thorens (5) described a dose-dependent hyperpolarization of up to 12 mV by NP in the rabbit main pulmonary artery.

In contraction and efflux studies, we have investigated whether NP might alter active and passive membrane processes possibly underlying drug-induced hyperpolarization of vascular smooth muscle (v.s.m.). GTN and isosorbide dinitrate were included in some of the experiments.

METHODS

In contraction experiments on rat aortae, the effects of temperature changes and ouabain on the action of NP, and of chloride substitution by isethionate on the action of NP, GTN, and isosorbide dinitrate were investigated by the methods described elsewhere (2).

To study ^{24}Na efflux, pieces of rabbit aorta were loaded with ^{24}Na, and their radioactivity measured continously during superfusion at 37°C with physiological salt solution (PSS) containing 10^{-7}M noradrenaline. 10^{-5}M NP was added after 50 min and withdrawn after 90 min.

To study ^{36}Cl efflux, pieces of rabbit aorta were loaded with ^{36}Cl for 90 min after an initial relaxation period for 2 h in normal PSS. Under constant rotation, the loaded preparations were transferred through a series of vials with 5 ml PSS at 37°C. 10^{-7}M noradrenaline was present from onset, and either NP or GTN were added after 20 min. In a control series, relaxation was induced by omission of calcium (+ 5 mM EGTA). Finally, the tissues were dissolved by Soluene and their radioactivity determined.

Rates of efflux were calculated as Δcpm/Δt/ total cpm at time t. In the study on ^{36}Cl efflux, cpm were added retrogradely, and by the method of least squares, the efflux equations $y = c \cdot e^{-kt}$ determined of the late linear phase of the ^{36}Cl efflux curve.

RESULTS

A. Contraction Studies: The reduction of temperature from 37°C to 27°C in the organbath did not alter the relaxant effect of NP on rat aortic strips which had been contracted previously by 10^{-8}M noradrenaline. Similarly, incubation with 10^{-4}M ouabain for 1 h had no obvious effects on the responsiveness of the preparations to NP.

Replacement of Cl^- by isethionate ions in the PSS almost completely abolished the vasodilator effect of NP (Fig.1a), isosorbide dinitrate (Fig.1b), and GTN in rat aortae contracted by high K^+_o, but not in preparations stimulated by noradrenaline. Also, replacement of Cl^- had no effect on the relaxant action of verapamil and papaverine.

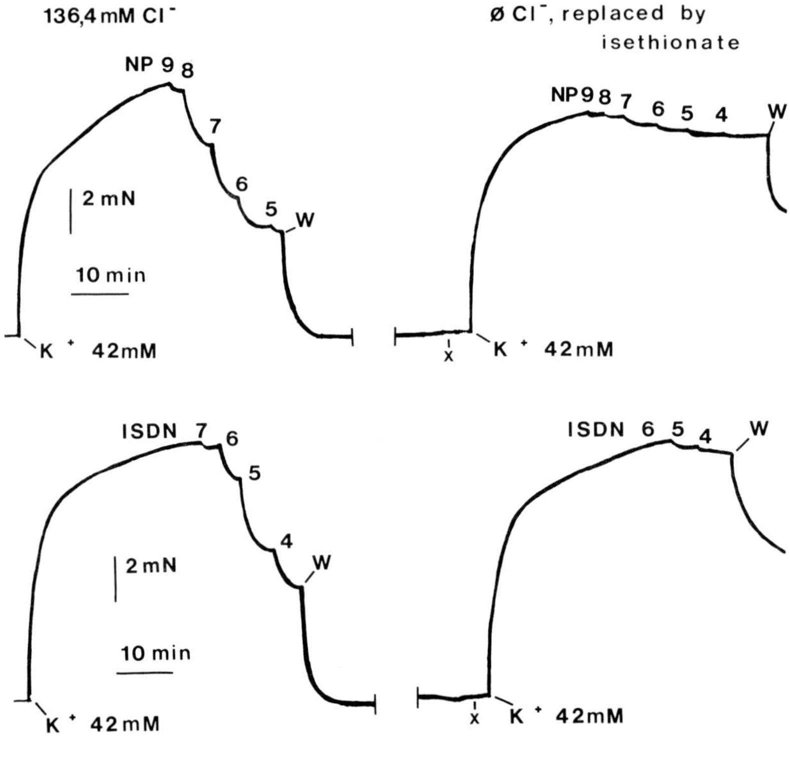

Fig.1. Effect of substitution of Cl^- by isethionate ions on the relaxant activity of a) sodium nitroprusside (NP, upper curves) and b) isosorbide dinitrate (ISDN, lower curves) on K^+-stimulated rat aortae; original recordings. Curves at left: controls, curves at right: replacement of Cl^- at x. Concentrations of drugs given as negative logarithms; W = washout of drugs and stimulant.

B. Efflux Studies: The efflux of ^{24}Na from rabbit aortae stimulated by 10^{-7}M noradrenaline was neither increased nor decreased during the period of superfusion with 10^{-5}M NP (Fig.2).

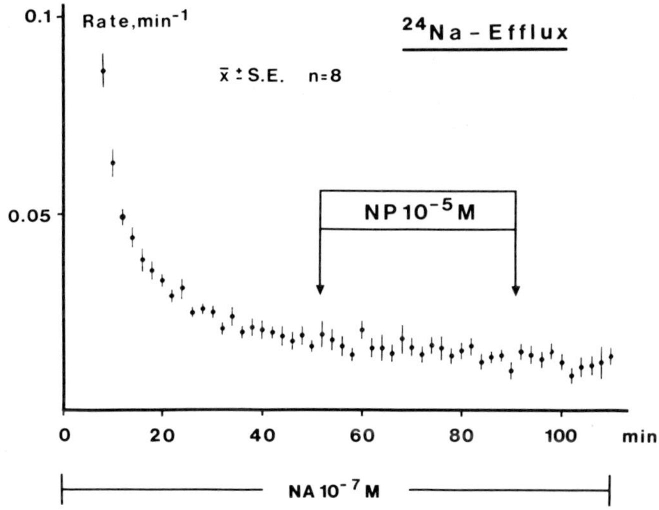

Fig.2. ^{24}Na efflux from rabbit aortae in the presence of noradrenaline (NA), and effect of addition of sodium nitroprusside (NP) to the superfusate at min 50 to min 90. Ordinate: rate of efflux = Δcpm/ Δt/total cpm at time t.

^{36}Cl efflux from rabbit aorta increased considerably with 10^{-7}M noradrenaline (rate coefficient k = 0.0713 $\pm$ 0.0018 min^{-1}; controls: 0.0399 $\pm$ 0.0031 min^{-1}; $\bar{x} \pm$ S.E., n = 10). Both, NP and GTN were found to be powerful inhibitors of the noradrenaline-induced increase of ^{36}Cl efflux (Fig.3). The rate coefficients k of the pertaining efflux equations obtained from 5 or 7 washouts, resp., fell from 0.0719 or 0.0724 min^{-1} in the presence of noradrenaline alone to 0.0424 and 0.0452 min^{-1} after addition of NP or GTN, resp. When instead of using NP or GTN, vasodilatation was induced by omission of calcium (+ 5 mM EGTA) from the PSS, no effect on ^{36}Cl efflux could be measured (Fig. 4).

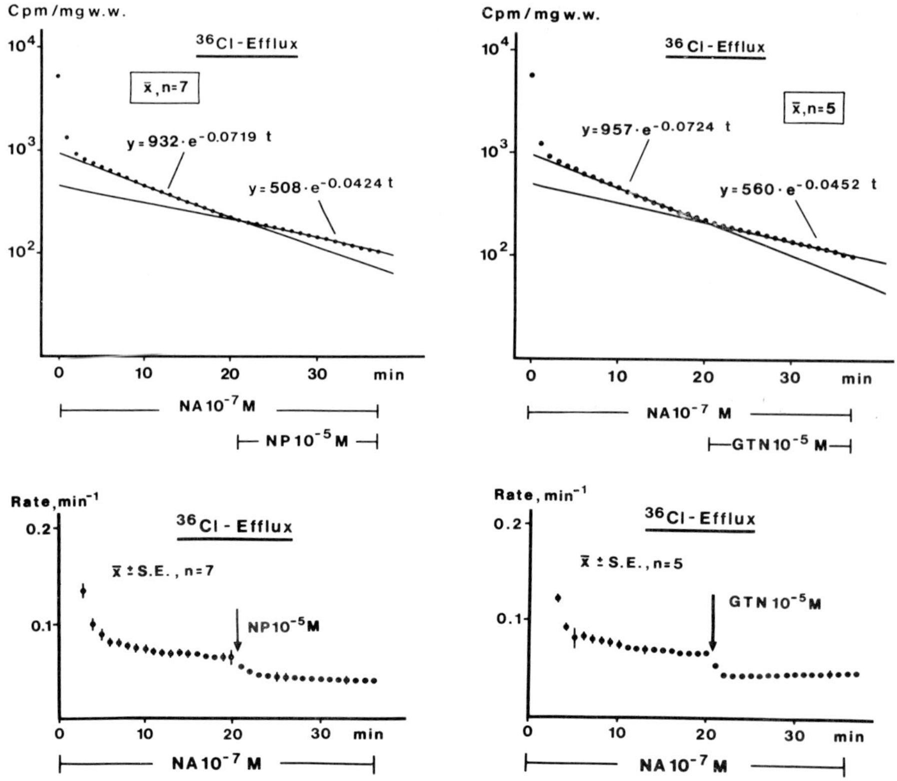

Fig.3. ^{36}Cl efflux from rabbit aortae in the presence of noradrenaline (NA), and effect of sodium nitroprusside (NP, at left) or nitroglycerine (GTN, at right) added at min 20. Ordinates of upper curves: cpm, logarithmic scale. Ordinates of lower curves: rate of efflux = Δcpm/ Δt/ total cpm at time t.

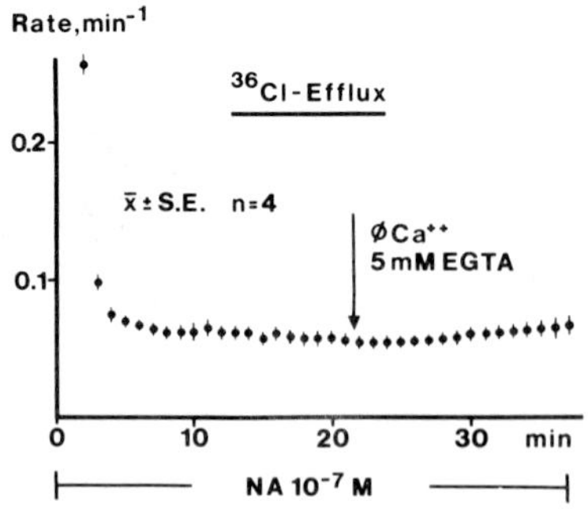

Fig.4. Effect of calcium withdrawal on ^{36}Cl efflux from noradrenaline-stimulated rabbit aortae. Ordinate: rate of efflux = Δcpm/ Δt/ total cpm at time t.

DISCUSSION

Vasodilator drugs may produce hyperpolarization of v.s.m. by 1.) activating an electrogenic Na pump, 2.) increasing K^+ permeability, 3.) reducing Na^+ permeability, and 4.) reducing Cl^- permeability as the equilibrium potential E_{Cl} (-20 to -30 mV) differs considerably from the resting potential E_m (-50 to -60 mV) in v.s.m. (6,7).

The finding that ^{24}Na efflux from rabbit aorta is not increased by NP, and that the vasodilator effect of NP is not influenced by alterations of temperature or by incubation with ouabain, seems to exclude the possibility that this drug hyperpolarizes v.s.m. by stimulation of an electrogenic Na pump.

Also, Na^+ permeability is probably not affected by NP either. Under steady state conditions, active Na^+ efflux should equal passive Na^+ influx (7), and a reduction of Na permeability should consequently have resulted in a decreased Na^+ efflux; this did not occur in our experiments.

In accordance with the observation of Wahlström (8), noradrenaline increased ^{36}Cl efflux from v.s.m. Relaxation of rabbit aorta by omission of extracellular calcium had no effect on ^{36}Cl efflux. On the other hand, relaxation of the noradrenaline-stimulated tissues by NP or GTN was paralleled by a great reduction of ^{36}Cl efflux. Since NP has no α-adrenergic blocking activity (2), and no such an effect has been claimed to be involved in the action of GTN, this observation may reflect a selective inhibition of Cl permeability by NP and GTN in v.s.m. Especially for NP, it is conceivable that this large anion may interfere with the kinetics of small anions at the cell membrane.

An estimate of chloride equilibrium potentials E_{Cl} based on the zero intercepts of efflux equations, a sorbitol space of 50 - 58 %, and a total tissue H_2O content of 78 % resulted in values between -30 to -40 mV. Although these are slightly higher than those reported for other types of v.s.m. (6,7,9), they are considerably lower than the resting potential E_m of the rabbit's large arteries which is about -60 mV (10,11). Hence, our results could explain the documented hyperpolarization of v.s.m. by NP, and they suggest that GTN likewise acts as a vasodilator by hyperpolarization.

No studies on the effect of NP and GTN on K^+ efflux have been performed by us as yet. Therefore, the possibility that an increase in K^+ permeability may be involved in the vasodilator action of these drugs, cannot be ruled out at present.

It is not clear why substitution of Cl^- by isethionate ions eliminates the vasodilator effect of NP and of the organic nitrates on K^+-

stimulated rat aortae. However, this finding suggests that these drugs act by the same cellular mechanism, and that the Cl^- gradient across the cell membrane may play an important role as a target for pharmacological actions in v.s.m.

ACKNOWLEDGMENTS

This study was supported by Deutsche Forschungsgemeinschaft within SFB 90 "Cardiovasculäres System". We are indebted to Dr.R. Gross, for having established the mathematical programmes on a Wang 2200 computer, and to Miss I. Villhauer for skilful technical assistance.

REFERENCES

1. Fermum,R., Klinner,U., Meisel,P. (1976) Acta biol. med. germ.35, 1347 - 1358
2. Kreye,V.A.W., Baron,G.D., Lüth,J.B., Schmidt-Gayk,H. (1975) Naunyn-Schmiedeberg's Arch.Pharmacol. 288, 381 - 402
3. Kreye,V.A.W., Lüth,J.B. (1976) Ionic actions on vascular smooth muscles. Ed. E.Betz, pp.145 -149, Springer, Berlin-Heidelberg-New York
4. Verhaege,R.H., Shepherd,J.T. (1976) J.Pharmacol.exp.Ther. 199, 269 - 277
5. Häusler,G., Thorens,S. (1976) Vascular neuroeffector mechanisms. Eds. J.A.Bevan et al., pp. 232 - 241, S.Karger, Basel
6. Siegel,G., Roedel,H., Nolte,J., Hofer,H.W., Bertsche,O. (1976) Physiology of smooth muscle. Eds. E.Bülbring and M.F.Shuba, pp. 19 - 39, Raven Press, New York
7. Wahlström,B.A. (1973) Acta physiol.scand. 89, 436 - 448
8. Wahlström,B.A. (1973) Acta physiol.scand. 89, 522 - 530
9. Jones,A.W. (1974) Circulation Res. 34/35, suppl.I, I-117 - I-122
10. Somlyo,A.V., Häusler,G., Somlyo,A.P. (1970) Science 169,490 - 491
11. Speden,R.N. (1970) Smooth muscle. Eds. E.Bülbring et al., pp.558-588, Edward Arnold (Publishers) Ltd., London

Excitation-Contraction Coupling in Smooth Muscle, R. Casteels et al. eds.
© 1977 Elsevier/North-Holland Biomedical Press

SUMMARY OF DISCUSSION ON "ION MOVEMENTS AND EXCITATION"

BY T. B. BOLTON (University of London)

Dr. N. Anderson reviewed some of the difficulties which must be surmounted if true voltage-clamp is to be achieved on a multifibre tissue such as smooth muscle. He drew attention to two problems : one was that when the membrane conductance is increased, the space constant falls so that to obtain uniform potential across the node (for example during the flow of early inward current) node width must be made very small. Another difficulty is the extracellular resistance which exists in series with the membrane resistance of cells, especially those deeper within the preparation. Because the potential recorded by the electrode consists of two components, one the true transmembrane potential, the other a potential proportional to current developing across this series resistance, poor control of the true transmembrane potential can occur. An important point which he stressed was that theoretical calculations have shown that merely from an examination of the form of the current records it is impossible to know if satisfactory voltage-clamp of a preparation has been achieved. Where the uniformity of nodal potential has been checked by inserting a microelectrode, inadequate voltage control has been shown to exist especially at early times after a step change in potential. Dr. Anderson was of the view that the best hope of achieving adequate voltage control lay in the use of single isolated smooth muscle cells, since he was pessimistic that adequate voltage control of multifibre preparations could be obtained.

Dr. Kao recounted some of the results he had obtained from smooth muscle of uterus and taenia using a double sucrose-gap technique in an attempt to achieve voltage-clamp conditions. He drew attention to early results which had shown a sagging of outward current, a feature that was almost absent in the records more recently obtained. He suggested that this indicated better voltage control.

Mironneau, Savineau and Kahéty gave a poster demonstration of results obtained on rat uterine smooth muscle using double sucrose gap "voltage-clamp". The tails of outward current they observed upon repolarization after a depolarizing step could be resolved into two components. One of these appeared to vary in size as the external calcium concentration was changed and they suggested that there was a component of outward (potassium) current which was affected by the influx of calcium during the early inward current.

In the interpretation of the results of Kao and Mironneau et al, the caveats of Anderson should be very much in our minds. The use of double sucrose gap technique to achieve voltage-clamp conditions has been the subject of much criticism in recent years. The effects of less than perfect voltage control by this technique being the main imponderable giving rise to uncertainty in the interpretation of the results.

BY E. BETZ (Universität Tübingen)

1) If one compares the effects of ions on the membranes of the smooth muscle cells of intestinal organs or myometrium and on those of the vascular smooth muscle cells - even in the same organs - one sees considerable differences in the mechanical or electrical responses during the same type of stimulus. During increased work of the smooth muscles of the organs, i.e. during an increase of muscular activities, there exists an increased demand for energy. It has been often measured that in this case the resistance vessels dilate, and their own smooth muscle cells relax. The smooth muscle cells of the vessels in an organ consisting of smooth muscles react in a manner opposite to the organ-smooth muscles. Since the fluid in the perivascular space of the arterioles is that of the extracellular space of the surrounding smooth muscle cells, the ionic stimulus for both types of smooth muscle cells is the same. From the papers and posters of this meeting it is difficult to understand why both types of smooth muscles react in such different ways to the same stimulus.

In other organs these problems have already been somewhat elucidated. It has been demonstrated in brain arteries and arterioles that an increase of perivascular Ca^{++} causes the vessels to constrict and a decrease of Ca^{++} causes a dilatation. For activation of the nerve cells Ca^{++} is also needed. If, now, the pCa^{++} is recorded with microelectrodes in the cortical interstitial space it can be observed that the Ca^{++} influx into the nervous tissue increases immediately on stimulation, thus causing a reduction of the interstitial Ca^{++}. The consequence of this reaction is a vessel dilatation (= relaxation of the vessel smooth muscle). This example shows that the electrical stimulus, which directly affects both types of tissue in the same way, causes a different effect on the smooth muscle tissue in vivo. In addition to the Ca^{++}-effects, H^+ and K^+ are released from brain cells during brain tissue activation thus increasing the H^+ and K^+ activities in the perivascular arteriolar space. Moderate increases of H^+ and K^+ are in turn dilatory stimuli thus coupling the vascular resistance to the demand for energy of the tissue.

This example may serve to point out that the functions of vascular smooth muscles can only be understood if the surrounding tissue is regarded simultaneously

- even if the surrounding tissue is smooth muscle.

2) The stimulus-response characteristics of smooth muscles do not only differ in various organs. It has been demonstrated in pial vessels that depletion of Ca^{++} in the perivascular space of thin arteries (20-80 µm diameters) causes a much more pronounced dilatation (in absolute and relative measure) than in arteries of 80-200 µm diameters. This demonstrates clearly the difficulty we are confronted with during this meeting when we try to compare the reactions of different smooth muscles.

Excitation-Contraction Coupling in Smooth Muscle, R. Casteels et al. eds.

PURIFICATION AND CHARACTERIZATION OF MICROSOMAL FRACTIONS FROM SMOOTH MUSCLE

Mary E. Carsten and Jordan D. Miller
Departments of Obstetrics and Gynecology and Anesthesiology
School of Medicine, University of California
Los Angeles, California 90024, U.S.A.

This presentation is aimed at summarizing the state of the art in isolating and characterizing subcellular components involved in contraction and relaxation of mammalian smooth muscles.

Requirement for intracellular calcium

Like skeletal and heart muscle, smooth muscle requires calcium for contraction, but unlike skeletal muscle, smooth muscle is more dependent on extracellular calcium. This difference parallels the difference in structure between skeletal and smooth muscles. Two lines of evidence establish the importance of intracellular calcium sources for smooth muscle activation. It was shown by Siegman and Gordon[1] that the amount of calcium that enters the cell with an action potential is not enough to account for the magnitude of the twitch tension produced. Van Breemen et al.[2] showed that with calcium efflux blocked, the contracted aorta was still capable of rapid relaxation. Thus, intracellular calcium binding lowers the cytoplasmic free calcium concentration to cause relaxation. In skeletal and heart muscle the sarcoplasmic reticulum (SR) has been identified as the intracellular calcium-accumulating site. Devine et al[3] have shown SR in a variety of smooth muscles. SR was identified as a calcium accumulating site in vascular smooth muscle by electron probe analysis[4].

Problems in isolating fractions from smooth muscle

In attempting to isolate an SR fraction from smooth muscles, procedures applicable to striated muscle SR have been used. It is, however, by no means certain that the same organelle has been isolated, because smooth muscle differs anatomically and functionally from skeletal and heart muscle. Indeed smooth muscles isolated from uterus, intestinal tract, and blood vessels may vary from one another. The relative paucity of smooth muscle SR established by electron microscopy[3] as compared to skeletal muscle creates special problems for those working with smooth muscle microsomal fractions.

TABLE I

COMPARISON OF SMOOTH MUSCLE MICROSOMAL FRACTIONS

	Carsten Miller[5] pregnant uterus bovine & human	Janis et al.[6,7] non-pregnant uterus rat & human	Hess, Ford[9] bovine aorta	Wei et al.[9] rat aorta	Moore et al.[10] rat aorta	Hurwitz et al.[11] guinea-pig ileum	Godfraind et al.[12] guinea-pig ileum
Ca binding nmol/mg protein							
ATP-no ATP		12.1 14.4		----12.5----	-----8----		---21.5---
Azide insens. -no ATP	23.8 52.2	6.9 7.5	-----10-----	----14.5----	----36----	-----9----	---21.1---
Oxalate -no ATP	-----33.8-----	32.8[x] 41.4[x]	-----35-----	----12.5[x]---	----36----	----20----	---72.9---
Ca conc. (μM)	-----20-------	100 17	-----45-----	----17------	----20----	----20----	---20-----
Protein conc. (μg/ml)	---300-500----	----10-50-----	---250-500--	---10-50----	---75-100--		---50-----
pH	------7.0-----	-----7.0------	-----7.0----	-----7.0----	-----6.8---	----7.4----	----6.6---
Temp. (°C)	-----37-------	----37--------	-----37-----	----37------	----37-----	---37------	---37-----
Time (min.)	------8-------	----10--------	------8-----	----10------	----20-----	---10------	---20-----
Fraction (g x 1000)	----15-40-----	----10-113----	----15-49---	---10-113---	---27-105--	--27-105---	---20-100-
Purification	35% sucrose	33% sucrose	80 mM KCl	33% sucrose	0.6 M KCl		
Anal. method	centrifugation	^{45}Ca filter	^{45}Ca filtrate	^{45}Ca filter	^{45}Ca filter	^{45}Ca filter	^{45}Ca filter

[x]Not corrected for azide

Preparation and characterization of microsomes

Various laboratories have prepared microsomal fractions by differential centrifugation, with a mitochondrial fraction being removed at 10,000, 15,000 or 20,000 g. The microsomal fraction, obtained by further centrifuging of the supernatant at high speed (between 40,000 and 110,000 g), may or may not be further purified on a sucrose density gradient or by washing with KCl. In Table I, results from several laboratories are listed for microsomes from bovine[5], rat[6] and human uterus[7,*], bovine[8] and rat aorta[9,10], and guinea-pig ileum[11,12]. In compiling the table, azide-sensitive and non-ATP-dependent calcium binding was subtracted. It appears that the ATP-dependent calcium binding is grossly similar for different smooth muscles, different laboratories, and different preparative and analytical procedures. It is disturbing that whether preparations are obtained between 10-100,000 or between 15-50,000 g only moderate purification is obtained, using the magnitude of ATP-dependent calcium binding as a criterion of purity. One possible interpretation is that ATP-dependent calcium binding is a property of many organelles. Nevertheless, the preservation of activity and/or purification appear to be greatest in our combination of differential centrifugation and sucrose density gradient, as seen in Table I. Though calcium binding in uterine microsomes[5] is almost as high as in cardiac SR[13], it must be remembered that only 17.5% of the cardiac SR protein was shown to be capable of accumulating calcium[14]. The Ca, Mg-ATPase activity, however, is lower in uterine microsomes[5] than in heart muscle SR[13], and the Ca-activated portion of the ATPase is relatively smaller.

Identification of microsomal fractions with cellular structures is important in order to allow us to assess the physiological significance of our findings for cellular function. Present methods are inadequate. Characteristics of SR are thought to be ATP-dependent calcium binding, increased calcium binding in the presence of oxalate, and Ca, Mg-ATPase. Markers for plasma membrane are ouabain-sensitive Na, K-ATPase, 5'-nucleotidase, phosphodiesterase, and adenylcyclase. However, all of these activities have been found to some extent in all microsomal fractions. If all these enzymes are true markers of plasma membrane, the contamination computed on the basis of enzyme activity should be the same from fraction to fraction for all enzymes. This is not the case, even though all studies were done by one investigator, presumably in one preparation[6].

*Carsten, M.E., unpublished

Therefore, we do not know whether a certain activity represents contamination with other components or is an inherent property. It follows that we do not know whether we have a pure fraction. Further purification may be limited by a gradual decline of specific activities with time.

A more promising approach is the use of chemical markers, whose specificity is based on their inability to penetrate through the cell membrane. The use of ligands such as wheat germ agglutinin (WGA), insulin or cholera toxin which specifically bind to receptors in the plasma membrane has been suggested[15]. The procedure requires either injection of the ligand into the blood stream, which assumes penetration of the ligand into the extracellular space of the target organ in an unaltered form, or application to tissue or homogenate. In the latter case, success depends on the presence of receptors only in plasma membrane. In our laboratory, attempts to label plasma membrane with wheat germ agglutinin were unsuccessful as all microsomal fractions were labeled.

Electron microscopy will give us only limited answers when applied to tissue fractions. For instance, it is extremely difficult to distinguish between vesicles formed from plasma membrane and SR. Furthermore, observed minor components may carry the major part of the biological activity.

Analytical procedures in calcium determination

The two methods used to determine calcium binding are 1) direct determination of calcium by atomic absorption spectroscopy after centrifugation, or 2) counting of tracer ^{45}Ca after filtration. The centrifugation method is more time consuming and requires more material, but gives unequivocal results, inasmuch as the calcium associated with the protein is directly measured. The filtration method is adapted from the method originally used in skeletal muscle to counting the filters instead of aliquots of the filtrate. This modification avoids some of the problems of the ^{45}Ca filtration method such as exchange of ^{45}Ca with intrinsic calcium. The latter can introduce large errors in the calculation due to a large intrinsic exchangeable calcium pool, a low free calcium concentration, and possible dependence on the size of the exchangeable calcium pool on either ATP (seen in skeletal muscle[16]) or added pharmacological agents[5,17]. Differences in exchangeable calcium may cause considerable error when comparing ATP-dependent calcium binding in different fractions of one preparation.

Other technical problems encountered are: 1) the necessity for using high radioactivity to an extent where the Ca in ^{45}Ca solutions has to be taken into account; and 2) the necessity of carefully washing the Millipore membrane filters. Using an ashing procedure, we found the filters to contain 650 nmol calcium (ave.). Six successive washes of the filters and prefilters with buffer removed on the average 56.0, 22.6, 8.0, 6.0, 5.4, 4.8 nmol calcium per ml, respectively, and when washing just the filters, 14.8, 4.1, 3.3, 2.5, 2.3 and 1.9 nmol calcium per ml. However, considerable variation was noted with different batches of filters. These findings eliminate the possibility of using the filtration method for the calculation of exchangeable calcium. The presence of calcium in the filter has the potential of forming calcium oxalate crystals in the filter (see below). Furthermore, the contact time of solutions with the filter is of importance. Passage of an incubation medium containing 20 nmol/ml calcium and ^{45}Ca can remove 1 nmol calcium from the incubation medium within 30 seconds, 1.5 nmol when 1 minute on the filter and more at longer contact time. This necessitates keeping filtration time to a minimum.

Role of calcium complexing anions

The physiological significance of oxalate in the calcium sequestering system is debatable. Oxalate was introduced into the system in the study of skeletal muscle[18] in order to increase the amount of calcium taken up by vesicular preparations and to exclusively measure calcium influx in contrast to an equilibrium of calcium binding and release. Recently, it has been suggested that calcium incorporation with and without oxalate are processes characteristic of two different organelles[6]. In the presence of oxalate, calcium transported across the membrane is precipitated inside the vesicles, a process distinctly different from calcium binding to specific sites in the vesicular membranes. In our experiments oxalate potentiation in uterus microsomes is small. There is a 30 to 50% increase in calcium incorporated at pH 7.0, a result confirmed by Godfraind[12] for intestinal-muscle preparations. At pH 6.6, calcium incorporation was found to be tripled[12]. In contrast, striated muscle shows a 40-100 fold increase in calcium incorporation with oxalate[13,19]. The physiological significance of the experiments using oxalate is not obvious, even supposing that there is a corresponding anion *in vivo*, because the velocity of contraction cannot be reconciled with the release of free calcium from a calcium precipitate.

TABLE II

CALCIUM BINDING IN THE PRESENCE OF Ca^{++}-EGTA BUFFERS

(8 minutes incubation at 37°C. 20 mM imidazole buffer, pH 7.0, 10 mM KCl, 2 mM ATP, 2 mM $MgCl_2$. Data are means of 4 determinations $\pm$S.E.M.)

Total Ca^{++} Concentration	Free Ca^{++} Concentration	EGTA Concentration	Intrinsic Ca^{x}	Total Ca bound	ATP-dependent Ca binding^{xx}
	(M)			(nmol/Ca/mg protein)	
1. 2×10^{-5}	2×10^{-5}	-----0-----	19.8 ± 2.7	44.3 ± 3.8	24.5 ± 2.3
2. 2×10^{-5}	2×10^{-7}	2.45×10^{-4}	10.0 ± 1.9	18.4 ± 3.9	8.4 ± 2.1
3. 2×10^{-4}	2×10^{-5}	1.80×10^{-4}	20.5 ± 2.4	60.4 ± 7.0	39.9 ± 5.7
4. 2×10^{-3}	2×10^{-5}	2.21×10^{-3}	17.6 ± 2.0	59.3 ± 7.6	41.7 ± 5.7

[x] The differences between 1, 3, and 4 not significant; between 1 and 2, $p < 0.002$

[xx] The difference between 1 and 2, $p < .002$; between 1 and 3, $p < 0.02$; between 1 and 4, $p < C.01$; between 3 and 4, $p < 0.2$

Role of EGTA

The amount of calcium bound is known to be a function of the free calcium concentration. In order to study calcium binding at concentrations which prevail inside the cell, one would have to use 10^{-7}M calcium. At this concentration, contamination by extraneous calcium from filters, glassware, water and chemicals are prohibitive. Therefore, EGTA buffers are used. We have previously reported that EGTA depletes the intrinsic calcium in smooth muscle microsomes in the absence of external calcium[20]. Because of the magnitude of the association constant of EGTA for calcium, EGTA appears to be able to partially deplete the intrinsic calcium even in the presence of external calcium as seen in Table II. At the same free calcium concentration, ATP-dependent calcium binding becomes a complicated function of the EGTA, total calcium, and free calcium concentrations. To evaluate the physiological significance of calcium binding in the presence of EGTA, one has to consider the calcium binding in relation to calcium association constants of the contractile proteins in smooth muscle, and these are still unknown. The situation is further complicated by the differences in Ca-EGTA association constants reported in the literature[21,22] and the dependence of these constants on pH, temperature, ATP and ions present. Thus, it is of primary importance for every investigator to clearly state not only which association constant was used in his computations, but also the final concentration of total calcium, of free calcium and of EGTA. For the results in Table II we used the association constant of Ogawa and calculated by the method of Katz et al.[23].

Physiological significance

A multitude of subcellular organelles, such as SR, plasma membrane, mitochondria, golgi apparatus, and lysozomes could be involved in calcium transport. Whether mitochondria can sequester calcium from very dilute solution at sufficient speed is under investigation in several laboratories[24,25]. Mitochondrial contamination in microsomal preparations can be easily assessed because of the azide-sensitivity of the mitochondrial calcium uptake. The function of smooth muscle SR in excitation-contraction coupling is deduced from striated muscle, where SR serves as a calcium sink and releases calcium on receiving a transmitted impulse. Can the SR function not only in excitation-contraction but also in pharmacomechanical coupling? The latter implies that a pharmacologic agent either penetrates the cell membrane and acts directly on the SR or that the agent interacts with the cell membrane and thereby releases a second messenger which in turn acts upon the SR. We have shown that agents such as prostaglandins (PG) have effects on microsomal calcium binding and release[5,26]. If the PG action were directed at cell membrane present in the microsomal preparation, the involvement of cyclic 3'-, 5'-adenosine monophosphate (cAMP) is an attractive hypothesis. So far it has not been documented that PGs act through a cAMP mechanism in this system. A third possibility is that the calcium binding mechanism is a property of cell membrane with direct PG action at the cell membrane. This implies that the microsomal preparation contains cell membrane, whose physiological function is to transport calcium from inside the cell to the extracellular fluid. In that case the microsomal fraction must contain cell membrane fragments turned inside out. In the course of future work it must be attempted to separate SR from cell membrane in order to answer these questions.

It is impossible to assess the physiological significance of calcium incorporation into smooth muscle microsomes on the basis of present results. We do not know 1) how much calcium is bound in contraction to actomyosin and/or troponin, 2) how much of these proteins occur in the respective smooth muscle, 3) what is the purity, percent recovery, and loss of activity of SR in a microsomal fraction, and what are the kinetics of initial calcium binding and calcium release.

To come back to our original goal, we query: Have we identified the intracellular calcium sequestering process? We can say that our knowledge about this process has grown. We still are not entirely certain that the isolated system is the system that can account for the observed physiological phenomena and thus the question of its physiological significance has to be left for future inquiry.

ACKNOWLEDGEMENTS

Supported by grant HD-00010, USPHS.

REFERENCES:

1. Siegman, M.J. and Gordon, A.R. (1972) Am. J. Physiol. 222, 1587-1593
2. Van Breemen, C. et al. (1973) Phil. Trans. R. Soc. Lond. 265, 57-71
3. Devine, C.E. et al.(1972) J. Cell Biol. 52, 690-718

4. Somlyo, A.P. et al. (1977) Calcium Compartmentalization in Vascular Smooth Muscle, in Biochemistry of Smooth Muscle (Stephens, N.L., ed.) pp 563-585, University Park Press, Baltimore
5. Carsten, M.E. and Miller, J.D. (1977) J. Biol. Chem. 252, 1576-1581
6. Janis, R.A. et al. (1977) Am. J. Physiol. 232, C50-C58
7. Janis, R.A. et al. (1976) Pflugers Arch. 365, 171-176
8. Hess, M.L. and Ford, G.D. (1974) J. Mol. Cell. Cardiol. 6, 275-282
9. Wei, J.W. et al. (1976) Circulation Res. 39, 133-140.
10. Moore, L. et al. (1975) Biochim. Biophys. Acta 413, 432-443
11. Hurwitz, L. et al. (1973) Science 179,384-386
12. Godfraind, T. et al. (1976) Biochim. Biophys. Acta 455, 254-268
13. Katz, A.M. and Repke, D.I. (1976) Circulation Res. 21, 153-162
14. Carsten, M.E. and Reedy, M.K. (1971) J. Ultrastruct. Res. 35, 554-574.
15. Chang, K.J. (1975) J. Biol. Chem. 250, 488-500
16. Carvalho, A.P. (1968) J. Gen. Physiol. 52, 622-642
17. Carsten, M.E. (1977) Can Hormones Regulate Myometrial Calcium Transport? in Biochemistry of Smooth Muscle (Stephens, N.L., ed.) pp 617-639, University Park Press, Baltimore
18. Hasselbach, W. and Makinose, M. (1961) Biochem. Z. 333, 518-528
19. Harigaya, S. and Schwartz, A. (1969) Circulation Res. 25, 781-794
20. Carsten, M.E. (1973) Gynecol. Invest. 4, 95-105
21. Ogawa, Y. (1968) J. Biochem. 64, 255-257
22. Owen, J.D. (1976) Biochim. Biophys. Acta 451, 321-325
23. Katz, A.M. et al. (1970) Biochim. Biophys. Acta 205, 473-490
24. Vallieres, J., et al. (1975) Arch. Biochem. Biophys. 170, 659-669
25. Wikstrom, M. et al. (1975) FEBS Letters 56, 120-123
26. Carsten, M.E. (1974) Prostaglandins 5, 33-40

Excitation-Contraction Coupling in Smooth Muscle, R. Casteels et al. eds.
© *1977 Elsevier/North-Holland Biomedical Press*

TIME-DEPENDENT CHANGES OF CALCIUM INFLUX AND EFFLUX RATES IN SARCOPLASMIC RETICULUM VESICLES*

Arnold M. Katz, Doris I. Repke+, Munekazu Shigekawa and Gary Fudyma
Department of Medicine, University of Connecticut
Farmington, Conn. 06032, USA

INTRODUCTION

Several investigators have shown that the rate of Ca efflux from sarcoplasmic reticulum vesicles increases with increasing $Ca^{§}_{o}$ (2-7). We have recently obtained evidence that the Ca permeability of these membranes is inversely proportional to Ca_i (6, 7), and that the Ca^{2+}-sensitive sites at the inside and outside surfaces of these membranes that influence Ca efflux interact independently with Ca^{2+} (8). A number of similarities between factors that influence Ca influx and efflux have been found, suggesting that the Ca pump may also participate in the control of Ca efflux (8, 9).

Careful studies of the changes in Ca content of SR vesicles during ATP-supported Ca uptake, in which Ca_i was maintained at a relatively constant level by the use of inorganic P to precipitate Ca within the vesicles, indicate that the filling of the vesicles does not follow a smooth curve. Instead, oscillations are seen as the Ca content of the vesicles reaches its maximum (7). Changes in Ca influx rate during the course of the reaction, such as noted by Barlogie (10), may partly explain these oscillations, as well as a slowing of Ca uptake velocity as the vesicles reach their "capacity" to store Ca.

---

*Supported by Research Grants HL-22135 and HL-21812 from the National Institutes of Health, and Research Grants from the Connecticut Heart Association and University of Connecticut Research Foundation.

+Current Address: Department of Medicine, Mount Sinai School of Medicine, New York, NY 10029, USA.

§Abbreviations: Ca_o: calcium ion concentration in the medium outside the vesicles; Ca_i: calcium ion concentration inside the vesicles, calculated as described previously (1); AcP; acetyl phosphate. As used in this article, calcium uptake and calcium release refer to the net gain or loss of calcium by the vesicles, calcium influx and calcium efflux refer to unidirectional calcium fluxes into and out of the vesicles, respectively.

The present study, which examines the rates of Ca influx and efflux during the Ca uptake reaction, indicates that this "capacity" is determined, at least in part, by time-dependent changes in both of these flux rates.

METHODS

Preparation of sarcoplasmic reticulum vesicles and measurements of calcium uptake, calcium release, calcium influx, and calcium efflux were as described previously (1, 6, 7, 9). Experiments were carried out in 0.12 M KCl, 5 mM $MgCl_2$, 40 mM histidine buffer (pH 6.8) using 5 mM M ATP or 10 mM AcP as energy donor. In studies using ATP, an ATP regenerating system (0.15 mg pyruvate kinase per ml, and 4.0 mM PEP) was used to prevent ADP formation and to maintain ATP concentration. Temperature was 25° C for all studies in which ATP was used as substrate, and 37° where AcP was substrate.

RESULTS AND DISCUSSION

A typical Ca uptake reaction, shown in Fig. 1, demonstrates that Ca content first rises and then falls after sarcoplasmic reticulum vesicles are added to a complete reaction mixture.

The fall in Ca content after 16 min. incubation (Fig. 1) cannot be attributed to substrate exhaustion as AcP is present in great excess. Instead, reduction in AcP concentration both delays and slows the Ca release phase (D. I. Repke _et al._, unpublished observations). Non-specific deterioration of the membranes can also be excluded as a cause for the Ca release phase as the transient calcium release caused by addition of a number of substances, including X537A, caffeine and low EGTA concentrations, markedly slows the spontaneous release of Ca (1, 7).

A simple explanation for the finding that Ca uptake ceases after 16 min. in the experiment shown in Fig. 1 is that the vesicles have become filled with an insoluble Ca precipitate. This explanation is unlikely, however, because Ca oxalate and Ca phosphate "capacities" are not constant, but depend on Ca_o when reactions are carried out with varying initial total $CaCl_2$/protein ratios (1, 2), and this "capacity" changes when the nature or concentration of the Ca-precipitating ion is altered (1).

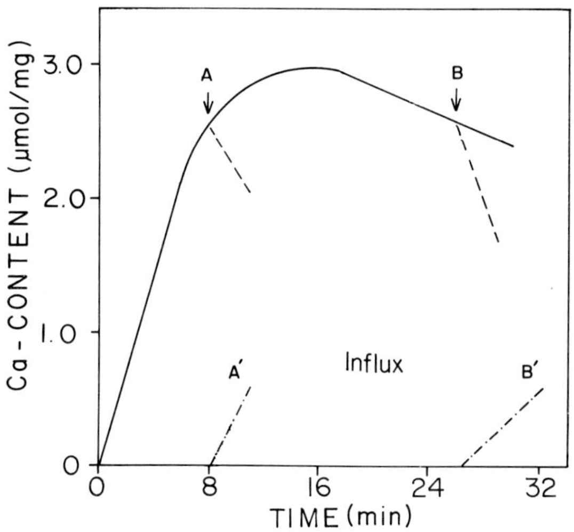

Figure 1: Time course of Ca uptake and release in a reaction carried out with 16 μg/ml protein, 60 μM^{45} $CaCl_2$, 10 mM AcP as substrate and 50 mM phosphate as Ca-precipitating anion (——→). Ca influx (—·—·→) was measured 8 and 26 min. after the start of the reaction (A' and B', respectively) by adding a small amount of carrier-free ^{45}Ca to an identical reaction mixture started with non-radioactive $^{40}CaCl_2$. Ca release (←— —) was also measured 8 and 26 min. after the start of the reaction (A and B, respectively) by adding EGTA (final concentration 3 mM) to an aliquot of the reaction mixture started with $^{45}CaCl_2$.

Measurements of Ca_o levels in experiments such as that shown in Fig. 1 demonstrate that the vesicles do not cease to accumulate Ca because of a decrease in Ca_o. In the experiment shown in Fig. 1, for example, Ca content reached its maximum and Ca release began when Ca_o was approximately 20 μM, well above the K_{Ca} for the uptake reaction in AcP (11, D. I. Repke <u>et al.</u>, unpublished observation). Thus, the decrease in Ca_o that accompanies Ca uptake, while contributing to the initial slowing of Ca influx, cannot explain the cessation of Ca uptake because the Ca pump remains active at the lowest levels of Ca_o seen in this study.

In order to test the hypothesis that the complex curve for Ca content shown in Fig. 1 resulted from time-dependent changes in Ca influx and efflux, Ca influx rates were measured at frequent intervals during the reaction. The rates of Ca influx 2 and 4 min. after the start of this reaction were 358 and 285 nmol/mg·min, respectively. These did not differ significantly from the concurrently measured rates of Ca uptake, indicating that Ca efflux rates are near zero during the initial phase of rapid Ca uptake. After 8 min., when Ca uptake rate had begun to decrease (Fig. 1), Ca influx slowed to 187 nmol/mg·min (A', Fig. 1). However, Ca uptake rate slowed even more, to 100 nmol/mg·min. The difference between influx and uptake rates, 87 nmol/mg·min. represents Ca efflux. While the initial slowing of Ca influx is due partly to a slight fall in Ca_o, Ca influx rate continued to show a marked time-dependent decrease until, after 26 min., it had fallen to 105 nmol/mg·min. (B', Fig. 1). As the vesicles at this time were releasing Ca at a rate of 43 nmol/mg·min., the Ca efflux rate at 26 min. was 148 nmol/mg·min. A progressive decrease in Ca influx rate coupled with a concurrent increase in Ca efflux rate thus accounts for the late rise and fall in Ca content shown in Fig. 1, and the calcium content at the time that Ca influx and efflux rates become equal is seen to represent the apparent "capacity" of the vesicles.

Confirmation of the finding that Ca efflux rate increased during the Ca uptake reaction shown in Fig. 1 was obtained in studies of the rate of Ca release after Ca influx was halted by the addition of EGTA to a final concentration of 3 mM. Virtually no EGTA-induced Ca release was seen during the initial phase of rapid Ca uptake. The rate of EGTA-induced Ca release increased progressively during the course of the reaction reaching 163 nmol/mg·min. after 8 min. (A, Fig. 1) and 280 nmol/mg·min. after 26 min. (B, Fig. 1).

Similar phenomena of Ca uptake followed by Ca release were seen in reactions in which ATP, instead of AcP, was the energy donor, except that the time-dependent changes in Ca efflux rates are more complex. For example, the initial peak of calcium content reached at high levels of Ca_o in reactions carried out with ATP was often followed by a secondary re-uptake phase that led to an even higher

Ca "capacity" (Repke, D. I. et al., unpublished observations).

We have shown previously that Ca efflux rates measured shortly after Ca uptake reactions reach a steady state of Ca content (i.e. at the time that "capacity" is maximal) increase with increasing Ca_o (6-8). It is possible that Ca efflux rate depends not on the level of Ca_o, but instead on Ca content, because the latter tends to increase at high levels of Ca_o. In several experiments, however, we observed that increasing Ca_o caused a decrease in Ca "capacity". In these experiments, Ca efflux rate was related to the level of Ca_o as shown previously (6-8), and not to the peak Ca content of the vesicles. The experiment shown in Fig. 1 provides additional evidence that factors other than Ca content influence Ca influx and efflux rates because at 8 and 26 min. after the start of the reaction, Ca contents were equal at 2.6 μmol/mg. Differences in Ca_o also cannot explain the different rates of Ca influx and efflux as Ca_o was 26 μM at these two times.

Several similarities between factors influencing Ca influx and efflux rates are consistent with the hypothesis that the Ca pump also participates in the control of Ca efflux from the sarcoplasmic reticulum (8, 9). The Ca efflux described in this, and previous studies cannot be attributed to complete reversal of the Ca pump (12-16) as the Ca efflux we have described increases with increasing Ca_o and ATP, both of which inhibit Ca pump reversal (12, 15, 16). Furthermore, the demonstration that Ca efflux can be promoted by Ca_o when AcP is substrate (8) cannot be explained by complete reversal of the Ca pump, which requires ADP. It is possible, however, that Ca efflux from the sarcoplasmic reticulum utilizes some of the reactions of the Ca pump. Such a mechanism, which would allow the Ca pump to serve also as a Ca "channel", can explain the present findings if the Ca pump sites shift from a "pump" mode to a "release" mode during the course of the reaction shown in Fig. 1. This behavior would account not only for the observed similarities between the control of Ca efflux and the Ca pump (8, 9), but also the tendency towards a reciprocal relationship between time-dependent changes in Ca influx and efflux rates. Further work will be necessary to demonstrate whether the phenomena documented in this report arise from such a mechansim, in which the Ca pump is utilized also to release Ca from the sarcoplasmic reticulum.

REFERENCES

1. Katz, A. M., Repke, D. I. and Hasselbach, W. (1977) J. Biol. Chem. 1938-1949.
2. Weber, A., Herz, R. and Reiss, I. (1966) Biochem. Z. 345: 329-369.
3. Makinose, M. (1973) FEBS Letters 37: 140-143.
4. Hasselbach, W., Fiehn, W., Makinose, M. and Migala, A. J. (1969) in The Molecular Basis of Membrane Function (D.C. Tosteson, Ed.) Prentice-Hall Inc., Englewood Cliffs, N. J. pp. 299-316.
5. Weber, A. (1971) J. Gen. Physiol. 57: 64-70.
6. Katz, A. M., Dunnett, J., Repke, D. I. and Hasselbach, W. (1976) FEBS Letters 67: 207-208.
7. Katz, A. M., Repke, D. I., Dunnett, J. and Hasselbach, W. (1977) J. Biol. Chem. 252: 1950-1956.
8. Katz, A. M., Repke, D. I., Fudyma, G. and Shigekawa, M. (1977) J. Biol. Chem. 252: 4210-4214.
9. Katz, A. M., Repke, D. I., Fudyma, G. and Shigekawa, M. (1977) in Calcium Binding Proteins and Calcium Function in Health and Disease, Elsevier-North Holland Publishing Company, In press.
10. Barlogie, B. (1972) Medical Thesis, University of Heidelberg, Fed. Rep. of Germany.
11. Hasselbach, W. (1972) in Molecular Bioenergetics and Macromolecular Biochemistry (H. H. Weber, Ed.) Springer-Verlag, Berlin, pp. 149-171.
12. Barlogie, B., Hasselbach, W. and Makinose, M. (1971) FEBS Letters 12: 267-268.
13. Makinose, M. (1971) FEBS Letters 12: 269-270.
14. Kanazawa, T., Yamada, S. and Tonomura, Y. (1970) J. Biochem. (Tokyo) 68: 593-595.
15. Panet, R. and Selinger, Z. (1972) Biochim. Biophys. Acta 235: 34-42.
16. Masuda, H. and de Meis, L. (1974) Biochim. Biophys. Acta 332: 313-315.

Excitation-Contraction Coupling in Smooth Muscle, R. Casteels et al. eds.
© 1977 Elsevier/North-Holland Biomedical Press

ATP DEPENDENT Ca^{2+} BINDING BY MITOCHONDRIAL AND MICROSOMAL FRACTIONS ISOLATED FROM DIFFERENT SMOOTH MUSCLES: INFLUENCE OF CYCLIC NUCLEOTIDES

ULRICH ZELCK and UWE KARNSTEDT
Institute of Pharmacology and Toxicology,
Medical Faculty of
the Wilhelm-Pieck-University Rostock,
25 Rostock,
Leninallee 70,
GDR

SUMMARY

Different concentrations of cyclic adenosine 3' , 5'-monophosphate and cyclic guanosine 3',5'-monophosphate failed to significantly influence the Ca^{2+} binding by mitochondria and microsomes isolated from pig coronary artery, guinea pig ileum guinea pig taenia coli and of microsomal membranes from rat uterus.

INTRODUCTION

Originating from the well known facts that calcium ions regulate the contraction and relaxation of the smooth muscle [1,2] and that cyclic adenosine 3',5'-mononophosphate (cAMP) is involved in the regulation of smooth muscle tone [3,4] , several investigations were undertaken to find an in vitro equivalent to the relaxing action brought about by this nucleotide in the physiological process. For this purpose the influence of cAMP on the Ca^{2+} uptake or Ca^{2+} binding by microsomal fractions from different smooth muscles was investigated by several laboratories [4-13]. In addition in some cases the influence of cAMP on the Ca^{2+} uptake and Ca^{2+} binding by mitochondrial smooth muscle fractions was investigated [4, 10-12, 14].

Since it has been observed that contraction of some kinds of smooth muscle is associated with increased levels of cyclic guanosine 3' ,5'-monophosphate (cGMP) [15,16] it was suggested that this nucleotide mediates contraction of the smooth muscle cell. In connection with this assumption Andersson et al. [4] and Nilsson and Andersson [25] reported on an enhanced ^{45}Ca release by a microsomal fraction from rabbit colon when 1 x 10^{-5} M cGMP were added. Furthermore, when cGMP (5x10^{-6} or 1 x 10^{-5} M) was combined with cAMP (1 x 10^{-5} M) the stimulating effect of cAMP was completely inhibited. Contrarily Clyman et al. [11] failed to show any effect of cGMP on the Ca^{2+} uptake by microsomes and mitochondria isolated from human umbilical artery. However, the results concerning the actions of cAMP and cGMP on the Ca^{2+} uptake or Ca^{2+} binding by subcellular fractions from different smooth muscles are conflicting.

To further knowledge in this field, we have investigated the influence of different concentrations of cAMP and cGMP on the Ca^{2+} binding and/or Ca^{2+} release by mitochondrial and microsomal fractions isolated from different smooth muscles (pig coronary artery, guinea pig ileum, guinea pig taenia coli and rat uterus) and statistically compared the results obtained over a period of about 4 years.

MATERIALS AND METHODS

Preparation of the subcellular fractions of smooth muscle

Just after killing the animals the pig coronary arteries, the guinea-pig ileum and the guinea pig taenia coli were removed in the cold, dissected free from adherent fat and loose connective tissue and cut into pieces of about 3-4 mm length (in some series of experiments adventitia-free medial-intimal layers of the pig coronary artery were prepared). The biological material was homogenized in icecold 0.25 M sucrose solution using an all-glass Potter-Elvehjem homogenizer. The resulting homogenate was filtered through three layers of gauze. The mitochondrial fractions were prepared by centrifugation of a 1000 x g (10 min) postnuclear supernatant for 20 min at 10,000 xg. The remaining supernatant was spun at 100 000 xg for 60 min to obtain the microsomal fraction. The preparation and characterization of the membrane enriched fraction of rat uterus, which was obtained after centrifugation of the crude microsomal pellet on a discontinuous sucrose gradient, was described in detail elsewhere [17]. All steps were carried out at 0-4°C and within 8 (pig coronary artery) or 4 h (guinea pig ileum and taenia coli) after the animals were sacrificed. The mitochondrial and microsomal pellets were suspended in 0.1 M KCl to give a protein concentration of about 1-5 mg/ml. Protein concentrations were measured by the method of Lowry et al. [18] using bovine serum albumin as a standard. Mitochondria which were found to be extremely unstable were always investigated immediately after isolation. The microsomal fractions were used within 4 h after preparation or (in the most cases) after storage at 4°C over night. The purity of the subcellular fractions was proven by electron microscopy and by cytochrome-C oxidase [19] and/or succinic dehydrogenase [20] estimation. No markable cross contamination was observed.

Ca^{2+} binding and Ca^{2+} release studies

Ca^{2+} binding and Ca^{2+} release was studied at 25 or 37°C in incubation mixtures (1000 or 400 µl) containing 100 mM KCl, 0.0035 - 5 mM $MgCl_2$, 0.0035 - 5 mM Na-ATP, 0.05 - 0.1 mM $CaCl_2$ (adjusted by atomic absorption measurement) with 2.5 - 5 µ Ci $^{45}CaCl_2$ and 5 mM histidine buffer, pH 7.0. Ca^{2+} release experiments were usually started after 20 min preincubation (at 25°C) and following dilution (1:25) with 100 mM KCl. Incubation mixtures were kept in a metabolic shaker. Ca^{2+} binding and release was determined by a microfiltration technique using Millipore (HAW:0.45 µm) or Sartorius (SM 11306; 0.45 µm) membrane filters. Blanks which lacked only the particulate material were filtered simultaneously. The amount of radioactive Ca^{2+} retained by the mitochondrial or microsomal protein was measured by liquid scintillation counting (LKB Wallac 81000) after filters had been dissolved in a formamide-dioxane scintillator solution. Atomic absorption measurements of the Ca^{2+} concentrations were carried out using a Varian-Techtron AA-1200 atomic absorption spectrophotometer. The electrometric equipment used for the measurement of Ca^{2+} binding with a calcium ion selective electrode was recently described in detail [21]. The pig heart and calf heart protein kinase was prepared by the method of Miyamato et al. (1969) and was a kind gift from the laboratory of Dr. E.-G. Krause and Dr. E. Will, Division of Cellular and Molecular Cardiology, Central Institute of Heart and Circulatory Regulation Research, Academy of Science of GDR.

Statistical methods. Values are presented as means ± S.E. of 3 determinations. The standard error (S.E.), for clarity not given in Figs., was about ± 5 % as an average. The significance of effects was checked by thestudent's t-test (paired samples).

Rejection criterion was chosen at the 1-5 % level of significance. Because the effects obtained from different experiments with the same subcellular preparation were always highly significant,only values obtained in experiments with different mitochondrial or microsomal fractions were used for the statistical analysis.

RESULTS

The results demonstrated in the following Figs. are more or less representative for a much larger (excluding Ca^{2+} release experiments) series of experiments obtained over a period of about 4 years. Although in these experiments usually no, or a rather low effect of cyclic nucleotides was observed, we should point out that a (low) number of experiments showed stimulation or reduction (larger number). However, trials to reproduce these effects were difficult. The results showed no statistically significant trends.

With respect to the extreme instability of the mitochondria* in the Ca^{2+} binding experiments with these organelles (using filter technique) controls and treated samples were always investigated at exactly the same moment using two sets of equipment. If this is omitted, extremely irritating results are obtained. Concerning microsomes this point was of less importance.

In Fig. 1 are shown examples demonstrating the effect of cAMP on the Ca^{2+} binding and Ca^{2+} release by mitochondria and microsomes isolated from pig coronary artery smooth muscle.

Fig. 2 demonstrates the influence of cAMP and cGMP at different ATP concentrations on the rate of Ca^{2+} binding by mitochondria and microsomes isolated from the

*For brevity mitochondrial and microsomal fractions are called mitochondria and microsomes.

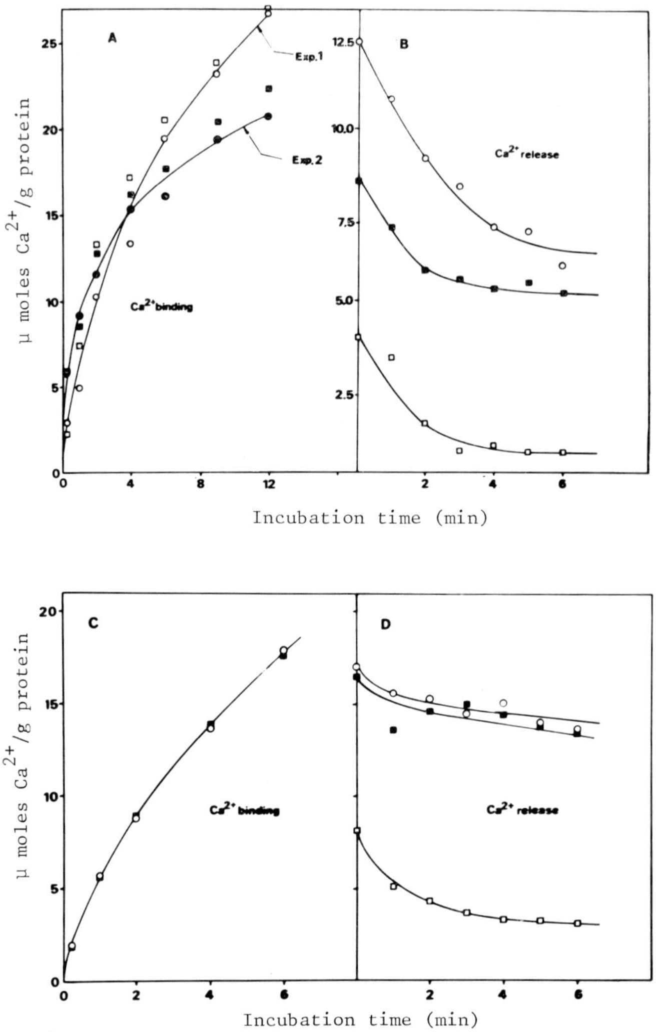

Fig. 1. Effect of cAMP (1 x 10^{-3} and 1 x 10^{-5} M) on the Ca^{2+} binding and Ca^{2+} release by mitochondria (A,B) and microsomes (C,D) from pig coronary artery at 25°C. Measurements of the control and the related treated samples were at all steps made simultaneously.

A. Exp. 1: control (o) + 1x10^{-3} M (□) cAMP; Exp. 2: control (●) + 1x10^{-5} M (■) cAMP. B: control (o) + 1x10^{-3} M (□) and 1 x 10^{-5} M (■) cAMP. C: control (o) + 1x10^{-5} M (■) cAMP. D: control (o) + 1 x 10^{-3} M (□) and 1 x 10^{-5} M (■) cAMP.

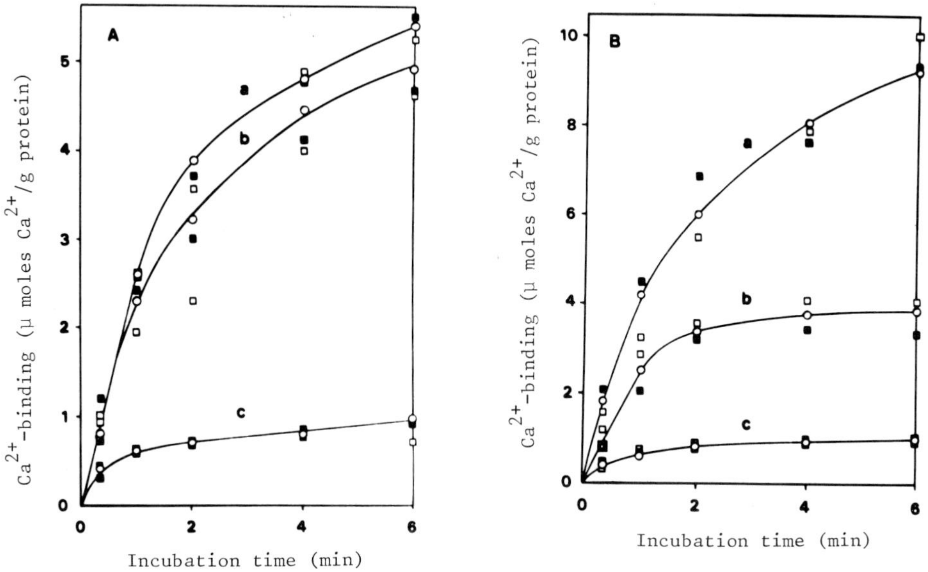

Fig. 2. Effect of 1 x 10^{-5} m cAMP (■) and 1 x 10^{-5} M cGMP (□) on the Ca^{2+} binding by microsomes from pig coronary artery at different ATP concentrations. A: a(5 mM) b(0.35 mM); c(0.035 mM). B. a(5 mM); b(0.1 mM); c(0.0035 mM).

medial-intimal layer of the pig coronary artery. In these experiments we examined the assumption made by Nilsson et al. [13] that the "rather high" ATP concentrations (3-5 mM) used by those investigators who have not observed any effect of cAMP on microsomal Ca^{2+} binding or Ca^{2+} uptake [5,8,9,11] is responsible for the negative effect obtained.

As to be seen, at the very low ATP concentrations used in these experiments (Andersson et al. [6] and Nilsson et al. [13] used 0.35 mM ATP) no markable effects could be demonstrated. This result will be underlined by experiments in which the measurement of Ca^{2+} binding was directly followed using a calcium ion sensitive electrode (Fig. 3). The results shown in Fig. 3 (A,B) clearly demonstrate that at an ATP concentration of 0.35 mM neither during the fast Ca^{2+} binding step nor in the saturation range were cAMP and cGMP able to produce any effect.

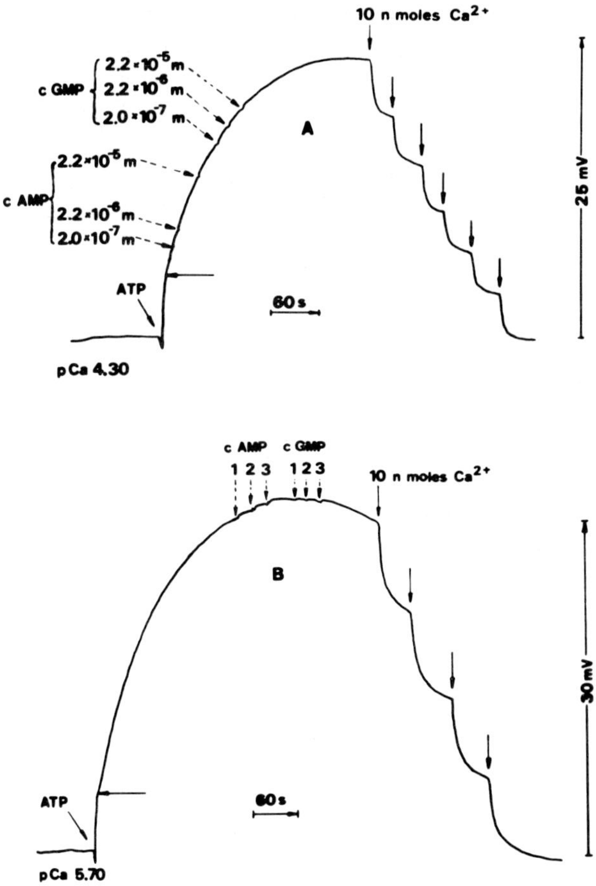

Fig.3. Effect of different concentrations cAMP and cGMP on the rate of Ca^{2+} binding by microsomes from pig coronary artery directly measured using a calcium ion sensitive electrode (K 4112, Radiometer) at 37°C. A. Influence of 2.0 x 10^{-7}, 2.2 x 10^{-6}, 2.2 x 10^{-5} M cAMP and cGMP, added at the fast Ca^{2+} binding range. B. Same concentrations cAMP and cGMP, added at the saturation level of Ca^{2+} binding (1= 2.0x10^{-7}, 2= 2.2x10^{-6}, 3= 2.2x10^{-5} M). ATP concentration: 0.35 mM. The magnitude of the self effect of ATP is marked by the horizontal arrow. The system was calibrated by stepwise addition of each 10 n moles Ca^{2+}, after the Ca^{2+} binding reaction has reached the saturation range.

In Fig. 4. The influence of cAMP and cGMP on the Ca^{2+} binding and Ca^{2+} release by mitochondria from guinea ileum is shown.
The result shown in Fig. 4 is generally consistent with that demonstrated for mitochondria from pig coronary artery (Fig. 1): The Ca^{2+} binding is only negligibly effected by 1 x 10^{-5} M cAMP and 1 x 10^{-5} m cGMP, whereas 1 x 10^{-5} and 1x10^{-3} M

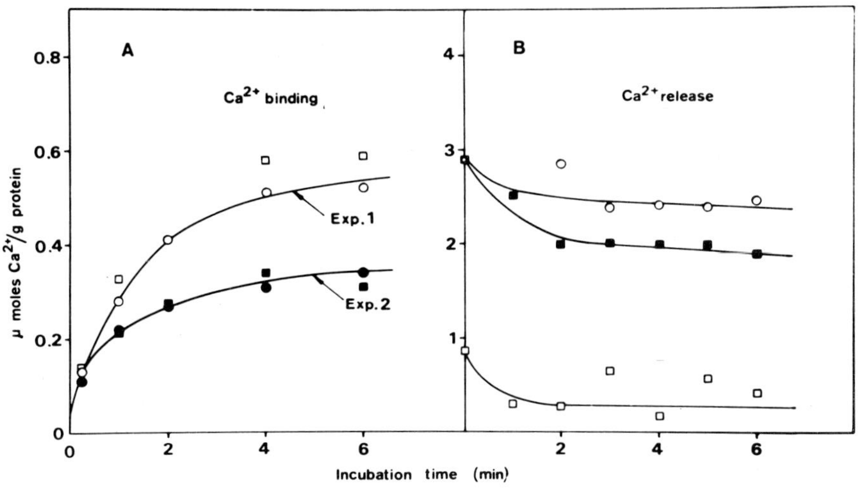

Fig. 4. Effect of cAMP and cGMP (both $1x10^{-5}$ M) on the Ca^{2+} binding (A), and of cAMP ($1x10^{-3}$ and $1x10^{-5}$ M) on the Ca^{2+} release (B) by mitochondria from guinea pig ileum at 25°C. A. Exp.1: control (o) + $1x10^{-5}$ M cAMP (□); Exp.2: control (●) + $1x10^{-5}$ M cGMP (■). B. control (o) + $1x10^{-5}$ (■) and $1x10^{-3}$ M cAMP (□).

cAMP caused a noticeably enhanced Ca^{2+} release, especially when the unphysiologically high cAMP concentration of 1 x 10^{-3} M was used.

Because the typical effects of cAMP and cGMP on the Ca^{2+} binding by microsomes from guinea pig ileum, mitochondria and microsomes from guinea pig taenia coli and the membrane enriched fraction from rat uterus are in general consistent with the results demonstrated for the effect of cAMP and cGMP on the Ca^{2+} binding by mitochondria and microsomes from pig coronary artery (Fig. 1 A,C; Fig. 2 A,B) figures are not here presented. Because the possibility had to be taken into account that in our experiments the cAMP dependent protein kinase activity of the subcellular fractions was lost during the preparation, additional experiments with added pig heart or calf heart protein kinase were performed on microsomes and mitochondria from pig coronary artery. Even in the presence of protein kinase cAMP and cGMP failed to produce any effect.

As pointed out before, in general the effects of different concentrations cAMP and cGMP on the Ca^{2+} binding by mitochondria and microsomes from the smooth muscles investigated were low or absent and not reproducible in separate series of experiments over a longer time period.
For this purpose we checked the data obtained on different microsomal or mitochondrial preparations statistically.

The following cAMP and cGMP concentrations were taken into account: $5x10^{-8}$, $1x10^{-7}$ $1x10^{-6}$, $3x10^{-6}$, $1x10^{-5}$ and $1x10^{-3}$ M. Each nucleotide concentration was tested at 5 or 7 incubation times (15 sec, 1,2,4,6,9 and 12 min).
Concerning mitochondria and microsomes from pig coronary artery, guinea pig ileum, guinea pig taenia coli and rat uterus (microsomal membranes only) the effect of the different concentrations of cAMP and cGMP on the rate of Ca^{2+} binding was not significantly different from the control.

DISCUSSION

Concerning Ca^{2+} binding or Ca^{2+} release by smooth muscle microsomes only a few laboratories have demonstrated a stimulation by cAMP (Anderson et al. 1972 on microsomes from rabbit colon, using 10^{-5} M cAMP; Baudouin-Legros and Meyer 1973 on microsomes from rabbit aorta, using 10^{-8} -10^{-4} M Dibutyryl cAMP; Webb and Bhalla 1976 on microsomes of rat aorta, using $5x10^{-6}$ M cAMP). In contrast Batra and Daniel 1971 (on microsomes from rat myometrium, using 10^{-3} - 10^{-6} M cAMP), Zelck et al. 1974 (on microsomes from pig coronary artery, using 10^{-5} M cAMP), Tomiyama et al. 1975 (on microsomes from guinea pig taenia caecum, using 10^{-5} M cAMP) and Clyman et al. 1976 (on microsomes from human umbilical artery) and Rangachari et al. 1976 (on microsomal membranes from rat myometrium using $5x10^{-6}$M cAMP or dibutyryl cAMP) failed to show any effect of the cyclic nucleotide. With respect to these conflicting results Nilsson et al. [13] tried to explain the negative findings by the "rather high" ATP concentrations (3-5 mM) used in these experiments, although Webb and Bhalla used a similar high ATP concentration (2 mM). As shown in Fig. 2 0.35 mM ATP (concentration used by Andersson et al. [6] and Nilsson et al. [13] as well as other low ATP concentrations (the physiological ATP concentration in the vascular smooth muscle is approximately 0.5 - 1.8 mM/kg wet weight [1]) used in the incubation mixture failed to produce any effect by cAMP. In this connection our results obtained by the ^{45}Ca filter technique are fully consistent with the absence of effect observed when the Ca^{2+} movement was measured using a calcium ion sensitive electrode (Fig. 3). In these experiments 0.35 mM ATP was used throughout.
Some remarks on the Ca^{2+} release studies (Fig. 1 B,D) : These data are not in accordance with the results of Ca^{2+} binding. Because these results represent only a few experiments they should be viewed critically. It seems necessary to test such results in a larger number of experiments under experimental conditions discriminating between active and passive release.

Concerning the influence of cGMP on microsomal Ca^{2+} binding of the smooth muscles investigated, the failing effect found is not consistent with findings of Andersson et al. [4] and Nilsson and Andersson [25] which demonstrated an enhanced Ca^{2+} release by microsomes from rabbit colon when 10^{-5} M cGMP was added. Our

results are in good agreement with findings of Clyman et al. [11] who failed to find any effect of cGMP on the Ca^{2+} uptake by mitochondria and microsomes from human umbilical artery.
The failure of a significant effect of cAMP on the Ca^{2+} binding found by us is in accordance with results recently published by Scarpa et al. [22] on the Ca^{2+} release by rat liver and rat and guinea pig heart mitochondria. These authors didn't find any stimulation of Ca^{2+} efflux when $5x10^{-7}$ to $1x10^{-5}$ M cAMP were added. In addition Borle has rescinded his previous findings [24] which demonstrated a cAMP stimulation of Ca^{2+} release from kidney, liver and heart mitochondria, and published a negative report [23].

Controversely Andersson et al. [4] reported on a significant reducing effect of 10^{-5} M cAMP on the Ca^{2+} binding capacity of mitochondria isolated from rabbit colon.
The rather low reducing effects previously found by us in single experiments on mitochondria of pig coronary artery [14] and guinea pig taenia coli [12] must be considered negligible in spite of the statistical results presented here.
As shown, cGMP also fails to effect the Ca^{2+} binding by mitochondria isolated from the different smooth muscles. This result is consistent with the findings of Clyman et al. [11].

Summarizing it should be pointed out that the lack of any significant effect of cAMP and cGMP on the Ca^{2+} binding of the different smooth muscles investigated is not in accordance with the assumption that the cyclic nucleotides cAMP and cGMP might act as intracellular mediators in the regulation of the free Ca^{2+} concentration of the smooth muscle cell.

ACKNOWLEDGEMENTS

The authors wish to thank Mrs. A. Saluski, Mrs. F. Pauli and Mrs. S. Schwieger for their skillful technical assistance. Thanks are further due to Dr. L. Konya (Central Res. Lab. Med. Univ. Debrecen, Hungary), and Dr. E. Albrecht (Inst. of Drug Res., Acad. of Sci. of GDR, Berlin) which were integrated in parts of this work.

REFERENCES

1. Somlyo, A.P. and Somlyo, A.V. (1968) Pharmac. Rev. 20, 197.
2. Somlyo, A.P. and Somlyo, A.V. (1970) Pharmac. Rev. 22, 249.
3. Andersson, R.G.G. (1972) Medical Dissertation No. 7 Linköping Sweden.
4. Andersson, R., Nilsson, K., Wikberg, J., Hohansson, S., Mohme-Lundholm, E. and Lundholm, L. (1975) Adv. in Cycl. Nucl. Res. 5, 491.
5. Batra, S.C. and Daniel, E.E. (1971) Comp. Biochem. Physiol. 38 A, 285.

6. Andersson, R. and Nilsson, K. (1972) Nature, New Biol. 238, 119.
7. Baudouin-Legros, M. and Meyer, P. (1973) Br. J. Pharmac. 47, 337.
8. Zelck, U., Konya, L. and Albrecht, E. (1974) Acta Biol. Med. germ. 32, K1.
9. Tomiyama, A., Takayanagi, I. and Takagi, K. (1975) Biochem. Pharmacol. 24, 9.
10. Webb, R.C. and Bhalla, R.C. (1976) J. Mol. Cell. Cardiol. 8, 145.
11. Clyman, R.J., Manganiello, V.C., Lovell-Smith, C.J. and Vaughan, M. (1976) Am. J. Physiol., 231, 1074.
12. Zelck, U. and Konya, L. (1977) Acta biol. med. germ. (accepted for publication)
13. Nilsson, K.B., Andersson, R., Mohme-Lundholm, E. and Lundholm, L. (1977) Acta pharmacol. toxicol. 40, Art. no. 2808.
14. Zelck, U., Karnstedt, U. and Jonas, L. (1976) Ergebn. exp. Med. 22, 55.
15. Schultz, G., Hardman, J.G. and Sutherland, E.W. (1973) Cyclic Nucleotides and Smooth Muscle Function. In: Asthma ed. by K.F. Austen and L.M. Lichtenstein, Academic Press, New York, pp. 123, 138.
16. Goldberg, N.D., Haddox, M.K., Hartle, D.K. and Hadden, J.W. (1973) Proc. 5th Int. Congr. Pharmacol. 5. 146.
17. Jankova, T., Paegelow, I., Reissmann, S. and Arold, H. (1975) Acta biol. med. germ. 34, 9.
18. Lowry, O.H., Rosebrough, N.J. Farr, A.L. and Randall, R.J. (1951) J. Biol. Chem. 193, 265.
19. Cooperstein, S.J. and Lazarov, A. (1951) J. Biol. Chem. 189, 665.
20. Arrigoni, O. and Singer, T.P. (1962) Nature 193, 1256.
21. Zelck, U., Karnstedt, U. and Lange, P. (1956) Acta biol. med. germ. 35, K 83.
22. Scarpa, A., Malmstrom, K., Chiesi, M. and Carafoli, E. (1976) J. Membrane Biol. 29, 205.
23. Borle, A.B. (1976) J. Membrane Biol. 29, 209.
24. Borle, A.B. (1974) J. Membrane Biol. 16, 221.
25. Nilsson, K.B. (1977) Acta physiol. scand. 99, 246.
26. Rangachari, P.K., Pernollet, M.G. and Worcel, M. (1976) Eur. J. Pharmacol. 40, 291.

Excitation-Contraction Coupling in Smooth Muscle, R. Casteels et al. eds

Characterization and Ca^{2+}-Accumulation by Membrane Fractions From Myometrium and Artery*

E.E. Daniel, C.Y. Kwan, M.A. Matlib
D. Crankshaw and A. Kidwai
Department of Neurosciences
McMaster University
Hamilton, Ontario
Canada L8S 4J9

INTRODUCTION

A standard procedure has been developed in this laboratory for isolation of subcellular membrane fractions from smooth muscle and applied to rat myometrium and mesenteric artery muscle. Our previous publications[1,2] did not deal directly with the contamination of membrane fractions by endoplasmic reticulum (ER) because no marker for this fraction was established in smooth muscle, also contamination by fragments mitochondrial membranes was not examined. Markers for ER have now been established; the full data will be published elsewhere**. In addition, lectin binding to plasma membrane (PM) and binding of ^{3}H-oxytocin to PM have been used to further characterize this fraction in myometrium. Some contamination of the fractions of rat arterial membranes by others from fat cells in the mesentery between blood vessels has also been eliminated.

METHODS

The membrane preparations of myometrium[1] and the mesenteric artery[2] have previously been described, and the same methods continue to be used with the following modifications: For both smooth muscles 0.25 M sucrose ±

---

* Supported by Medical Research Council of Canada and Ontario Heart Foundation

** Matlib *et al.* (in preparation)

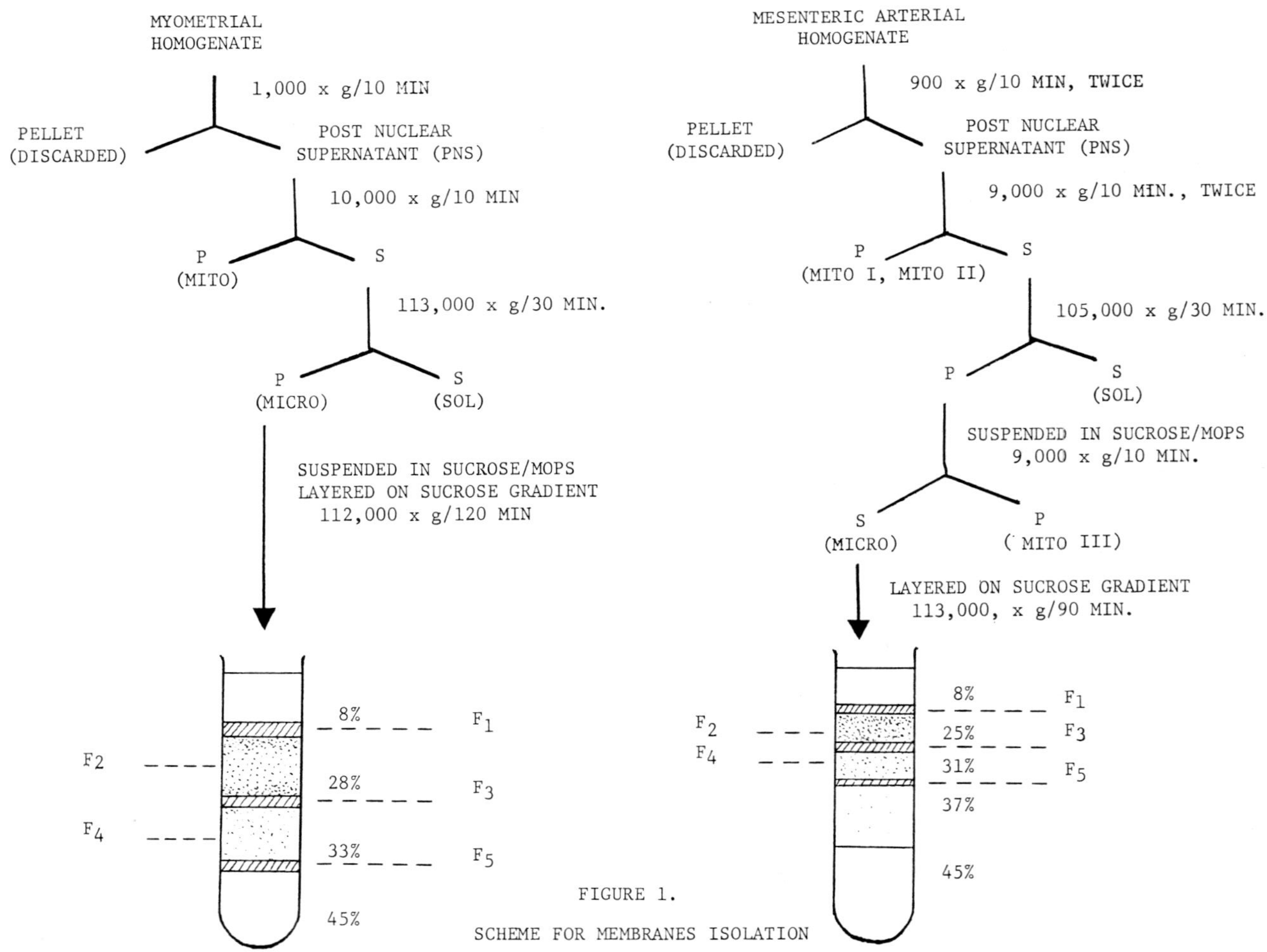

FIGURE 1.

SCHEME FOR MEMBRANES ISOLATION

with 10-25 mM MOPS has been substituted for sucrose ± histidine. The isolation schemes now used are shown in Figure 1. As markers for PM, 5'-nucleotidase[3], alkaline phosphatase[4] as well as binding of ^{3}H-wheat germ agglutinin[5] (WGA) and of ^{3}H-oxytocin[6] were studied. The lectin was bound to tissues incubated in glucose-free Krebs solution prior to homogenization; the oxytocin was bound to isolated fractions. For ER, we used distribution of ^{3}H-leucine incorporation into proteins after 2 minutes[7] and NADPH-cytochrome c reductase[8]. For mitochondrial inner membrane, we used cytochrome c oxidase[9] and succinate-cytochrome c reductase[10]; for outer membrane of mitochondria we tried to use NADH-cytochrome c reductase[8] (rotenone-insensitive). Ca^{++} uptake was measured by Millipore filtration as described previously[1,2]. Protein was determined according to Lowry et al.[11].

RESULTS

Myometrium membrane fractions and the distribution of marker enzyme activities are shown in Fig. 2. Figure 2A shows that 5'-nucleotidase activity, oxytocin and WGA binding go in parallel in different fractions. Our unpublished[6] studies have established that oxytocin binding occurs at its receptors in the PM. The present data clearly establish that 5'-nucleotidase activity is an acceptable marker for PM in smooth muscle and that this membrane is greatly enriched in fractions F1 and F2 and is present in lesser amounts in other fractions.

Figure 2B shows the distribution of ^{3}H-leucine incorporation into proteins as well as NADPH-cytochrome c reductase activity. In comparison it shows that leucine incorporation usually parallels the distribution of NADPH-cytochrome c reductase in membrane fractions. These data and others suggest that NADPH-cytochrome c reductase is an adequate marker for ER in this tissue.

Figure 2C shows the distribution of mitochondrial marker enzymes activity. Mitochondria and fraction F4 contain the highest activity of succinate-cytochrome c reductase activity. Rotenone-insensitive

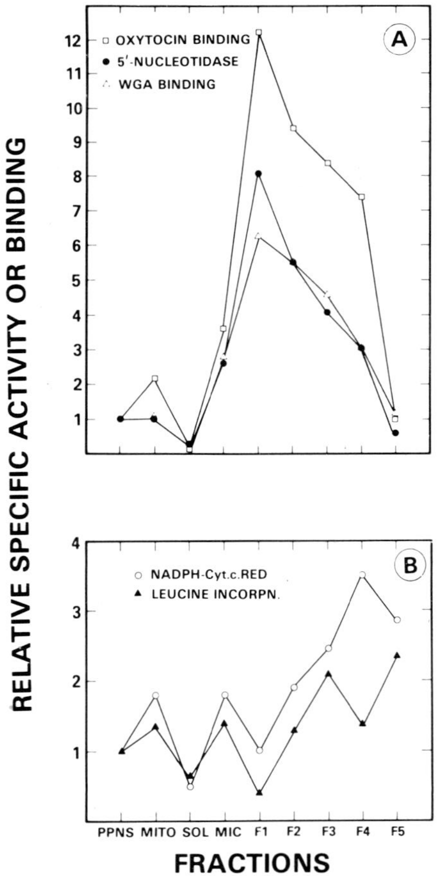

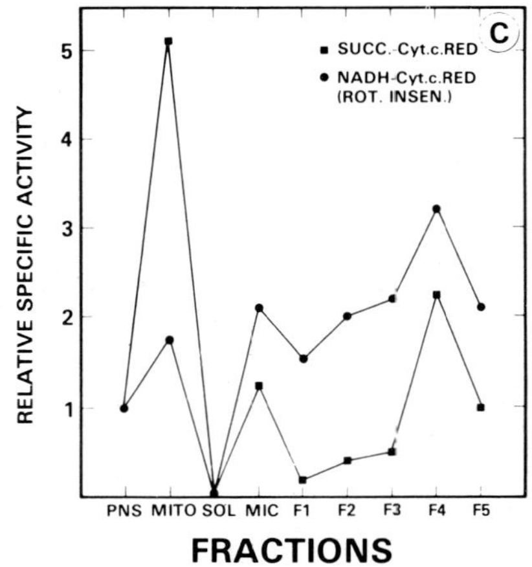

DISTRIBUTION AND YIELD OF PROTEINS IN FRACTIONS

(D)

FRACTIONS	TOTAL PROTEIN (mg)	PROTEIN YIELD mg/g Tissue
PNS	98.90	31.90
MITO	12.10	3.90
MIC	14.60	4.71
SOL	67.40	21.74
F1	1.85	0.59
F2	1.40	0.45
F3	0.62	0.20
F4	0.50	0.16
F5	6.90	2.22

FIG. 2 DISTRIBUTION OF MARKERS AND PROTEINS IN DIFFERENT FRACTIONS OF RAT MYOMETRIUM

NADH-cytochrome c reductase, being present in both mitochondrial outer membrane and ER, may also be present in PM. Fraction F1, contained about 70-80% PM, had very little mitochondrial inner membranes, but did contain some ER (about 20%). Fractions F2 and F3 contained appreciable PM (about 50%), but also had considerable ER (about 50%). Fraction F5 had ER (about 40%), considerable mitochondrial inner membranes and probably some contractile proteins, with very little PM. Fraction 4 was similar to fraction 5, but contained more PM and ER. These results fully justify our contention[1] that F1 is mainly PM and that PM carries ATP-dependent Ca^{2++}-transport (20-25 umoles Ca^{2++}/g protein/10 min).

ARTERY

We discovered that an appreciable part of the membrane fractions we had been studying derived from fat and other cells in the mesentery between arteries. Results of one separate analysis of these components are given in Table 1.

TABLE 1

PROPERTIES OF ISOLATED ARTERIES AND OTHER MESENTERY CELLS*

Properties	Arteries	Mesentery Cells
Wet weight (gm)	2.6	1.2
Total protein of PNS (mg)	18.9	8.6
Total protein of F_1 (mg)	0.7	0.2
5'-nucleotidase of F_1 (u moles/mg/hr)	50.0	238.4
Alkaline phosphatase of F_1 (u moles/mg/hr)	38.8	5.2
Ca^{2+} uptake of F_1 (u moles/gm/10 min.**		
Passive uptake	2.0	3.6
Active uptake	12.2	10.0

* From 12 Wistar male rats

** Free Ca^{2+} concentration in the reaction medium was 17 uM. Passive uptake - in the absence of ATP; active uptake - in the presence of 5 mM ATP.

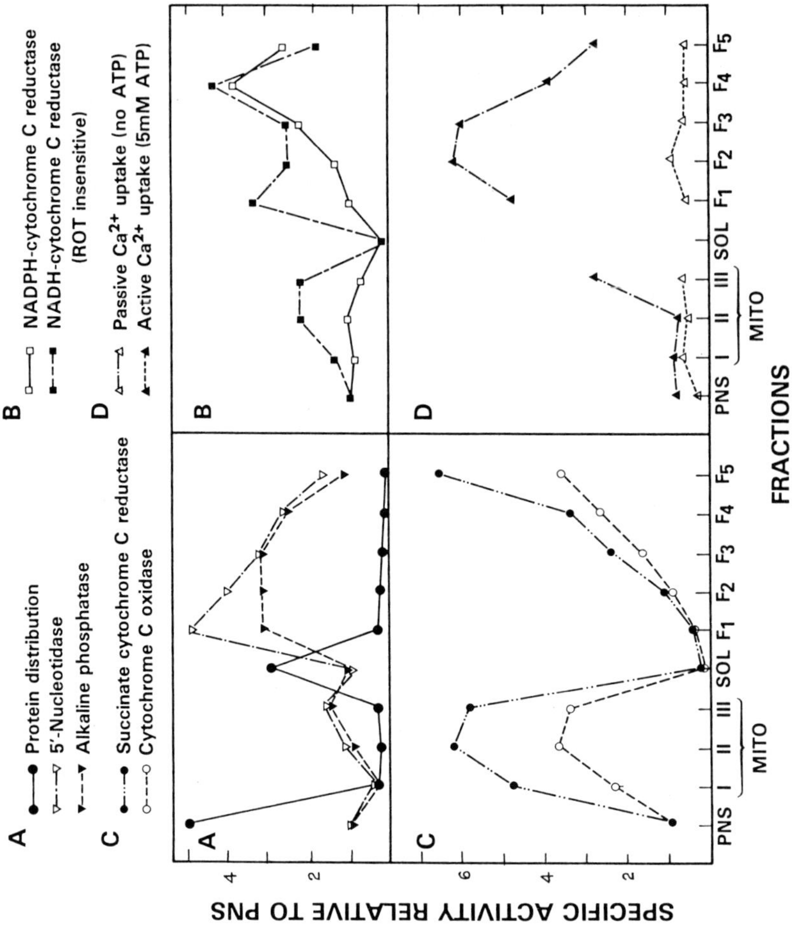

FIG. 3 ENZYME DISTRIBUTION AND CALCIUM UPTAKE OF ARTERIAL MEMBRANES

Obviously, the mesentery cells contribute up to 20-30% of the total protein. The specific activities of 5'-nucleotidase and alkaline phosphatase in these components are different, while their Ca^{2+} uptake capacities are quite comparable. The distribution of enzymes and Ca^{2+} uptake in fractions free of mesentery cells is shown in Fig. 3. Fraction 1 contains mostly PM (70-80%), some ER (20%). The amount of cytochrome c oxidase and succinate-cytochrome c reductase (inner membrane) are small. The substantial amounts of NADH-cytochrome c reductase (rotenone insensitive) seems too large to reflect simply outer membrane of mitochondria plus ER. This enzyme may be a component of smooth muscle PM, but further study is required. Clearly, F_1 contains a capacity for ATP dependent Ca^{2+} accumulation; a portion of this as shown previously is prevented by Ca^{2+} ionophores such as X-537A or A23187. This is in contrast to myometrium in which virtually all was prevented. Presumably substantial ATP-dependent Ca^{2+} binding to PM sites as well as Ca^{2+} transport occur. F_2 is similar to F_1 but has slightly more ER and mitochondrial membrane. F_4 contains mostly ER, but appreciable quantities of PM and mitochondrial membrane.

DISCUSSION

These studies clearly establish 5'-nucleotidase as a marker for smooth muscle PM, strongly suggest the validity of NADPH-cytochrome c reductase as an ER marker and raise the question of whether rotenone-insensitive NADH-cytochrome c reductase is present throughout all smooth muscle membrane. These studies also show conclusively that PM is capable of accumulating Ca^{2+} by transporting and binding it. No possible contamination of PM by another fraction could account for its Ca^{2+} accumulation. ER and mitochondria also accumulate Ca^{2+}, but to a lesser extent. These data support our previous analysis showing that Ca^{2+} transport and binding by PM play a major role in regulating intracellular levels of Ca^{2+}. We are now in a position to examine further mechanisms of Ca^{2+} binding and transport as well as their changes under physiological and pathophysiological conditions.

SUMMARY

Membranes of myometrium and mesenteric artery have been separated and characterized by marker enzymes, receptor binding and lectin binding, 5'-nucleotidase has been established as a marker enzyme for plasma membrane. NADPH-cytochrome *c* reductase for endoplasmic reticulum and other markers exist for mitochondria. We still require a suitable marker for outer membrane of mitochondria. The purity of plasma membrane fraction is sufficient to establish the possible role of this membrane in Ca^{2+} binding and transport.

REFERENCES

1. Janis, R.A., Crankshaw, D.J. and Daniel, E.E. (1977). Amer. J. Physiol. 232: C50-C58.

2. Wei, J.W., Janis, R.A. and Daniel, E.E. (1976). Blood Vessels 13: 279-292.

3. Song, C.S. and Bodansky, O. (1967). J. Biol. Chem. 242: 694-699.

4. Bessey, O.A., Lowry, O.H. and Brock, M.J. (1946). J. Biol. Chem. 164: 321-329.

5. Garfield, R.E. and Somlyo, A.P. (1977). Exptl. Cell Res. (in press).

6. Crankshaw, D. *et al*. (in preparation).

7. Kidwai, A.M., Radcliffe, M.A., Lee, E.Y. and Daniel, E.E. (1973). Biochim. Biophys. Acta. 298: 593-607.

8. Sottocasa, G.L., Kuylenstierna, B., Ernster, L. and Bergstrand, A. (1967). J. Cell Biol. 32: 415-438.

9. Cooperstein, S.J. and Lazarow, A. (1951). J. Biol. Chem. 189: 665-670.

10. King, T.E. (1967). Methods in Enzymology, Vol. 10, pp. 216-217. Academic Press, New York.

11. Lowry, O.H., Rosebrough, N.J., Farr, A.L. and Randall, R.J. (1951). J. Biol. Chem. 193: 265-275.

Excitation-Contraction Coupling in Smooth Muscle, R. Casteels et al. eds.

Ca^{2+}-BINDING, CYCLIC NUCLEOTIDE METABOLISM AND MEMBRANE PHOSPHORYLATION OF SUBCELLULAR STRUCTURES FROM RABBIT INTESTINE

Karin B. Nilsson, Rolf G.G. Andersson, Sverker Börjesson, Ludmila Mackerlova,
Ella Mohme-Lundholm and Lennart Lundholm
Department of Pharmacology, Linköping University
Sweden

INTRODUCTION

Contraction and relaxation of smooth muscle is often associated with changes of the level of cyclic nucleotides of the intact muscle but the role of these changes for the muscle function is under discussion[1,2,3,4,5,6]. It has been suggested that relaxation caused by β-adrenoceptor agonists and phosphodiesterase (PDE) inhibiting drugs is mediated by an increased level of cyclic AMP[1,2,3,4], whereas contraction following stimulation of cholinergic receptors is mediated either by a reduced level of cyclic AMP[1] and/or an increased level of cyclic GMP[7]. Low concentrations of isoproterenol can relax rat uterus with no or only slight increase of the cyclic AMP level[8], however. The relaxing action of nitroglycerin in the same muscle was associated with a marked increase of the cyclic GMP level[6]. Contraction of smooth muscle by carbachol and other drugs is also sometimes associated with a marked increase of the cyclic AMP level[1]. Some investigators have therefore suggested that changes of cyclic nucleotides do not necessarily mediate contraction and relaxation in smooth muscle[5,6]. An alternative explanation is that there exist different pools of cyclic nucleotides with alternative functions[2] in smooth muscle.

The contraction-relaxation cycle in smooth muscle is regulated by changes of the concentration of free Ca^{2+} in the myoplasm[9]. An influence of cyclic nucleotides on the contraction-relaxation cycle will therefore probably be associated with changes in the myoplasmic Ca^{2+} concentration[1,2]. This concentration can principally be influenced in two ways, either by an action on influx or efflux of Ca^{2+} through the plasma membrane (A) or by translocation of Ca^{2+} within the cell (B)[10]. Subcellular structures in the smooth muscle cell which are able to bind or release Ca^{2+} are[11]: 1. the plasma membrane including the invaginations of the membrane called "surface vesicles"; 2. the sarcoplasmic reticulum; 3. the mitochondria. Drugs may preferentially influence the Ca^{2+} metabolism and the contraction-relaxation cycle either via the A or via the B mechanism[10]. The structures listed under B may be of varying importance for Ca^{2+} translocation in different kinds of muscle or following different contracting or relaxing drugs[12]. The existence of several pools of calcium with different localization and regulatory mechanisms in the smooth muscle cell is therefore probable, even if more conclusive evidence is rather scanty. Below we will present some findings which indicate the existence of several Ca^{2+}-pools in rabbit colon muscle. They are influenced differentially by cyclic nucleotides and by some drugs. We also present some studies of the effect of cyclic AMP on the phosphorylation of membrane proteins and its possible significance for Ca^{2+} binding.

MATERIALS AND METHODS

Subcellular fractions from rabbit colon muscle were prepared as described earlier[11]. The tissue was homogenized in a solution of 0.3 M sucrose, 0.002 M ascorbic acid and 0.02 M Tris HCl, pH 7.2. After differential centrifugation of the homogenate, the pellet sedimenting between 17.300 x g for 20 min and 40.000 x g for 90 min was resuspended in 0.08 M NaCl and 0.005 M Na_2-oxalate pH 7.0. It was placed on a sucrose density gradient consisting of 35 %, 45 % and 55 % sucrose layers and centrifuged at 50.000 x g for 2 hrs. Two protein fractions were obtained, one from the 35 % sucrose layer, and another between the 35 and 45 % sucrose layers. The biochemical and morphological characteristics of these two fractions have been studied[11]. Ca^{2+}-binding and -release were studied with the Millipore filtration technique. Adenylate cyclase, PDE and ATPase activities were determined as described by Nilsson[11]. Protein kinase from the colon homogenate was partially purified by DEAE-cellulose chromatography. The enzyme was eluted with a linear gradient (0-0.4 M) of NaCl (Fig. 4). The fractions were assayed for protein kinase as described by Corbin et al.[13] in the absence and presence of 2 μM cyclic AMP. Histone or microsomal proteins were used as substrates. Other biochemical methods have been used as described by Nilsson[11].

RESULTS

Ca^{2+}-binding and -release and cyclic nucleotides in different subcellular fractions.

The "plasma membrane" (35-45 % sucrose) fraction contained morphological fragments of the sarcolemma and some damaged mitochondria but only a few vesicles from the sarcoplasmic reticulum. Biochemically the cholesterol/phospholipid ratio was high in this fraction whereas the ability to incorporate [^{3}H]-leucine into proteins was absent. [^{3}H]-ouabain was bound to a moderate degree and cytochrome c oxidase activity was present[11] (Fig. 1). The fraction contained adenylate cyclase, guanylate cyclase and phosphodiesterase activity. It contained about 20 nmol/mg protein of bound Ca^{2+} [14]. The Ca^{2+} binding was increased in the presence of 0.35 mM ATP with about 5 nmol/mg protein. Most of the binding took place within the first 2-3 min (Fig. 2). The binding was inhibited by 3 mM salyrgan but not influenced by 5 mM sodium azide or by cold storage for 24 h. Cyclic AMP (1×10^{-5}M) stimulated the Ca^{2+} binding in the presence of 0.35 mM ATP (Fig. 2).

Isoproterenol (2×10^{-6}M) stimulated both Ca^{2+} binding and formation of cyclic AMP in the presence of 0.35 mM ATP (Fig. 3). In the presence of 5 mM ATP and especially when combined with 5 mM sodium oxalate the Ca^{2+} binding was much higher

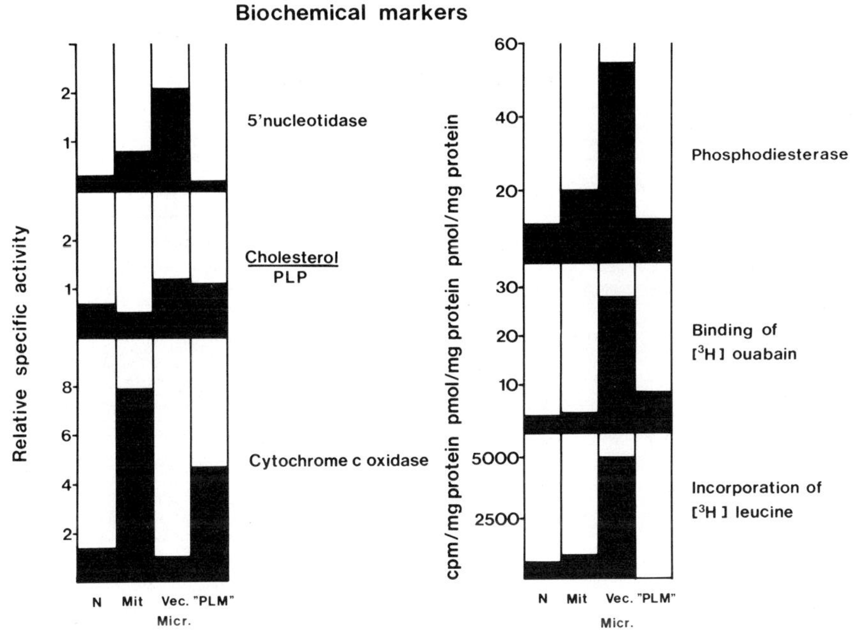

Fig. 1. The distribution of some biochemical markers in subcellular structures of rabbit colon smooth muscle, i.e. from left to right: nuclei (N), mitochondria (Mit), vesicular microsomes (Vec. Micr.) and "plasma membrane" ("PLM"). The data represent the mean of 4-8 different experiments $\pm$S.E.M.

and more prolonged[11]. The formation of cyclic AMP was also much higher at 5 mM ATP (Fig. 3) and probably maximal as the cyclic AMP formation was not further stimulated by isoproterenol (Fig. 3). Addition of Ca^{2+} did not stimulate the basal ATP hydrolysis but the increase of Ca^{2+} binding following isoproterenol or cyclic AMP was associated with an increased ATPase activity[11]. All action of isoproterenol was inhibited by a β-adrenoceptor agonist sotalol. Papaverine markedly inhibited the PDE-activity of the fraction and stimulated the Ca^{2+}-binding[11]. Acetylcholine or carbachol reduced the adenylate cyclase activity by about 40 % and released Ca^{2+} from the fraction preloaded with Ca^{2+}. Both effects were inhibited by atropine. The Ca^{2+} release was also partly counteracted when cyclic AMP (1×10^{-5}M) was added. Imidazole stimulated the PDE-activity and released Ca^{2+} from the 35-45 % fraction preloaded with Ca^{2+} [11]. Cyclic GMP (1×10^{-5}M) released Ca^{2+} from the fraction preloaded with Ca^{2+}. The nucleotide was, however, most potent in inhibiting the stimulating action of cyclic AMP on Ca^{2+} binding (with up to 55 %). Already in a concentration of 10^{-7}M did cyclic GMP reduce the effect of cyclic AMP (1×10^{-5}M) by 30 %[15].

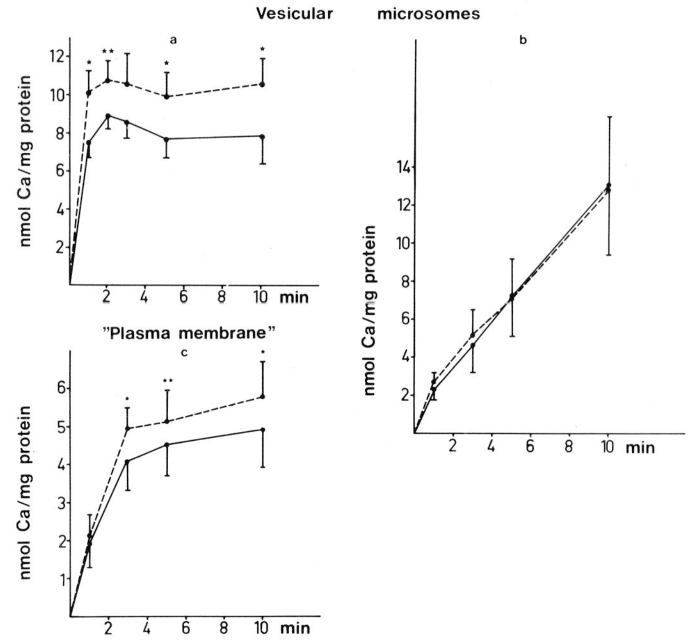

Fig. 2. The effect of 10^{-5} cyclic AMP (·---·) on the Ca binding of the vesicular microsomes and "plasma membrane" fractions in the presence of 0.35 mM Mg ATP (a,c) and in the presence of 5 mM Mg ATP (b). Mean ± S.E.M. (n = 5-8). The effect of cyclic AMP was significantly separated from the control curve (·——·) * = $p < 0.05$; ** = $p < 0.01$.

Morphologically the vesicular 35 % sucrose fraction consisted of two populations of vesicles with diameters of 400-900 A and 1000-2000 A respectively. To some of the vesicles ribosomes were attached. Biochemically the fraction had both a high potency to bind [^{3}H]-ouabain and incorporate [^{3}H]-leucine in proteins but had almost no cytochrome c oxidase activity (Fig. 1). The ATPase activity in the presence of Mg^{2+}, Na^+ and K^+ was high. The fraction probably consisted mostly of surface vesicles derived from the plasma membrane and vesicles from the sarcoplasmic reticulum[11].The vesicular fraction showed both similarities and dissimilarities in comparison with the plasma membrane fraction with regard to Ca^{2+} binding and cyclic nucleotide metabolism. It contained adenylate- and guanylate cyclase and phosphodiesterase activity. The calcium content was about 20 nmol/mg protein[14]. In the presence of 0.35 mM ATP exogenous cyclic AMP (1 x 10^{-5}M) stimulated the Ca^{2+} binding (Fig. 2).

In the presence of 5 mM ATP when both the intrinsic cyclic AMP formation and Ca^{2+} binding were higher (Fig. 3), exogenous cyclic AMP had no effect on Ca^{2+} binding (Fig. 2). Ca^{2+} binding was associated with a stimulation of the ATPase activity. Isoproterenol had no effect on Ca^{2+} binding or cyclic AMP formation in the vesicular fraction (Fig. 3) and the action of carbachol on these parameters were weaker and of shorter duration than in the 35-45 % fraction[11]. In a concentration of 10^{-7} - 10^{-5}M cyclic GMP reduced the stimulating effect of cyclic AMP (1 x 10^{-5}M) on Ca^{2+} binding by 25-30 %. Cyclic GMP did not release Ca^{2+} from the Ca^{2+}-preloaded vesicular fraction (in contrast to its effect on the 35-45 % fraction)[15].

<u>The mitochondrial fraction</u> consisted, to about 60 %, of whole or somewhat damaged mitochondria. It also consisted a "clumpy material" probably of nuclear origin and some vesicles. The cytochrome c oxidase activity of this fraction was high and the cholesterol/phospholipid quotient was low. It bound little [^{3}H]-ouabain and incorporated little [^{3}H]-leucine in proteins, indicating low contaminations with elements from the plasma membrane or endoplasmic reticulum[11]. The fraction bound Ca^{2+} in the presence of 0.35 mM ATP. The Ca^{2+}-binding capacity was reduced by sodium azide and after cold storage for 24 h (in contrast to the microsomal fraction). Cyclic AMP (1 x 10^{-5}M) did not influence the Ca^{2+} binding of the mitochondrial fraction. Adenylate cyclase and PDE activity was present in the crude "mitochondrial" fraction. The formation of cyclic AMP in this fraction was not influenced by isoproterenol but was increased by carbachol by 300 % which was in contrast to the decreasing action of cholinergic drugs in the 35-45 % fraction.

<u>Effect of cyclic AMP on phosphorylation of histones and microsomal proteins catalyzed by protein kinases.</u>

The cyclic AMP-dependent protein kinases are mediators of many of the effects of cyclic AMP in several tissues and have been suggested to be involved in the effect of cyclic AMP on Ca^{2+} binding[16,17]. An increased protein kinase activity in response to smooth muscle relaxing drugs such as isoproterenol and 1-methyl-3-isobutylxanthine was observed in the intact rabbit colon muscle by us. It was also observed that protein kinase from skeletal muscle stimulated the Ca^{2+} binding of microsomal fractions from rabbit colon[14].

When protein kinase from colon muscle was partially purified by chromatography on a DEAE-cellulose column two peaks of cyclic AMP-dependent activity were present. The first peak (peak I) was eluted at 0.09 M NaCl, while the main peak (peak II) was eluted at approximately 0.2 M NaCl (Fig. 4).

The two different peaks were concentrated and their phosphorylating activity was tested on the membrane proteins of the Ca^{2+}-binding microsomal fractions from colon muscle. The phosphorylation of the colon microsomal subfractions was catalyzed by the peak II enzyme in the presence of 10^{-5}M cyclic AMP (Fig. 5). The enzyme

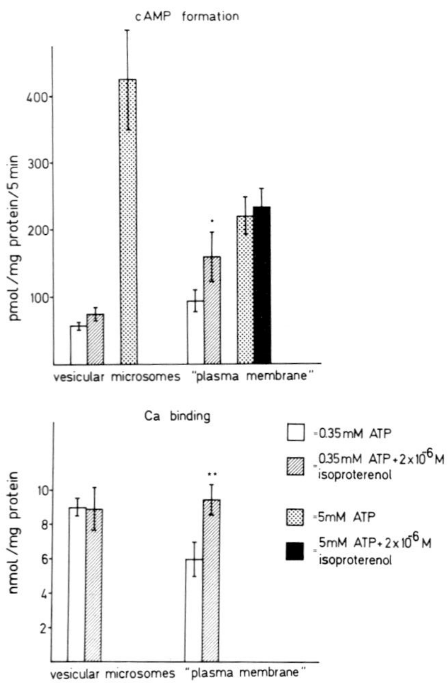

Fig. 3. Cyclic AMP formation and Ca^{2+} binding of vesicular microsomes and a "plasma membrane" fraction in the absence and presence of isoproterenol (2×10^{-6}M). Mean $\pm$ S.E.M. (n = 5-9). * = $p < 0.05$; ** = $p < 0.01$.

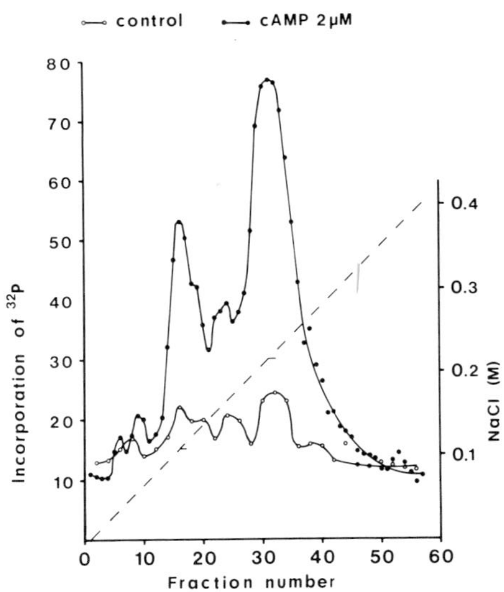

Fig. 4. Separation of cyclic AMP dependent protein kinases from crude extracts of rabbit colon by chromatography on a DEAE-cellulose column.

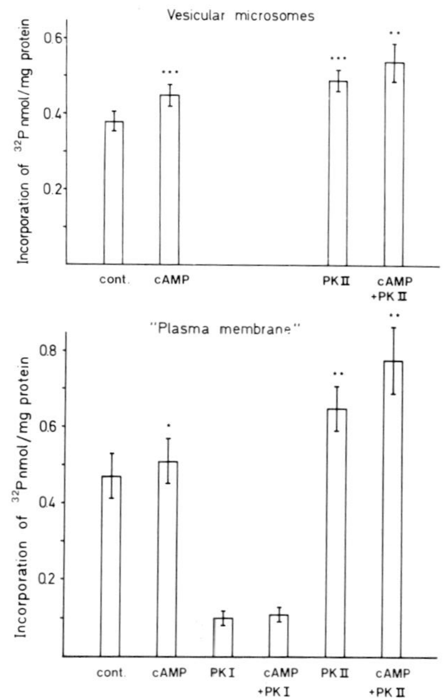

Fig. 5. Phosphorylation of vesicular microsomes (92 μg) and "plasma membrane" proteins (56 μg) by protein kinases (PK I and PK II, 40 μg) in the absence and presence of 10^{-5}M cyclic AMP. Mean $\pm$ S.E.M. (n = 3-7). * = $p < 0.05$; ** = $p < 0.01$; *** = $p < 0.001$.

preparation obtained from peak I did not catalyze the phosphorylation of the microsomal proteins, instead it inhibited the phosphorylation of the microsomes. Cyclic AMP was not able to counteract this effect (Fig. 5). The properties of the enzyme of peak I thus differed markedly if histone or microsomal proteins were used as substrates. The phosphorylation process of microsomal proteins was also investigated with the ESCA-technique[18] (electron spectroscopy for chemical analyses). By this method it was shown that phosphate groups were incorporated into the microsomal proteins. The increase of the phosphate groups was demonstrated even after washing the microsomes. An increased binding of Ca^{2+} to the microsomal membranes was also demonstrated with the ESCA-technique in parallel with the incorporation of the phosphate groups. Studies on the two peaks of protein kinases are in progress to correlate their effects on phosphorylation of the microsomal proteins with that on Ca^{2+} binding.

DISCUSSION

Formation of cyclic nucleotides and Ca^{2+} binding.

The effects of the cyclic nucleotides on Ca^{2+} binding as the influence of drugs on the formation of cyclic AMP was different in the subcellular fractions. In both the "plasma membrane" and "vesicular" fractions cyclic AMP stimulated the Ca^{2+} binding but had no effect on the Ca^{2+} binding of the mitochondrial fraction. There were several indications of a close association between Ca^{2+} binding and cyclic AMP formation especially in the "plasma membrane" fraction both under basal conditions in presence of ATP (different ATP concentrations) and under the influence of drugs (isoproterenol, cholinergic drugs, papaverine, imidazole). In both the microsomal fractions cyclic GMP counteracted the effect of cyclic AMP on Ca^{2+} binding. In the 35-45 % fraction cyclic GMP also released Ca^{2+} from the fraction preloaded with Ca^{2+}.

Regarding factors influencing the cyclic AMP formation isoproterenol only increased the formation in the 35-45 % fraction but not in the 35 % microsomal or the mitochondrial fraction. Cholinergic drugs reduced cyclic AMP formation in the 35-45 % fraction but increased it in a crude mitochondrial fraction. Regarding the formation of cyclic GMP only Ca^{2+} in a rather high concentration was found to stimulate the formation in the 100.000 x g supernatant fraction[15].

These results indicate that determination of the total changes of the cyclic AMP content in the whole muscle probably gives limited information of its functional significance since the formation may either take place in a compartment where the Ca^{2+} metabolism is influenced (plasma membrane, sarcoplasmic reticulum) or where it has no effect (mitochondrial fraction). A drug as isoproterenol may stimulate the cyclic AMP-formation in only one compartment essential for Ca^{2+} binding while cholinergic drugs may have quite contrasting effects on cyclic AMP formation in compartments where Ca^{2+}-binding is or is not influenced by the nucleotide.

Ca^{2+} binding and phosphorylation of microsomal membranes.

Cyclic AMP increased the Ca^{2+}-binding capacity of microsomal proteins and a maximum was reached within 2-4 min in the presence of 0.35 mM ATP. The nucleotide did not stimulate the more marked and prolonged uptake of Ca^{2+} which occurred in the presence of 5 mM ATP + sodium oxalate. This indicates that cyclic AMP probably increased the binding sites for Ca^{2+} in the microsomes rather than stimulated a Ca^{2+} pump of the microsomal vesicles. As a working hypothesis we suggest that cyclic AMP activates protein kinase and thereby phosphorylate microsomal proteins. The number of phosphate groups able to bind Ca^{2+} is thus increased. The action of the peak I protein kinase indicates, however, the presence of a factor in the microsomal fraction that was able to reduce the extent of phosphorylation. This factor may be a phosphoprotein phosphatase.

It has been reported that prostaglandins are able to activate a phosphoprotein phosphatase[19]. The possibility has to be considered that some contracting drugs stimulate the release of Ca^{2+} by reducing the degree of phosphorylation of the microsomal proteins. This effect may either depend on a decrease of the cyclic AMP level and a reduction of the protein kinase activity or an activation of a phosphoprotein phosphatase.

REFERENCES

1. Andersson, R., Nilsson, K., Wikberg, J., Johansson, S., Mohme-Lundholm, E. and Lundholm, L. (1975) Adv.in Cycl.Nucl.Res.(Drummond, G.I., Greengard, P. and Robison, G.A., eds.) Vol. 5, pp. 491, Raven Press, N.Y.
2. Andersson, R.G.G. and Nilsson, K.B. (1977) Biochemistry of Smooth Muscle (Stephens, N., ed.) p. 263, University Park Press, Baltimore.
3. Bär, H.B. (1974) Adv. in Cycl.Nucl.Res., Vol. 4, p. 195, Raven Press, N.Y.
4. Vulliemoz, Y., Verosky, M. and Triner, L. (1977) Biochemistry of Smooth Muscle (Stephens, N., ed.) p. 293, University Park Press, Baltimore.
5. Batra, S.C. and Daniel, E.E. (1971) Comp.Biochem.Physiol. 38A:285.
6. Diamond, J. (1977) Biochemistry of Smooth Muscle (Stephens, N., ed.) p. 343, University Park Press, Baltimore.
7. Lee, T.P., Kuo, J.F. and Greengard, P. (1972) Proc.Nat.Acad.Sci. 69:3287.
8. Nesheim, B.I., Osnes, J.B. and Øye, I. (1975) Br.J.Pharmacol. 53:403.
9. Rüegg, J.C. (1971) Physiol. Rev. 51:201.
10. Hurwitz, L. and Suria, A. (1971) Annu. Rev. Pharmacol. 11:303.
11. Nilsson, K.B. (1977) Linköping University Med.Diss. 49:1.
12. Golenhofen, K. (1975) The Smooth Muscle of the Artery, pp. 46, Plenum Press, N.Y.
13. Corbin, J.D., Soderling, T.R. and Park, C.P. (1973) J.Biol.Chem. 248:1813.

14. Andersson, R.G.G., Djärv, L.,Nilsson, K. and Wikberg, J. (1976) Stimulus secretion coupling in the gastrointestinal tract (Case, R.M. and Goebell, H., eds.) pp. 1, Medical technical Publ. Co., Lancaster.
15. Nilsson, K.B. and Andersson, R.G.G. (1977) Acta physiol.scand. 99:246.
16. Alexander, R.N., Fontana, J.A. and Lovenberg, W. (1973) Fed. Proceed. 32:711.
17. Sands, H., Meyer, T.A. and Rickenberg (1973) Biochim.Biophys.Acta 302:267.
18. Andersson, R.G.G., Nilsson, K.B., Magnusson, K.-E. and Johansson, L. (1976) Fifth Int.Congr. Histochemistry and Cytochemistry, p. 22, Bucharest.
19. Layne, P., Constantopoulos, A., Judge, J.F.X., Rauner, R. and Najjar, V.A. (1973) Biochem. Biophys. Res. Commun. 53:800.

Financial support was provided by Swedish State Medical Research Council (04X-101-13B; 04X-2080-11B; 04X-4498-03).

Excitation-Contraction Coupling in Smooth Muscle, R. Casteels et al. eds.

SOME PROPERTIES OF CHEMICALLY SKINNED SMOOTH MUSCLE FIBERS

M. Endo, T. Kitazawa, S. Yagi, M. Iino and Y. Kakuta*
Department of Pharmacology and *First Department of Internal Medicine, Tohoku University School of Medicine, Sendai 980, Japan.

INTRODUCTION

Skinned skeletal muscle fibers, originally introduced by Natori[1], have made a substantial contribution to the elucidation of the mechanisms of muscle contraction and of its control. Skinned fibers are superior to glycerinated fibers in that (1) all the contractile and regulatory proteins are kept in the physiological state, while denaturation or removal of some of the proteins during glycerination cannot be easily excluded, and (2) the functions of the sarcoplasmic reticulum (SR) are preserved, which is usually not the case in glycerinated fibers. Similar preparations of smooth muscle, therefore, are undoubtedly very useful, but it is clear that mechanical skinning of smooth muscle cells, is very difficult because of their small size. Ohtsuki et al[2] have recently established a method of making holes in the cell membrane by using saponin. Ohtsuki further suggested us that since saponin acts on cholesterol molecules in the membrane (cf. ref. 2) and since the content of cholesterol in the SR membrane is much less than that of the surface membrane[3,4], saponin may specifically act on the surface membrane but not on the SR. We followed his suggestion and could indeed demonstrate that a certain concentration of saponin can make the surface membrane of skeletal muscle fibers freely permeable to various substances, without impairing the functions of the SR and of the contractile system[5], and that the same applies also to the cardiac muscle fibers[6,7]. We have, therefore, tried to make skinned smooth muscle fibers by using saponin[8], and examined some of the properties of the saponin-treated smooth muscle fibers. A similar effort has recently been reported by another group[9].

MATERIALS AND METHODS

Taenia caecum of the guinea-pig was isolated and small bundles with a diameter of 100-200 μm were dissected under a stereomicroscope. Alternatively, strips of guinea-pig pulmonary artery were used. These preparations were tied at each end with silk or nylon thread, and suspended in a trough through which solutions could be perfused rapidly. Isometric tension was recorded with either a straingauge transducer (U-gauge), a mechanoelectronic transducer (RCA 5734) or a photoelectric transducer by R. A. Meiss[10]. Experiments were conducted at 25°C, by circulating water at the temperature underneath the trough.

Normal external solution contained 150 mM NaCl, 4 mM KCl, 2 mM Ca-methanesulfonate, 5 mM N-2-hydroxyethyl-piperazine-N'-2-ethanesulfonic acid (HEPES) neutralized with Tris(hydroxymethyl)aminomethane (pH 7.4) and 5.6 mM glucose. High potassium solution was obtained by replacing all of NaCl in the normal solution with KCl and adding 44 mM sucrose to prevent swelling[11]. The composition of normal relaxing solution was 115 mM K-methanesulfonate, 4 mM Mg-methanesulfonate, 4 mM ATP-Na_2, 2 mM ethyleneglycol-bis-(β-amino ethylether)-N-N'-tetraacetic acid (EGTA), 20 mM piperazine-N-N'-bis [2-ethanesulfonic acid] (PIPES), brought to pH 6.8 with KOH. The concentration of EGTA was altered when necessary. In activating solutions, 10-20 mM total EGTA was used and a specified amount of Ca-methanesulfonate was added. Free Ca ion concentration was calculated by assuming an apparent association constant of CaEGTA, $5 \times 10^5 M^{-1}$[12]. In some cases, concentrations of free Mg ion and/or MgATP in relaxing and activating solutions were altered by changing concentrations of Mg-methanesulfonate and ATP-Na_2. The levels of free Mg and other constituents were calculated by assuming an association constant of MgATP 10^4 M^{-1}. In all of the above alterations from the normal relaxing solution, ionic strength was kept constant by adjusting the concentration of K-methanesulfonate.

Saponin obtained from Merck & Co., Inc. was dissolved usually in normal relaxing solution at each time shortly before the use. Caffeine was dissolved in a suitable solution.

RESULTS

The effect of saponin on smooth muscle cell membrane : In each preparation, responses of intact smooth muscle fibers to high K solutions were first examined. The preparation was then immersed in a normal relaxing solution and treated with saponin. Fig. 1 shows time course of the effect of 50 µg/ml saponin. As is seen, 10^{-4} M Ca produced only a small and slow tension development before saponin treatment, but a rapid tension development after the treatment. The tension was completely abolished when the activating solution was replaced by a relaxing solution. Removal of Mg or ATP induced a slight but definite rigor tension after the saponin treatment, which was abolished immediately when Mg or ATP was reapplied. These results indicate that the surface membrane of the smooth muscle fiber became permeable to substances such as Ca, Mg, EGTA and ATP after the saponin treatment.

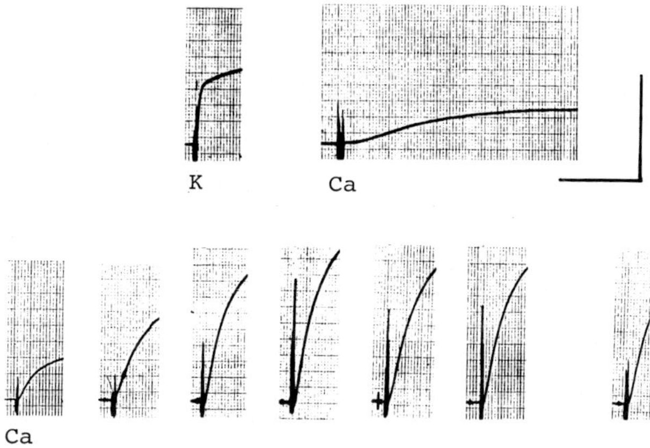

Fig. 1 Initial part of tension development of a small fiber bundle from taenia caecum. Top row : before saponin treatment. The bundle was activated with a high K solution (K) or by transferring from a relaxing solution to an activating solution containing 10^{-4} M free Ca ion (Ca). Similar solution without ATP produced a similar response. Bottom row : after saponin treatment. The bundle was repeatedly activated with the activating solution containing 10^{-4} M Ca. At each interval including that between the last in the upper row and the first in the bottom row, the bundle was made relax by applying a relaxing solution and then treated with 50 μg/ml saponin for 5 min, except the last interval where 300 μg/ml saponin was used. Calibration, tension: 20 mg, time: 1 min.

As seen in Fig. 1, with an increase in the total duration of the saponin treatment, the rate of tension development on application of Ca became faster and faster, until maximum effect was reached at total about 20 min of the treatment. The time course was similar with 100 μg/ml saponin. Therefore, we adopted 25 min treatment with 50-100 μg/ml saponin as a standard procedure of chemical skinning for taenia. It should be noted, however, that if bundles were thick, a much higher concentration of saponin was required.

In strips of pulmonary artery, it was our impression that the higher concentrations of and/or a long treatment with saponin might have a deleterious effect on the tension development of skinned smooth muscle fibers. Therefore, we tentatively adopted 4 min treatment with 500 μg/ml saponin for this preparation.

Maximum tension obtained with 10^{-4} M Ca after the saponin treatment was usually greater than the tension produced by a high K solution applied to the same bundle before the skinning (Fig. 2), suggesting that all fibers in the bundle were probably skinned.

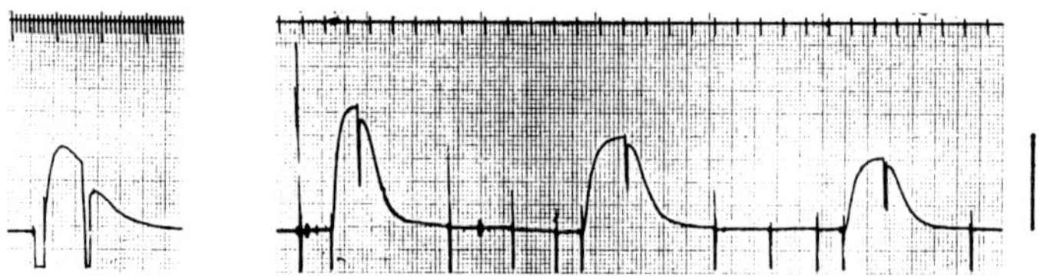

Fig. 2 Comparison between K-contracture of a bundle from taenia caecum and Ca-responses after saponin treatment. K: high K solution, Na: normal external solution, Ca: activating solution containing 10^{-4} M free Ca ion. After K-contracture, the bundle was treated with 50 μg/ml saponin for 25 min. Before each Ca contraction, the bundle was equilibrated with a relaxing solution containing 0.5 mM EGTA. After the Ca contractions it was washed with a relaxing solution containing 10 mM EGTA three times. Deterioration here was stronger than the average. Tension calibration: 30 mg, time: 1 sec per division for K-contracture and 1 min for Ca-responses.

Reversibility of responses of chemically skinned smooth muscle fibers: As shown in Figs. 1 and 2, the contractions of skinned taenia produced by a high concentration of Ca could be repeated in the same preparation at least several times with a minor deterioration. However, contractions with lower activating concentrations of Ca was more strongly depressed, especially after the preparation was activated by high Ca. As a result, the pCa-tension relation of a preparation gradually shifted to the right on repeating the determination (Fig. 3). To obtain physiologically meaningful pCa-tension

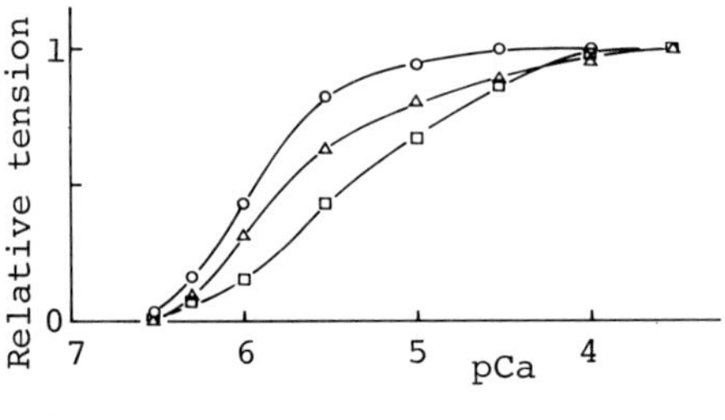

Fig. 3 Deterioration of Ca sensitivity of a skinned taenia caecum. pCa-tension relations were determined repeatedly in the same preparation. Symbols (○ , △ , □) show the first, the second and the third determination respectively.

relation, therefore, we applied activating solutions with various free Ca levels cumulatively, i.e., successively in the order of increasing concentration as in Fig. 4, and only the data of the first of these cumulative trials of each skinned preparation were accumulated in the following.

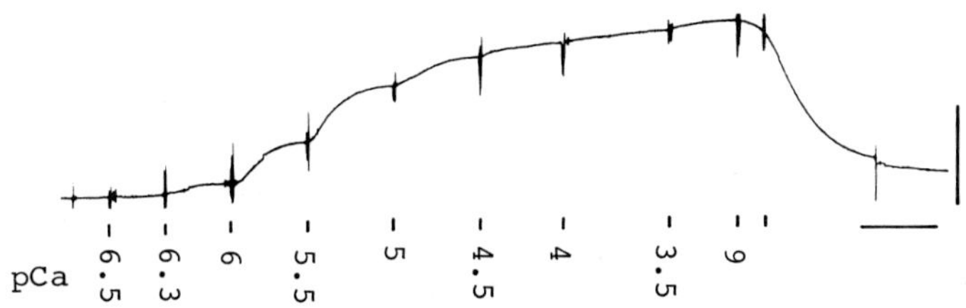

Fig. 4 An example of records used to determine pCa-tension relation of skinned smooth muscle fibers. Figures indicate pCa of the solution applied. Calibration, tension: 20 mg, time: 2 min.

The reversibility of skinned pulmonary artery preparations was worse than that of taenia. This might partly be due to an insufficient supply of ATP to the contractile system because of thickness of the preparation. It was found that when a higher concentration of ATP was used, the deterioration was slower. Therefore, 10 mM ATP, together with 8.6 mM total Mg, was regularly used for skinned pulmonary artery preparations.

The pCa-tension relation: (1) Difference between two kinds of smooth muscle. Fig. 5 shows the pCa-tension relations of skinned fibers of the two kinds of smooth muscle. In taenia caecum, small contractions were usually observed at pCa 6.5. Half maximum and maximum tension was observed at about pCa 5.6 and 4 respectively. This relation is not so different from that of skeletal and cardiac muscles. Skinned pulmonary artery, however, showed a significantly higher Ca-sensitivity: about 30% of maximum tension was obtained already at pCa 6.5 and half maximum at pCa 6.3. A higher MgATP concentration was used in pulmonary artery than in taenia but the conceivable contribution of the MgATP concentration to the high Ca sensitivity was excluded by the fact that with similarly high MgATP concentration (crosses in Fig. 5) taenia caecum still showed the lower Ca sensitivity.

(2) The absence of effect of stretch. In skinned skeletal[13] and cardiac fibers[6,14,15], stretching the muscle increased an apparent Ca sensitivity. Therefore, it is interesting to see whether or not it also applies to the smooth muscle. This has been examined and the results are shown also in Fig. 5. In taenia caecum, the length-tension relation of K-contracture of intact preparation was similar to that already reported[16]. The pCa-tension relations were, therefore, determined at the length that gave the maximum tension (open symbols), which is near the slack length, and at stretched lengths

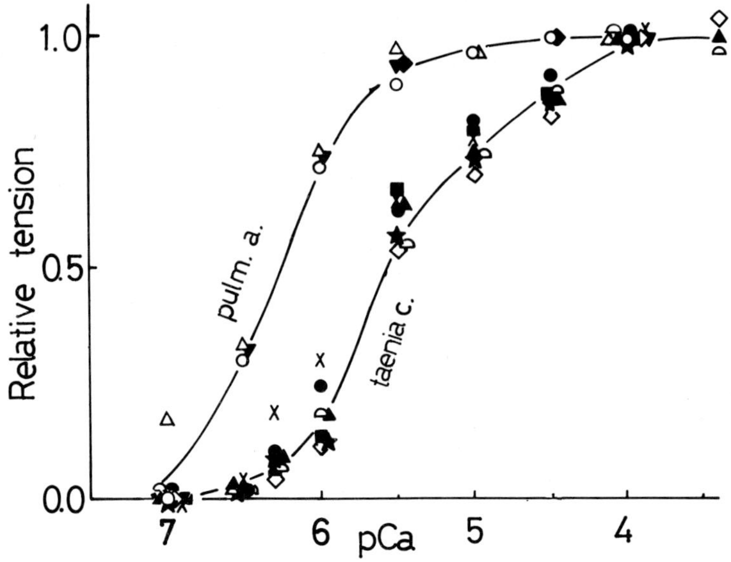

Fig. 5 pCa-tension relation of skinned taenia caecum and pulmonary artery at different lengths. Each symbol indicates a different bundle. Not all our data but only those of the series of experiments for length effect were plotted. Solutions contained 0.9 mM free Mg, 7.7 mM MgATP and 20 mM total EGTA in the case of pulmonary artery, and 0.9 mM free Mg, 3.1 mM MgATP and 10 mM total EGTA for taenia caecum, except crosses which were done in the presence of 1 mM free Mg and 8 mM MgATP. Open and filled symbols show that the relation was determined at slack and stretched length respectively. In taenia caecum, the dissection of the bundles was made either in the presence (◇,⌓,●,▲) or absence (■,★) of 10^{-6} g/ml adrenaline. Some preparations dissected in the absence of adrenaline developed maximum tension not at the slack length but at a stretched length. They were not included in the figure. Further explanation, see text.

that gave 50-60% maximum tension (filled symbols). In pulmonary artery, the pCa-tension relation was also determined at slack length (open symbols) and after about 40% stretch (filled symbols). In this preparation, peak tension was obtained at the stretched length. It is clear that in both kinds of smooth muscle skinned fibers, stretch did not affect the pCa-tension relation at all.

(3) The influence of Mg and MgATP. A change in MgATP concentration between 3 and 8 mM did not appreciably alter the pCa-tension relation of skinned taenia caecum as already described (Fig. 5). An increase in free Mg ion concentration tended to shift the pCa-tension relation slightly to the right. The effect, however, appears to be much smaller than that in skeletal and cardiac muscle[17-19].

The effect of Mg on high Ca contraction: Filo et al.[20] demonstrated that unlike skeletal muscle, glycerinated vascular smooth muscle could not contract in a low free Mg solution even if the level of free Ca was sufficiently high. This is in accord with biochemical studies in which the requirement of free Mg for contractile reactions was clearly demonstrated[21]. However, as shown in Fig. 6, skinned smooth muscle fibers developed a good tension in the presence of a low free Mg. Although the tension tended to decrease with the decrease in free Mg below 1 mM, free ATP concentration increased at the same time under these conditions. Since free ATP in this concentration range inhibits tension production at least in skeletal muscle[22], the reduction of tension below 1 mM free Mg might well be explained by the effect of free ATP.

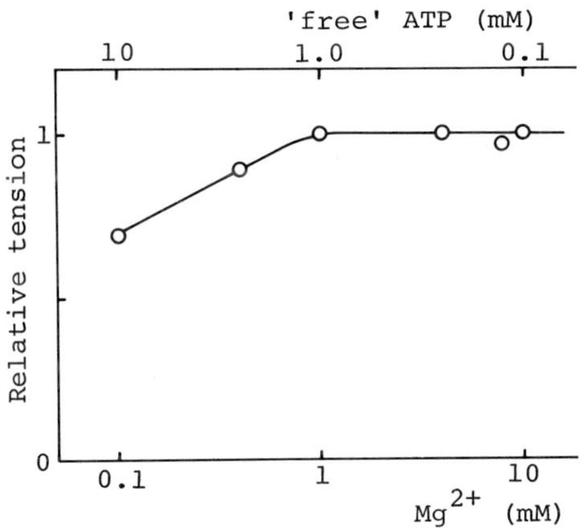

Fig. 6 Effect of changing free Mg ion concentration on tension of a skinned taenia caecum produced by 10^{-4} M Ca. The preparation started to be in an altered free Mg solution a few min before the application of Ca. Since MgATP concentration was always kept at 4 mM, the decrease in free Mg was accompanied by an increase in ATP unbound with divalent cations ('free' ATP).

Possible responses of the SR: (1) Caffeine contracture of intact fibers of taenia caecum. As shown in Fig. 7, in a thin bundle of taenia caecum, 25 mM caffeine produced a contracture almost as large as K-contracture, in the presence of 4 mM EGTA, although pretreatment with the EGTA solution slowly decreased the caffeine response. Similar caffeine contracture could be obtained in completely depolarized fibers in a high K solution, as shown at the bottom of the figure. These results suggest the presence of an appreciable Ca store, probably SR, in these smooth muscle cells under normal condition.

(2) Caffeine responses of skinned taenia caecum. In taenia caecum skinned with 50 μg/ml saponin for 25 min, a simple application of 30 mM caffeine did not cause an appreciable contraction (Fig. 8a). However, if the skinned fiber was pretreated with 10^{-6} M of free Ca

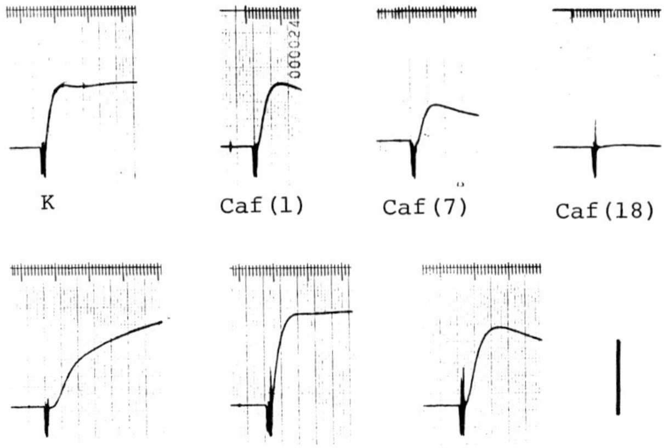

Fig. 7 Caffeine contractures of an intact fiber bundle from taenia caecum. Solutions applied are as follows. Top row K: high K solution. Caf(1), Caf(7), Caf(18): 25 mM caffeine in modified external solution containing 4 mM EGTA and no Ca. Figures in the parentheses indicate time in minute between removal of Ca by the modified external solution and application of caffeine. Between each caffeine responses the bundle was returned to and kept in normal external solution until normal K-contracture was obtained. Bottom row K(1), K(7): high K solution. Figures in the parentheses are time in minute after the bundle was returned to the external solution containing normal Ca. Caf(2, K): 25 mM caffeine in modified high K solution containing 4 mM EGTA and no Ca. The bundle was equilibrated in the Ca-depleted K solution for 2 min. Tension calibration: 50 mg, time: 1 sec per small division.

ion for several min, the same concentration of caffeine produced a transient contraction (Fig. 8 b and d). If the duration of pretreatment with Ca was short, no response was obtained by caffeine (Fig. 8c). These results suggest that the transient contractions induced by caffeine is due to a release of Ca from some store, possibly the SR, which slowly accumulates Ca during the course of several min in 10^{-6} M Ca.

The caffeine responses of pulmonary arteries have not yet been studied.

DISCUSSION

It is clear that by the use of saponin smooth muscle fibers could be skinned chemically within a short time. This is an additional advantage of the saponin method over the glycerinated fiber method, since the former allows us to make a direct comparison between intact and skinned fiber responses of the same preparation very easily. As regards the size of the holes made by saponin, Ohtsuki et al[2]

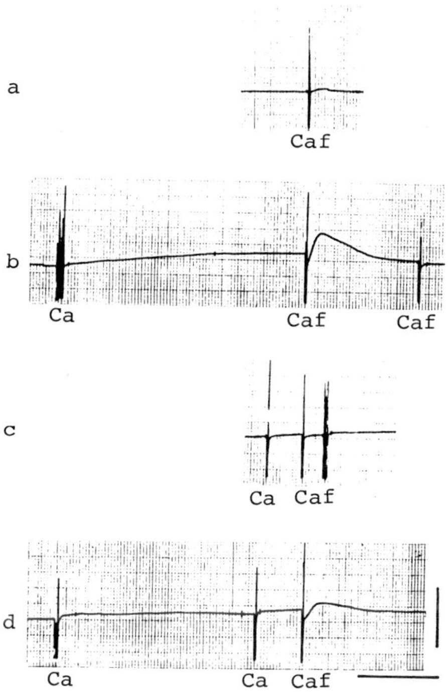

Fig. 8 Effect of caffeine on a skinned taenia caecum. Ca: an activating solution containing total 0.1 mM EGTA and 10^{-6} M free Ca. Caf: 30 mM caffeine in Ca described above. Calibration: 10 mg and 1 min. For further explanation, see text.

have shown that at least in some types of the cells, molecules as large as ferritin can freely pass through the saponin-treated membrane.

Reversibility of saponin-treated smooth muscle fibers were rather poor as far as Ca-sensitivity is concerned (Fig. 3). Clearly an improvement is necessary for the method, although some information could be obtained with confidence by using only first contractions of each preparation as done in this report. In the case of pulmonary artery, an effort to obtain thinner sheets of smooth muscle cells might largely improve the responsiveness and reversibility of the preparation.

A higher Ca-sensitivity was obtained in our pulmonary artery preparation. Similar results on pig pulmonary artery were found by using glycerinated preparation[23]. However, whether or not this is a general property of vascular smooth muscle is not clear yet. The pioneering work by Filo et al.[20] showed that Ca-sensitivity of glycerinated rabbit vascular smooth muscle was similar to that of skeletal muscle. Apparently further investigations are necessary.

In striated muscles, the longer the muscle length, the greater the apparent Ca-sensitivity of the contractile system. This was clearly demonstrated in the region of the descending limb of the length-tension curve[6,13-15], but the same tendency continues to exist in the region of the ascending limb of the curve[24]. On the other hand, a substantial change in muscle length in smooth muscle did not alter the pCa-tension relation (Fig. 5). What is implied in this difference, however, is not clear, especially since the mechanism of increase in the apparent Ca-sensitivity by stretch in striated muscles is totally unknown.

In our skinned smooth muscle fibers, more than 50% of maximum tension was developed by applying Ca in the presence of low free Mg (Fig. 6). This appears to disagree with biochemical experiments in which Mg reguirement of millimolar range for actomyosin ATPase and superprecipitation was demonstrated[21]. Recently Nonomura et al.[25] have shown that Mg ions are necessary for the formation of myosin filaments. Ohashi and Nonomura[26] have also shown that in skinned smooth muscle fibers treated with a low free Mg solution for a sufficient period, no thick filaments were observed. However, it is conceivable that disintegration of myosin filaments in skinned fibers might take time and not occur within a few minutes, a duration used in our experiments. In any case our results appears to suggest that the contractile reaction of the smooth muscle <u>per se</u> does not require a millimolar range of free Mg. The effect of long immersion in a low free Mg solution is under investigation.

One of the biggest advantages of the saponin method is possible preservation of the SR. In frog skeletal muscle fibers, 30 min treatment with 50 μg/ml saponin did not affect the function of the SR[5]. Similar treatment in guinea-pig taenia caecum also appears to preserve at least some of the functions of its SR (Fig. 8). Further studies along this line are now being undertaken in the hope that more information might be obtained about physiological role played by the smooth muscle SR.

SUMMARY

1. Chemically skinned smooth muscle fibers were successfully made by using saponin and their properties were examined.
2. Ca-sensitivity of the contractile system of taenia caecum was similar to those of striated muscles, but that of pulmonary artery was significantly higher.
3. Length change did not affect the Ca-sensitivity of smooth muscle,

unlike striated muscles.

4. Skinned smooth muscle fibers developed respectable tension even in the presence of low free Mg.
5. Possible SR responses were obtained in skinned taenia caecum.

REFERENCES

1. Natori, R., 1954, Jikei. Med. J. 1, 18-28
2. Ohtsuki, I., Palade, G.E. and Jamieson, D.J., 1977, J. Cell Biol. in press
3. Martonosi, A., 1968, Biochim. Biophys. Acta 150, 694-704
4. Waku, K., Uda, Y. and Nakazawa, Y., 1971, J. Biochem. 69, 483-491
5. Endo, M., 1976, Folia pharmacol. japon. 72, 9P-10P (in Japanese)
6. Endo, M., 1976, J. Physiol. Soc. Jap. 38, 157 (in Japanese)
7. Kitazawa, T., 1977, Jap. J. Pharmacol. in press (abstract)
8. Kitazawa, T. and Endo, M., 1977, Blood Vessels in press (abstract)
9. Saida, K. and Nonomura, Y., 1977, Folia pharmacol. japon. in press (abstract, in Japanese)
10. Meiss, R.A., 1974, J. Appl. Physiol. 37, 459-463
11. Jones, A.W., Somlyo, A.P. and Somlyo, A.V., 1973, J. Physiol. 232, 247-273
12. Ogawa, Y., 1968, J. Biochem. 64, 255-257
13. Endo, M., 1972, Nature (New Biol.) 237, 211-213
14. Winegrad, S. Personal communication
15. Fabiato, A.. Personal communication
16. Åberg, A.K.G. and Axelsson, J., 1965, Acta physiol. scand. 64, 15-27
17. Ebashi, S. and Endo, M., 1968, Progr. Biophys. Mol. Biol. 18, 123-183
18. Kerrick, W.G.L. and Donaldson, S.K.B., 1972, Biochim. Biophys. Acta 275, 117-122
19. Fabiato, A. and Fabiato, F., 1975, J. Physiol. 249, 497-517
20. Filo, R.S., Bohr, D.F. and Ruegg, J.C., 1965, Science 147, 1581-1583
21. Murphy, R. A., Bohr, D.F. and Newman, D.L., 1969, Am. J. Physiol. 217, 666-673
22. Sawada, E. and Endo, M. Unpublished observation
23. Kuriyama, H. Personal communication
24. Endo, M., 1975, Seibutsubutsuri (Biophysics) 15, 247-259 (in Japanese)
25. Nonomura, Y., Mikawa, T., Toyooka, T. and Ebashi, S., J. Mol. Biol. Submitted
26. Ohashi, M. and Nonomura, Y., J. Mol. Biol. Submitted

Excitation-Contraction Coupling in Smooth Muscle, R. Casteels et al. eds.

EGTA-STIMULATED AND OUABAIN INHIBITED ATPase OF VASCULAR SMOOTH MUSCLE

Julius C. Allen and Charles L. Seidel
Department of Medicine, Section of Cardiovascular Sciences
Baylor College of Medicine and the Fondren-Brown Cardiovascular
Research and Training Center, The Methodist Hospital
Houston, TX 77030

INTRODUCTION

The role of the sarcolemmal surface in the regulation of cytoplasmic Ca^{++} in vascular smooth muscle has not been clarified. Devine et al. (1) have suggested that internal membrane structures, rather than cell membrane structures may relate to a given tissue's ability to utilize intracellular Ca^{++}. Hurwitz et al. (2) however, suggested that while Ca^{++} binding coincided with a cell membrane fraction on sucrose gradient in guinea pig intestinal smooth muscle, it did not so correlate with a cell membrane fraction from rabbit aorta. These latter workers used 5'-nucleotidase as a cell membrane marker with rabbit aorta, since "Na^+,K^+ dependent ATPase activity is not found in this tissue," quoting the work of Verity and Bevan (3). However, it should be noted that Wolowyk et al. (4) earlier showed that such Na^+,K^+-ATPase activity could be demonstrated in rabbit aorta with a special treatment of an isolated fraction.

Since that time, papers have appeared indicating that at least in some vascular smooth muscle preparations, Na^+,K^+-ATPase is present. Hess and Ford (5) demonstrated such activity in bovine aorta (activity = 10 µmoles P_i/mg pro/hr) Wei et al. (6) in rat mesenteric arteries (activity = 15.1 µmoles P_i/mg pro/hr), and Limas and Cohn (7) in dog mesenteric arteries (activity ∿7.8 µmoles P_i/mg pro/hr). Preiss and Banaschak (8) have attempted to characterize a very low activity enzyme (∿1.0 µmoles P_i/mg pro/hr) from sheep carotid artery.

Vascular smooth muscle then, does contain a Na^+,K^+-ATPase activity which in all probability, is the enzymatic manifestation of the sarcolemmal Na^+ pump as in other tissue (9). Therefore, previous observations (2) of lack of Ca^{++} binding to sarcolemma may reflect the absence of an active Na^+,K^+-ATPase, rather than a characteristic of the sarcolemma. However, since the exact nature of the regulation of cytoplasmic Ca^{++} in vascular smooth muscle is still not known, and may be different for different muscles, the sarcolemma and the Na^+ pump may still be involved.

MATERIALS AND METHODS

1) Isolation: Microsomal fractions from canine, rabbit and rat aorta were isolated by the method described previously (10) involving differential centrifugation in .25 M Sucrose-Tris, pH 7.4. A single dog aorta was used, while at least 2 rat or rabbit aortae were pooled. In each case, media was removed from the rest of the artery. Final microsomal pellets were suspended in 0.5-1.0 ml of isolation medium.

2) Assay: All activities were assayed in a Gilford Model 250 recording spectrophotometer using the pyruvate kinase-lactate dehydrogenase linked enzyme system as previously described, except the total volume was 1.0 ml (11).

3) SDS Treatment of Microsomal Fractions: Final microsomal pellets were divided in half for treatment and control. SDS was added to the experimental tube at a final concentration of .2 mg/ml. The final volume was adjusted so that a ratio of SDS to original tissue weight ranged from 0.2-0.5 mg/g.

4) Protein Determination: Protein was determined by the Lowry method (12) or micro-Kjeldahl.

RESULTS

Dog Aorta: Freshly isolated dog aortic microsomes show no ouabain sensitive ATPase activity, but do show EGTA inhibited Ca^{++} sensitive activity (10) suggesting the presence of Ca^{++} binding membranes.

However, if these microsomal preparations are treated with SDS, the addition of relatively large amounts of EGTA (10 mM) then stimulates the ATPase activity, which can in turn be inhibited by ouabain (Figure 1).

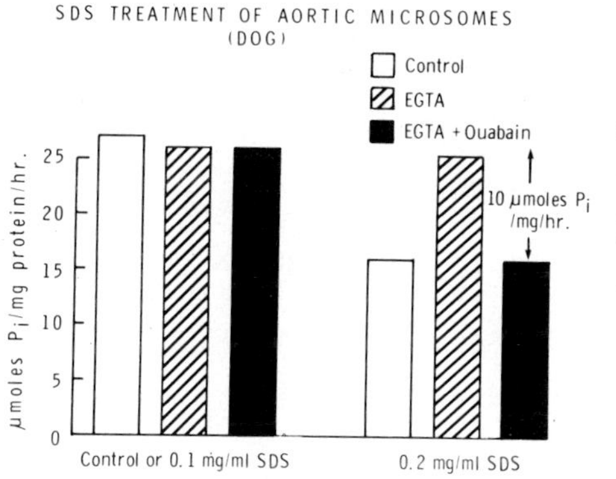

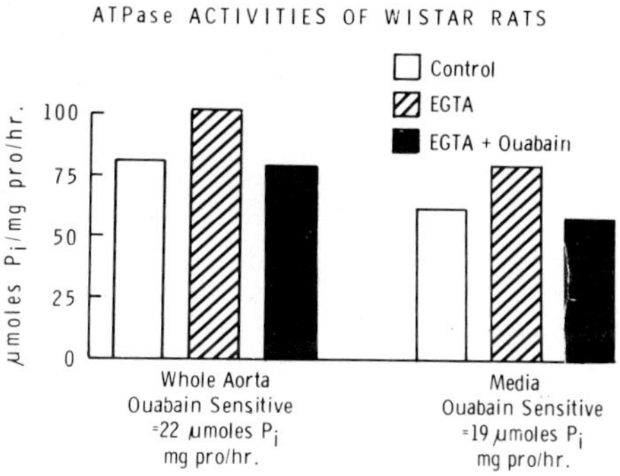

Fig. 1. ATPase activities of dog aortic microsomes, showing basal activity (open bars), activity in the presence of EGTA (hatched bars), and activity in the presence of EGTA + ouabain (dark bars). Left side bars indicate control or ineffective SDS treatment (.1mg/ml).

Fig. 2. ATPase activities of microsomal fractions from Wistar rat aortae. Left side, whole aorta; right side, media only.

Rat Aorta: Rat aortic microsomes, on the other hand, do not require SDS treatment to demonstrate EGTA stimulation and subsequent ouabain inhibition (Figure 2). It is also of considerable interest that the ouabain sensitive EGTA stimulated component is much greater in aorta of SHR than in the age-matched controls (Figure 3).

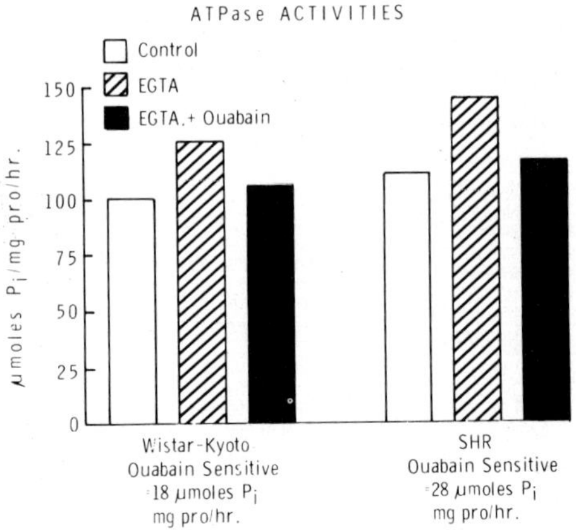

Fig. 3. ATPase activities of SHR and age-matched Wistar-Kyoto controls (media only).

Rabbit Aorta: Preliminary work with rabbit aortic microsomal preparation indicate that they also can be stimulated by EGTA after SDS treatment (Table 1).

TABLE 1

Effects of Ouabain and EGTA on control and SDS-treated Rabbit Aortic Microsomes

	Control		SDS-Treated	
	-Ouabain	+Ouabain	-Ouabain	+Ouabain
-EGTA	100*	85	100*	82
+EGTA	85	76	141	90

*In each case (i.e., control and SDS treated) 100% = the total enzyme activity in the absence of added ouabain or EGTA. These absolute values are 25% lower in the SDS treated microsomes than in the control. Additional % values are based on each separate control.

DISCUSSION

It is clear from these data, as well as the data presented earlier (1-8) that Na^+,K^+-ATPase exists in smooth muscle as it does in skeletal and cardiac muscle. Thus, there is sufficient evidence to suggest that mechanisms of active transport in these tissues are the same as in other tissue. Na^+,K^+-ATPase is the enzymatic manifestation of the Na^+ pump.

The integrated role of the Na^+ pump in cellular Ca^{++} regulation has been proposed for many tissues, but has not been clarified for either cardiac or skeletal muscle. For example, there are significant differences between Na^+,K^+-ATPase isolated from cardiac and skeletal muscle, but none of these differences have been suggested to account for the differences in Ca^{++} requirements of these two tissues (13).

Furthermore, Drummond and coworkers (14,15) have isolated sarcolemmal membranes from both cardiac and skeletal muscle, which show Ca^{++} binding and uptake capacity. How these data may relate to cation transport or muscle contractility remains unknown. However, these preparations could be phosphorylated by protein kinase, and could be stimulated by c-AMP. In addition, the phosphorylated membranes accumulated more Ca^{++} than the control preparations.

In vascular smooth muscle, mechanisms controlling cytoplasmic (activator) Ca^{++} are even less well documented than in the other muscle types, and as a result, sarcolemmal membranes may well constitute part of such regulation. The fact that Ca^{++} binding can occur to a membrane fraction from aortae that also contains Na^+,K^+-ATPase activity, may not simply reflect heterogeneity of the fraction, but in view of the above work (14,15) may well mirror a functional characteristic of the cell membrane.

The microsomal fractions under study are clearly very heterogeneous. However, other fractions from this procedure sedimented at different forces show no EGTA stimulation or ouabain sensitivity, suggesting the relatively specific nature of the treatment, and/or the specific constituents of the microsomal fractions.

Gel electrophoresis demonstrated the presence of a band of 68-100,000 daltons consistently in membrane fractions from all aortic preparations, but functional involvement of these bands has not been determined.

We have demonstrated with canine aorta that microsomal fractions bind Ca^{++}, but do not respond to oxalate (10). Wei et al. have also shown that while Ca^{++} sequestration to sarcoplasmic reticulum membranes from rat mesenteric artery can be stimulated by oxalate, such stimulation does not occur when Ca^{++} sequestration is to sarcolemmal membranes (6). The Ca^{++} binding that occurs to canine aortic microsomes may then be largely due to cell membrane material reflecting then, a lack of oxalate stimulation. The presence of the ouabain sensitive ATPase activity is consistent with this interpretation.

The present microsomal fractions contain both Ca^{++} binding and Na^{+},K^{+}-ATPase activities. The latter occurs only when EGTA is added in relatively high amounts. A possible mechanism is that the EGTA chelates an endogeneous metal (Ca^{++}?) allowing an appearance of the Na^{+},K^{+}-ATPase activity, which can then be inhibited by ouabain. Such a situation implies that there may be a relationship between Ca^{++}, Na^{+} and K^{+} transport sites in sarcolemma of vascular smooth muscle. Such interaction may have significant implications regarding ionic transport mechanisms and the alternate regulation of activator Ca^{++}.

We have shown earlier (10) that while the preparation we are studying is phosphorylated by protein kinase and c-AMP, this phosphorylation does not affect Ca^{++} sequestration. Other workers, however (16) have suggested that such phosphorylation of vascular smooth muscle membranes may well affect Ca^{++} sequestration. The role of c-AMP in vascular smooth muscle Ca^{++} regulation is not at all clear (17) and further suggestions as to its specific mechanism await further purification.

The data from this report show that Ca^{++} binding, EGTA stimulation and Na^{+},K^{+}-ATPase activity all occur in the same fraction. Such interaction may well be a reflection of functional interactions in vascular smooth muscle.

REFERENCES

1. Devine, C.E., Somlyo, A.V. and Somlyo, A.P. (1973) J. Cell. Biol. 52:690.
2. Hurwitz, L., Fitzpatrick, D.F., Debbas, G. and Landon, E.J. (1973) Science 179:384.
3. Verity, M.A. and Bevan, J.A. (1969) Biochem. Pharm. 18:327.
4. Wolowyk, M.W., Kidwai, A.N. and Daniel, E.E. (1971) Can. J. Biochem. 49:376.
5. Hess, M.L. and Ford, G.D. (1974) J. Molec. Cell. Cardiol. 6:275.
6. Wei, J.W., Janis, R.A. and Daniel, E.E. (1976) Blood Vessels 13:279.
7. Limas, C.J. and Cohn, J.N. (1974) Circ. Res. 35:601.
8. Preiss, R. and Banaschak, H. (1976) Acta Biol. Med. Germ. 35:453.
9. Schwartz, A., Lindenmeyer, G.E. and Allen, J.C. (1975) Pharmacological Rev. 27:3.
10. Allen, J.C. (1977) Blood Vessels 14:91.
11. Schwartz, A., Allen, J.C. and Harigaya, S. (1969) J. Pharmacol. Exp. Ther. 168:31.
12. Lowry, O.H., Rosebrough, N.J., Farr, A.L. and Randall, R.J. (1951) J. Biol. Chem. 193:265.
13. Pitts, B.J.R., Wallick, E.T., Van Winkle, W.B., Allen, J.C. and

Schwartz, A. (1977) Arch. Biochem. Biophys. (In press).

14. Sulakhe, P.V., Drummond, G.I. and Ng, D.C. (1973) J. Biol. Chem. 248:4150.

15. Hui, C.W., Drummond, M. and Drummond, G.I. (1976) Arch. Biochem. Biophys. 173:415.

16. Webb, R.C. and Bhalla, R.C. (1976) J. Molec. Cell. Cardiol. 8:145.

17. Diamond, J. (1977) Evidence for disassociation between cyclic nucleotide levels and tension in smooth muscle, In: The Biochemistry of Smooth Muscle, (N.L. Stephens, ed.), University Park Press, pp. 343-362.

ACKNOWLEDGEMENTS

These studies were supported by HL 16947, and the American Heart Association, Texas Affiliate. Dr. Allen is an Established Investigator of the American Heart Association.

Excitation-Contraction Coupling in Smooth Muscle, R. Casteels et al. eds.

MICROSOMAL AND MITOCHONDRIAL CALCIUM PUMPS IN SMOOTH MUSCLE

Norbert VERBEKE, Nicole MOREL and Théophile GODFRAIND
Laboratoire de Pharmacodynamie Générale et de Pharmacologie
Université Catholique de Louvain
Avenue E. Mounier 73 UCL 7350
1200 Bruxelles Belgique

INTRODUCTION

We have reported that microsomes prepared from the longitudinal smooth muscle of the guinea pig ileum did accumulate calcium from the medium, when it was supplemented with Mg-ATP. Oxalate stimulated this uptake in a way depending on the pH of the medium. Biochemical evidences indicated that the calcium incorporation with and without oxalate were the result of two processes related to distinct subcellular components[1].

In the present study, we have performed a morphological analysis of the microsomal fraction in order to assess the proposed heterogeneity of the components involved in cellular calcium homeostasis. Furthermore, as mitochondria have been claimed to play an important role in the cytosolic calcium regulation in smooth muscle[2], a comparison was made between the calcium uptake by the microsomal and the mitochondrial fraction of intestinal smooth muscle.

MATERIALS AND METHODS

■ Preparation of the subcellular fractions

The longitudinal muscle of the guinea pig ileum was prepared and homogenized as described previously[1]. Microsomes were obtained by 3 successive centrifugation steps : 500 g x 10 min - 20 000 g x 30 min - 100 000 g x 60 min. A mitochondrial enriched fraction was pelleted twice by a 15 000 g x 15 min spun of the 1 000 g x 10 min supernatant. Subcellular fractions were kept at 0°C and used within 6 hours after isolation.

■ Morphological studies

Smooth muscle microsomes were examined by electron microscopy after incubation either in the presence of calcium and oxalate or in the presence of strontium. Experimental conditions are outlined in the text. Incubations were stopped by quick cooling to 0°C and centrifugation at 300 000 g for 1 hour.

Pellets were suspended in chilled 4% glutaraldehyde in 0.1 M phosphate buffer (pH 7.4) and spun at 300 000 g for 2 hours.
Samples were postfixed in 1% OsO_4 in 0.1 M phosphate buffer(pH 7.4) dehydrated through an ascending series of alcohols and embedded in Epon. Sections (0.04 μm) were stained with uranyl acetate and lead citrate according to Reynolds[3].

■ Calcium incorporation studies

Smooth muscle microsomes and mitochondria were incubated, under continuous magnetic stirring, at 37°C, in a solution containing 20 mM Tris/maleate (pH 6.6), 3 mM Tris/ATP, 5 mM $MgCl_2$, 100 mM KCl 0 or 10 mM NaN_3 and ^{45}Ca (25 mCi/mol). The free Ca^{2+} concentration was controlled using Ca-EGTA buffers[1].
Microsomal and mitochondrial protein concentration was 0.05 mg/ml. ^{45}Ca binding by the subcellular components was measured by the Millipore filtration method as described elsewhere[1].

RESULTS

■ Morphological analysis

The microsomal fraction was constituted of smooth, empty vesicles of 0.1 to 0.4 μm in diameter. After an incubation in the presence of calcium and oxalate, some vesicles were found to contain a precipitate sticking to the internal wall (Fig.1). It is likely that this precipitate was constituted by calcium oxalate. Such an observation was also made by de Meis et al[4] for striated muscle microsomes treated in the same way.
Strontium has been used as a marker to localize subcellular calcium sequestering sites in smooth muscle [5,6]. We incubated microsomes in the presence of strontium instead of calcium. As Fig.2 illustrates, most of the microsomal vesicles were unstained,while others were stained by strontium bound to their membranes and a few presented an accumulation of strontium inside their lumen and a staining of their membranes.

Fig.1 Electron micrograph of microsomes after incubation, at 37° C, in the presence of 20 mM Tris/maleate (pH 6.6), 5 mM $MgCl_2$, 3 mM ATP, 100 mM KCl, 0.02 mM $CaCl_2$ and 5 mM Tris/oxalate. Ca-oxalate precipitates can be noticed inside some vesicles (1). The horizontal bar represents 0.5 μm.

Fig.2 Electron micrograph of microsomes after incubation, at 37° C, in the presence of 20 mM Tris/maleate (pH 6.6) 5 mM $MgCl_2$, 3 mM ATP, 100 mM KCl, and 0.1 mM $SrCl_2$. Strontium stains some vesicular membranes (1), while others are filled by strontium precipitate (2). The horizontal bar represents 0.5 μm.

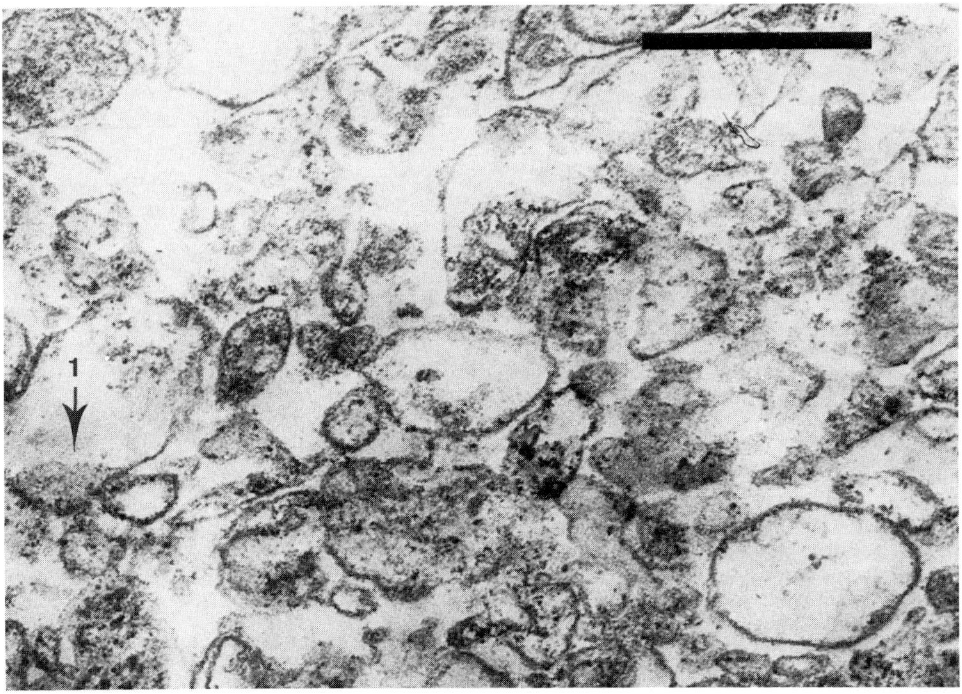
1

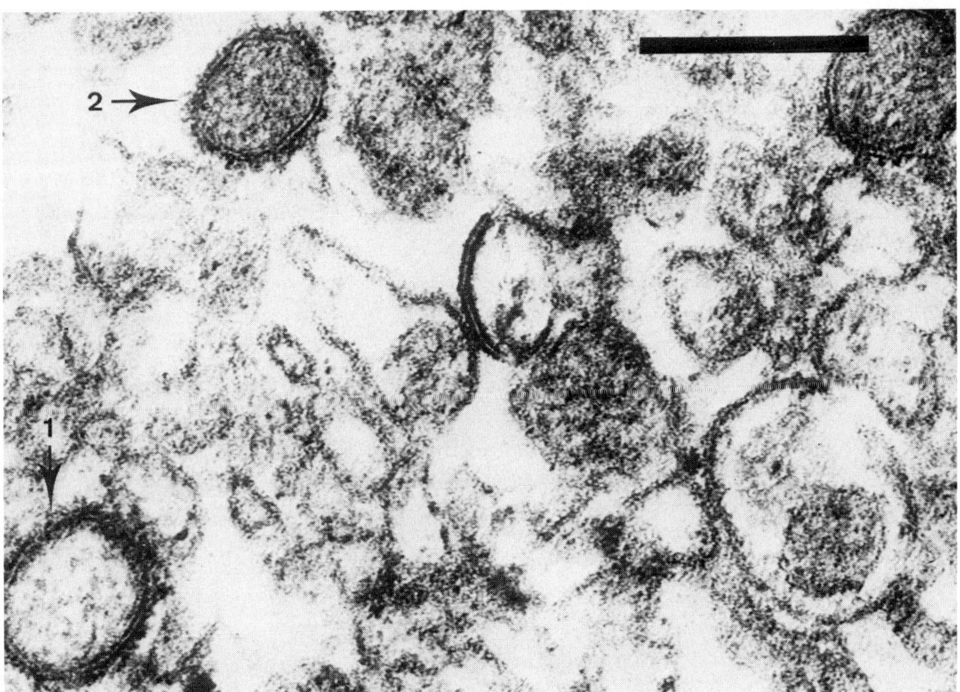
2
1

■ Microsomal and mitochondrial calcium uptake

Calcium accumulation by smooth muscle microsomes has been extensively described elsewhere[1]. The present experiments were devoted to the comparison of calcium binding by the mitochondria and microsomes of the same tissue, studied in identical conditions.

Mitochondria took up calcium from the incubation medium when it was supplemented with Mg-ATP. Succinate exerted a stimulatory action but was unable to support calcium uptake by its own. Inorganic phosphate led to a massive calcium storage. Azide inhibited almost completely calcium sequestration in the mitochondrial fraction.

The affinity of microsomes and mitochondria for calcium was compared by measuring calcium binding at equilibrium, in both fractions, at various free calcium concentrations ranging from 10^{-7}M to 10^{-4}M. In microsomes, the threshold concentration was comprised between 10^{-7}M and 5 x 10^{-7}M. In mitochondria, the threshold concentration was tenfold higher. At calcium concentrations higher than 3 x 10^{-5} M, mitochondria bound more calcium than microsomes (Fig.3).

DISCUSSION

From a biochemical study on smooth muscle microsomes[1], we came to the conclusion that calcium binding without oxalate is due to a mechanism different from the one responsible for calcium uptake enhanced by oxalate. This was shown by the observation that changes in factors such as pH, pCa and monovalent cations affected differently the two processes. Furthermore, it was proposed that the two calcium sequestering activities were linked to distinct subcellular components, as their distribution profiles were different. This hypothesis is strengthened by the results of the present morphological analysis of microsomes after incubation in the presence of strontium. Among the few microsomal vesicles responsible for divalent cations storage, some show only a binding of the cations to their membranes and others present a further accumulation inside their lumen.

The relative importance of microsomal components and mitochondria in cytosolic calcium homeostasis is a matter of controversy [2,7-9]. When assayed at high calcium concentrations, mitochondria seem indeed to be the major calcium sink among smooth muscle subcellular components. However, their affinity for calcium is lower than that of microsomes. At physiological relevant calcium concentrations ($\pm$ 10^{-6}M), microsomal components are likely to be responsible for cytoplasmic calcium regulation in smooth muscle.

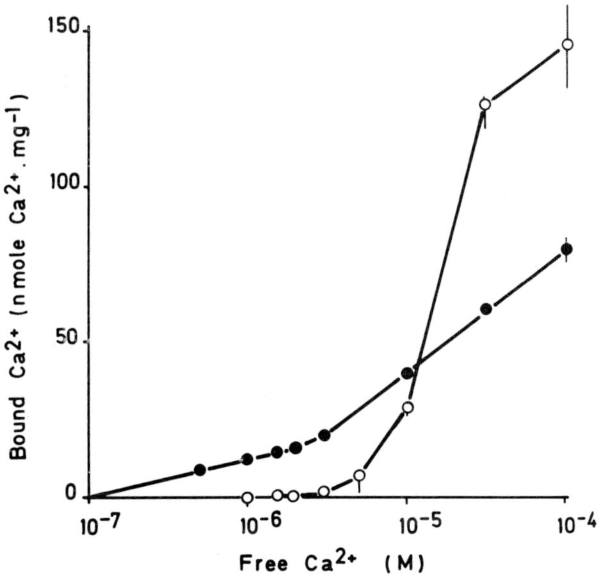

Fig. 3 Microsomal and mitochondrial calcium binding as a function of free Ca^{2+} concentrations in the medium.
Microsomal ^{45}Ca content (closed symbols) was measured after 30 minutes incubation, at 37°C, in the presence of 20 mM Tris/maleate (pH 6.6), 5 mM $MgCl_2$, 3 mM ATP, 100 mM KCl and various concentrations of Ca^{2+} buffered by Ca-EGTA.
Each result is the mean of six determinations ± S.E.M.
Mitochondrial ^{45}Ca content (open symbols) was estimated as the azide sensitive calcium binding of the mitochondrial fraction. It was measured after 30 minutes incubation, at 37°C, in the presence of 20 mM Tris/maleate (pH 6.6), 5 mM $MgCl_2$, 3 mM ATP, 100 mM KCl, 0 or 10 mM NaN_3 and various concentrations of Ca^{2+} buffered by Ca-EGTA.
Each result is the mean of nine determinations ± S.E.

Acknowledgements: This work has been supported by F.R.S.M. (grant n° 20121). Norbert VERBEKE is Aspirant du F.N.R.S. Nicole MOREL is Boursier de l'IRSIA. We thank Dr X. STURBOIS for valuable discussion and J. MICHOTTE for skilful technical assistance.

REFERENCES

1. GODFRAIND, T., STURBOIS, X. & VERBEKE, N. (1976) Biochim. Biophys. Acta, 455, 254

2. BATRA, S.C. (1973) Biochim. Biophys. Acta 305, 428

3. REYNOLDS, E.W. (1963) J. Cell. Biol. 17, 258

4. DE MEIS, L. (1970) J. Biol. Chem. 245, 1883

5. DEVINE, C.E., SOMLYO, A.V. & SOMLYO, A.P. (1972) J. Cell. Biol. 52, 690

6. GODFRAIND, T., STURBOIS, X. & VERBEKE, N. (1973) Arch. Int. Physiol. Biochem. 81, 786

7. HESS, M.L. & FORD, G.D. (1974) J. Mol. Cell. Cardiol. 6, 275

8. VALLIERE, J., SCARPA, A. & SOMLYO, A.P. (1975) Arch. Biochem. Biophys. 170, 869

9. RAEYMAEKERS, L., WUYTAK, F. & CASTEELS, R. (1976) Arch. Int. Biochem. Physiol. 84, 355

Excitation-Contraction Coupling in Smooth Muscle, R. Casteels et al. eds.

THE IMPORTANCE OF CALCIUM BINDING BY SUBCELLULAR COMPONENTS OF SMOOTH MUSCLE IN EXCITATION CONTRACTION COUPLING

Satish Batra
Department of Pharmacology and the
Department of Obstetrics and Gynecology,
University of Lund, Lund, Sweden

The role of calcium in the regulation of muscular activity is widely recognized. Calcium is required in the process of excitation contraction coupling and must accordingly be withdrawn from the contractile proteins for the ensuing relaxation. The source of activator Ca in smooth muscle could, besides the extracellular Ca, originate from several intracellular stores as well as from the sarcolemma. In fact, any vesicular organelle in the cell, or sarcolemma itself, if it is capable of transporting and concentrating Ca, is clearly a potential candidate for storing activator Ca.

From the biochemical data obtained in recent years, mitochondria, reticulum as well as sarcolemma have been implicated as sources of activator Ca in smooth muscle. In the case of uterine smooth muscle, however, the data on the kinetics of Ca uptake by mitochondria particularly the Ca uptake capacity and concentrating ability, obtained in this laboratory, led to the proposal that this organelle could adequately fulfil the requirements for acting as Ca "sink"[1]. This proposal is further supported by the recent results of Wikström _et al._[2] and Janis _et al._[3]. However, other sites for Ca storage such as the sarcolemma and sarcoplasmic reticulum cannot be ruled out simply on the basis of these biochemical data. The need for a more critical examination of these sources of activator Ca, for reasons given previously, must again be emphasized[1,4].

In an attempt to resolve these questions further, kinetics of mitochondrial and microsomal fractions isolated from rabbit myometrium were examined. Furthermore, the influence of hormonal exposure on the endogenous Ca concentration of the myometrium as well as Ca uptake by the isolated subcellular fractions was studied.

At a concentration of 10^{-6} M Ca which is approximately the threshold for contractile activity, the initial rate of mitochondrial Ca uptake (Fig 1) was an order of magnitude higher than that of the microsomal Ca uptake (Fig 2).

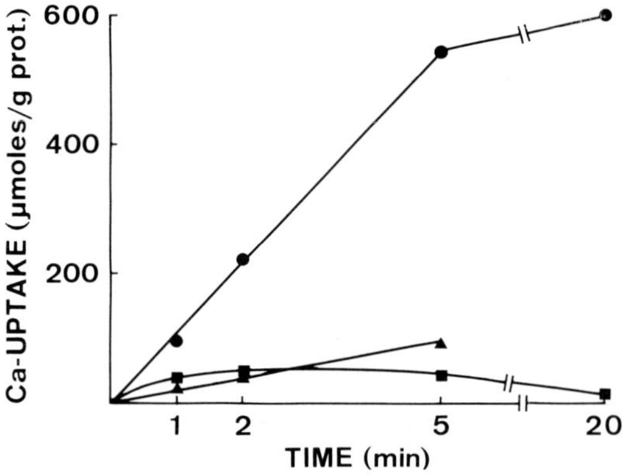

Fig 1. Ca uptake by rabbit myometrial mitochondria from a medium containing 10^{-6} M Ca^{++} (adjusted with Ca-EGTA buffer). Other constituents of the incubation medium were: (●) 100 mM KCl, 5 mM ATP, 5 mM $MgCl_2$, 10 mM succinate and 30 mM Hepes buffer, pH 7.0, (▲) ATP from the medium was omitted; 3 µM rotenone and 5 mM K-phosphate were included, (■) ATP, $MgCl_2$ and KCl were omitted; 3 µM rotenone, 5 mM K-phosphate and 250 mM sucrose were included.

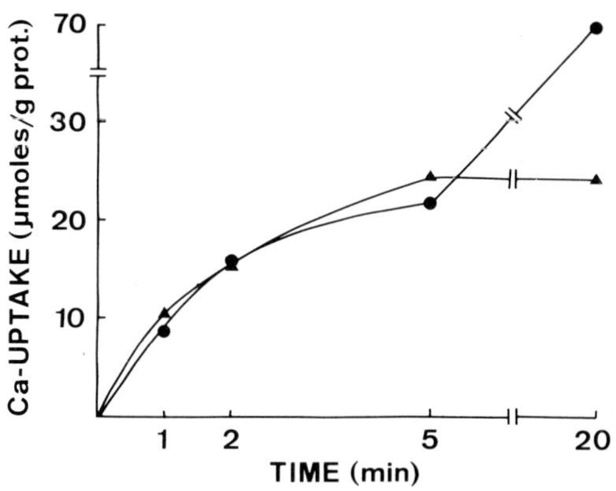

Fig 2. Ca uptake by rabbit myometrial microsomes from a medium containing 10^{-6} Ca^{++} (adjusted with Ca-EGTA buffer). Other constituents of the incubation medium were: (▲) 100 mM KCl, 5 mM ATP, 5 mM $MgCl_2$, 5 mM NaN_3 and 30 mM Hepes-buffer, pH 7.0, (●) with 5 mM K-oxalate added to the medium.

The presence of oxalate although increased Ca uptake capacity, measured after 20 min, it had no effect on the initial rates of Ca uptake. In certain other smooth muscles oxalate has been shown to increase both the rate and capacity of Ca uptake substantially[5] but -

caution is necessary in interpreting data on Ca uptake measured in the presence of oxalate.

The initial rates of Ca uptake by mitochondria and by microsomes were also measured at concentrations lower than 10^{-6} M. With 10^{-7} M Ca in the medium the rates of uptake by the two fractions were very low (2-3 µmoles Ca/g protein/min) and not significantly different from each other. However, when the concentration of Ca exceeded 3×10^{-7} M the rate of Ca uptake by mitochondrial fraction was significantly higher than that by the microsomal fraction. These results are similar to those reported previously on the human myometrium[1].

<u>Hormonal control of myometrial calcium.</u> Available evidence indicates that progesterone domination in the rabbit uterus results on the whole in reduced contractile activity (or inhibition) whereas progesterone withdrawal has the opposite effect[6]. In fact, progesterone withdrawal in the rabbit appears to be a prerequisite for the evolution of myometrial activity at parturition[6]. We have obtained a direct evidence for the inhibitory effect *in vivo* of progesterone on the rabbit myometrial activity.

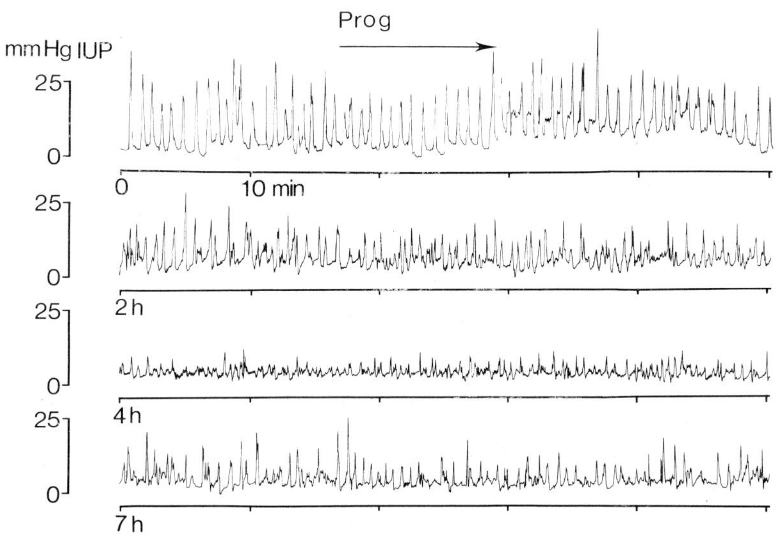

Fig 3. Effect of progesterone given as intravenous infusion (2 µg/h for 4 h) on the intrauterine pressure (IUP)[7] in a nonpregnant rabbit. Note that in the lower tracing, obtained 3 hours after the termination of the infusion, the uterine contractility has partly resumed its original (nonpregnant type) pattern.

An intervenous infusion of progesterone diminishes the spontaneous activity of the myometrium rather rapidly, and on cessation of the progesterone infusion the activity tends to resume its original pattern of contraction marked by high amplitude (Fig 3).

An analysis of tissue Ca concentration showed that progesterone withdrawal leads to a greater Ca concentration in the myometrium (Fig 4).

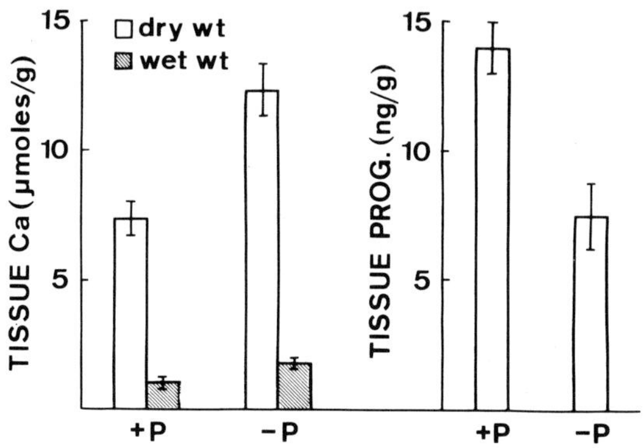

Fig 4. Effect of progesterone treatment (+ P) and subsequent withdrawal (- P) on myometrial Ca and myometrial progesterone concentration. Estrogenized rabbits were treated daily with progesterone for 3 days and used on the 4th day (+ P) or on the 6th day (- P).

A measurement of the progesterone concentration[8] in the myometrium clearly showed that progesterone withdrawal was effective in reducing myometrial progesterone substantially (Fig 4). Mitochondria isolated from these myometria showed a good respiratory control, and no significant difference in this respect between the two groups of rabbits was observed (Fig 5).

Ca uptake capacity measured in the presence of both succinate and ATP in the medium (Fig 6) as well as the rate of Ca uptake, supported by respiration (Fig 7), was significantly lower in mitochondria isolated from progesterone-dominated myometrium than those isolated from myometrium subjected to progesterone withdrawal. This observation also suggests that the lower tissue Ca in progesterone-dominated myometrium was not the result of decreased Ca entry into the myometrial cell but due to lower accumulation by the subcellular structures. Interestingly, progesterone *in vitro* has little effect on both myometrial contractions[9] and mitochondrial Ca uptake[10].

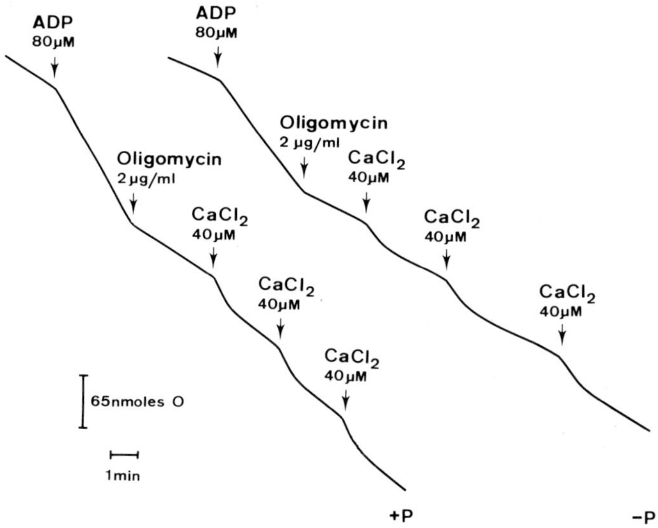

Fig 5. Respiratory activity of mitochondria isolated from the two groups of rabbits shown in Fig 4. Incubation medium contained 200 mM sucrose, 20 mM KCl, 5 mM $MgCl_2$, 5 mM K-phosphate, 10 mM succinate, 12 µM rotenone, 20 mM Hepes-buffer, pH 7.2 and 1.7 mg (+ P) and 1.4 mg (- P) mitochondrial protein/ml.

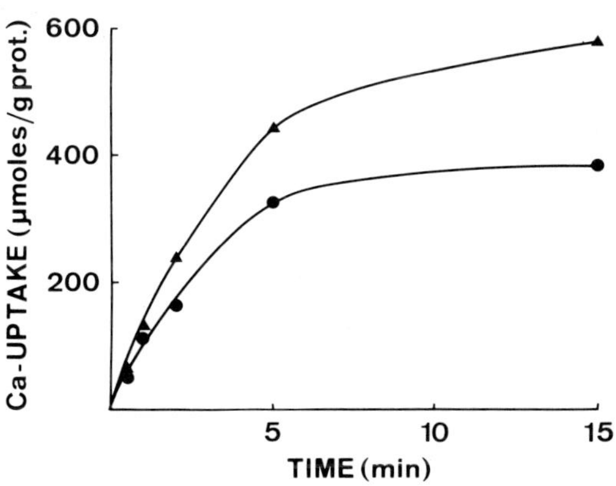

Fig 6. Effect of progesterone administration (●) and subsequent withdrawal (▲) on Ca uptake capacity of mitochondria isolated from rabbit myometrium (See Fig 4). Ca uptake was measured in the presence of both ATP and succinate in the medium.

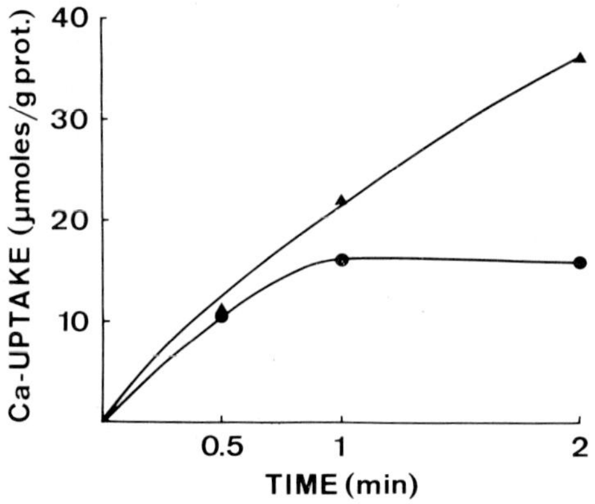

Fig 7. Effect of progesterone administration (●) and subsequent withdrawal (▲) on respiration-supported Ca uptake (see Fig 4). Succinate (10 mM) was used as substrate and rotenone 3 µM was included in the medium.

The effect of estrogen on myometrial Ca was not examined since this hormone appears to have little significance in the reproductive physiology of the rabbit. Although progesterone levels in the blood change greatly during pregnancy and before parturition, estrogen levels, in this species, do not[6].

The effect of progesterone on Ca binding by the subcellular fractions of the myometrium was further studied by using pseudopregnant rabbits. This condition in which endogenous progesterone is produced, lasts for about 18 days. Uterine progesterone concentration is maximal around day 12 and declines to very low levels again by day 18 of pseudopregnancy[11].

Preliminary data of these studies indicated that the rate of Ca uptake by both mitochondria and microsomes was lower in progesterone dominated (12-day pseudopregnant) rabbits than those relieved of progesterone domination (18-day pseudopregnant).

On the basis of data presented here, one cannot resolve the question of the precise source of activator Ca. Nevertheless, the findings elucidate an important aspect of excitation-contraction coupling in uterine smooth muscle namely, the hormonal influence on Ca uptake kinetics. In the rabbit, progesterone is the most important hormone for maintaining pregnancy and for the control of parturition[6].

It appears that this hormone can reduce the uptake of Ca by myometrial mitochondria *in vivo* as evidenced by measurements of tissue Ca and Ca uptake by mitochondria isolated from these tissues.

SUMMARY AND CONCLUSIONS

Data are now accumulating in different laboratories indicating that the relative importance of different intracellular organelles on one hand and the extracellular Ca on the other may vary from one type of smooth muscle to another. In fact there is some evidence showing that the Ca uptake ability of the intracellular organelles in the same type of smooth muscle could vary from one species to another[12]. It is not surprising then to find that divergent results have been reported from different laboratories and a debate on this issue continues.

The present data on Ca uptake kinetics support our previous proposal[4] that mitochondria in uterine smooth muscle could make a very significant contribution in the regulation of myoplasmic Ca. The contribution by sarcoplasmic reticulum or sarcolemma although difficult to fully acknowledge, cannot, however, due to their strategic location, be ignored[1].

Progesterone withdrawal which is considered to be a prerequisite for parturition in the rabbit led to an increase in Ca accumulation by mitochondria. Assuming that Ca so stored is, or can be, utilized for the evolution of myometrial activity, it would then be available in higher amounts after progesterone withdrawal when myometrium develops maximal contractile force for the expulsion of the fetus.

ACKNOWLEDGEMENTS

Lena Timby provided excellent technical assistance. Our work is supported by the Ford Foundation and the Swedish Medical Research Council (project No. 4504). A part of this work was done in collaboration with Prof. N.-E. Saris (University of Helsinki) through an award to the author of a short-term EMBO fellowship.

REFERENCES

1. Batra, S. (1975) in Calcium Transport in Contraction and Secretion (Carafoli, E., Clemente, F., Drabikowski, W. and Margreth, A., eds.), pp. 87-94, North-Holland, Amsterdam
2. Wickström, M., Ahonen, P. and Luukkainen, T. (1975) FEBS Letters 56, 120-123

3. Janis, R.A., Lee, E.Y., Allan J. and Daniel, E.E. (1976) Pflügers Arch. 365, 171-176

4. Batra, S. (1973) Biochim. Biophys. Acta 305, 428-432

5. Raeymaekers, L., Wuytack, F., Batra, S. and Casteels, R. (1977) Pflügers Arch. 368, 217-223

6. Thorburn, G.D., Challis, J.R.G. and Robinson, J.S. (1977) in Biology of the Uterus (Wynn, R.M., ed.) pp. 653-732, Plenum Press, New York

7. Laudański, T., Åkerlund, M. and Batra, S. (1977) J. Reprod. Fert. (in press)

8. Batra, S. and Bengtsson, L.Ph. (1976) J. Steroid Biochem. 7, 599-603

9. Batra, S. and Bengtsson, B. (1977) J. Physiol. (in press)

10. Batra, S. (1973) Biochem. Pharmacol. 22, 803-809

11. Thorbert, G., Batra, S., Owman, Ch., Rosengren, E. and Sjöberg, N.-O. (1976) Endocrinology 99, 1207-1212

12. Janis, R.A. and Daniel, E.E. (1977) in The Biochemistry of Smooth Muscle (Stephens, N.L., ed.), pp. 653-671, University Park Press, Baltimore

Excitation-Contraction Coupling in Smooth Muscle, R. Casteels et al. eds.

COMPARISON OF THE Ca ACCUMULATION BY DIFFERENT SUBCELLULAR FRACTIONS OF SMOOTH MUSCLE.

L. Raeymaekers, F. Wuytack and R. Casteels
Laboratorium voor Fysiologie
Campus Gasthuisberg
B-3000 Leuven
Belgium

SUMMARY :

The energy-dependent Ca^{2+} uptake by a mitochondrial and by a microsomal fraction of the guinea-pig taenia coli was studied in solutions with Ca^{2+} concentrations buffered by Ca-EGTA mixtures. The experiments were performed at 25°C and at pH 7.1. The Ca^{2+} accumulation by the mitochondria was measured in the presence of both succinate (5 mM) and ATP (5 mM). A correction for passive binding was calculated from the simultaneous measurement of Ca accumulation in the presence of a specific mitochondrial inhibitor (5 mM azide or 5 μM FCCP). In experiments on the microsomal fraction, oxalate (5 mM) and a mitochondrial inhibitor were always added. The active Ca^{2+} uptake in the microsomal vesicles was assumed to correspond to the difference in the amount of ^{45}Ca taken up in the presence and in the absence of ATP (5 mM). By using these procedures, the necessity of extensive purification of the isolated fractions was avoided.

The maximum rate of energy-dependent Ca uptake in the mitochondria amounted to 40 μmoles Ca^{2+}/g protein.min. The maximum rate was independent of the free Mg^{2+} concentration between 0.5 and 5 mM. Because 5 mM ATP was present, these values correspond to a total Mg concentration of 5.2 and 10 mM respectively. The rate of uptake was half-maximum at 6 μM Ca^{2+} when the free Mg^{2+} concentration was 5 mM. The apparent K_m decreased to 2.5 μM at a free Mg^{2+} concentration of 0.55 mM. At all Ca^{2+} concentrations the rate of energy-dependent ^{45}Ca uptake decreased when the total Mg concentration was lowered below 5 mM. The lowest Ca^{2+} concentration at which energy-dependent ^{45}Ca uptake occurred, was not clearly affected by Mg^{2+}. This lowest Ca concentration was about 3×10^{-7} M.

The maximum rate of ATP-stimulated ^{45}Ca accumulation by the microsomes was 9 μmole Ca^{2+}/g. protein.min. The ^{45}Ca uptake at all Ca^{2+} concentrations was inhibited by 5 mM free Mg^{2+}, but was unaffected by Mg^{2+} in the range 0.3 to 1 mM. The apparent K_m was 7×10^{-7} M Ca^{2+} and the lowest Ca^{2+} concentration at which active ^{45}Ca uptake occurred was about 4×10^{-8} M.

These results suggest that under physiological conditions the cell organelles corresponding to the microsomal fraction are more important for the regulation of the cytoplasmic Ca^{2+} than the mitochondria.

INTRODUCTION :

The relaxation of a smooth muscle is probably brought about by the removal of Ca from the cytoplasm by an energy-requiring transport process. Energy-dependent Ca uptake has been shown to occur in mitochondria (1-3) and membranes (4-5) isolated from smooth muscle. K_m values for Ca uptake were reported for microsomal membranes from vascular smooth muscle (5) and for isolated mitochondria from vascular smooth muscle (2) and from uterus (6). The isolated membranes have a higher affinity for Ca ($K_m = 10^{-6}$ M) than the mitochondria (K_m = 17 µM or higher). In the present study, a mitochondrial and a microsomal fraction was isolated from the guinea-pig's taenia coli. The affinity for Ca and the rate of energy-dependent Ca uptake in both fractions were studied under similar conditions and in the range of the physiologically important Ca concentrations. The Ca concentrations were buffered with mixtures of Ca and EGTA. These experiments allow a more direct comparison of the two Ca transport systems.

METHODS :

The experimental procedures have been described in detail elsewhere (7). Briefly, the tissues were treated with collagenase and then homogenized in modified Chappell-Perry medium with the aid of a Potter-Elvejhem homogenizer. A mitochondria enriched and a microsomes enriched fraction were prepared by differential centrifugation and used for the Ca uptake experiments without further purification. The solution used for the study of the Ca uptake contained (mM) : sucrose, 125, KCl, 50, imidazole, 20, and variable concentrations of EGTA, $CaCl_2$ and $MgCl_2$. For experiments on the mitochondrial fraction this solution contained also (mM) : Na-ATP, 5, Na-succinate, 5, and for experiments on the microsomal fraction : K-oxalate, 5, Na-azide and ATP, when added, 5. The pH was 7.1 and the temperature 25°C. In order to calculate the free Ca^{2+} and Mg^{2+} concentration, the reactions of ATP^{4-}, $HATP^{3-}$, $EGTA^{4-}$, $HEGTA^{3-}$ with H^+, Ca^{2+} and Mg^{2+} were considered. For the association constants for ATP and for references see ref. (8). The association constant used for the $CaEGTA^{2-}$ complex was $10^{10.42}$ M^{-1} (9). By giving a certain value to (H^+), (Ca^{2+}), (Mg^{2+}), $(ATP)_{tot}$ and $(EGTA)_{total}$, the total concentrations of Mg and Ca that have to be added to the solution can be calculated from this set of 12 equations.

The ^{45}Ca uptake was stopped by filtration through a Millipore filter of pore size 0.45 µ and by washing the filter with 2 ml buffered sucrose solution containing

10 mM EGTA.

RESULTS :

1. The functional integrity of the isolated mitochondria.

O_2 consumption measurement

The rate of O_2 consumption of mitochondria isolated under optimal conditions increases 2.5 to 4 times on addition of ADP, showing that respiration is coupled to oxidative phosphorylation. Addition of Ca also induces a stimulation of respiration, giving a Ca : O ratio of 4. When the mitochondria were prepared less carefully, e.g. when albumin was omitted from the homogenization medium, the O_2 consumption measurements showed that phosphorylation in these mitochondria was uncoupled.

Dependence of Ca uptake on the method of preparation and on the presence of ATP

It was observed that the mitochondria took up Ca at a faster rate in the presence of ATP than in the presence of an oxidizable substrate (5 mM succinate). The uptake was fastest, however, when both ATP and succinate were added. The Ca uptake in the presence of succinate alone was highly dependent on the method of preparation of the isolated mitochondria (e.g. with or without fatty acid free albumin in the homogenization medium), whereas the Ca uptake in the presence of ATP did not vary under these different conditions. Therefore all further experiments were done in the presence of both succinate and ATP. Moreover, these conditions resemble more closely the in vivo situation.

2. Estimation of the amount of mitochondrial and microsomal protein in the intact tissue.

From the homogenate four fractions were obtained by differential centrifugation. These fractions are the nuclear, mitochondrial, microsomal fractions and the final supernatant. The mitochondrial and the microsomal fraction each contained 4.3 % of the total protein of the original homogenate.
The Ca uptake in the mitochondria and in the microsomal vesicles is expressed as the uptake per g of "mitochondrial" or "microsomal" protein. These terms do not mean "pure mitochondrial protein" or "pure microsomal protein". They only refer to the protein mixture as it is present in the mitochondrial or in the microsomal fraction. Only a limited and unknown amount of this protein is pure mitochondrial

For the mitochondria, we measured in every fraction the cytochrome c oxidase activity. In order to estimate the amount of microsomal membranes, the difference in the amount of Ca taken up in the presence and in the absence of oxalate was measured, as suggested by Ford and Hess (5). As expected, the cytochrome c oxidase activity was highest in the mitochondria enriched fraction. The activity, however, was greater in the nuclear fraction, because of the high protein content of this fraction. The total oxalate-stimulated Ca uptake was also largest in the nuclear fraction, amounting to 488 $\pm$ 87 (n=4) µmole Ca.kg w. weight^{-1}, whereas the corresponding values for the mitochondrial fraction and for the microsomal fraction were 181 $\pm$ 15 (n=3) and 155 $\pm$ 16 (n=4) respectively. This observation indicates that the mitochondrial fraction is contaminated with microsomal vesicles. This was confirmed by electronmicroscopical observations. The majority of impurities in the microsomal fraction were myofilaments.

The calculated values for the amount of mitochondrial and microsomal protein in the intact tissue were 7.8 and 18.7 g.kg w. weight^{-1} respectively. This last value is probably an underestimation, because vesicles present in the final supernatant, large membrane fragments in the "nuclear" fraction, and right-side-out vesicles are not detected.

3. The rate of the energy-dependent Ca uptake as a function of the Ca concentration. Effect of Mg^{2+}.

The maximum rate of azide-sensitive Ca uptake in the mitochondrial fraction amounted to 40 µmole.g protein^{-1}.min^{-1}. This maximum rate was independent of the total Mg concentration in the range 5 to 10 mM. Since 5 mM ATP was always present, these total Mg concentrations correspond to free Mg^{2+} concentrations of 0.4 and 5 mM respectively. The Ca^{2+} concentration for half-maximal rate of Ca uptake was 2.5 µM at a total Mg concentration of 5.2 mM. The apparent K_m value was shifted to higher values when the amount of Mg was either increased or decreased (fig. 1). The lowest Ca^{2+} concentration at which energy-dependent Ca uptake occurred was not clearly affected by Mg. This lowest Ca^{2+} concentration was about 3×10^{-7} M.

The maximum rate of ATP-dependent Ca uptake in the microsomal fraction was 9 µmole Ca.g protein^{-1}.min^{-1}. At all Ca concentrations, the rate of uptake was independent of the Mg concentration in the range 5 to 6 mM (corresponding to free Mg^{2+} concentrations of 0.4 to 1.15 mM respectively). The K_m value was 0.7 µM. The lowest Ca concentration at which ATP-dependent Ca uptake occurred was about 2×10^{-8}M. 5 mM free Mg^{2+} inhibited the rate of uptake at all Ca concentrations tested (up to 10^{-5} M Ca) (fig. 1).

DISCUSSION :

Because the relaxation of intestinal smooth muscle is probably induced by remo-

or microsomal protein. It is possible to determine the total amount of the mitochondria or of the microsomal membranes in the tissue (T g protein. kg w. weight^{-1}) by measuring a specific characteristic of the mitochondria or of the microsomes in the corresponding fraction (C units of activity.g protein^{-1}) and by comparing it to the activity in the homogenate or to the sum of the activities in all isolated fractions (S units of activity.kg w. weight^{-1}). Then the total amount $T = \frac{S}{C}$.

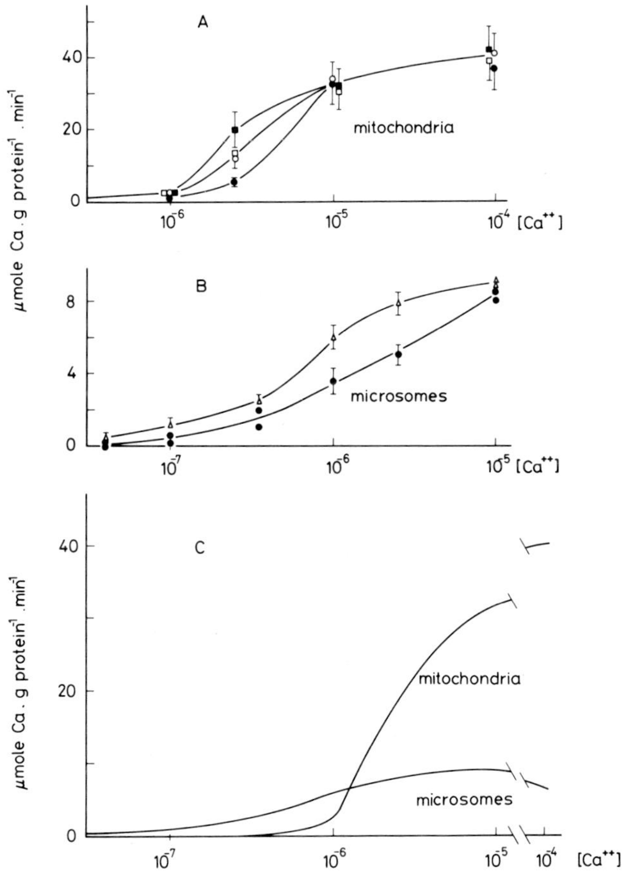

Fig. 1.

The rate of the energy-dependent Ca uptake (μmole Ca.g protein^{-1}.min^{-1} on the ordinate) in mitochondria (A) and in microsomes (B) isolated from the guinea-pig taenia coli as a function of the Ca^{2+} concentration (on the abscissa on a logarithmic scale). The different symbols refer to different Mg concentrations.
(●) 5 mM free Mg^{2+} or 10 mM Mg_{tot}
(o) 1.15 mM free Mg^{2+} or 6 mM Mg_{tot} (■) 0.55 mM free Mg^{2+} or 5.2 mM Mg_{tot} (□) 0.4 mM free Mg^{2+} or 5 mM Mg_{tot} (Δ) 0.8 mM free Mg^{2+} or 5.5 mM Mg_{tot}.

The curve in A represented by the filled squares (■) and the curve in B represented by the open triangles (Δ) are plotted on the same scale in C.
For further details see text.

val of Ca from the cytoplasm by an energy-dependent transport process, we have measured the rate of the energy-dependent Ca uptake in mitochondria and in microsomal vesicles. Moreover, the present methods allow us to discriminate between Ca uptake occurring in the mitochondria and that occurring in the microsomes without isolation of pure fractions.

It was shown that the Ca transport system in the microsomal membranes has a higher affinity for Ca than the mitochondrial system. The apparent K_m value for Ca uptake is about two times lower, and the lowest Ca concentration at which energy-dependent uptake occurs is about ten times lower in the microsomes then in the mitochondria. This observation strongly suggests that the Ca concentration in the cytoplasm of the relaxed muscle is determined by the microsomal Ca transport system. This Ca transport system could be localized in the surface membrane or in the sarcoplasmic reticulum, or in both. In the latter case, the observed Ca transport in the microsomal fraction could be the sum of the uptake in two different systems. In taenia coli, however, the sarcoplasmic reticulum is poorly developed and internal Ca stores, as they occur in vascular smooth muscle, are not present (10). The microsomal Ca uptake therefore may correspond to transport of Ca across the surface membrane. Since the maximum rate of Ca uptake in the microsomal fraction at 25°C is 9 µmole Ca.g protein^{-1}.min^{-1}, and since the intact tissue contains at least 18.7 g microsomal protein.kg w. weight^{-1}, the rate of Ca uptake in the microsomal membranes is at least 170 µmole Ca.kg w. weight^{-1}.min^{-1}. This values should be multiplied by about 2 to obtain the rate of uptake at 35°C (unpublished observation). The microsomal membranes could therefore have a Ca transport system which is sufficiently fast to bring about relaxation of the taenia coli without any Ca uptake by other cellular organelles, such as the mitochondria. These latter organelles more likely play a role in buffering cytoplasmic Ca when the muscle is strongly stimulated.

REFERENCES :

1. Batra, S. (1973) Biochim. Biophys. Acta, 305, 428-432.
2. Vallières, J., Scarpa, A. and Somlyo, A.P. (1975) Arch. Biochem. Biophys. 170, 659-669.
3. Wikström, M., Ahonen, P., Luukkainen, T. (1974) in : Physiology and genetics of reproduction, Vol 4, 177-188, Plenum Press, New York.
4. Hurwitz, L., Fitzpatrick, D.F., Debbas, G. and Landon, E.J. (1973) Science 179, 384-386.
5. Ford, G.D. and Hess, M.L. (1975) Circulat. Res. 37, 580-587.
6. Batra, S. (1975) in : Calcium transport in contraction and secretion, 87-94, North Holland Publishing Company, Amsterdam.
7. Raeymaekers, L., Wuytack, F., Batra, S. and Casteels, R. (1977) Pflügers Arch. 368, 217-223.
8. Vianna, A.L. (1975) Biochim. Biophys.Acta 410, 389-406.
9. Godt, R.E. (1974) J. gen. Physiol. 63, 722-739.
10. Gabella, G. and Raeymaekers, L. (1975) J. Physiol. 249, 28-29P.

Excitation-Contraction Coupling in Smooth Muscle, R. Casteels et al. eds.

SUMMARY OF DISCUSSION ON "ROLE OF SUBCELLULAR STRUCTURES IN Ca HOMEOSTASIS"

BY S. EBASHI (University of Tokyo)

What organelle should be responsible for the regulation of intracellular Ca ion concentration of smooth muscle has been a debated question, which should be solved to understand the mechanism of excitation-contraction coupling in smooth muscle. However, there are at least two obstacles which must be overcome : i) preparing a homogeneous subcellular fraction retaining a high physiological activity, and ii) identifying the fraction with a specified structure of the cell. Carsten, a real pioneer in this field, has reviewed the present situation of the studies along this line. She gave a serious warning against hasty conclusions, e.g. identifying the origin of a subcellular fraction by using only one or two 'markers'.

A majority of investigations, as represented by Raeymaekers et al., were of the opinion that the Ca uptake of the microsomal fraction should play a physiological role, although as suggested by Batra and some other investigators the role of mitochondrial Ca uptake must not be disregarded. In view of the results obtained with skeletal muscle, it is quite natural to assume that the microsomal fraction might be mainly represented by the sarcoplasmic reticulum (SR), but before reaching this conclusion there remain many things to be settled.

Verbeke et al. have shown an interesting observation that the oxalate-stimulated Ca accumulation and Ca 'binding' in the absence of oxalate are related to different structures of microsomal fraction. In this connection, I should like to point out an observation that Ca uptake of very fresh microsomal preparations of skeletal muscle is only weakly stimulated by oxalate and with the lapse of time the stimulation becomes prominent as observed with ordinary preparations.

Since the ionic processes at the surface membrane of smooth muscle are quite different from those of striated muscle, it is very plausible to consider that the surface membrane of smooth muscle would play a crucial role in regulating intracellular Ca ion concentration. Daniel et al. presented the results of their elaborate efforts to prepare a homogeneous surface membrane fraction. As they themselves have stated, this is certainly the first step to our final goal. Allen's work as to the presence of Na, K-ATPase in the cell membrane fraction may be another approach to the problem of surface membrane.

A new type of chemically skinned fibers shown by Endo et al., i.e., treated with saponin, may deserve our keen attention. We hope that this new device will be as fruitful as the use of mechanically skinned fibers in linking the studies of subcellular fractions with the events of living muscle fibers.

Ford posed many questions as to whether or not the data obtained with biochemical preparations could explain physiological events. Indeed, the most important aspect of such biochemical research is to provide fundamental knowledge which should underlie physiological mechanisms. However, people are often very enthusiastic in pursuing the 'basic' or molecular mechanism, rather ignoring the physiological meaning of the matter. For instance, although beautiful schemes have been formulated for the Ca pumping mechanism, the question of whether or not the rates of Ca uptake of the fragmented SR so far reported can explain the rate of relaxation of skeletal muscle has not yet been seriously addressed (in our calculation, it explains only about one fifth of the required velocity). Since the situation of smooth muscle is like that of skeletal muscle SR, I feel that we smooth muscle investigators should keep in mind our physiological concern in future research.

It will perhaps be a long time before we can present such sophisticated work as described in this session by Prof. A. Katz.

BY A. M. KATZ (University of Connecticut)

We have had considerable experience with a stimulating effect of cyclic AMP on the calcium pump of cardiac sarcoplasmic reticulum (1). From these experiences, it can be stated that failure to see an effect of cyclic AMP on a smooth muscle calcium pump may be overlooked, even if a stimulatory effect exists, for several reasons. These can be grouped as follows :

1. Failure to phosphorylate a regulatory protein
   a. Not enough protein kinase
   b. Presence of a protein kinase inhibitor
   c. High phosphoprotein phosphatase activity
   d. Active phosphodiesterase activity
   e. Use of an inhibitory, high, cyclic AMP concentration

2. Failure to start with a dephosphorylated regulatory protein.
   High endogenous adenylate cyclase activity could make enough cyclic AMP to phosphorylate a regulatory protein, even in the absence of added cyclic AMP.

3. Failure to examine the regulated aspect of calcium transport.
   Based on our experiences with cardiac sarcoplasmic reticulum, this requires :
   a. That transport <u>rate</u>, not capacity ("binding" or with calcium pre-

cipitating anion) be measured.

b. That Ca^{2+} concentration be in range around the K_{Ca} for calcium transport as in cardiac sarcoplasmic reticulum, phosphorylation of phospholamban appears to change the Ca^{2+}-sensitivity of the pump, not its maximum velocity.

4. Artefact. High (> 0.2 mg/ml) concentrations of soluble proteins can increase the permeability of millipore filters, thereby allowing Ca-filled vesicles to pass into the filtrate. This artefact reduces apparent calcium uptake.

It is possible, of course, that an effect of cyclic AMP is really absent !

1. Katz, A.M., Tada, M. and Kirchberger, M.A. (1975). Control of calcium transport in the myocardium by the cyclic AMP-protein kinase system. _Adv. Cyclic Nucleotide Res._ 5 : 453-472.

Excitation-Contraction Coupling in Smooth Muscle, R. Casteels et al. eds.
© 1977 Elsevier/North-Holland Biomedical Press

THE LANTHANUM METHOD

C. van Breemen, O. Hwang and B. Siegel
Dept. of Pharmacology, University of Miami School of Medicine, Miami, Fla., USA.

The La-method was designed to allow measurement of fluctuations in Ca content of smooth muscle cells (1). Fig. 1 illustrates diagrammatically both the need for and the principle of quenching cellular Ca. When a smooth muscle preparation is equilibrated in a ^{45}Ca labelled physiological salt solution (PSS) approximately 95% of the radioactivity will be located extracellularly. This results from a large extracellular space containing many Ca binding sites (2). Upon transfer of the preparation to non-labelled PSS, ^{45}Ca washes out at a continually deminishing rate making it difficult to define efflux of cell ^{45}Ca as a unique exponential process. Thus in order to measure Ca uptake associated

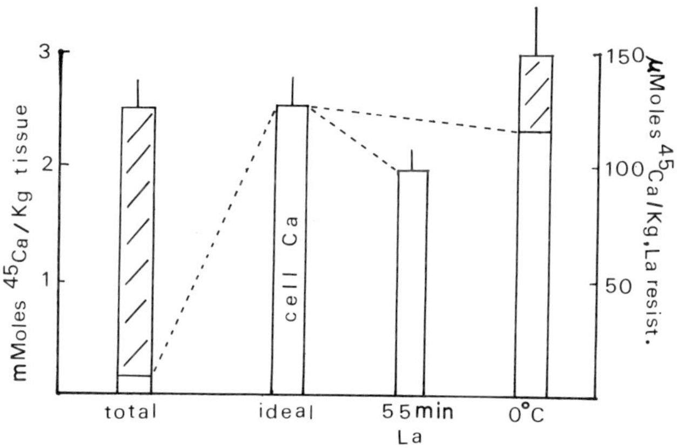

Fig. 1: Schematic representation of exchangeable Ca distribution in the aorta before (total) and after perfect quenching (ideal), 55 min wash in 10 mM La, 0Ca PSS (Ca free physiological salt solution) or a 55 min wash in the same solution at 0° C. For the last 3 bars use units on right ordinate.

with myofilament activation it is necessary to first remove the extracellular Ca with minimal loss of cellular Ca. To this effect it was proposed to soak the tissue in the presence of 2-10 mM La and absence of Ca. The high affinity of La for Ca binding sites (3) favors displacement of extracellular Ca and La occupation of Ca transport sites at the cell membrane. Ideally only cellular Ca should remain for analysis but in fact some Ca is lost from the cells before analysis or some extracellular Ca remains. The latter leads to greater variability of the Ca measurements.

Displacement of extracellular Ca:

Fig. 2A shows that addition of 2 mM $LaCl_3$ to Ca free tris buffered PSS causes displacement of an additional .7 mMoles Ca/Kg aorta. When smooth muscle is exposed to La first, Ca binding and influx appear blocked such that ^{45}Ca uptake can be accounted for by free diffusion into the extracellular space (Fig. 2B). Weiss and Goodman (5) and Godfraind (6) have observed Ca displacement by La in the longitudinal muscle of the ileum and rat aorta. To obtain effective dis-

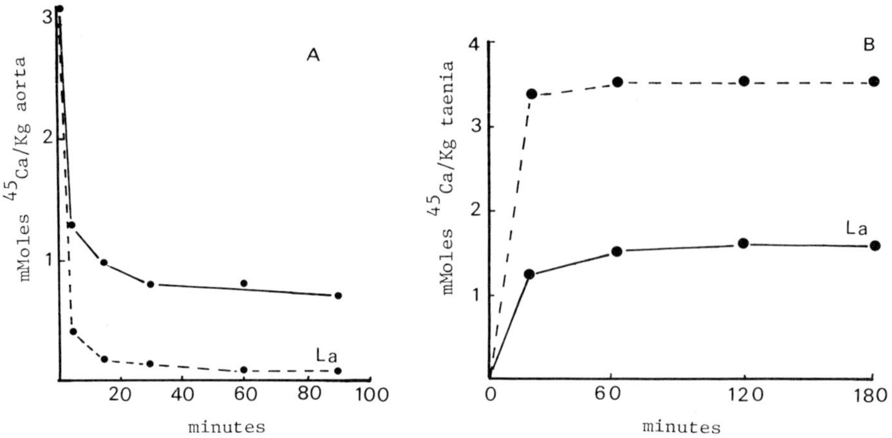

Fig. 2A: The loss of labelled Ca from aortic strips which had been exposed to ^{45}Ca for 20 minutes, to either OCa, PSS or OCa, PSS with 2 mM $LaCl_3$ (La). Fig. 2B: Labelled Ca uptake by guinea pig taenia coli from PSS containing 4.5 mM Ca, labelled with ^{45}Ca, or the same solution plus 10 mM $LaCl_3$ (La). (from ref. 4).

placement of extracellular bound Ca during the La wash, care must be taken to prevent Ca trapping. Phosphates and carbonates must be replaced by organic buffers and tissue damage kept to a mimimum. Damaged cells will leak phosphates which precipitate with La in the extracellular space resulting in variable Ca trapping. For example the guinea pig ceacum must be continually superfused with PSS during dissection of the taenia coli in order to obtain reproducible results with the La-method.

La inhibition of Ca efflux:

The possible sites at which La could act to inhibit Ca efflux are illustrated in Fig. 3A. Good evidence exists that La blocks all passive membranes Ca transport. A millipore filter impregnated with mixed phospholipids and cholesterol and suspended between aqueous solutions is traversed by narrow negatively charged pores (7). Exposure to La arrests all Ca flux across such a membrane (Fig. 3B) while the fixed negative charge is reversed as witnessed by reversal of the diffusion potential. Thus La binds to the negative sites changing the

Fig. 3A: A diagrammatic representation of various Ca transport mechanisms: a) diffusion through negatively charged pore b) ionophore mediated Ca transport c) Na-Ca exchange d) Ca exchange diffusion and f) Ca-ATPase-Ca pump. Fig. 3B: Ca flux across a millipore filter impregnated with phospholipids and cholesterol. At the arrow 0.5 mM $LaCl_3$ is added to the aqueous solutions on both sides of the membrane (from ref. 4). Fig. 3C: Potential difference across the same membrane as it separates 50 mM $CaCl_2$ from 5 mM $CaCl_2$. At the arrow 1 mM $LaCl_3$ is added (+ scope input connected to sol. of lower conc.) (from ref 8). Fig. 3D: ^{45}Ca uptake by smooth muscle of the rabbit aorta from 80 mM K PSS in absence and presence of 2 mM $LaCl_3$ (from ref. 4). Fig. 3E: ^{45}Ca efflux from the squid axon. A the arrows $LaCl_3$ is added to final concentrations of .2 mM, 1 mM and 5 mM. Fig. 3F: Stimulation of ^{45}Ca efflux from the squid axon by 2 mM cyanide and subsequent inhibition by 5 mM $LaCl_3$ (from ref. 9).

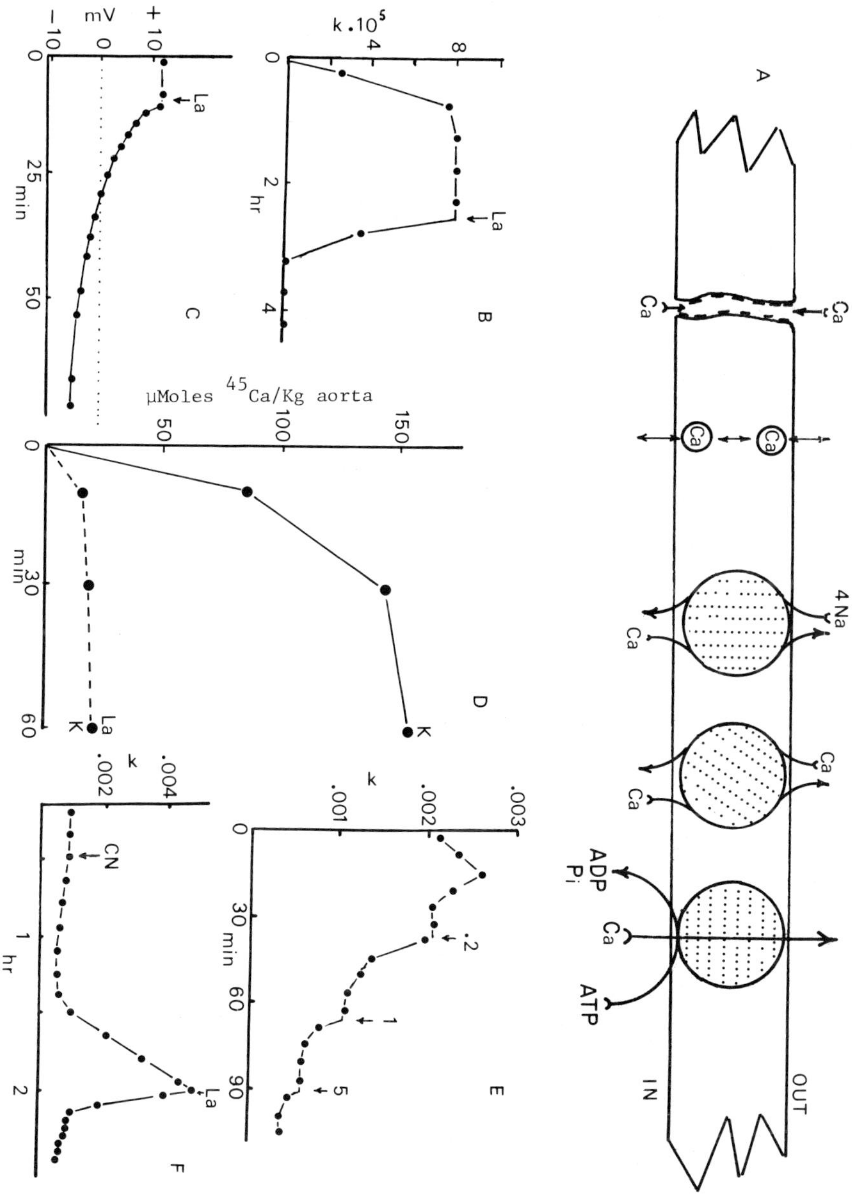
A
Ca
Ca
4Na
Ca
ADP
Pi
ATP
OUT
IN
B
k.10^5
hr
La
C
mV
min
La
D
µMoles 45Ca/Kg aorta
min
K
La
K
E
k
min
.2
1
5
F
k
hr
CN
La

Figure 3

fixed surface charge to positive. Ionophore mediated Ca fluxes are also inhibited by La (see below). Fig. 3D shows that the high K stimulated cellular Ca uptake by rabbit aortic smooth muscle is also blocked by La. Half maximal inhibition of this process occurs at 10^{-4} mM La. Fig. 3E shows that ^{45}Ca efflux from an internally injected squid axon is more than 80% inhibited by 5 mM extracellular La. ^{45}Ca efflux from an axon under normal physiological conditions has components dependent on ATP, extracellular Na and Ca. When the axon is poisoned by CN^- most of the Ca efflux depends on Na influx and this fraction is again inhibited by 5 mM La (3F). The La affinity of the Na-Ca exchange pump appears to be lower than the other Ca transport sites (10). It is more difficult to obtain

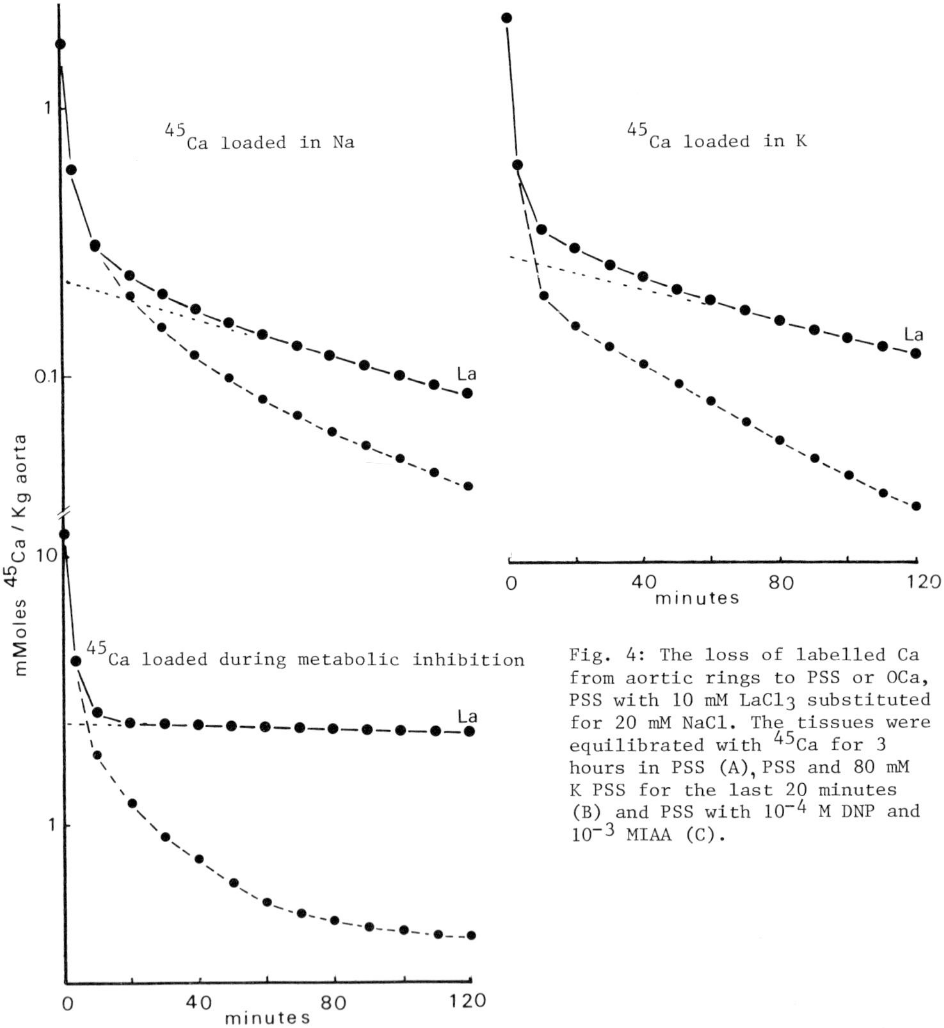

Fig. 4: The loss of labelled Ca from aortic rings to PSS or 0Ca, PSS with 10 mM $LaCl_3$ substituted for 20 mM NaCl. The tissues were equilibrated with ^{45}Ca for 3 hours in PSS (A), PSS and 80 mM K PSS for the last 20 minutes (B) and PSS with 10^{-4} M DNP and 10^{-3} MIAA (C).

similar information on ^{45}Ca efflux from smooth muscle. La displacement of extracellularly bound Ca has an effect on ^{45}Ca washout opposite to its inhibition of cellular ^{45}Ca efflux. However, 10 mM La is still seen to inhibit ^{45}Ca washout from aortic rings which were previously equilibrated 3 hours in ^{45}Ca labelled PSS (in mM: NaCl 140, KCl 5.0, $MgCl_2$ 1.0, dextrose 10.0, Hepes 5.0, $CaCl_2$ 1.5, pH 7.2 bubbled with O_2 at 37° C) 10 mM $LaCl_3$ was added to the loading solution for the last 5 minutes. The rings were then washed out in PSS or Ca free PSS + 10 mM $LaCl_3$ (substituted 20 mM $NaCl_2$). Control tissues were washed out in PSS.

The final portion of the log ^{45}Ca decay curve in the presence of La becomes linear. If the extrapolation of this linear portion to t=0 is assumed to represent the cellular Ca fraction then a regular 55 minute La wash will underestimate this fraction by 36%. The La inhibition is accentuated by increasing the cellular ^{45}Ca fraction by high K depolarization during the last 20 minutes of loading (Fig. 4B). The La block is nearly complete if the cells were ^{45}Ca loaded during metabolic inhibition and elevated extracellular Ca. However in this last case we are only testing passive transmembrane Ca fluxes.

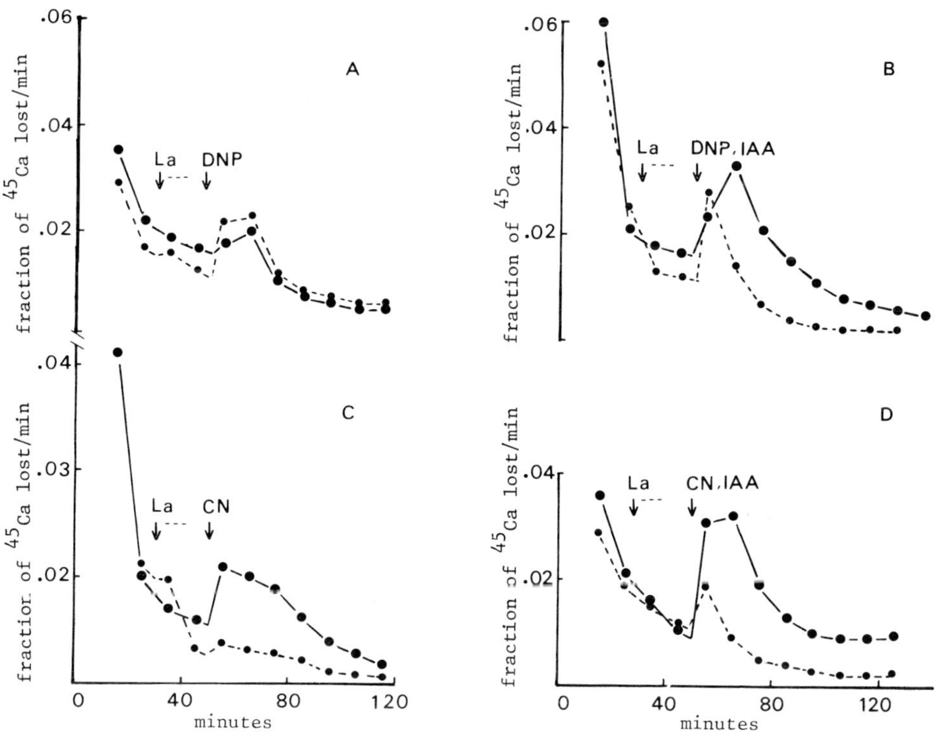

Fig. 5: The effect of 10^{-4} DNP (A), 10^{-4} M DNP + 1 mM IAA (B), 2 mM CN (C) and 2 mM CN + 1 mM IAA on the rate of ^{45}Ca efflux from rabbit aorta into PSS with (dotted line) or without 10 mM La.

Mitochondrial inhibitors and or idoacetic acid (IAA) cause transient stimulation of cellular ^{45}Ca efflux which is independent of Ca influx. This presents an ideal opportunity for studying La inhibition of ^{45}Ca efflux devoid of extracellular ^{45}Ca washout. However, Fig. 5A and B show a lack of La inhibition in the presence of dinitrophenol, as was recently reported by Deth (12).

To determine if this was due to inability of La to inhibit Ca extrusion or to a peculiar property of DNP, two milder ways of mitochondrial inhibition were tested (Fig. 5C and 6A). Both the CN and rotenone stimulated Ca fluxes were about 80% inhibited by 10 mM La. Neither 2 mM CN nor 10^{-5} M rotenone affected high K^+ or norepinephrine induced contractions whereas 10^{-4} M DNP caused variable suppression of these parameters. Mitochondrial inhibition causes only partial ATP depletion (13). The remaining ATP appears sufficient to fuel the Ca pumps since the aortic smooth muscle cells remained relaxed and showed no net Ca gain

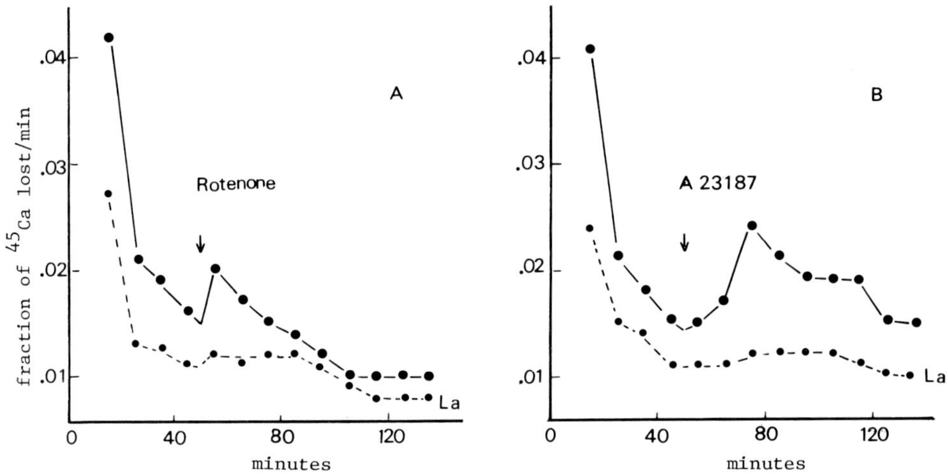

Fig. 6: The effect of 10^{-5} M rotenone or 10^{-6} M A23187 on ^{45}Ca efflux from rabbit aorta into PSS with (dotted line) or without 10 mM La.

during 30 minutes CN exposure. We thus conclude that 10 mM La markedly inhibits active Ca extrusion from smooth muscle. Half maximal inhibition of CN stimulated ^{45}Ca efflux occurred between 0.1 and 0.2 mM La which is similar to 0.1 mM La for 50% inhibition of high K^+ contractions. The Ca ionophore A23187 markedly stimulated the rate of ^{45}Ca efflux across the cell membrane. La almost completely inhibited this mode of Ca transport. A control experiment showed that La did not transfer A23187 from an organic hydrocarbon phase (30% butanol/70% toluene) into an aqueous phase. Thus it appears that La blocks the ionophore mediated Ca efflux at its membrane location. High concentrations of La thus rather nonspecifically block a variety of Ca transport systems.

NE stimulated ^{45}Ca efflux is partially inhibited by La, whereas caffeine stimulated Ca release is not affected by La. The mechanism whereby DNP and caffeine (12) bypass La inhibition is not clear at present.

Reversibility of La inhibition:

Fig. 7A shows that La inhibition of a high K^+ contraction cannot be reversed

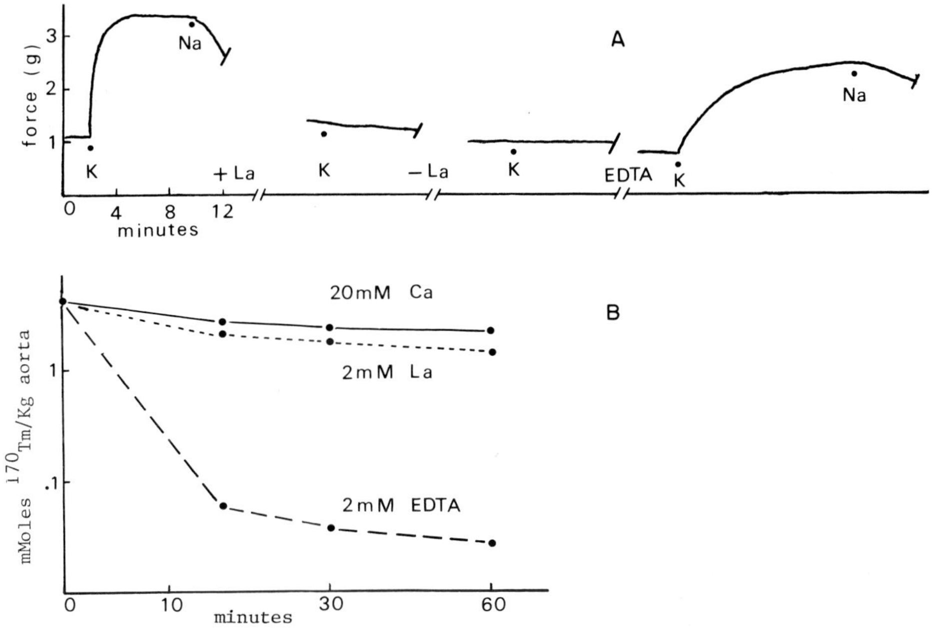

Fig. 7A: An isolated ring of rabbit aorta is stimulated with 80 mM K. 10 minutes exposure to 2 mM $LaCl_2$ blocks the next high K contracture. Removal of La for 30 minutes does not remove the La block. 10 minutes exposure to 2 mM EDTA followed by 15 minutes in normal PSS does restore the high K response.
Fig. 7B: Rings of rabbit aorta were exposed to PSS with 2 mM labelled Tm for 5 minutes. The ^{170}Tm was subsequently washed out in the presence of 20 mM Ca, 2 mM La or 2 mM EDTA.

by removal of La. 10 minutes of EDTA followed by normal PSS will however return the high K^+ response. In accordance with this EDTA reversal of the La block Fig. 7B shows that ^{170}Tm can only be removed from the aorta by EDTA. Tm is a lanthanide with exactly the same Ca blocking potency as La. 0.1 mM Tm gives half maximal inhibition of an 80 mM K induced contraction. It is used here because ^{140}Tm is a more convenient tracer. The aortae in Fig. 8B were loaded with ^{170}Tm for only 5 minutes before washing out the label in a series of test tubes containing PSS. Neither 20 mM Ca nor 2 mM La was able to displace the bound Tm. Thus EDTA chelates Tm or La which is bound to the cell membrane to restore Ca permeability.

La entry into smooth muscle cells:

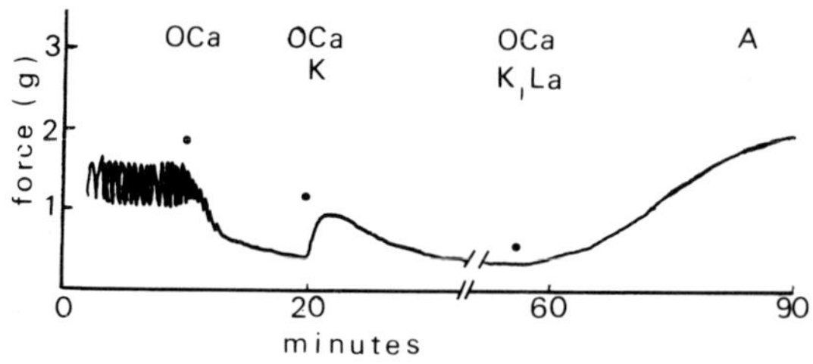

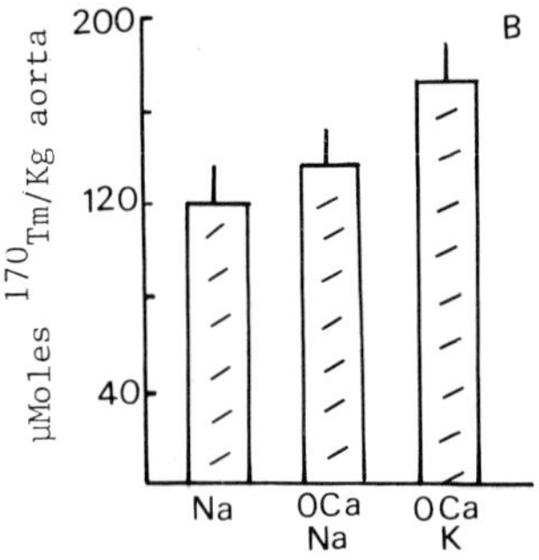

Fig. 8A: Tension recording of isolated guinea pig taenia coli suspended in PSS. Ca is removed from the solution and 10 minutes later 80 mM K substituted for Na. During the break in the curve (K) is returned to normal. Subsequent elevation of K in the presence of 2 mM La causes a substantial contraction.
Fig. 8B: ^{170}Tm uptake during 10 minutes in control, Ca free or Ca free 80 mM K solutions containing 2 mM labelled Tm. The pieces of taenia coli were washed for 55 minutes in PSS + 2 mM EDTA before analysis.

Although electron microscopy indicates that La is distributed extracellularly (14) it is conceivable that low concentrations escape detection. Fig. 8 provides evidence that high K depolarization of the guinea pig taenia coli stimulates the entry of La. Under Ca free conditions high K induces only a small transient contraction. However in the presence of La a slow irreversible contraction develops. La is thought not to directly activate the myofilaments but on entry to displace Ca into the cytoplasm (4). In Fig. 8B ^{170}Tm is used as a marker for La. Strips of taenia coli are incubated for 10 minutes in ^{170}Tm labelled PSS containing 2 mM Tm or the same solution without Ca, or ^{170}Tm labelled OCa, 80 mM K, PSS with 2 mM Tm following 10 minutes in OCa PSS. They were then washed in Ca free PSS + 2 mM EDTA for 55 minutes ashed and the radioactivity counted in a liquid scintillation counter. All tissues accumulated ^{170}Tm which was not removed by EDTA and high K stimulated this ^{170}Tm uptake. Fig. 8 thus provides evidence that 170 Tm and in analogy La can penetrate the membrane of the taenia coli during high K^+ depolarization. This should not interfere with the La-method for quenching Ca, if simultaneously the Ca efflux is inhibited.

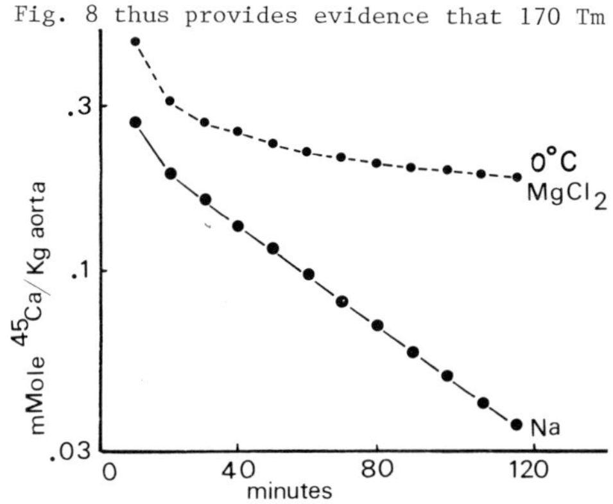

Fig. 9: Loss of ^{45}Ca from rabbit aorta into PSS or buffered isotonic $MgCl_2$ at 0° C.

Improvements on the La-method:

Casteels has suggested metabolic inhibition during the La wash. In this case use of DNP should be avoided. Complete metabolic inhibition causes membrane leakiness which could lead to La phosphate precipitates and extracellular Ca trapping. This may explain why in our hands this modification led to greater variability in the Ca determinations. A useful imporvement would be a cellular Ca quenching method which does not require omission of physiological phosphate and carbonate buffers. Such an attempt was made by the ^{45}Ca washout in isotonic $MgCl_2$ buffered to pH 7.2 with 5 mM Hepes and cooled to 0° C (Fig.9). However, again we had more scatter in the ^{45}Ca uptake data then with our use of the La-method. It is hoped that improvements can be made to reduce the partial loss of cellular Ca during the La wash period. It must be stressed however that a consistent approximation of cellular Ca even if it underestimates is extremely useful in elucidating Ca fluxes associated with excitation contaction coupling and relaxation.

Ca fluxes and c-AMP mediated relaxation in the canine coronary artery:

Although the physiological stimuli regulating the coronary blood flow are much debated, there is considerable evidence that elevated c-AMP is causally related to relaxation. Figure 10 correlates the contractile activity of rings cut

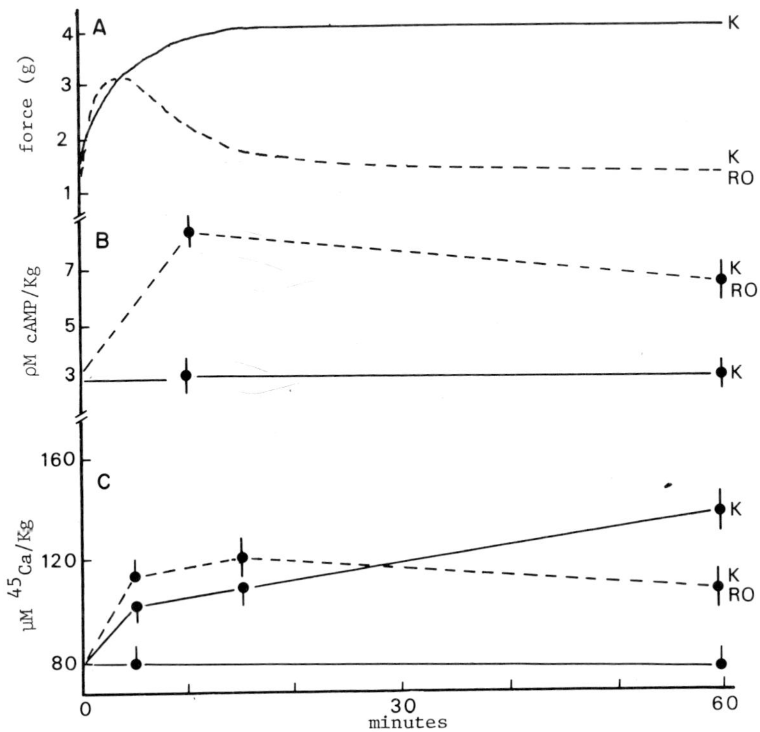

Effects of 1 mM RO20-1724 on tension induced by 80 mM K (A), c-AMP concentration (B)and exchangeable cellular Ca (C) in isolated dog coronary artery

from isolated dog coronary with their cyclic AMP content and exchangeable cellular Ca (measured by the La-method) during high K contraction and phosphodiesterase induced relaxation. The dog coronary artery rings were first equilibrated with ^{45}Ca labelled PSS for 80 minutes and then changed to 80 mM K PSS (equimolar substitution of KCl for NaCl) with or without 1 mM RO20-1724 labelled at the same specific activity. Some tissues were left in the labelled PSS. At the times indicated 10 mM $LaCl_3$ was added. The tissues were left in the labelled La solution for 5 minutes and then transferred to the 10 mM La PSS without Ca for a 55 minute wash before analysis for ^{45}Ca content. c-AMP was measured by means of the Schwarz-Mann radioimmune assay kit from extracts derived from tissues frozen in acid alcohol and homogenized by polytron.
The results indicate that the coronary arterial smooth muscle relaxes when c-AMP is elevated, and the cellular Ca is still elevated. Since the total Ca is the same with or without c-AMP and the free Ca^{2+} is lower with c-AMP, c-AMP must stimulate Ca^{2+} accumulation into a subcellular store, probably the SR. It is however very interesting that after relaxation is complete the elevated c-AMP appears to be associated with a net extrusion of Ca. These results are consistent with the hypothesis that c-AMP induces relaxation by stimulating the SR Ca pump (15),which dominates initially,but also aids in maintaining low cellular Ca by stimulation of the active Ca extrusion pump.

REFERENCES

1. van Breemen, C. and Mc Naughton, E. (1970). Biochem. Biophys. Res. Comm. 39, 569-574.
2. Goodford, P.J. and Wolowyk, M.W. (1972) J. Physiol. 224, 521-535.
3. Lettvin, J.Y., Pickard, W.F., Mc Cullock, W.S. and Pitts, W. (1964) Nature Lond. 202, 1338-1339.
4. van Breemen, C., Farinas, B.R., Casteels, R., Gerba, P., Wuytack, F. and Deth, R. (1973) Phil. Trans. R. Soc. Lond. B. 265, 57-71.
5. Weiss, C.G. and Goodman, F.R. (1969). J. Pharmac. Exp. Ther. 169, 40-55.
6. Godfraind, T. (1976) J. Physiol. 260, 21-35.
7. Mikulecky, D.C. and Tobias, J.M. (1964). J. Cell and Comp. Physiol. 64, 151.
8. van Breemen, C. (1968) Biochem. Biophys. Res. Com. 1968, 977-983.
9. van Breemen, C. and de Weer, P. (1970) Nature 226, 760-761.
10. Katzung, B.C., Reuter, H. and Pazig, H. (1973) Experientia 29, 1073.
11. van Breemen, C., Wuytack, F. and Casteels, R. (1975) Pflügers Arch. 359, 183-196.
12. Deth, R.C. (1977). Fed. Proc. 36, 4030.
13. Casteels, R., van Breemen, C., and Wuytack, F. (1972) Nature New Biol. 239, 249-251.
14. Langer, C.A. and Frank, J.S. (1972) J. Cell. Biol. 54, 441-455.
15. Anderson, R.G.G. and Nilsson, K.B. (1977). in The Biochemistry of Smooth Muscle, Ed.,Newman L. Stephens, M.D. University Park Press, Baltimore, 263-292.

Excitation-Contraction Coupling in Smooth Muscle, R. Casteels et al. eds.

APPROACHES TO DELINEATION OF DIFFERING CALCIUM BINDING SITES IN SMOOTH MUSCLE

George B. Weiss
Department of Pharmacology
University of Texas Health Science Center
Dallas, Texas, U. S. A.

The overall objective of our studies has been to increase understanding of the ionic basis underlying drug-induced stimulatory or inhibitory responses in smooth muscle. The general approach employed has been, in most part, a kinetic and isotopic one. Tension changes have been correlated with ionic fluxes and content in isolated tissues and the changes observed have been related to a common series of events preceding and/or leading to smooth muscle contraction or relaxation. More specifically, the working hypothesis employed is that effects of stimulatory and inhibitory agents on smooth muscle responsiveness are a direct function of the amount of intracellular Ca^{++} available and this, in turn, depends upon variations in the uptake, distribution, binding and release of Ca^{++} in each type of smooth muscle in a given species.

Today, I would like to discuss three aspects of this rather general problem. First, I will describe the basic techniques (and assumptions) we have employed to delineate ^{45}Ca movements in isolated smooth muscle tissues. Second, I will review some of the basic concepts we have derived from tension and ^{45}Ca flux studies with divalent and trivalent ions employed as partial antagonists of Ca^{++}. Finally, I will consider the question of how to quantify both the affinity of Ca^{++} for membrane sites and the number of Ca^{++}-binding sites. I will present a new approach we have derived for providing additional information about the number of binding sites and the affinity of Ca^{++} for these sites in an isolated vascular smooth muscle preparation.

A more experimentally-oriented understanding of the complexity of Ca^{++} movements in smooth muscle has emerged in recent years. Previously, it had been assumed that three Ca^{++} compartments (extracellular, membrane, intracellular) existed and interchanged with each other in a reversible manner. This formulation soon proved untenable. The primary problem was that stimulatory agents acting to release membrane-bound ^{45}Ca did not produce any detectable changes in

^{45}Ca efflux. To postulate an inward unidirectional Ca^{++} movement requires a more complex structural basis for Ca^{++} translocations. Thus, a more detailed general model seems appropriate. The basis for this model is an increased number of equilibrating and interchanging Ca^{++} compartments and sites (or stores). Those which exist and are of varying significance in different smooth muscle preparations include extracellular space, superficial (including stabilizing) binding sites, sequestered (less superficial) binding sites, intracellular space, intracellular storage sites (endoplasmic reticulum, mitochondria), and outward Ca^{++} transport. An example helpful in visualizing how a stimulatory agent would alter intracellular Ca^{++} levels without affecting ^{45}Ca efflux might be that of, say, norepinephrine on sequestered Ca^{++}. In this instance, release of sequestered Ca^{++} would increase the intracellular Ca^{++} concentration (and elicit a contractile response) and this, in turn, would be mainly rebound at intracellular storage sites and sequestered sites. Little if any additional extracellular ^{45}Ca efflux would be observed. However, it is important to recognize that measurement of ^{45}Ca fluxes does not yield any direct information concerning specific anatomical location of ^{45}Ca sites.

One way to obtain information about the binding of ^{45}Ca at cell fiber sites is to incubate isolated muscle preparations in Ca^{++}-deficient solutions. The resulting increased binding of ^{45}Ca at Ca^{++}-depleted sites effectively magnifies the bound ^{45}Ca fraction of the total ^{45}Ca uptake. This is best observed when the ^{45}Ca uptake is obtained in muscles incubated in Ca^{++}-free solutions and is calculated as a tissue/medium ratio. For example, the ^{45}Ca tissue/medium ratios for dog aorta, carotid and terminal mesenteric arteries are, respectively, 7.24, 40.64 and 17.78 ml/g[(1)]. Thus, we can conclude that, of these three tissues, carotid has the highest Ca^{++}-binding capacity and aorta has the least.

The movement or loss of ^{45}Ca can most effectively be detected when the data obtained are plotted as desaturation or rate coefficient plots. In these cases, counts of ^{45}Ca efflux per unit time are plotted as a percentage of the total ^{45}Ca originally taken up or as the fraction of that ^{45}Ca in the muscle emerging per unit time. In either case, the important point is to portray the ^{45}Ca

efflux as a known fraction of that ^{45}Ca present in the muscle. Other approaches have been used but are not as quantitatively precise. If ^{45}Ca efflux counts are plotted directly, rate changes observed may well involve non-specific release of proportionately small or even indeterminate fractions of Ca^{++} of questionable functional significance. If a ^{45}Ca efflux pattern is constructed from a series of residual ^{45}Ca uptakes (determined by analysis of several muscles incubated in ^{45}Ca and subsequently soaked for varying time intervals in a nonradioactive solution), then the plot is much less precise and the values represent an indirect efflux measurement.

Now, based upon these experimental approaches, we have investigated the effects of a number of divalent and trivalent ions and of various types of drugs. With ionic effects, the objective is to delineate different Ca^{++}-binding sites by differentially affecting these sites.

The actions of Sr^{++} provide an appropriate example. When 1.5 mM Sr^{++} is substituted for equivalent concentrations of Ca^{++}, then the contractile response of rabbit aortic smooth muscle to norepinephrine is inhibited to a greater degree than that to high K^+[2,3]. This differs from the effects obtained with rabbit aorta in a Ca^{++}-free solution when the response to norepinephrine persists long after that to high K^+ is abolished[4]. Further information about the mechanism of the differential responses obtained can be elicited from ^{45}Ca uptake and efflux studies. The ^{45}Ca tissue/medium ratio in rabbit aortic smooth muscle incubated in a Ca^{++}-free solution is approximately 25-30 ml/g[5]. In the presence of 1.5 mM nonradioactive Ca^{++}, this ratio decreases to approximately 1.5 ml/g. In the presence of 1.5 mM Sr^{++}, the tissue/medium ratio of approximately 4.0 ml/g indicates a marked effect but somewhat less than that of Ca^{++}. However, the effect of Sr^{++} on the slow component of ^{45}Ca efflux is the same as that of Ca^{++}[5]. On the basis of these and other experimental determinations (including the findings that Sr^{++} can substitute for Ca^{++} as a direct activator of the smooth muscle contractile mechanism[6] and that, in rabbit aorta, norepinephrine induces contractions by utilizing Ca^{++} from more sequestered sites than does high K^+[4]), we can derive some specific information about the mechanism by which Sr^{++} acts in aortic smooth muscle. Even though Sr^{++} displaces Ca^{++} from sequestered sites, stimulatory agents cannot utilize Sr^{++} from these sites in the same manner as Ca^{++}. However, K^{+}-induced responses, which depend upon uptake of Ca^{++} (or Sr^{++}) rather than release from sequestered sites, are not as inhibited as are those with norepinephrine in Sr^{++}-containing solutions.

The actions of the trivalent lanthanides in vascular smooth muscle can be considered as well-established(7). These ions (La^{+++}, Nd^{+++}, Eu^{+++}, Lu^{+++}) block ^{45}Ca uptake though the actions of Lu^{+++} are somewhat weaker than the others(8). Indeed, a high enough concentration of La^{+++} will block essentially all ^{45}Ca uptake and extracellular binding and the ^{45}Ca tissue/medium ratio will approximate the extracellular tissue space in both vascular(9) and intestinal(10) smooth muscle. However, they do not block the efflux of ^{45}Ca and, in the presence of high concentrations of La^{+++} (10 mM), virtually all of the ^{45}Ca present is depleted within approximately 20 minutes(9). This lanthanide-induced depletion is primarily an indirect effect (inhibition of uptake) because measurements with ^{147}Pm indicated that this lanthanum analog did not accumulate intracellularly to any significant extent and that the rate-limiting factor for ^{147}Pm efflux was desorption from the cell membrane(11).

The question of a more precise quantitation of Ca^{++} binding in terms of affinity of Ca^{++} for binding sites and of number of binding sites for Ca^{++} first arose during a recent study of the mechanism of the potentiating action of phosphatidyl serine in aortic smooth muscle(12). On the basis of the finding that responses to stimulatory agents other than high K^{+} were potentiated by phosphatidyl serine, it was postulated that phosphatidyl serine might act by increasing the effectiveness of Ca^{++} at more sequestered sites. More specifically, it was found that phosphatidyl serine decreased the exchangeability of Ca^{++} at sequestered sites and, in this manner, augmented the release of sequestered Ca^{++} by stimulatory agents and potentiated the resulting contractile response. However, these experiments, as designed, could not differentiate between an increase in affinity of sequestered sites for Ca^{++} or an increase in the number of sequestered binding sites for Ca^{++}.

A method designed to distinguish between these two possibilities has been developed by measuring the binding of Ca^{++} at membrane sites in muscles incubated for 30 minutes or 60 minutes with ^{45}Ca at different extracellular Ca^{++} concentrations. The ^{45}Ca uptakes, expressed as tissue/medium ratios, are summarized in table 1. By adjusting these values for the contents of the extracellular space

TABLE 1

TISSUE/MEDIUM RATIOS FOR ^{45}Ca IN RABBIT AORTIC SMOOTH MUSCLES

Extracellular Ca^{++}	Incubation in ^{45}Ca	^{45}Ca Tissue/Medium Ratios
mM	min	ml/g ± S.E.
0.004	30	46.587 ± 2.302
0.011	30	35.541 ± 2.169
0.031	30	24.916 ± 1.017
0.101	30	11.598 ± 0.914
1.501*	30	2.259 ± 0.115
1.501*	30	2.207 ± 0.187
5.001	30	1.377 ± 0.060
0.004	60	54.046 ± 5.942
0.011	60	42.341 ± 2.030
0.031	60	28.926 ± 0.957
0.101	60	12.531 ± 0.764
1.501*	60	2.346 ± 0.098
1.501*	60	2.702 ± 0.308
5.001	60	1.448 ± 0.067
15.001	60	0.888 ± 0.029

*Values derived from sets obtained at two different times.

and the specific activity of the incubation solution, Ca^{++} uptake values, expressed as μmol/g tissue, can be obtained in all cases. These values and the ratio of Ca^{++} bound/Ca^{++} free are summarized in table 2. Further information about the relationships between these uptake values can be obtained with Scatchard-type plot(13). This plot, designed to determine affinity of ions for various protein-binding sites, can also provide analogous information in more complex systems of the type employed in this study. A Scatchard plot of the data in table 2 is included in figure 1 (some additional points obtained at 0.2 and 0.3 mM Ca^{++} are included in the figure but not in the tables). From this plot, we can conclude that the Ca^{++} taken up is bound at two sites, a high affinity

TABLE 2

UPTAKE OF ^{45}Ca IN RABBIT AORTIC SMOOTH MUSCLES

Extracellular Ca^{++}	Incubation in ^{45}Ca	Calculated Ca^{++} Uptake**	Ratio of Bound Ca^{++}/Free Ca^{++}
mM	min	μmol/g	
0.004	30	0.185 ± 0.009	46.250
0.011	30	0.386 ± 0.024	35.091
0.031	30	0.760 ± 0.032	24.516
0.101	30	1.130 ± 0.092	11.188
1.501*	30	2.775 ± 0.173	1.849
1.501*	30	2.697 ± 0.280	1.797
5.001	30	4.836 ± 0.300	0.967
0.004	60	0.214 ± 0.024	53.500
0.011	60	0.461 ± 0.022	41.909
0.031	60	0.884 ± 0.030	28.516
0.101	60	1.224 ± 0.077	12.119
1.501*	60	2.906 ± 0.147	1.936
1.501*	60	3.440 ± 0.462	2.292
5.001	60	5.191 ± 0.335	1.038
15.001	60	7.170 ± 0.435	0.478

*Values derived from sets obtained at two different times.

**Product of tissue/medium ratio minus extracellular space (0.41 ml/g) and extracellular Ca^{++} concentration ("free Ca^{++}").

site which is clearly present at low Ca^{++} concentrations and a low affinity site which is delineated at higher Ca^{++} concentrations. Furthermore, the relative number of binding sites (n) present per unit tissue (given as μmoles Ca^{++} bound per gram tissue) can be estimated from the x-axis intercept. These values as well as calculation of K_D, the apparent dissociation constant (K_D = n/y intercept) values are included in figure 1. It is clear that the affinities for Ca^{++} and apparent number of low affinity binding sites for Ca^{++} are similar after both 30 and 60 minutes of exposure to ^{45}Ca. However, the apparent number of high affinity binding sites is less after 30 minutes of ^{45}Ca incubation than after 60 minutes. This indicates that binding of Ca^{++} at high affinity sites is not complete (equilibrated) after 30 minutes exposure to ^{45}Ca. Possibly, high affinity Ca^{++} sites are less accessible than low affinity Ca^{++} sites.

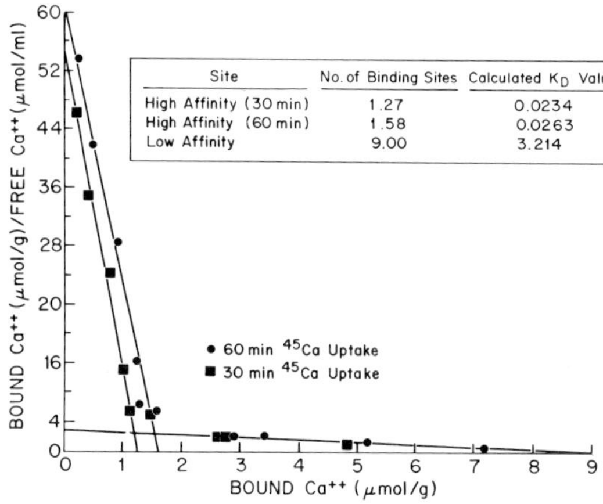

Fig. 1. Scatchard plot of ^{45}Ca uptake from solutions containing different extracellular Ca^{++} concentrations into rabbit aortic smooth muscle.

More important, it is clear that all of the Ca^{++} taken up is present at two specific membrane sites. It is also obvious that actions of agents which inhibit Ca^{++} uptake and/or binding might be more specifically described in terms of effects on either affinity of Ca^{++} for sites or numbers of available sites for either high or low affinity sites. Indeed, preliminary data indicate that both Sr^{++} and Lu^{+++} block uptake of high affinity Ca^{++}, whereas only Lu^{+++} has a significant inhibitory effect on low affinity Ca^{++}. On the basis of these actions as well as differing actions on ^{45}Ca efflux (Sr^{++} displaces ^{45}Ca as well as increases the maintained rate of washout, whereas Lu^{+++} only increases the maintained rate of washout), it is possible to conclude that Sr^{++} displaces high affinity Ca^{++} but not low affinity Ca^{++}, whereas Lu^{+++} blocks uptake of Ca^{++} to high affinity sites. By utilizing similar approaches, it should also be possible to examine the effects of specific drugs in terms of clearly delineated effects on high and low affinity sites.

REFERENCES

1. Goodman, F.R., Adams, H.R. and Weiss, G.B., 1975, Blood Vessels 12: 248-260.

2. Hudgins, P.M. and Weiss, G.B., 1969, Fed. Proc. 28: 541.

3. Hudgins, P.M., 1969, J. Pharmacol. Exp. Ther. 170: 303-310.

4. Hudgins, P.M. and Weiss, G.B., 1968, J. Pharmacol. Exp. Ther. 159: 91-97.

5. Hudgins, P.M. and Weiss, G.B., 1969, Am. J. Physiol. 217: 1310-1315.

6. Bohr, D.F., 1964, Canad. Med. Assoc. J. 90: 174-179.

7. Weiss, G.B., 1974, Annu. Rev. Pharmacol. 14: 343-354.

8. Weiss, G.B. and Goodman, F.R., 1975, J. Pharmacol. Exp. Ther. 195: 557-564.

9. Goodman, F.R. and Weiss, G.B., 1974, Proc. 11th Rare Earth Conf. 2: 716-722.

10. Burton, J. and Godfraind, T., 1974, J. Physiol. (London) 241: 287-298.

11. Weiss, G.B. and Goodman, F.R., 1976, J. Pharmacol. Exp. Ther. 198: 366-374.

12. Goodman, F.R., Weiss, G.B. and Goth, A., 1976, J. Pharmacol. Exp. Ther. 198: 168-175.

13. Scatchard, G., Coleman, J. S. and Shen, A.L., 1957, J. Am. Chem. Soc. 79: 12-20.

Excitation-Contraction Coupling in Smooth Muscle, R. Casteels et al. eds.
© 1977 Elsevier/North-Holland Biomedical Press

DISSOCIATION OF CALCIUM DEPENDENT REACTIONS AT DIFFERENT SITES: LANTHANUM BLOCK OF CONTRACTION AND RELAXATION IN A MOLLUSCAN SMOOTH MUSCLE

Betty M. Twarog
Department of Anatomical Sciences
Health Sciences Center
State University of New York at Stony Brook
Stony Brook, New York 11794
U. S. A.

I. NEURAL CONTROL OF MOLLUSC CATCH MUSCLE

The anterior byssus retractor muscle (ABRM) of the mollusc, *Mytilus edulis*, is a smooth muscle which resembles the smooth muscles of vertebrates in both structure and function (1, 2, 3, 4). Figure 1 summarizes current knowledge and speculations concerning the roles of neural mediators and the mechanisms which activate phasic and catch contractions and control relaxation of catch. Cholinergic nerves cause excitation and contraction. The released acetylcholine activates a depolarizing receptor causing an increased conductance to Na^+, K^+ and Ca^{2+} ions (5, 6). This depolarization brings the membrane to threshold for the generation of a calcium dependent spike (4, 7) and Ca^{2+} flux into the muscle cell increases abruptly. The depolarization and/or the Ca^{2+} influx during the spike trigger release of Ca^{2+} from internal stores (8) and at an intracellular pCa between 7 and 6, contraction is activated (9, 10, 11). The spike-induced tension is maximal when the Ca^{2+} concentration of the bathing medium is 10 mM and it does not increase if the Ca^{2+} concentration is increased (4, 12). This contraction is termed catch because the tension developed persists after activation ceases. Acetylcholine can be applied to ABRM depolarized in K^+, or to a muscle in which depolarization is prevented by omitting Na^+ from the perfusing medium (12, 13, 14). In both of these cases no spike is fired but influx of Ca^{2+} into the cell leads to a slow activation of the contractile system. The degree of activation is proportional to Ca^{2+} in the bathing medium, tension is barely detectable at 10 mM Ca^{2+}, and appreciable only above 50 mM Ca^{2+}. Relaxation ensues as soon as Ca^{2+} is removed from the external bathing medium. This contraction is of the phasic type.

Tryptaminergic nerves release serotonin, which specifically terminates catch contraction. When tryptaminergic nerves are firing, all responses to excitation are phasic (15). Serotonin binds at a specific relaxing receptor (16). No changes in membrane polarization precede or accompany the relaxing action of serotonin (17). Evidence has accumulated that the primary effect of serotonin is to increase the concentration of intracellular cyclic AMP (18, 19, 20, 21).

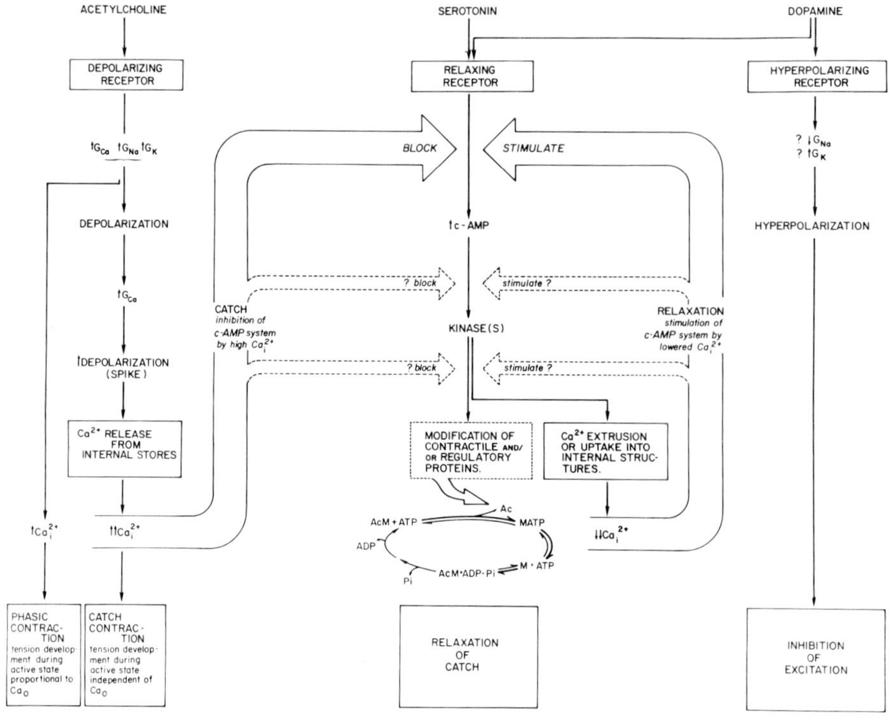

Although we proposed in 1966 (22) that relaxation of catch depends on regulation of intracellular Ca^{2+} levels we now incline towards an older speculation that the contractile system is itself directly affected (4, 23). There is now evidence that increased concentrations of cyclic AMP induce phosphorylation of the contractile or regulatory proteins (24, 25). The result could be a decreased affinity of the contractile proteins for Ca^{2+} (26), increased affinity for ATP, or a conformational change that interferes with the interaction of actin and myosin. However, despite the probability that the serotonin-cyclic AMP system directly modifies the contractile mechanism, there is considerable evidence that serotonin also decreases intracellular Ca^{2+} concentrations (4). Cyclic AMP and phosphorylations may mediate this action by increasing Ca^{2+} uptake into internal stores or by stimulating active Ca^{2+} extrusion across the cell membrane. Phosphorylation of elements in the microsomal fraction has been reported (27).

Our most recent hypothesis, set forth in Figure 1, suggests that the high intracellular Ca^{2+} concentrations produced by release of intracellular Ca^{2+} stores not only activate the contractile mechanism but also inactivate the serotonin-cyclic AMP triggered relaxing system. Thus the activating Ca^{2+} initiates catch by "turning off" the relaxing mechanism. There is probably a Ca^{2+} threshold for catch (i.e. turning-off of relaxation) which differs from the Ca^{2+} threshold for maximal activation. Observations by Marchand-Dumont and Baguet (26) and by Achazi (24) permit a hypothesis that relaxation may be inhibited by high levels of Ca^{2+}. The hypothesis is also consistent with observations that firing of the cholinergic excitatory nerves produces a catch contraction unless the relaxing nerves are also active (15). The hypothesis thus states that the primary relaxing process induced by serotonin is a cyclic-AMP mediated modification of the contractile system as suggested by Marchand-Dumont and Baguet (26) and by Achazi (24), but it further postulates that serotonin-induced lowering of internal Ca^{2+} concentrations plays a modulating role, relieving Ca^{2+}-induced inhibition of the relaxing system, and thus speeding relaxation.

Dopamine can activate the relaxing receptor (28, 29). However, there is strong evidence now that serotonin is the mediator released by relaxing nerves, although essential structural similarities shared by dopamine and serotonin enable the former to combine with the serotonin receptor (15, 16). Dopamine, however, may be a transmitter at inhibitory nerve endings (29). Dopamine combines with a receptor which triggers hyperpolarization of the membrane. The hyperpolarization can be blocked with ouabain, or cold (-2°C). Block of hyperpolarization by ouabain in no way alters the effect of dopamine on the relaxing receptor nor does ouabain change serotonin action.

II. LANTHANUM ION (LA^{3+}) AS A TOOL TO DISSOCIATE CA^{2+} DEPENDENT REACTIONS

A. Application of La^{3+} to normally polarized and depolarized ABRM

The speculation that catch is initiated by increased levels of intracellular Ca^{2+} and that relaxation is primarily effected by cyclic-AMP controlled alteration of the contractile mechanism, and secondarily modulated via the intracellular Ca^{2+} concentration prompted our use of methods which could selectively influence Ca^{2+} binding and/or movements at different sites. In our studies, we used La^{3+} as a Ca^{2+} antagonist in a molluscan smooth muscle in order to analyze the relative roles of Ca^{2+} influx and release from internal sites under different stimulating conditions (4, 30). We also compared the effects of La^{3+} as applied to normally polarized and depolarized muscle because La^{3+} applied to a depolarized muscle might reach sites which were beyond the superficial or which were selectively exposed by depolarization. Our interpretations of the results took into consideration that La^{3+}, under the conditions in which we used it, might not only interfere with Ca^{2+} movements but also affect sites on Ca^{2+} controlled enzymes.

B. Physiological and pharmacological observations

The results and experimental details of these investigations will be published elsewhere (30).

We studied contraction and relaxation in response to a variety of stimuli, each of which is distinct in its influence on Ca^{2+} movements. Addition of 5 mM La^{3+} to Ca^{2+}-containing ASW was equivalent, in general, to exposing the muscle to Ca^{2+}-free ASW. The evidence supports the hypothesis that La^{3+}, bound at a superficial site on the ABRM membrane interferes with entry of Ca^{2+} from the external medium. La^{3+} applied to a depolarized muscle apparently reaches sites beyond the superficial and blocks processes which are not dependent on the presence of Ca^{2+} in the external medium, for example, relaxation.

Acetylcholine (ACh) contraction

In 5 mM La^{3+}, catch tension in response to ACh is usually blocked within 25 minutes. Although the extent of ACh-induced Ca^{2+} influx in the ABRM is disputed (31, 32), the Ca^{2+}-spike of ABRM is known to be blocked in 0 Ca^{2+} ASW and in La^{3+}-ASW. Since in normally polarized ABRM, the spike apparently triggers Ca^{2+} release from internal stores (4, 8) 0 Ca^{2+} and La^{3+} probably act by inhibiting the triggering process. When spiking is blocked, the increased Ca^{2+} conductance induced by ACh permits entry of Ca^{2+} down its concentration gradient activating a phasic contraction which is readily blocked by 0 Ca^{2+} and La^{3+}-ASW.

High K^+ Contraction

On exposure to 5 mM La^{3+}, the response to K^+ ASW is rapidly blocked. K^+ ASW causes net Ca^{2+} influx in the ABRM (33). The response to K^+ ASW is blocked in 0 Ca^{2+} ASW. It is probable that La^{3+} blocks activation of K^+ contraction by blocking Ca^{2+} influx.

Caffeine Contraction

The first caffeine response after 60 minutes in 0 Ca^{2+} ASW or La^{3+} ASW remains at the control level but subsequent responses under these conditions are blocked. The results reinforce the hypothesis that caffeine contraction depends entirely on release of Ca^{2+} from internal stores, while replenishment of Ca^{2+} available for release at these sites depends on influx of Ca^{2+} from the external medium (4, 34). When La^{3+} is applied to a depolarized muscle, the caffeine response is unaltered, indicating that under these conditions La^{3+} does not block caffeine release of Ca^{2+} or Ca^{2+} activation of the contractile mechanism.

Serotonin and Dopamine Relaxation

The relaxing response to serotonin (and dopamine) persists during hours of exposure to Ca^{2+}-free ASW or in 5 mM La^{3+} ASW. However, depolarization of the ABRM by K^+ ASW or ACh for 2 to 5 minutes in the presence of La^{3+} completely blocks relaxation. It seems that block of relaxation requires that La^{3+} penetrate beyond superficial sites. It may be that Ca^{2+} sequestering sites in the membrane are required for relaxation. Alternatively La^{3+} may bind at the inner surface of the membrane, thus blocking outward Ca^{2+} transport as in erythrocytes (35). Outward Ca^{2+} transport has been suggested by us (4) and by Bloomquist and Curtis (32, 36, 37) who have reported that active Ca^{2+} extrusion is stimulated by relaxing mediators. It is also possible that La^{3+} blocks Ca^{2+} uptake into sarcoplasmic reticulum or mitochondria. Hodgson *et al*, (38) have observed La^{3+} accumulation in mitochondria of smooth muscle. Our morphological studies suggest La^{3+} accumulation at the inner membrane leaflet but have not revealed accumulation in mitochondria.

However, function might be affected by levels of La^{3+} too low to be visualized. A quite different explanation of La^{3+} block of relaxation follows from the speculation that Ca^{2+} modulates the formation of cyclic-AMP, or phosphorylation of contractile and/or regulatory proteins. Thus Ca^{2+} sites on adenylate cyclase or kinases would be sensitive to inhibition by La^{3+}. La^{3+} has been reported to block adenylate cyclase activity (39). It will be most interesting to test the effect of La^{3+} on adenylate cyclase activity and phosphorylation in ABRM.

C. Morphological Observations

With Professor Maynard M. Dewey and Mr. David Colflesh, observations were made of muscles soaked in 10 mM La^{3+} in a normal bathing medium (ASW) and in a medium rich in K^+ (high K^+ ASW). After treatment with 10 mM La^{3+} the control and the test (high K^+ ASW) muscles were fixed overnight at room temperature in 6% glutaraldehyde with 0.2 M cacodylate buffer containing 5% sucrose and .002 M $CaCl_2$ at pH 7.4 or 6.5. The muscles were then washed in the same buffer overnight at low temperature and post-fixed in equal parts of 4% osmium tetroxide and buffer for about 2 hours, then washed, dehydrated and embedded for observation by electron microscopy.

Plate 1A is an electronmicrograph including a nexus in a control ABRM incubated for 30 minutes in ASW. Plate 1B and C display areas of nexus after incubating ABRM in 10 mM La^{3+} ASW. There is an electron dense region spanning the extracellular leaflets and gap of the nexal (gap) junctions. In no other area was an electron dense deposit noted which was specifically attributable to pretreatment with La^{3+}. No changes were observed in cell organelles even after an hour soaking in La^{3+}. La^{3+} accumulates solely at the nexal junction and does not disrupt any cell structure which we have visualized. Plate 1D displays ABRM after exposure to high K^+ ASW for 5 to 20 minutes in the presence of 10 mM La^{3+}. Electron dense areas can be seen at the inner surface of the cytoplasmic membrane leaflet, perhaps indicating specific binding sites of La^{3+} in depolarized muscle. Oschman and

---

Plate 1. Electronmicrographs of ABRM. (A) Incubated in ASW; (B, C) Incubated in 10 mM La^{3+} ASW and (D) Incubated in high K^+ ASW containing 10 mM La^{3+}.

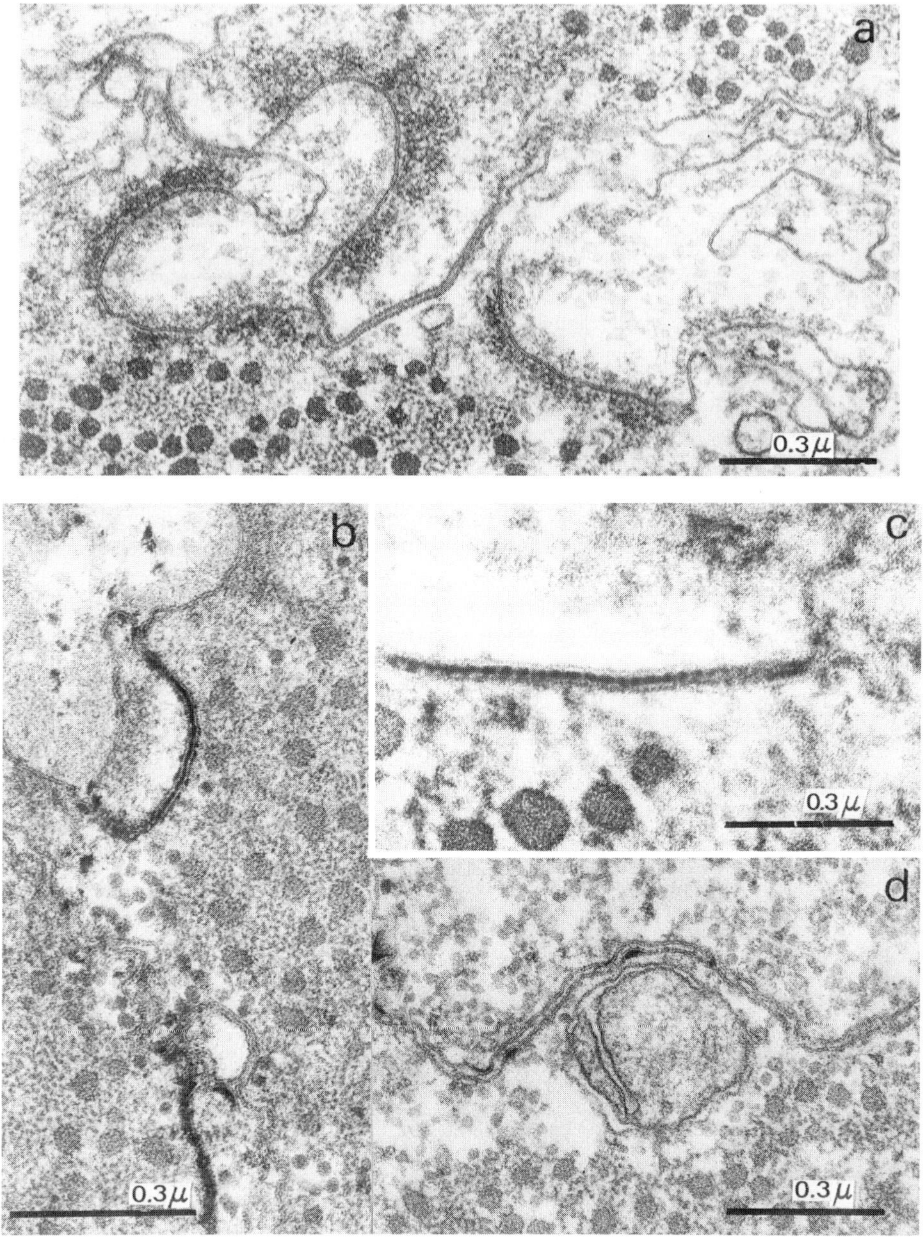
a
0.3 μ
b
0.3 μ
c
0.3 μ
d
0.3 μ

Wall have attributed similar densities in intestinal membranes to Ca^{2+} in fixing solutions (40). No densities are seen on the mitochondrial membrane. In high K^+-ASW, La^{3+} accumulates at areas of nexus, but this is not displayed here.

Brink, Kensler and Dewey have completed a study of the effects of La^{3+} on the nexus using freeze fracture techniques (41, 42). Freeze-fracture replicas display extensive regions of nexus. They are characterized by particles in the PF face (9.5 to 11 nm diameter) and pits in the EF face (3.5 to 4.5 nm diameter). The localization of the particles and pits is similar to that observed in vertebrate heart muscle and differs from the arrangement in arthropods where the particles are found in the EF face and pits in PF. In the control (incubated in ASW) the pits and particles are spaced at random. After incubation for 20 minutes in 5 mM La^{3+} ASW certain pits at the periphery of the nexus become highly ordered with respect to one another. Optical diffraction indicates a hexagonal array of pits in these ordered areas. The particles do not display ordering under these conditions. After longer times of soaking the EF face consistsentirely of hexagonally ordered pits while the particles remain randomly distributed. When muscles were incubated in high K^+ ASW containing 5 mM La^{3+}, the hexagonal ordering of the pits proceeded more rapidly. The particles remained randomly distributed. However, in high K^+ ASW containing La^{3+}, after the pits achieve maximum order they then begin to smooth out and disappear. Within an hour, the EF face is devoid of pits, although the particles in the PF face remain and their order seems unaltered. This effect may signal structural and functional disruption of the nexus. Disappearance of pits has been observed only when La^{3+} is applied in high K^+ ASW or after glycerol treatment. The interpretation of this phenomenon is of considerable interest. La^{3+} is known to "imitate" high Ca^{2+} as a stabilizer (43). A decrease in membrane fluidity caused by La^{3+} could lead to ordering of the pits. It may be that La^{3+} affects the cell membrane generally, but the effect is visualizable by morphological methods only in the area of the nexus where pits and particles provide a sensitive indicator of underlying structural changes. It would seem that in depolarized and glycerinated muscle La^{3+} combines at sites which were not exposed in the control and thus directly affects pit structure. It seems less likely that La^{3+} disrupts the relationship between pits and particles since the ordering of pits and particles is consistently independent (42).

In summary: the electron microscopic observations do not conclusively establish differences in La^{3+} localization in controls and muscles depolarized in high K^+ ASW, although there are indications that electron dense deposits at the inner membrane leaflet are observed only after depolarization in La^{3+}. On the other hand, the freeze-fracture studies of Brink, Kensler and Dewey clearly show that La^{3+} treatment of depolarized ABRM produces at areas of nexus a disruption of pit structure not seen in control muscles, which may well reflect penetration of La^{3+} to a specific site, not exposed in the control.

D. Observations of ^{45}Ca efflux

In experiments in collaboration with Professor P. Bianchi and Mr. S. Narayan ABRM was equilibrated overnight at 19° C in aerated ASW containing 0.1 to 10 microcurie ^{45}Ca. The muscles were removed from the equilibrating solution, rinsed four times in 0 Ca^{2+} ASW, placed in a chamber containing 2.5 ml 0 Ca^{2+} ASW and collections made at 2.5, 5 and 10 minutes and thereafter at 10 minute intervals for 150 minutes. One ml was analyzed for ^{45}Ca and 1.5 ml was analyzed for ^{40}Ca. At the end of the experiment, the muscles were analyzed for electrolyte content. Calcium, magnesium, sodium and potassium were measured by atomic absorptiometry. The control muscle was washed in the 0 Ca^{2+} ASW. The test muscle was washed in 0 Ca^{2+} ASW containing 10 mM La^{3+}.

The rate constants of ^{45}Ca efflux were calculated. There was no difference in rate constants of efflux from normally polarized muscles under control conditions and in solutions containing La^{3+}. The electrolyte analysis showed no significant difference between control and La^{3+} treated muscles with respect to Ca, Mg or K content. However, the sodium content was lower after washout in 0 Ca^{2+} ASW containing La^{3+}.

During depolarization by K^+ ASW in the presence of La^{3+}, the rate of efflux of ^{45}Ca increased. Following K^+ depolarization the rate of ^{45}Ca efflux was the same as in normally polarized ABRM. The Ca and Mg content of the muscles was unchanged, while the K content increased and the Na content decreased.

It is concluded that in ABRM, 10 mM La^{3+} does not alter the rate of ^{45}Ca efflux, whether it is applied to normally polarized or depolarized muscles. This result may help to resolve the question of how La^{3+}, applied to a depolarized muscle, specifically blocks relaxation. Of the three explanations set forth in II C, the first, that La^{3+} blocks outward transport of Ca^{2+}, seems to be ruled out. Thus La^{3+} inhibition of intracellular Ca^{2+} translocation or of adenylate cyclase or a kinase are more probable explanations.

REFERENCES

1. Twarog, B. M., 1976, Physiol. Rev. 56, 829-838
2. Twarog, B. M., 1971, Proc. Interntl. Congr. Physiol. Sci. 9, 572
3. Twarog, B. M., Dewey, M. M. and Hidaka, T., 1973, J. Gen. Physiol. 61, 207-221
4. Twarog, B. M. and Muneoka, Y., 1973, Cold Spring Harbor Sym. 37, 489-504
5. Twarog, B. M., 1967, J. Physiol. 192, 857-868
6. Hidaka, T. and Twarog, B. M., 1977, Gen. Pharmac. 8, 83-86
7. Twarog, B. M. and Hidaka, T., 1971, J. Gen. Physiol. 57, 252
8. Sugi, H. and Yamaguchi, T., 1976, J. Physiol. London 257, 531-548
9. Baguet, F., 1973, Pflugers Arch. 340, 19-34
10. Baguet, F., and Marchand-Dumont, G., 1974, Pflugers Arch. 354, 75-85
11. Nichols, M. R., 1975, Calcium and Contraction in a Molluscan Smooth Muscle. Ph. D. Thesis. Tufts University, Medford, Mass.
12. Muneoka, Y., 1966, Biol. Bull. Hiroshima Univ. 33, 25-33
13. Muneoka, Y., 1973, Comp. Gen. Pharmac. 4, 277-284
14. Muneoka, Y., 1974, Comp. Biochem. Physiol. 47A, 61-70
15. Twarog, B. M., Muneoka, Y. and Ledgere, M., 1977, J. Pharmac. Exptl. Therap. 201, 350-356
16. Twarog, B. M. and Cole, R. A., 1972, Comp. Biochem. Physiol. 43A, 331-335
17. Muneoka, Y. Cottrell, G. A. and Twarog, B. M., 1977, Gen. Pharmac. 8, 93-96
18. Cole, R. A. and Twarog, B. M., 1972, Comp. Biochem. Physiol. 43A, 321-330
19. Dolling, B., Achazi, R. K., Zebe, E., and Ahlert, U., 1972, Naturwissenschaften 59, 313

20. Achazi, R. K., 1974, Z. Naturforsch 29, 451-452

21. Achazi, R. K., Dolling, B. and Haakshorst, R., 1974, Pflugers Arch. 349, 19-27

22. Twarog, B. M., 1966, Life Sciences 5, 1201-1213

23. Twarcg, B. M., 1954, J. Cell Comp. Physiol. 44, 141-163

24. Achazi, R. K., 1977: Zur 5-HT-induzierten Erschlaffung von glatten Mollusken-musken. Habilitationsschrift, Westfalischen Wilhelms-Universitat, Munster

25. Achazi, R. K., 1976, Verh. Dtsch. Zool. Ges. 1976, 90

26. Marchand-Dumont, G. and Baguet, F., 1974, Pflugers Arch. 354, 87-100

27. Kohler, G., Heilmann, C., Nickel, E. and Florey, E., 1976, Pflugers Arch. 365

28. Hidaka, T., 1969, American Zoologist 9, 251

29. Hidaka, T., Yamaguchi, H., Twarog, B. M. and Muneoka, Y., 1977, Gen. Pharmac. 8, 87-91

30. Muneoka, Y. and Twarog, B. M., 1977, J. Pharmac. Exptl. Therap. (in press)

31. Sugi, H. and Tameyasu, T., 1975, J. Physiol. Soc. Japan 37, 215

32. Bloomquist, E. and Curtis, B. A., 1975, Amer. J. Physiol. 229, 1244-1248

33. Hagiwara, E. and Nagai, T., 1970, Japan. J. Physiol. 20, 72-83

34. Muneoka, Y., and Mizonishi, T., 1969, Zool. Mag. 78, 101-107

35. Weiner, M. L. and Lee, K. S., 1972, J. Gen. Physiol. 59, 462-475

36. Bloomquist, E. and Curtis, B. A., 1972, J. Gen. Physiol. 59, 476-485

37. Bloomquist, E. and Curtis, B. A., 1975, Amer. J. Physiol. 229, 1237-1243

38. Hodgson, B. J., Kidwai, A. M. and Daniel, E. E., 1972, Canad. J. Physiol. Pharmac. 50, 730-733

39. Nathanson, J. A., Freedman, R. and Hoffer, B. J., 1976, Nature 261, 330-332

40. Oschman, J. L. and Wall, B. J., 1972, J. Cell Biol. 55, 58-73

41. Kensler, R., Brink, P. R. and Dewey, M. M., 1977, Biophys. J. 17, 163a

42. Kensler, R., Brink, P. R. and Dewey, M. M., 1977 (manuscript)

43. Weiss, G. B., 1974, Ann. Rev. Pharmac. 14, 343-354

ACKNOWLEDGMENTS

The work mentioned in this review was supported in part by Grant NS 12857 from the U.S.P.H.S. Thanks are due to Professor M. M. Dewey, Dr. P. R. Brink and Mr. R. Kensler, who permitted me to discuss their unpublished data, and to Dr. R. K. Achazi, who provided me with a copy of his Habilitationsschrift.

Excitation-Contraction Coupling in Smooth Muscle, R. Casteels et al. eds
© *1977 Elsevier/North-Holland Biomedical Press*

^{45}CA-UPTAKE OF STOMACH SMOOTH MUSCLE DURING DIFFERENT MODES OF ACTIVATION

Ernst Lammel
Department of Physiology, University of Marburg,
Deutschhausstraße 2, D-3550 Marburg, W.-Germany

Studying the effect of so-called calcium antagonistic substances on the activity of different types of smooth muscle it was concluded that two mechanisms for raising the intracellular Ca concentration and inducing contraction exist in this tissue: A P system which can be blocked by verapamil, D 600 (methoxyverapamil, Knoll AG, Leverkusen, West Germany), and nifedipine (BAY a 1040, Bayer, West Germany), and a T system which is resistant to this group of substances (1,2,3). On the basis of this model the finding shown in Fig. 1 can be interpreted by the assumption that in fundus muscle of the guinea-pig stomach both systems contribute to physiological activation, while exposure to high potassium solution induces tension by only acting via the P-system: The P-antagonistic agent nifedipine (10^{-6} mol/l) produces only a partial suppression of the contraction force induced by acetylcholine (ACH). However, it totally blocks the potassium contracture ($[K^+]_o$ = 120 mmol/l) leaving only a small initial contraction transient.

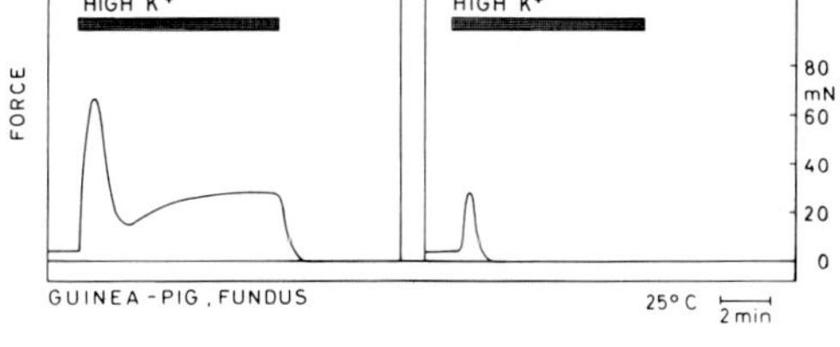

Fig. 1. Effect of nifedipine (10^{-6} mol/l) on an acetylcholine (ACH, $2 \cdot 10^{-5}$ g/ml) induced contraction (A) and on a potassium contracture (B, $[K^+]$ = 120 mmol/l) in isolated preparations of the fundus region of the guinea-pig stomach.

Electrophysiological as well as flux measurements indicate that P-antagonists act by blocking the transmembrane Ca influx related to the process of activation (4,5,6,7). The present studies were carried out in order to test first whether this mechanism also holds for the action of nifedipine in stomach muscle, and secondly whether the nifedipine-resistant component of an ACH-response is connected with a nifedipine-resistant influx component or if this component has to be attributed to release of Ca from cellular stores.

From the time course of ^{45}Ca-content of fundus preparations after incubation in radioactive solution (Fig.2) it can be seen that both kinds of stimulation considered lead to an increase of the rate of uptake of ^{45}Ca as well as of the amount of Ca which could be labelled in the tissue. (The ^{45}Ca uptake was measured in isolated circular strips according to the "lanthanum method" as described by van Breemen et al. (6) using a hepes-buffered modified Krebs solution of 25°C: Variable time of incubation in the ^{45}Ca-labelled experimental solution, 3 min 10 mmol/l $LaCl_3$ in the presence of ^{45}Ca, 50 min washing in a Ca-free solution containing 10 mmol/l $LaCl_3$.)

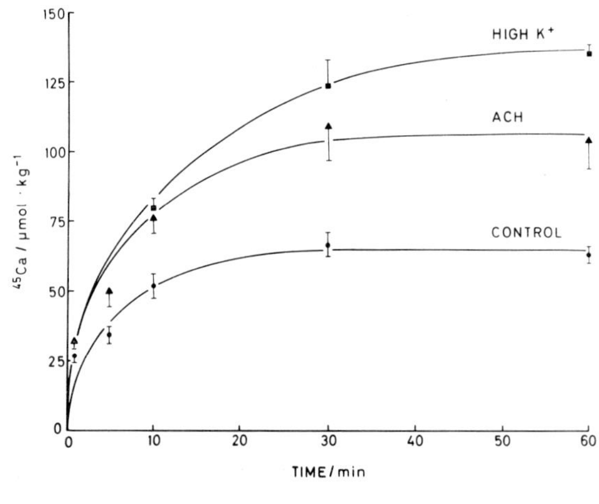

Fig.2. Time course of ^{45}Ca content (μmol/kg wet weight) as determined with the lanthanum method during exposure to labelled control solution (control), solution containing $2 \cdot 10^{-5}$ g/ml acetylcholine (ACH), and 120 mmolar K^+ solution (high K^+).

The effect of P-blockade by nifedipine on the ^{45}Ca-uptake was investigated in comparative experiments on pairs of preparations dissected from the same height of the fundus using a radioactive incubation period of 10 min. Fig.3 illustrates the influence of nifedipine on the uptake component related to K^+ depolarization. It is evident that this component was totally blocked during the time period following the initial contraction transient (r.h.s. groups of bars). When the contraction transient was included in the radioactive exposure time a small but not statistically significant uptake increment of the depolarized preparations was found (center group of bars).

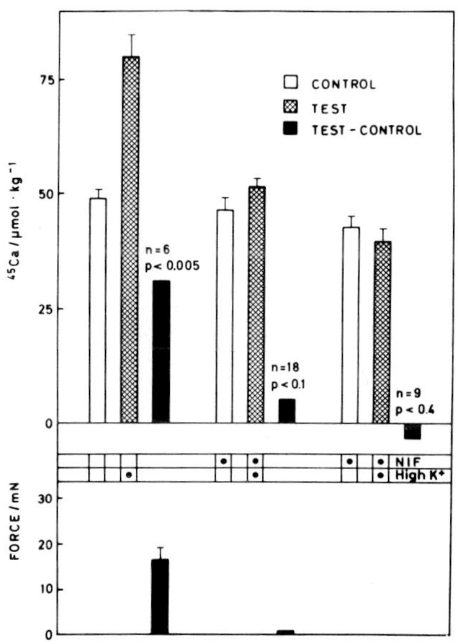

Fig.3. Effect of nifedipine (10^{-6} mol/l, NIF) on ^{45}Ca-uptake and contracture force induced by 120 mmolar K^+ solution (high K^+).
Upper part: ^{45}Ca-content after 10 min of radioactive incubation. Treatment of the preparation as marked below the bars. Black (test-control): Difference between the ^{45}Ca-content of test and control preparations. Center group of bars: Results of a series in which the initial contraction transient (compare Fig.1) was included in the incubation period. R.h.s. group: Results of a series in which the preparations were transferred to the labelled solution after the contraction transient.
Lower part: Mean value of force developed by the activated preparations during the test period.

In contrast to the findings with K^+ activation there was still an increased ^{45}Ca-uptake of preparations activated by ACH if both the control and the test preparations were treated with nifedipine, as can be seen from the results represented by the center bars of Fig.4. The incremental uptake amounted to about 50 % of the corresponding component in the absence of nifedipine (compare black bars of the center and l.h.s. group in Fig.4). The reduction of ^{45}Ca-uptake due to P-blockade was accompanied with a decrease of the ACH induced force development of 25 % (see lower part of Fig.4).

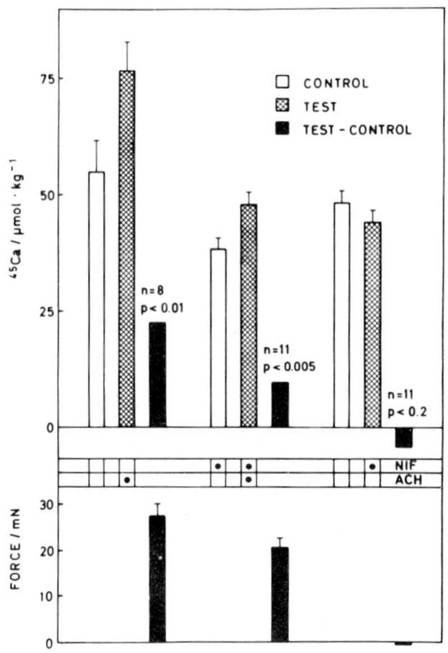

Fig.4. Effect of nifedipine (10^{-6} mol/l, NIF) on the ^{45}Ca-uptake induced by acetylcholine ($2 \cdot 10^{-5}$ g/ml, ACH). Details as described in Fig.3.

Nifedipine did not alter the basic Ca exchange unrelated to activity: A small but not significant decrease of ^{45}Ca-uptake of preparations under nifedipine was found in comparison with untreated strips (see r.h.s. bars of Fig.4). This result corresponds to the very low spontaneous activity of the preparations and to the accordingly small tension decrease producable by nifedipine under the experimental conditions (hepes buffer, 25°C).

Conclusion: Summarizing the results it can be concluded that an activation of fundus muscle induced by ACH, like a K^+ contracture, is connected with an increased transmembrane Ca influx. However, in contrast to the influx due to K^+ depolarization the influx induced by ACH consists of two components, one, which can be blocked by nifedipine, and another one, which is resistant to this P antagonist. This finding is consistent with the observation that contractions of fundus preparations under P blockade are still dependent on the presence of external Ca (2). It indicates the existence of a Ca transport mechanism in the membrane related to physiological activation which is resistant to P antagonists.

REFERENCES

1. Golenhofen, K. (1976) In: Physiology of smooth muscle, ed. by E.Bülbring and M.F.Shuba, p.197-202. New York: Raven Press
2. Boev, K., Golenhofen, K., and Lukanow, J. (1976) In: Physiology of smooth muscle, ed. by E.Bülbring and M.F.Shuba, p.203-208
3. Golenhofen, K., Wagner, B., and Weston, A. This volume
4. Van Breemen, C., Farinas, B.R., Casteels, R., Gerba, P., Wuytack, F., and Deth, R. (1973) Phil.Trans.R.Soc.Lond.B., 265, 57-71
5. Kohlhardt, M., Bauer, B., Krause, H., and Fleckenstein, A. (1972) Pflügers Arch., 335, 309-322
6. Mayer, C.J., van Breemen, C., and Casteels, R. (1972) Pflügers Arch., 337, 333-350
7. Kroeger, E.A., Marshall, J.M., and Bianchi, C.P. (1975) J. Pharmac.exp.Ther. 193, 309-316.

Excitation-Contraction Coupling in Smooth Muscle, R. Casteels et al. eds.

ION BINDING PROPERTIES OF VASCULAR CONNECTIVE TISSUE

G. Siegel[1], R. Ehehalt[1], H. Gustavsson[2], and L.-Å. Fransson[3]

[1]Institute of Physiology, Biophysical Research Group, The Free University of Berlin, D-1000 Berlin 33, Germany; [2]Departments of Physical Chemistry 2 and [3]Physiological Chemistry 2, Chemical Center, The Lund Institute of Technology, S-22007 Lund 7, Sweden

SUMMARY

1) Two new approaches to the study of ion binding properties in isolated vascular connective tissue as well as in mucopolysaccharides have been made using atomic absorption and nuclear magnetic resonance spectroscopy.

2) The special binding characteristics of K^+ and Ca^{++} ions in their dependence on the pH value indicate that these ion species play a regulating role during excitation and contraction processes.

INTRODUCTION

From earlier experiments it is known that the connective tissue structures of blood vessels have an excessive binding capacity for mono- and divalent cations[1,2]. Especially it has been proven that glycosaminoglycans have a cooperative interaction with Ca^{++} ions and bind a considerable number of these ions at specific sites and diffusely[3]. Further, it has long been known from light-microscopical and electron-optical experiments that the vascular smooth muscle cells are surrounded by a basal lamina and fitted with a fine network of connective tissue fibres like a glove. Due to this anatomical peculiarity, narrow tissue clefts between these extracellular connective tissue structures and the cell membrane are separated from the rest of the extracellular space. It could be possible that a change in the ion binding results from a change in the physical state functions of the extracellular space and/or a change in conformation of the proteoglycans, and this could lead to alterations in the ion concentrations in the narrow tissue clefts around the cell membrane. That would be a possible way for the extracellular connective tissue compartment, due to its structurally bound ions, to influence the excitation and contraction processes of smooth muscle cells. The following experiments show a new approach to study the release or binding mechanisms of cations and anions to molecular structures in vascular connective tissue.

MATERIALS AND METHODS

Experimental Preparations and Solutions: Common carotid arteries were surgically removed from anaesthetized dogs within 5 min. Adventitial connective tissue was isolated by microdissection and equilibrated in a modified Krebs solution having the following composition: Na^+ 151.16; K^+ 4.69; Ca^{++} 2.52; Mg^{++} 0.11; Cl^-

140.79; HCO_3^- 16.31; $H_2PO_4^-$ 1.38; glucose 7.77 mM (37°). The pH value of these solutions was varied between 4.0 and 9.0 whereby the pH steps 6, 7 and 8 were attained by changing the O_2/CO_2 ratio of the gas mixture with which the solutions were aerated. The pH steps 4 and 5 were arrived at by titration of the Krebs solutions with 65 % HNO_3, the pH step 9 by titration with 25 % NH_4OH. The preparations were incubated for 3 hours.

Determination of the Ion Concentrations: The total ion concentrations of the isolated tissue were determined using an atomic absorption spectrometer (Perkin-Elmer model 403). After determination of wet and dry weight, the preparation was placed in an ice-cold quartz Erlenmeyer flask containing pure oxygen and Pt catalyst and burned ash-free. The gaseous products were dissolved in a defined volume of 1.3 % HNO_3. Atomic absorption measurements were made directly from this fluid for Na, K and Mg; for Ca after addition of a 2 % lanthanum nitrate solution; and for Cl by measurement of Ag after quantitative precipitation with $AgNO_3$. The total method error is < ± 5 %.

^{23}Na Nuclear Magnetic Resonance Measurements: A modified Varian XL-100 Fourier transform spectrometer operating at a resonance frequency of 26.47 MHz was used for the ^{23}Na NMR transverse relaxation rate, $R_2 = 1/T_2$, determinations. Each spectrum is the result of at least 200 accumulations with an acquisition time of 0.4 s, and each reported value is the average of at least 3 separate determinations. R_2 is calculated from $R_2 = 1/T_2 = \pi \cdot \Delta\nu_{1/2}$, where $\Delta\nu_{1/2}$ is the line width at half height of the observed resonance signal.

The longitudinal relaxation rate, $R_1 = 1/T_1$, measurements were performed on a Bruker BK 322s NMR spectrometer at 23.812 MHz, using the 180°-t-90° plot method. The corresponding T_2's were obtained using the Meiboom-Gill-Carr-Purcell sequence with 8 ms between the 180° pulses. In both cases, the signals were averaged using a Varian CAT-1024 time average computer.

The samples were prepared by dissolving an amount of Chondroitin sulphate·Na salt (SERVA, Heidelberg), corresponding to 0.1 m concentration in both carboxylate and sulphate groups in 2 g of 0.1 m aqueous NaOH solution. The so-called chondroitin sulphate was found to be composed of a mixture of 95 % chondroitin sulphate and 5 % protein. Therefore the SERVA substance may be classified as a chondroitin sulphate-peptide (CS-P). In addition to the CS-P, experiments were made with pure chondroitin sulphate without any protein (CS)[4]. The *pH titrations* were performed by adding small volumes of 1 m aqueous HCl. The degree of neutralization, α, is defined as follows: When $\alpha = 1$, the excess of hydroxide has been titrated; $\alpha = 0$ corresponds to a further addition of 1 eq hydrochloric acid per eq substance.

For the *dilution* experiments the samples were prepared by dissolving an amount of CS-P, corresponding to 0.15 m concentration in both carboxylate and sulphate groups in 2 g H_2O (pH = 5.7). The initial Na concentration was thus as in the pH titration experiments 0.3 m. To lower the pH value to about 3.6 (cf. Fig. 2A, lower plateau of the pH titration curve[5]), a small volume of 1 m aqueous HCl was added. The samples were then diluted with a defined amount of water, whereby the ratio of the molar concentrations of the substance to the metal ions remained constant ($A = \frac{c_S}{c_L} = 0.5$). The pH value of the probes showed practically no change during the dilution.

For the *competition* experiments a 0.1 m CS-P solution in 2 g H_2O was made and the pH was adjusted to 3.6 by the addition of a small amount of 1 m aqueous HCl. Then defined amounts of cristalline NaCl, KCl or $CaCl_2 \cdot 2H_2O$ were added to the samples. The pH value of the samples remained unchanged with these additions.

RESULTS AND DISCUSSION

Ion Binding to Connective Tissue

Fig. 1 shows the Na^+ and Cl^- binding on isolated adventitial connective tissue of the A. carotis as a function of the pH value of the Krebs solution. We see that,

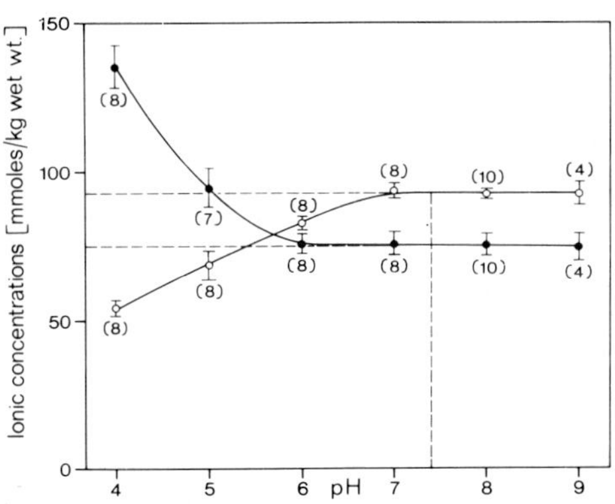

Fig. 1. Na (o) and Cl binding (•) of isolated adventitial connective tissue from the dog A. carotis as a function of the pH of the Krebs solution. Dashed lines indicate characteristic physiological pH and concentration values.

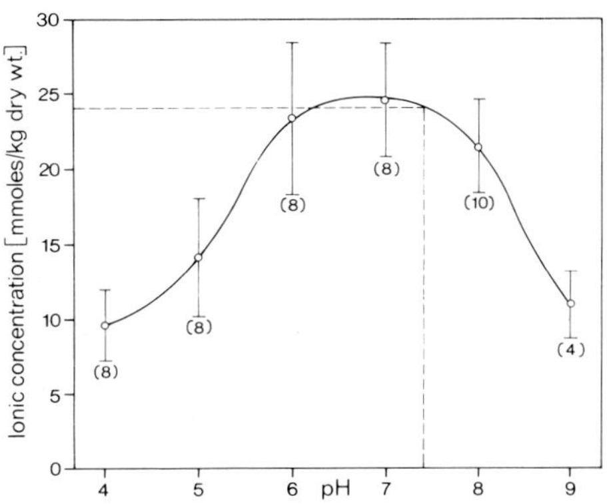

Fig. 2. K binding of isolated adventitial connective tissue from the dog A. carotis as a function of the pH of the Krebs solution.

as expected, the Na concentration increases with increasing pH up to a saturation value, the Cl concentration on the other hand decreases. The ion concentrations are given per kg of tissue wet weight, whereby the tissue water constitutes approximately 62 % of the total weight. To calculate the ionic fractions bound to the connective tissue structures, the amount of ions in the tissue water should be subtracted from the total amount and the results adjusted to the tissue dry weight; largely negative values result. That does not justify the conclusion that Na^+ and Cl^- ions cannot also be bound. Obviously the two compartment model used here is an oversimplification. From the literature we know that water in and around connective tissue structures could exist in various forms which are called "structured", "free" and "solidlike"[6,7]. Thus it is possible that the ion concentrations in the immediate neighbourhood of connective tissue macromolecules are different than in the surrounding water. In Fig. 1 we see further that in the physiological pH range the Cl concentration is 18 mmoles/kg wet weight lower than the Na concentration. The electroneutrality is supported by the other negative ions of the Krebs solution (HCO_3^-; $H_2PO_4^-$). Finally it must be asked if the Na^+ and Cl^- ions bound to connective tissue could play a regulating role in membrane excitation processes during a pH change. The curves in Fig. 1 indicate not, because the Na as well as the Cl bindings have reached a stationary level in the physiological pH range.

Starting at pH = 4 the K binding, similarly to the Na binding, increases with increasing pH (Fig. 2). The K binding attains a maximum of 25 mmoles/kg dry weight at pH = 6.8 and decreases with basic pH. This decrease of the K binding with increasing basicity of the pH is an extremely unusual result for a pH dependent cation binding curve, which in principle is similar to a pH titration curve. An explanation for this behaviour is possible when one considers the strong increase in Ca binding in the basic pH range (cf. Fig. 3). Clearly the K^+ and Ca^{++} ions compete for the same binding sites on connective tissue structures. Because of the higher affinity of Ca^{++} ions the K^+ ions are expelled from the binding sites (cf. Fig. 7). The actual K binding curve shows that by a displacement of the pH value from 7.4 towards acidity the K^+ ions are bound to connective tissue, by a displacement towards basicity they are released. The physiological meaning of this mechanism could be that the result of an acidosis would be the decrease of extracellular K^+ concentration in the small pockets between the cell membrane and the basal lamina, the result of an alkalosis would be the increase of the extracellular K^+ concentration in the neighbourhood of the cell membrane. The first effect will lead to membrane hyperpolarization with relaxation, the second effect to depolarization with contraction.

The possible existence of an autoregulatory mechanism for local blood perfusion of a tissue area has been suggested in an earlier paper[5]. It was speculated, that changing the pH to tissue acidosis leads to a release of Ca^{++} ions from connective tissue proteoglycans, especially from the basal lamina of the vascular smooth muscle cells[8]. The released Ca^{++} increases the Ca^{++} concentration in the narrow clefts between the cell membrane and the basal lamina, thus possibly effecting a membrane stabilization with hyperpolarization and/or a shift of the threshold for initiation of contraction to a greater depolarization[9,10]. The very steep Ca binding curve in Fig. 3 confirms that even a very small pH change in the physiological range can lead to a considerable release or binding of Ca^{++} ions. In principle the same argument holds for Mg^{++} ions but their binding curve is flatter. However, it must be noted that the Mg^{++} concentration in our modified Krebs solution was only 1/10 that

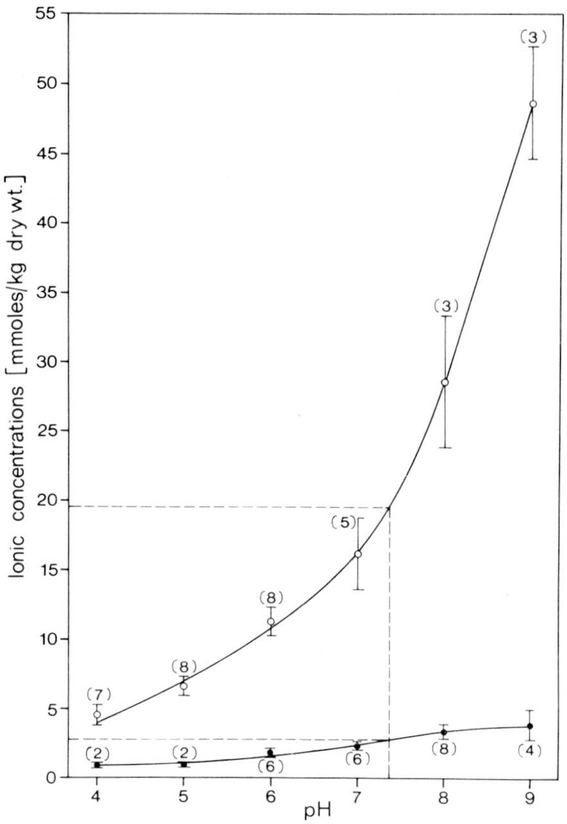

Fig. 3. Ca (o) and Mg binding (•) of isolated adventitial connective tissue from the dog A. carotis as a function of the pH of the Krebs solution.

of a normal blood plasma. Finally it can be stated that in spite of their antagonistic competition for binding sites in vascular connective tissue K^+ and Ca^{++} ions can work synergistically in their physiological effects: tissue acidosis could lead to a membrane hyperpolarization and relaxation in vascular smooth muscle cells, tissue alkalosis to depolarization and contraction.

Ion Binding to Mucopolysaccharides

A quantitative description of the ion binding properties of polyelectrolytes such as connective tissue structures demands the determination of affinity constants for the various ion species. The tissue that was used for the experiments described above consists of at least three very complicated macromolecules (collagen, elastin, proteoglycan). In addition to this in a Krebs solution the various ion species compete for specific and diffuse binding sites on the macromolecules. Therefore we decided to study the ionic interactions in substances derived from connective tissue using ^{23}Na NMR relaxation rate and chemical shift measurements[5]. To obtain a good basis for interpretation it is useful to begin on relatively simple systems like chondroitin and dermatan sulphate and hyaluronic acid. Fig. 4 shows the relations between the degree of neutralization and the $^{23}Na^+$ relaxation

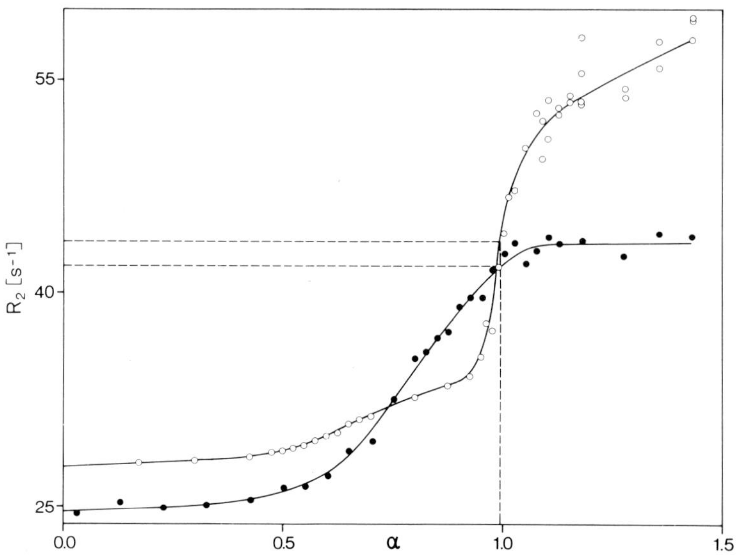

Fig. 4. The $^{23}Na^+$ relaxation rate, $R_2[s^{-1}]$, as function of the degree of neutralization, α, for aqueous solutions of CS-P (○) and CS (•) at 0.1 m concentration calculated on carboxylic acid content. The total Na^+ ion concentration is 0.3 m.

rate for CS-P and CS. For CS-P it is not possible to discuss these results in full but some interesting observations can be made: i) With the usual fast exchange assumption it is suggested from comparisons of our R_1 and R_2 results that the change in R_2 at high α (and pH) values is dependent on a correlation-time change and thus some conformational alteration. The change in R_2 at neutral pH and below is at least partly due to a release of Na^+ ions with decreasing pH. ii) For the purified chondroitin sulphate (CS) a simple sigmoid titration curve was obtained in studies of R_2 as a function of α. Apparently the peptide content of CS-P has a marked effect on the ionic interactions, and cooperative effects probably have to be invoked to explain the differences[5]. It is especially remarkable that the ^{23}Na relaxation rate of CS-P changes fastest around the physiological pH value, while this does not hold for CS (cf. dashed lines of Fig. 4). The parallel to the Ca^{++} and Mg^{++} binding properties of the total vascular connective tissue around the physiological pH is obvious.

Fig. 5 shows that the $^{23}Na^{+}$ relaxation rate decreases with continuing dilution (equal for CS-P and Na) until it reaches the value for free Na^{+} ions. The purpose

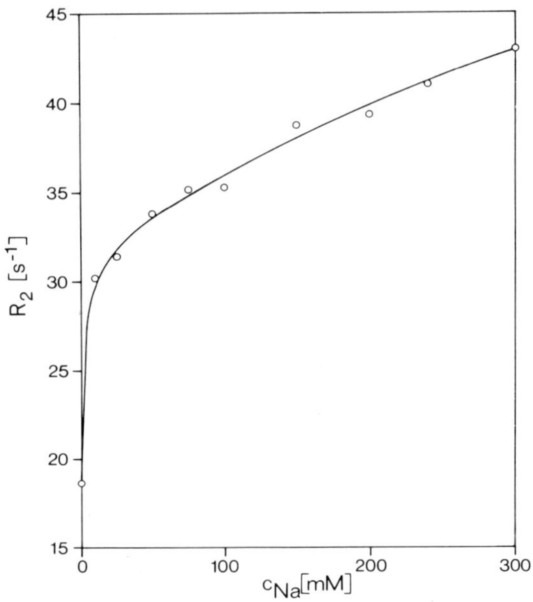

Fig. 5. The $^{23}Na^{+}$ relaxation rate, $R_2[s^{-1}]$, as function of the Na concentration in aqueous solutions of CS-P during a dilution experiment.

of such a dilution experiment is to be able to determine the specific Na affinity constants and relaxation rates. These data are required for a parameter reduction in the calculations of affinity constants, e.g. for K^{+} and Ca^{++}, in the much more complicated competition experiments (cf. Fig. 7).

The observed $^{23}Na^{+}$ relaxation rate, R_{obs}, depends on the relaxation at the various binding sites which the ligand Na^{+} can occupy. With a fast exchange between the binding sites R_{obs} is the weighted average of the individual relaxation rates; the weighting factors are the fractional concentrations

$$R_{obs} = p_f R_f + \sum_i p_i R_i \ , \tag{1}$$

where R_f = relaxation rate of the free ligand; R_i = relaxation rate for binding on site i; p_f = fractional concentration of free ligands; p_i = fractional concentration of the ligand bound to site i. If one takes a monomeric unit of chondroitin sulphate shown schematically in Fig. 6 then the ligand can associate with the sub-

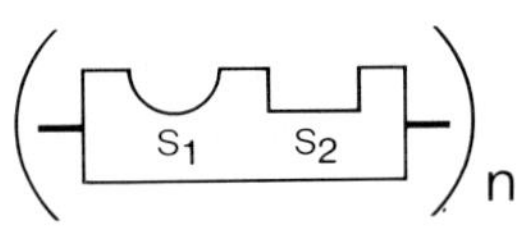

Fig. 6. Scheme of a monomeric unit of chondroitin sulphate with two binding sites S_1 and S_2.

strate at two binding sites (carboxylate and sulphate groups) with relaxation rates R_1 and R_2.

$$R_{obs} = \frac{[L]}{c_L} \cdot R_f + \frac{[LS_1]}{c_L} \cdot R_1 + \frac{[LS_2]}{c_L} \cdot R_2 \tag{2}$$

where $[L]$ = concentration of the free ligand; $[LS_1]$, $[LS_2]$ = concentrations of the ligands bound to sites 1 and 2; c_L = total concentration of the ligand. The concentrations of the free and bound ligands are connected through their dissociation equilibria

$$K_1 = \frac{[L][S_1]}{[LS_1]} \; ; \qquad K_2 = \frac{[L][S_2]}{[LS_2]} \tag{3}$$

and

$$c_L = [L] + [LS_1] + [LS_2] \tag{4}$$

$$c_S = [S_1] + [LS_1] \tag{5}$$

These relations can be utilized to find $[LS_1]$ and $[LS_2]$ as functions of $[L]$ and the dissociation constants. From these equations one gets an equation of the third order in $[L]$

$$[L]^3 + [L]^2\{c_S + (c_S-c_L) + (K_1+K_2)\} + [L]\{(c_S-c_L)(K_1+K_2) + K_1K_2\} - c_LK_1K_2 = 0 \tag{6}$$

Writing the solution to this equation as $[L] = f(c_L, K_1, K_2)$ and considering also that during a dilution experiment the concentration ratio $\frac{c_S}{c_L} = A$ remains constant we arrive at the relaxation rate $R_{obs}(c_L)$

$$R_{obs} = \frac{f(c_L,K_1,K_2)}{c_L} R_f + \frac{f(c_L,K_1,K_2)\cdot A}{f(c_L,K_1,K_2)+K_1} R_1 + \frac{f(c_L,K_1,K_2)\cdot A}{f(c_L,K_1,K_2)+K_2} R_2 \tag{7}$$

After making a nonlinear least square fit to the curve $R_{obs}(c_L)$ in Fig. 5, the ratio of the dissociation constants results as $K_1 : K_2 = 1 : 1000$.

These dissociation constants will be required for the calculation of K^+ and Ca^{++} dissociation constants in competition experiments. The results of such experiments are shown in Fig. 7. One recognizes that with the increasing concentration

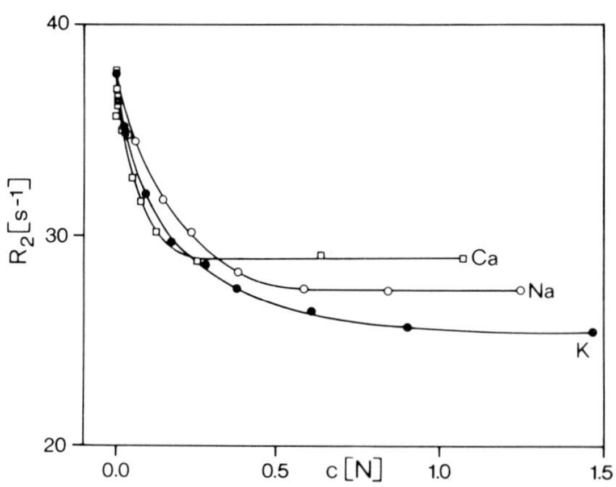

Fig. 7. $^{23}Na^+$ relaxation rate, R_2, as a function of Na, K and Ca concentrations in the presence of a 0.1 m CS-P solution at pH = 3.6.

of Na, K and Ca in a 0.1 m CS-P solution the ^{23}Na relaxation rate decreases to differing but approximately constant end values, whereby none of these reaches the R_f of Na^+. This is clearly due to the fact that Na^+ ions are expelled from their specific binding sites upon addition of, e.g. Ca^{++} ions in a range of low Ca concentrations, but at higher Ca concentrations this is not the case. It appears that in addition to the specific site binding there is also a diffuse binding for the ion species studied here. If one looks at the initial parts of the curves in the range $0 < c < 0.2$ N, which include also the physiological ion concentrations, then one notices that the ^{23}Na relaxation rate decreases fastest upon addition of Ca^{++} ions, slower upon addition of K^+ ions and slowest upon addition of Na^+ ions. This means that due to the high affinity of Ca^{++} ions for the binding sites of CS-P, this ion species expels most of the Na^+ ions when all three ion species are at the same normality. The affinity constants decrease in the sequence: Ca > K > Na. On the one hand, it can be seen that, as in isolated connective tissue, Ca^{++} ions can drive out K^+ and Na^+ ions from CS-P binding sites (cf. Figs. 2 and 3). On the other hand, the competition experiments with CS-P completely confirm the finding from isolated connective tissue that Ca^{++} ions are bound most extensively, K^+ ions less extensively and Na^+ ions least of all. This holds even with the differing concentrations of the various ion species in a Krebs solution. In the future it will require somewhat strenuous efforts to clarify the ion binding properties of such complex polyelectrolyte structures as connective tissue. Probable conformational changes and allosteric effects can make the task more difficult.

ACKNOWLEDGEMENTS

We are grateful to Professors H.P. Koepchen, B. Lindman and S. Forsén for helpful and invaluable suggestions and discussions. We thank Mrs. Ch. Fuhrmann for her excellent technical assistance and Messrs. H. Dannenberg, J. Ofner and W. Slania from the precision workshop for their help in the development of new analytical methods. Thanks are also due to Miss M. Schindler for her outstanding work in preparing the illustrations and to Miss E. Gaebel for the editorial elaboration and writing of the manuscript.

This work was supported by the Deutsche Forschungsgemeinschaft (Si 182/1-4) within its overall program "Structure and Function of Biological Membranes", the Stiftung Volkswagenwerk (Az. 111 327), the Erwin Riesch-Stiftung (ERS 7/8) and the Swedish Natural Science Research Council.

REFERENCES

1. Siegel, G., Jäger, R., Nolte, J., Bertsche, O., Roedel, H. and Schröter, R., in Cervós-Navarro, J., *Pathology of Cerebral Microcirculation*, Walter de Gruyter, Berlin and New York, 1974, p. 96.
2. Siegel, G., Roedel, H., Nolte, J., Hofer, H.W. and Bertsche, O., in Bülbring, E. and Shuba, M.F., *Physiology of Smooth Muscle*, Raven Press, New York, 1976, p.19.
3. Rees, D.A., in Whelan, W.J., *Biochemistry of Carbohydrates*, Vol. 5, MTP Int. Rev. Sci., Butterworths and University Park Press, London and Baltimore, 1975, p. 1.
4. Fransson, L.-Å., Biochim. Biophys. Acta, 437 (1976) 106.
5. Siegel, G., Gustavsson, H., Ehehalt, R. and Lindman, B., in Lewis, D.H., *Proc. IXth World Conf. Eur. Soc. Microcirculation*, S. Karger, Basel, München, Paris,

London, New York, Sydney, 1977.
6. James, Th.L., *Nuclear Magnetic Resonance in Biochemistry*, Academic Press, New York, San Francisco, London, 1975.
7. Berendsen, H.J.C., Fed. Proc., 25 (1966) 971.
8. Catchpole, H.R., Joseph, N.R. and Engel, M.B., A.M.A. Arch. Pathol., 61 (1956) 503.
9. Bohr, D.F., Science, 139 (1963) 597.
10. Somlyo, A.P., Somlyo, A.V., Shuman, H. and Garfield, R.E., in Betz, E., *Ionic Actions on Vascular Smooth Muscle*, Springer-Verlag, Berlin, Heidelberg, New York, 1976, p. 17.

Excitation-Contraction Coupling in Smooth Muscle, R. Casteels et al. eds.

THE LANTHANUM-RESISTANT CALCIUM FRACTION

Théophile GODFRAIND
Laboratoire de Pharmacodynamie Générale et de Pharmacologie
Université Catholique de Louvain
Avenue E. Mounier 73 UCL 7350
1200 Bruxelles Belgique

In smooth muscle, most of the calcium exchange determinations are obscured by exchange occurring at extracellular sites. As an example, table 1 reports data obtained in the rat aorta. The contraction of this muscle in response to noradrenaline shows a component sensitive to Ca_o (fig. 1), this suggests that noradrenaline stimulates Ca influx. As shown on table 1, the rate of ^{45}Ca uptake by the tissue does not seem to be influenced by noradrenaline.

Several attemps have been made to estimate cytoplasmic Ca changes occurring during excitation of smooth muscle. It has been shown that lanthanum replaces Ca on superficial binding sites and that it does not penetrate the cell (Weiss & Goodman, 1969). On the basis that it might block transmembrane fluxes of calcium, it has been proposed that the Ca content of a muscle washed in La solution was an estimate of cellular Ca (van Breemen & al, 1973).

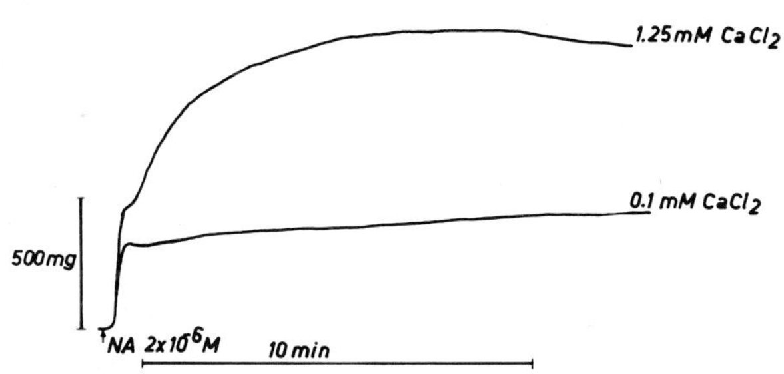

Fig.1. Influence of extracellular Ca concentration on the tonic contraction evoked by noradrenaline in rat aorta (Kaba, 1973)

TABLE 1. ^{45}Ca content of rat aorta (m-mole/kg wet wt) incubated for various times in radioactive physiological solution at 37°C containing noradrenaline 10^{-5}M which evokes the maximum contraction (Godfraind, 1976).
Note the absence of action of noradrenaline on total ^{45}Ca content and on its rate of uptake.

Time in radioactive solution (min)	Physiological solution	Time in the presence of noradrenaline (min)	Physiological solution + noradrenaline 10^{-5}M
2	1.62 ± 0.055 (6)	-	-
5	2.08 ± 0.046 (6)	0	-
7	2.17 ± 0.022 (6)	2	2.19 ± 0.047 (4)
9	2.07 ± 0.030 (6)	4	2.22 ± 0.047 (6)
13	2.21 ± 0.052 (6)	8	2.31 ± 0.040 (6)
20	2.22 ± 0.025 (20)	15	2.24 ± 0.051 (14)
21	2.15 ± 0.033 (6)	-	-
120	2.31 ± 0.022 (28)	-	-
240	2.43 ± 0.023 (8)	-	-

The evidence that La blocks both Ca binding to superficial sites and Ca influx is given by the observation that the calcium diffusion space in the presence of lanthanum is similar to the inulin diffusion space (fig.2). However, La does not block completely Ca efflux as shown in several smooth muscles (table 2) (Burton & Godfraind, 1974; Godfraind, 1976).

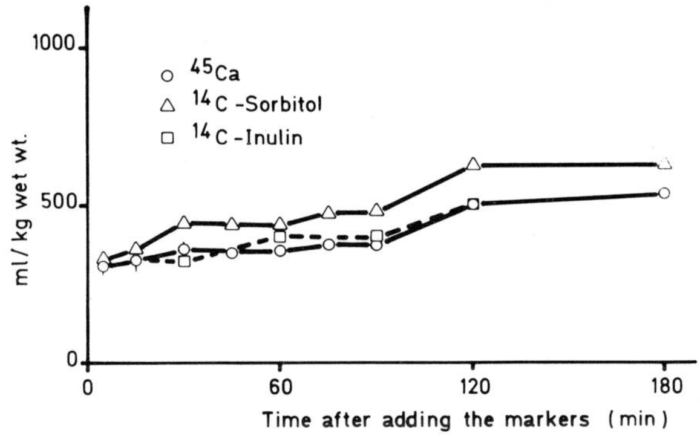

Fig. 2. Diffusion space of calcium and of the extracellular markers inulin and sorbitol in intestinal smooth muscle bathed in La-solution (Burton & Godfraind, 1974). Note their similarity, indicating an absence of Ca binding and influx.

An attempt has been made in order to estimate Ca fluxes across the smooth muscle cell membrane by measuring the Ca turnover in the La-resistant Ca fraction (Godfraind, 1976). This fraction was defined as the one which was not displaced when the tissue was washed in 50 mM $LaCl_3$. In this solution, most of Ca displacement occurred during the first 5 minutes and Ca efflux from intracellular stores appeared to be delayed (table 2). This allowed an estimation to be made of the La-resistant Ca fraction.

Noradrenaline increased the rate of uptake of ^{45}Ca into this Ca fraction. In resting rat aorta, this uptake was equal to 6 µmole $^{45}Ca\ kg^{-1}min^{-1}$. In the presence of noradrenaline $10^{-5}M$, it increased to 36 µmole $^{45}Ca\ kg^{-1}min^{-1}$ (fig.3). There was no net gain in tissue Ca and Ca efflux increased in a similar manner (fig. 4).

Table 2. ^{45}Ca content (mmole/kg wet wt) of rat aorta incubated for 20 min in ^{45}Ca-physiological solution and washed either in physiological solution or in 50 mM-La solution (Godfraind, 1976). In La-solution, ^{45}Ca efflux was faster than in physiological solution during the first 5 min of washout, thereafter, it was lower but not completely absent.

^{45}Ca content of controls before washing : 2.27 ± 0.045 (6)

^{45}Ca content of aortae washed in non-radioactive solutions :

Time in non-radioactive solution (min)	Physiological solution (4)	50 mM-La solution (4)
2	0.49 ± 0.01	0.25 ± 0.03
5	0.17 ± 0.01	0.10 ± 0.01
10	0.08 ± 0.005	0.08 ± 0.009
15	0.064± 0.004	0.069± 0.006
30	0.033± 0.001	0.043± 0.004
60	0.017± 0.001	0.028± 0.002

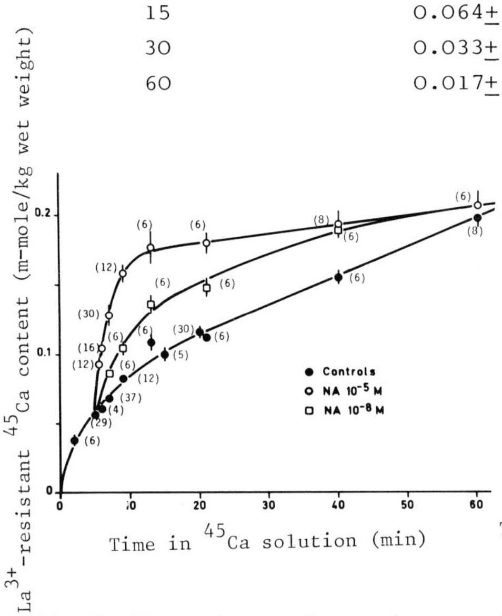

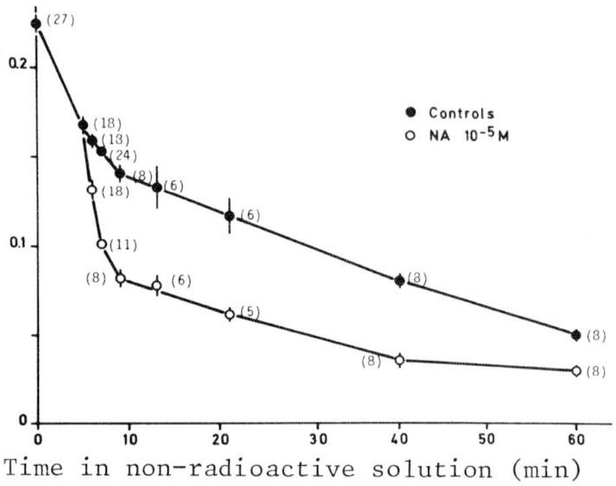

Fig.3. The action of noradrenaline on the uptake of ^{45}Ca into the La-resistant Ca fraction of rat aorta (Godfraind, 1976 modified).

Fig.4. The action of noradrenaline on the rate of ^{45}Ca efflux from the La-resistant Ca fraction (Godfraind, 1976 modified)

Note : the experiments were performed as follows : after an incubation in the physiological solution with or without noradrenaline, the aortae were washed for 5 min in the 50 mM La solution and their residual content in ^{45}Ca was estimated by liquid scintillation counting. Changes in ^{45}Ca content are related to transmembrane Ca fluxes.

The increase in Ca uptake was dose dependent, a response of 50% of the maximum being produced by 2×10^{-8}M noradrenaline as for the contraction. In the presence of phentolamine, the dose-effect curves for the action of noradrenaline on ^{45}Ca uptake were displaced in a manner characteristic of competitive antagonism (fig.5). The pA_2 for phentolamine was 7.8 as for the contraction. In view of the similarity of pA_2 when phentolamine was used as an antagonist, it appears that the activation of alpha-adrenergic receptors was responsible both for the contraction and for the increase rate of Ca exchange in the La-resistant Ca fraction. The rate of exchange from this fraction appears to be the biologically important component of Ca exchange across smooth muscle cell membrane.

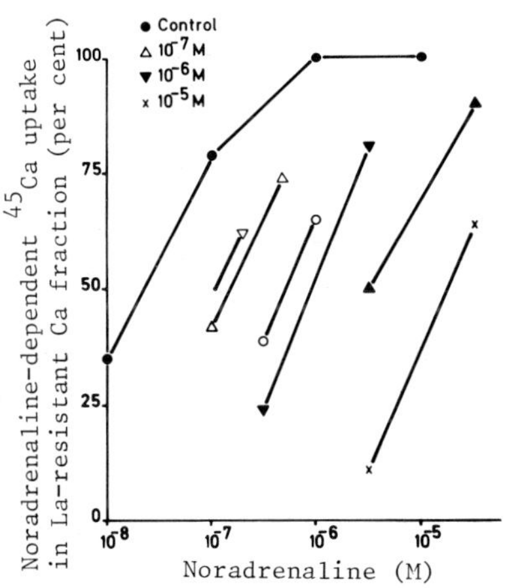

Fig.5. Noradrenaline dose-effect curve on the ^{45}Ca influx in the presence and absence of phentolamine. (Godfraind, 1976a modified)

In recent experiments, the rate of calcium exchange into the La-resistant Ca fraction has been studied in the presence of cinnarizine.

Cinnarizine is a non competitive antagonist for noradrenaline in rat aorta : when the preparation was preincubated in the presence of the drug, the maximum response was depressed, whereas the ED_{50} was unaffected. Cinnarizine inhibits the component of the contraction sensitive to extracellular calcium. This has been observed in muscles stimulated by noradrenaline and in depolarized preparations in which the contraction was evoked by increasing the concentration of extracellular calcium. In the latter, the action of cinnarizine was observed down to 10^{-9}M (Godfraind & Kaba, 1969 and 1972). Cinnarizine had no effect on the ionic composition of the tissue. At concentrations lower than 10^{-5}M, it did not reduced the ^{45}Ca content of the La-resistant fraction of aortae washed at 37°C in ^{45}Ca physiological solution for 5 min or longer. However, concentrations comprised between 10^{-8} and 10^{-5}M reduced the noradrenaline-sensitive ^{45}Ca uptake into the La-resistant fraction. The depression of the rate of ^{45}Ca influx was parallel to the depression of the tonic contraction sensitive to extracellular calcium. The concentration of cinnarizine producing a 50 per cent inhibition of the maximum was equal to 7.9 x 10^{-7}M for the two processes. At cinnarizine 10^{-5}M, both processes were nearly abolished. The noradrenaline-increased rate of calcium efflux from the La-resistant Ca fraction has been measured for cinnarizine concentrations comprised between 3 x 10^{-7}M and 10^{-5}M. For cinnarizine 3 x 10^{-7}M, both efflux and influx were reduced to the same extend whereas for cinnarizine 10^{-5}M the influx was blocked and the reduction of the efflux was not greater than with cinnarizine 3 x 10^{-7}M. This indicates that the calcium fraction contributing to the efflux corresponded to intracellular calcium released by noradrenaline. This observation is consistent with the one done in Ca-free EGTA solution, in which noradrenaline 10^{-5}M evoked a Ca efflux equal to 50% of the efflux observed in physiological solution. This efflux was not affected by cinnarizine.

The action of noradrenaline 10^{-5}M on Ca fluxes in isolated rat aorta has been studied not only at 37°C but also at 27°C, 17°C and 4°C. Noradrenaline sensitive Ca influx into the La-resistant fraction was more affected by temperature changes than was noradrenaline sensitive Ca efflux. At 27°C, the influx was reduced to 38% of controls and at 17°C, it was completely blocked. As far as the efflux is concerned, the remaining rates were respectively equal to 76 % of control at 27°C and to 42 % at 17°C; at 4° C, the noradrenaline-sensitive Ca efflux was abolished.

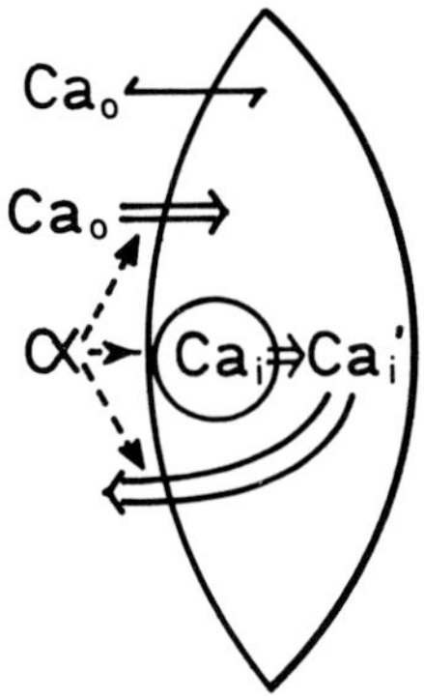

Fig. 6.

Schematic representation of the transmembrane fluxes of calcium and of their sensitivity to alpha-adrenergic stimulation.
As indicated by the dotted line, the stimulation of the alpha-adrenergic receptor influences Ca influx, Ca efflux and Ca_i release. These three actions are likely the consequence of three distinct cellular processes. The calcium exchange which is not sensitive to noradrenaline represents the basal Ca-Ca exchange.

The study of the calcium exchange into the La-resistant Ca fraction gives quantitative informations on the Ca pools mobilized by the alpha adrenergic receptor activation. The observations here summarized indicate that the alpha adrenergic receptor is coupled to at least three calcium controlling processes (fig. 6) which are involved in the release of intracellularly stored calcium, in the increase in the rate of Ca influx and in the increase in the rate of Ca efflux.

Acknowledgement : This work has been supported by F.R.S.M. (grant n° 20121).
I thank Mrs FANARD for typing this paper.

REFERENCES

BREEMEN, C. van, FARINAS, B.R., CASTEELS, R., GERBA, D., WUYTACK, F. & DETH, R. (1973) : Phil. Trans. R. Soc., 265, 57
BURTON, J. & GODFRAIND, T. (1974) : J. Physiol. (London), 241, 287
GODFRAIND, T. (1976) : J. Physiol. (London), 260, 21
GODFRAIND, T. & KABA, A. (1969) : Brit. J. Pharmacol., 36, 549
GODFRAIND, T. KABA, A. (1972) : Arch. Int. Pharmacodyn. Ther., Suppl., 196, 35
KABA, A. (1973) : Contribution à l'étude de calcium dans la pharmacologie du muscle lisse vasculaire. Vander (Louvain)
WEISS, G.B. & GOODMAN, F.R.(1969): J.Pharmacol. Exp. Ther., 169, 46
WEISS, G.B. (1974) : Ann. Rev. Pharmacol., 14, 343.

Excitation-Contraction Coupling in Smooth Muscle, R. Casteels et al. eds.

THE LANTHANUM-RESISTANT FRACTION OF CALCIUM IN THE RAT TAIL ARTERY

Vladimir Palatý
Department of Anatomy
The University of British Columbia
Vancouver, B.C., V6T 1W5, Canada

INTRODUCTION

For reasons discussed in detail by van Breemen et al. (1), satisfactory separation of the cell Ca from extracellular Ca in preparations of the vascular wall can be achieved only if the rate of translocation of Ca across the sarcolemma has been sufficiently and selectively reduced. The best solution of this problem available at present is based on the property of La^{3+} and its congeners to block transmembrane fluxes of Ca^{2+} in both directions (2,3). 'The La method' (3) has been criticized (4) because of the more or less convincing evidence that La^{3+} may cross the membrane and displace some of the intracellular bound Ca. In the absence of a better alternative, the La method was adopted to study the distribution and transport of Ca in the title tissue.

METHODS

Preparation: 25 - 30 mg. segments of the artery were obtained and treated as described elsewhere (5), exposed to 0.5 μM monensin for 15 minutes in order to deplete endogenous catecholamines and allowed to recover for 3 hours in aerobic Krebs solution.

Determination of the size of the La-resistant fraction: The slow decline in the residual tissue ^{45}Ca activity observed if the La wash at 37 °C was extended beyond the usual 1 hour period (3) was virtually absent if the tissue was washed at 2 °C. Under the latter conditions, the activity level attained after 1.5 hours remained constant for the subsequent 4 hours. A 2 hour wash at 2 °C was used, therefore, in all the experiments. Lowering of the temperature of the La wash, which has been mentioned also by van Breemen (6), makes the La method applicable also to those experiments in which the tissue has been exposed to a *mobile* ionophore, such as X-537A, capable of transporting La^{3+}. Washed and dried samples were solubilized with Protosol (NEN) and counted in the usual manner.

Labeling of Ca in the La-resistant fraction: As the commonly used 2 hour loading period was found to be too short to permit uniform labeling of Ca in the fraction, a different procedure was eventually developed. After recovery from the monensin treatment, the segments were rinsed with HEPES-buffered solution, and equilibrated with the ^{45}Ca-containing, HEPES-buffered solution first for 18 hours at 2 °C and then for 3 hours at 37 °C. This avoided using an impracticable loading period such as 8 to 12 hours at 37 °C.

Solutions and drugs: The composition of bathing solutions was based on a Krebs formula (in m-moles per litre): NaCl 115, $NaHCO_3$ 25, NaH_2PO_4 1.2, KCl 5, $CaCl_2$ 1.7, $MgSO_4$ 1.2, D-glucose 11. A HCO_3^-- and P_i-free version used in the isotope experiments contained the same concentrations of cations and HEPES in the concentration of 10 m*M*. It was aerated with O_2 and its pH at 37 °C was 7.2. The La wash solution had the same composition except that it also contained $LaCl_3$ in the concentration of 10 m*M*. Monensin, A 23187 (Eli Lilly), X-537A (Hoffman-LaRoche), valinomycin, and carbonyl cyanide *m*-chlorophenyl hydrazone ("CCCP") were added in the form of stock solutions in $(CH_3)_2SO$. The final concentration of the solvent in the bath did not exceed 0.1 % (v/v).

RESULTS

The course of exchange of ^{45}Ca in the La-resistant fraction for external ^{40}Ca under steady-state conditions at 37 °C is illustrated in Fig. 1. It could be approximated by the function: $Ca_t = 1.0 \exp(-2.9t) + 0.3 \exp(-0.24t)$, where Ca_t is the unexchanged portion of the fraction (in m-moles of Ca/kg dry tissue) after time t in the isotope-free solution (in hours). Identical results were obtained using segments that had not been treated with monensin.

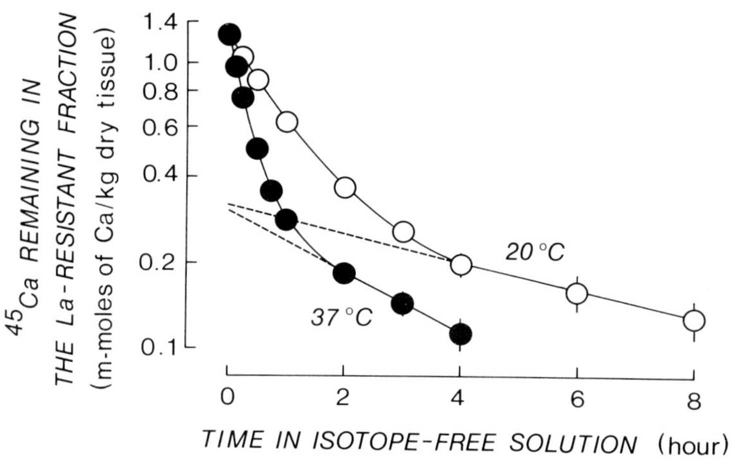

Fig. 1. Exchange of ^{45}Ca in the La-resistant fraction for external ^{40}Ca under steady-state conditions. In all the figures, the points represent means of 6 individual values. The limits of S.D. are indicated by vertical lines.

Measurement of the exchange at 37, 30, 25 and 20 °C made it possible to estimate the activation energies of the rapidly- and slowly-exchanging components in that range as approximately 11.4 and 9.1 kcal/mole, grad, respectively. The rates of exchange at temperatures below 10 °C were too slow (the apparent overall k at 2 °C 0.09 hr^{-1}) to allow but a rough estimate of the activation energy of the rapidly-exchanging component as not less than 19.5 kcal/mole, grad.

The steady-state exchange was measured also under the conditions of complete dissipation of the transmembrane Na and K gradients resulting from inhibition of the Na pump with 1 m*M* ouabain. To avoid any indirect effect of ouabain due to release of endogenous catecholamines (7,8), it was essential to use catecholamine-depleted segments. As illustrated in Fig. 2, there was a highly significant increase of about 0.4 m-mole/kg dry tissue in the size of the rapidly-exchanging component, but the two rate constants and the size of the slowly-exchanging component remained essentially unaltered.

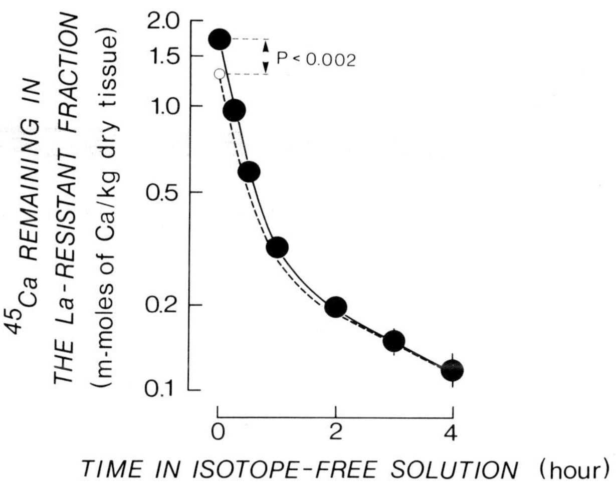

Fig. 2. Steady-state exchange under conditions of complete dissipation of transmembrane Na and K gradients. Ouabain was present in both the loading and isotope-free solutions in the concentration of 1 m*M*. The dashed curve represents the exchange in normal segments illustrated in Fig. 1.

As shown in Fig. 3, addition of norepinephrine (NE) to the loading solution increased the size of the rapidly-exchanging component by about 0.2 m-moles/kg dry tissue. Because of a change in the tissue H_2O associated with the NE contracture, it was important to express the results on the dry tissue basis. In another experiment, NE was added to the isotope-free solution only. The marked increase in the rate of exchange during the initial 7.5 minutes could be attributed at least in part to release of Ca from the NE-mobilizable stores (9,10), the size of which could be estimated as about 0.2 m-moles/kg dry tissue. It was also found that exposure of strips to either 5 μ*M* A 23187 or 10 m*M* caffeine (10) for 15 minutes was accompanied by abolition of the phasic response to NE in Ca-free solution. Under these conditions, the size of the La-resistant fraction was reduced by a similar amount, i.e. about 0.2 m-moles/kg dry tissue.

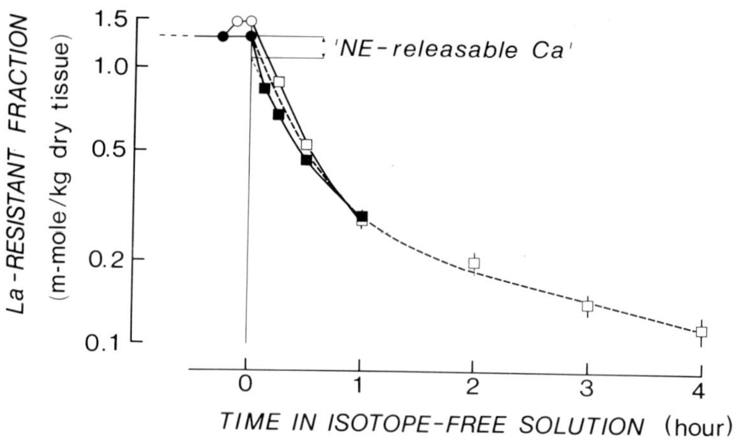

Fig. 3. The effect of norepinephrine on the La-resistant fraction. NE was added either to both the loading (○) and isotope-free (□) solutions or to the isotope-free solution only (■) in the concentration of 20 μ*M*.

Arterial strips could be exposed to 1 m*M* NaCN or 0.5 μ*M* CCCP for as long as 1 hour without any appreciable changes in tension, the phasic response to NE, and the tissue Na and K. Respiration of segments was blocked completely by 1 m*M* NaCN, while 0.5 μ*M* CCCP increased the O_2 consumption from 0.30 to 0.75 μl/mg, hr. As shown below, both NaCN and CCCP brought about a decrease in the size of the La-resistant fraction of some 0.4 m-moles/kg dry tissue, most likely by releasing Ca accumulated actively by mitochondria. In ouabain-treated segments, NaCN reduced the fraction by about 0.35 m-moles/kg dry tissue.

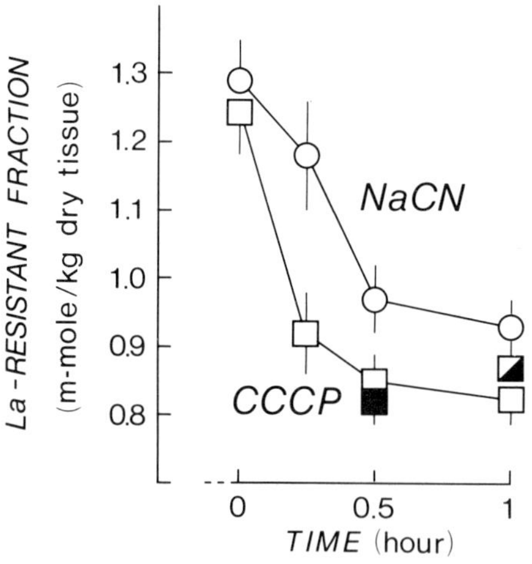

Fig. 4. The effects of NaCN and uncouplers on the La-resistant fraction. At the end of the standard loading period, NaCN (○) or CCCP (□) were added to the loading solution in the final concentrations of 1 m*M* and 0.5 μ*M*, respectively. There was no significant enhancement of the effect of CCCP if valinomycin was also present in the final concentration of 2 μ*M* (■). The effect of 0.5 μ*M* monensin (◪) was comparable to that of CCCP.

If added in the final concentration of 4 μ*M*, the ionophore X-537A abolished the phasic response to NE in Ca-free solutions, increased the O_2 consumption by about 70 %, and brought about rapid dissipation of the Na and K gradients, but it did not affect the resting tension. As illustrated in Fig. 5, X-537A reduced the size of the rapidly-exchanging component by about 0.4 m-moles/kg dry tissue. It will be noted that the latter value is considerably lower than the amount of Ca that, on the basis of previous results, would be expected to be present in the NE-releasable stores and in mitochondria. X-537A did not alter significantly the size of the slowly-exchanging component or the two rate constants.

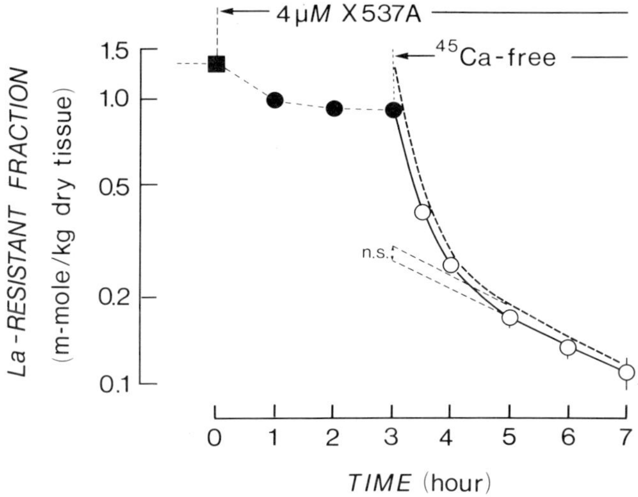

Fig. 5. The effect of X-537A on the La-resistant fraction. X-537A was added at the end of the loading period in the concentration of 4 μ*M*.

DISCUSSION

The course of exchange of ^{45}Ca in the La-resistant fraction for external ^{40}Ca reveals that the fraction consists of at least two kinetic components. The rapidly-exchanging component was found to respond to various experimental conditions in a manner expected from the cell Ca if translocation of Ca across the sarcolemma were the main rate-limiting step. The estimates of the sizes of two distinct Ca pools, the NE-releasable stores and the Ca accumulated actively by mitochondria, are in reasonable agreement with the data obtained by similar techniques for the rabbit aorta (9,10).

The increase in the size of the rapidly-exchanging component associated with dissipation of the Na and K gradients cannot be attributed to release of endogenous catecholamines (7,8) and it does not seem to be due to enhanced active accumulation of Ca by mitochondria. Since neither ouabain nor X-537A increased the tension of catecholamine-depleted strips, it is rather unlikely that dissipation of the Na and K gradients caused any major increase in $[Ca^{2+}]_i$.

The finding of a slowly-exchanging component of the La-resistant fraction was somewhat surprising. The component is not an artefact of the loading procedure, since it was found also in a control experiment in which we used the standard loading period of 2 hours at 37 ^{o}C. It may be significant that the size of the component cannot be altered even by ionophores capable of transporting Ca^{2+} such as X-537A or A 23187. The results of some preliminary experiments are not incompatible with the opinion of Mayer et al. (11) that an appreciable quantity of extracellular Ca may be trapped in the tissue upon addition of La.

This research has been supported by a grant from the Medical Research Council of Canada. Monensin and A 23187 were kindly provided by Eli Lilly, X-537A by Hoffman-LaRoche.

REFERENCES

1. van Breemen, C., Farinas, B.R., Casteels, R., Gerba, P., Wuytack, F. and Deth, R. (1973) Phil. Trans. Roy. Soc. Lond. B 265, 57-71
2. Weiss, G.B. (1974) Ann. Rev. Pharmacol. 14, 343-354
3. van Breemen, C., Farinas, B.R., Gerba, P. and McNaughton, E.D. (1972) Circulation Res. 30, 44-54
4. Daniel, E.E. and Janis, R.A. (1975) Pharmac. Therap. 1, 695-729
5. Palatý, V. (1974) J. Physiol. Lond. 242, 555-569
6. van Breemen, C. (1976) in Vascular Neuroeffector Mechanisms (Bevan, J.A., Burnstock, G., Johansson, B., Maxwell, R.A. and Nedergaard, O.A., Eds.), pp. 67-79, Karger, Basel.
7. Bonaccorsi, A., Hermsmeyer, K., Smith, C.B. and Bohr, D.F. (1977) Amer. J. Physiol. 232, H140-H145
8. Karaki, H. and Urakawa, N. (1977) Eur. J. Pharmacol. 43, 65-72
9. Deth, R. and Casteels, R. (1977) J. Gen. Physiol. 69, 401-416
10. Deth, R. and van Breemen, C. (1977) J. Membrane Biol. 30, 363-380
11. Mayer, C.J., Ölscher, J. and Riemer, J. (1976) Pflügers Arch 362, R 26

Part II : Erwin-Riesch Symposium, Heidelberg, 15-16 July, 1977

BIOCHEMICAL ASPECTS OF EXCITATION-CONTRACTION COUPLING IN SMOOTH MUSCLE

Excitation-Contraction Coupling in Smooth Muscle, R. Casteels et al. eds.
© 1977 Elsevier/North-Holland Biomedical Press

THE CONTRACTILE AND CYTOSKELETAL ELEMENTS OF VERTEBRATE SMOOTH MUSCLE.

J. V. Small
Institute of Molecular Biology of the Austrian
Academy of Sciences Billrothstr.11,
5020 Salzburg, Austria.

INTRODUCTION

From X-ray diffraction and electron microscope investigations it is now fairly well established that the contractile apparatus of vertebrate smooth muscle consists of filaments of myosin and of actin oriented more or less parallel with respect to the cell axis (see review by Small, 1977a). Such an arrangement would suggest that contraction is brought about, as in striated muscle, via a process of shearing, or relative sliding between the thick and the thin filaments. But the means by which such a sliding process takes place in smooth muscle to produce a net contraction remain unclear. Disagreement still exists over the structural organisation of the thick filaments and the general organisation of the thin filaments with respect to sites of anchorage and polarity change within the cytoplasm. In the present report I shall not be able to shed much new light on the contractile process in smooth muscle but shall attempt to summarise the current state of our knowledge of the structure of the contractile apparatus. This will, I hope, provide some indication of those areas warranting more critical and further investigation.

The thick myosin filaments

Determining the in vivo form of myosin is not alone a matter of fixing muscles under "physiological conditions" (Somlyo et aL, 1973; Ashton et al., 1975) but requires the use of independent means of controlling the state of the contractile apparatus during chemical fixation. In our earlier studies (see review by Small, 1977a) we used the X-ray diffraction method as one means of monitoring the state of the myosin component (see also Lowy et al., 1973). However, under the conditions needed to observe the myosin 143Å reflection in the living muscle and subsequently after chemical fixation, there occurred an aggregation of the myosin filaments into wider ribbon like structures, of which we were at that time not aware. In various other electron microscope studies in which conventional preparative procedures were used cylindrical myosin filaments could be observed (see e.g. Rice et al., 1970; Somlyo et al., 1973) and these were considered from a conservative viewpoint to be the more likely form of myosin in vivo.

Direct evidence against the presence of "ribbons" under normal conditions and in favour of the cylindrical filaments came from studies of isolated and demembranated smooth muscle cells (Small 1977a, b). In these studies in which the state of the contractile apparatus could be controlled via the external medium thick filaments of about 160Å in diameter were observed both under conditions of rigor and relaxation (Figs. 1 and 2).

More recently, we have linked detergent-extracted smooth muscle strips to a tension transducer and tested these for their contractility and Ca-sensitivity. With freshly extracted taenia coli muscle strips we consistently observed a Ca-sensitivity with respect to tension development in the order of about 50%. Following chemical fixation during relaxation at 30°C (in the presence of Mg ATP and EGTA) cylindrical thick filaments were again observed in the cells but their number and distribution has so far proved somewhat variable.

By gentle homogenisation the thick filaments could be separated from the isolated cells (Small 1977b; Fig. 3a) and showed a structure essentially indistinguishable from that exhibited by synthetic smooth muscle myosin filaments (Figs. 7 & 8; Sobieszek, 1972, 1977; Small, 1977b). The important feature of these filaments was the absence of a central bare zone, as seen on native filaments from striated muscle and a continuous distribution of cross-bridges along the filament length. The same type of organisation is also observed on filaments assembled from various non-muscle myosins (Hinssen et al., in preparation). The length of the isolated native thick filaments was very variable and occasionally very long filaments, up to about 8 μ long were observed (Small, 1977b).

The thin filaments

The thin filaments of vertebrate smooth muscle lack any troponin-like components and are composed only of actin and tropomyosin (Sobieszek & Bremel, 1975; Sobieszek & Small, 1976). From the densitometry of stained polyacrylamide gels of whole muscle we found that

Fig. 1. Cross section of a smooth muscle cell from the guinea pig taena coli after treatment with Triton X-100 (rigor state). The actin and myosin filaments appear in groups of variable shape and size; the 100Å filaments and dense bodies occur between these groups. (From Small & Sobieszek, 1977). Bar, 0.1 μ.

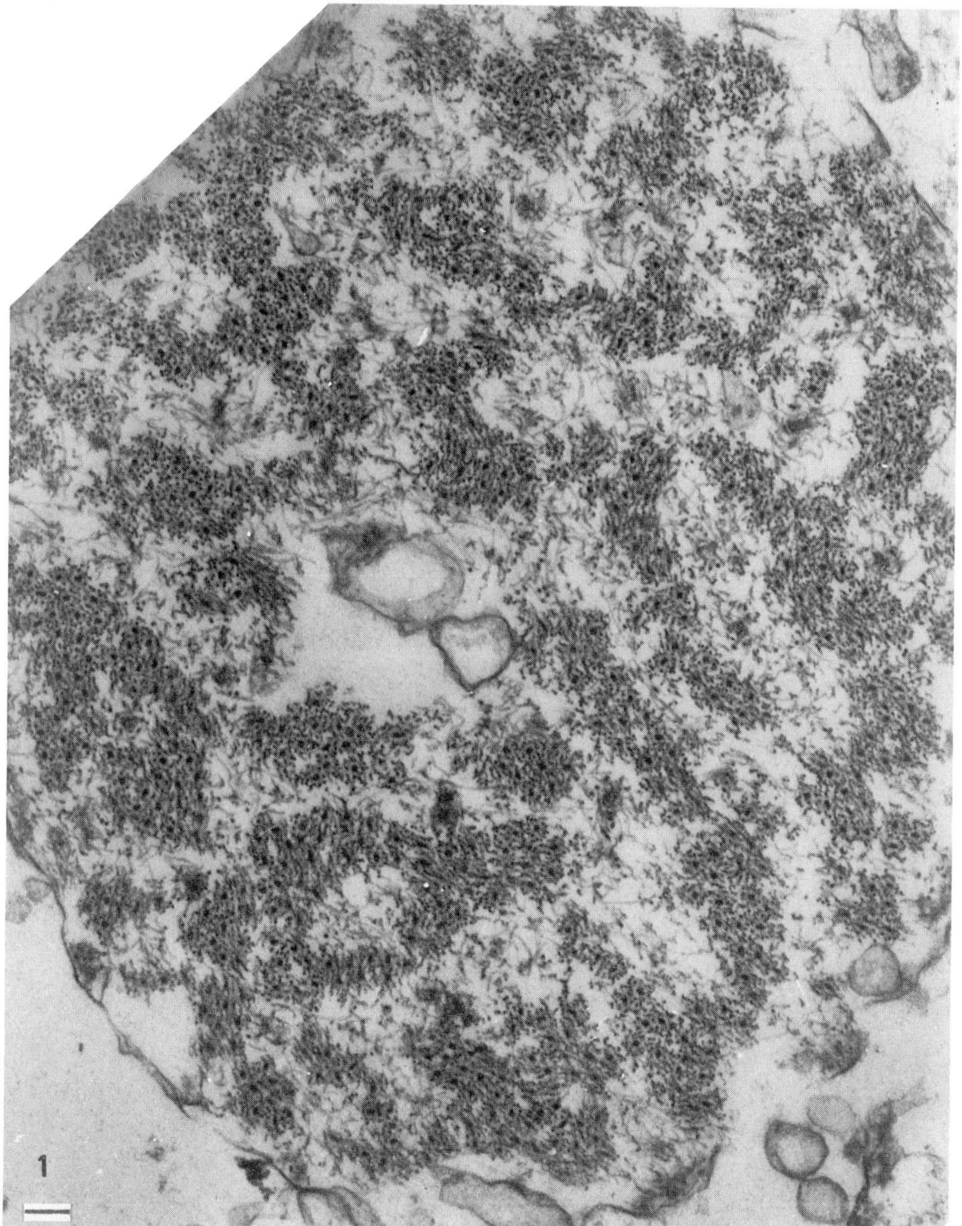
1

the actin to myosin mass ratio in different smooth muscles varies considerably, from about 1:0.6 in hog carotid to 1:1 in guinea pig taenia coli and 1:1.6 in guinea pig vas deferens (Small & Sobieszek, 1977). While comprehensive data is at present lacking these differences may be correlated with the different ratios of actin to myosin filaments obtained from thin transverse sections of smooth muscle. Accordingly, for suitably preserved material counts with taenia coli gave an actin to myosin filament ratio of 10 to 13:1 (Small, 1977b) and with vascular smooth muscle about 15:1 (Somlyo et al., 1973).

At present there is no direct information available on the length of the thin filaments in smooth muscle. Likewise, their precise longitudinal organisation with respect to the thick filaments remains unclear. The view still persists that the intracellular dense bodies observed in smooth muscle, or possibly "some proportion of them" serve as pseudo Z-lines giving rise to mini-sarcomeres within the cell (Heumann, 1970; Ashton et al., 1975). But even the recently presented evidence for the attachment of thin filaments to these dense bodies (Ashton et al., 1975) is unconvincing and other data points more directly to the association of the dense bodies only with the 100Å filament cytoskeleton (see below). The densely staining plaques observed on the plasmalemma of smooth muscle cells appear to be the only anchorage points for the thin filaments and the attachment of actin to these sites is indicated in both longitudinal sections (see e.g. Ashton et al., 1975) and transverse sections (Figs. 3 & 4).

The 100Å filament cytoskeleton

The 100Å filaments of smooth muscle form a complex network throughout the smooth muscle cell which has connections to the cytoplasmic dense bodies and to the plaques on the plasmalemma (Cooke, 1976; Small & Sobieszek, 1977). The latter plaques as such appear to be anchoring sites for both actin and the 100Å filaments (see also Figs. 3 & 4). The 100Å filament network is shown most clearly following the repeated extraction of actomyosin from isolated smooth muscle cells or from smooth muscle strips (Small & Sobieszek, 1977 and Fig. 5). Isolated cells extracted in this way appear as "ghost cells" which still retain the general cell form. From this result it was concluded that the 100Å filaments perform a cytoskeletal role in smooth muscle (Small & Sobieszek, 1977).

Fig. 2. Taenia coli cell fixed in and after exposure to a relaxing medium containing Mg ATP and EGTA for 1 min at R.T. (see Small, 1977). Bar, 0.1 μ

Figs. 3,4. Association of the thin filaments and 100Å filaments with the dense plaques on the plasmalemma. Bars, 0.1 μ.

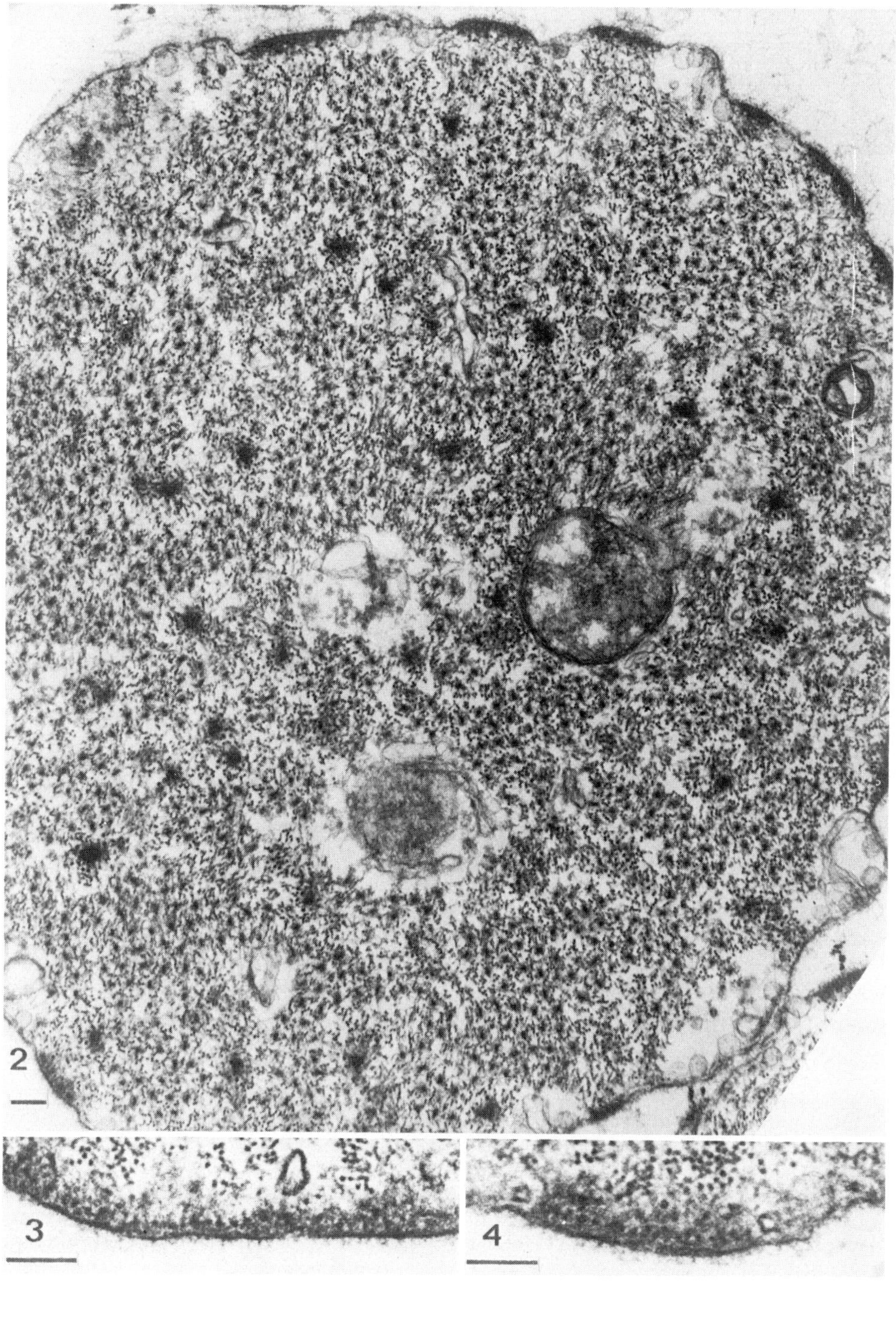
2
3
4

While resistant to extraction in solutions of high ionic strength the 100Å filaments are readily degraded by proteolytic enzymes and may be removed from smooth muscle cells without affecting the general integrity of the actin and myosin filaments. This is shown most readily in triton-extracted smooth muscle cells in rigor. In this state (Fig. 1) the actin and myosin filaments tend to form groups within the cell which may be identified with the birefringent fibrils recognised in the light microscope (Small, 1974; 1977a). The 100Å filaments are observed between these groups, either free or in association with the dense bodies. After digestion of such cells with crude collagenase the 100Å filaments and eventually the dense bodies are no longer seen but the general organisation of the actin-myosin filament groups remains intact (Fig. 6). Cells treated in this way also contract, albeit more slowly than normal, in the presence of Mg ATP (Small & Sobieszek, 1977). These data then point to the independence of the 100Å filament-dense body network from the contractile apparatus proper of the cell. We may also note that cells lacking the 100Å filaments tend to lose their nucleus rather readily and are also more easily dissociated into myofilaments under relaxing conditions, properties which might be expected from the loss of an intracellular cytoskeleton. Data obtained from studies of other cell types indicated that the 100Å filaments may serve generally as cytoskeletal components (see Small & Sobieszek, 1977).

As we have shown elsewhere (Small & Sobieszek, 1977) the protein of the 100Å filaments, which we have termed "skeletin", has a polypeptide molecular weight of 55,000 and an amino acid composition closely similar to that deduced for vertebrate neurofilaments. This protein is not present in smooth muscle actomyosin and has no effect on the Ca-sensitivity of the actomyosin complex.

The organisation of the contractile apparatus

We are as yet rather far from a description of the contractile apparatus of smooth muscle as detailed as that long established for skeletal muscle. However, the information now available allows us to consider some important constraints in devising a model. One rather

Fig. 5. A cell from a taenia coli muscle strip extracted to remove actomyosin (see Small & Sobieszek, 1977). Only the 100Å filaments and dense bodies may be recognised. Bar, 0.1 μ.

Fig. 6. A cell from a taenia coli muscle strip extracted first with Triton X-100 and then digested with collagenase (1mg/ml, 1h 37°C). The 100Å filaments and dense bodies have been removed but the actin-myosin filament bundles retain their integrity and orientation. Bar, 0.1 μ.

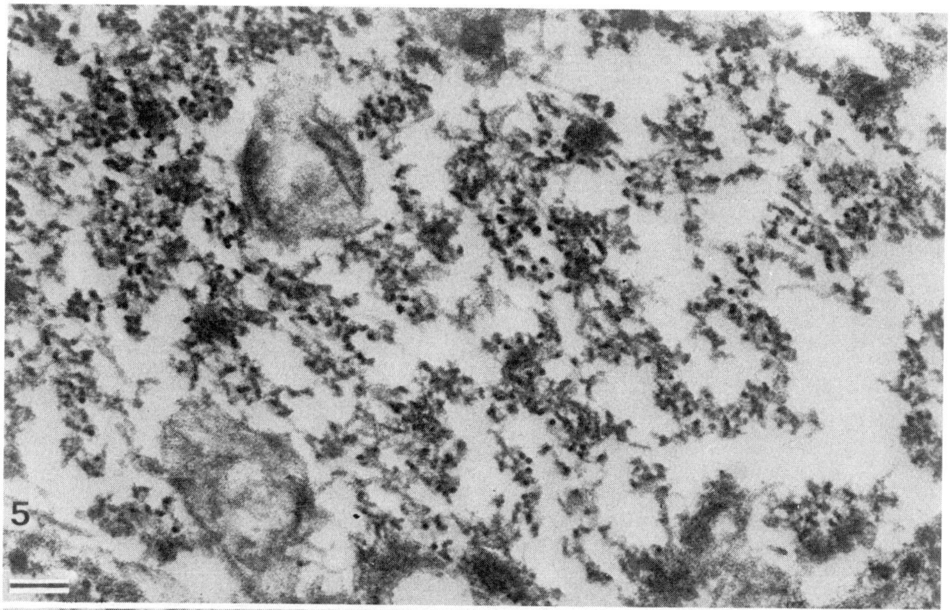

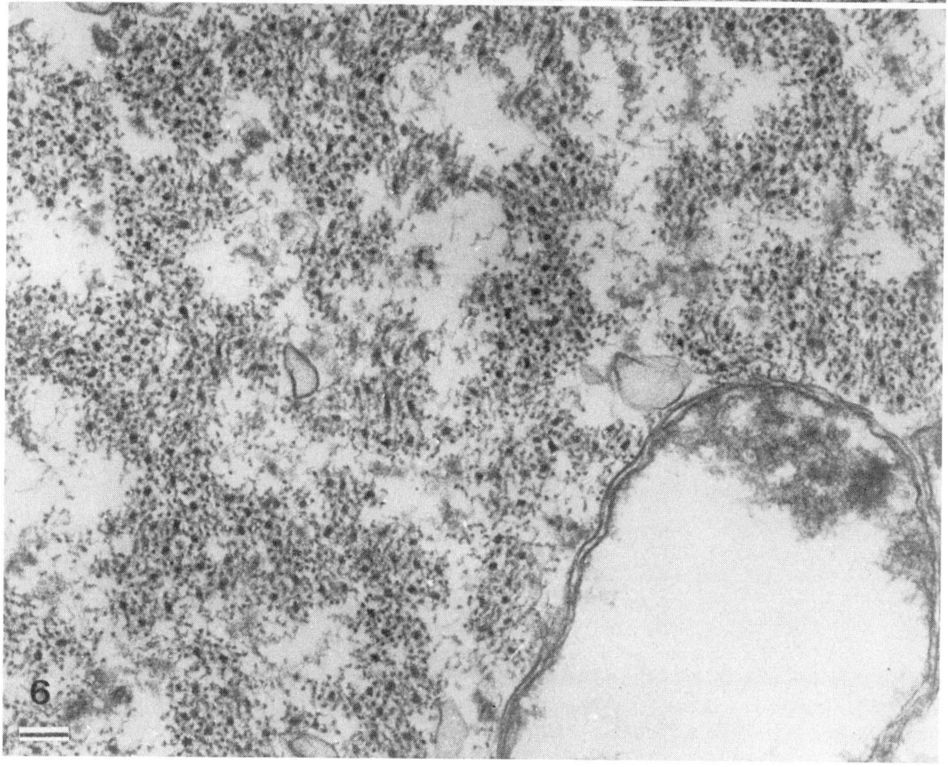

schematic model which incorporates some of the features that have already been discussed is shown in Fig. 10 (Small, 1977a).

At the core of the problem is the structural organisation of the myosin filaments. Since these filaments lack the bipolarity shown by the native filaments of skeletal muscle the filament arrangement must be fundamentally different from that of a sarcomere. While optical diffraction analysis has already yielded information about the symmetry of the cross bridge organisation on smooth muscle myosin filaments (Sobieszek, 1972; 1977; Small 1977b) it is not yet clear how the polarity of the myosin molecules may be arranged with respect to, for example their position in one helical strand in the filament surface or perhaps their localisation on one side of the filament. We may only suggest that the continuous distribution of cross bridges along the filament length likely allows interaction of one thin filament with the whole length of a thick filament.

In the absence of any structures analagous to Z-lines it is apparent that the shearing process is constrained only by the polarity and geometry of organisation of the myofilaments. Whether or not there is some type of longitudinal grouping of the thin filaments remains unclear. In transverse sections of extended cells in rigor we have been struck by the apparent uniformity in the ratio of thick and thin filaments in the actin-myosin filament groups; we have found no clear evidence for the presence of only actin or myosin filaments. Ashton et al. (1975) claimed to identify some filament grouping from studies of semi-thick transverse sections but the evidence is not too compelling. Clearly, the use of antibodies to smooth muscle myosin (Groschel-Stewart et al., 1975) with isolated, differentiated smooth muscle cells should help to clarify this point.

A further unavoidable feature of the smooth muscle cell is the centrally placed nucleus, around which the contractile machinery must

Fig. 7 Native thick filaments seperated from isolated taenia coli cells under relaxing conditions (see Small, 1977b). A periodicity of about 140Å arising from the myosin cross bridges occurs along the entire filament length. Bar, 0.2 μ.

Fig. 8. Synthetic myosin filament formed from purified smooth muscle myosin. Cross bridges are seen along the entire filament, as with the native filaments. Bar, 0.2 μ.

Fig. 9. Phase contrast micrograph of shortened, Triton-extracted taenia coli cell in rigor. Obliquely arranged fibrils are visible in the cytoplasm. Bar, 10 μ.

Fig. 10. Tentative model of the contractile apparatus (see Small, 1977b). "Contractile units" (cylinders) are arranged so as to join opposing anchorage sites on the plasmalemma. The 100Å filaments form a network between the contractile units.

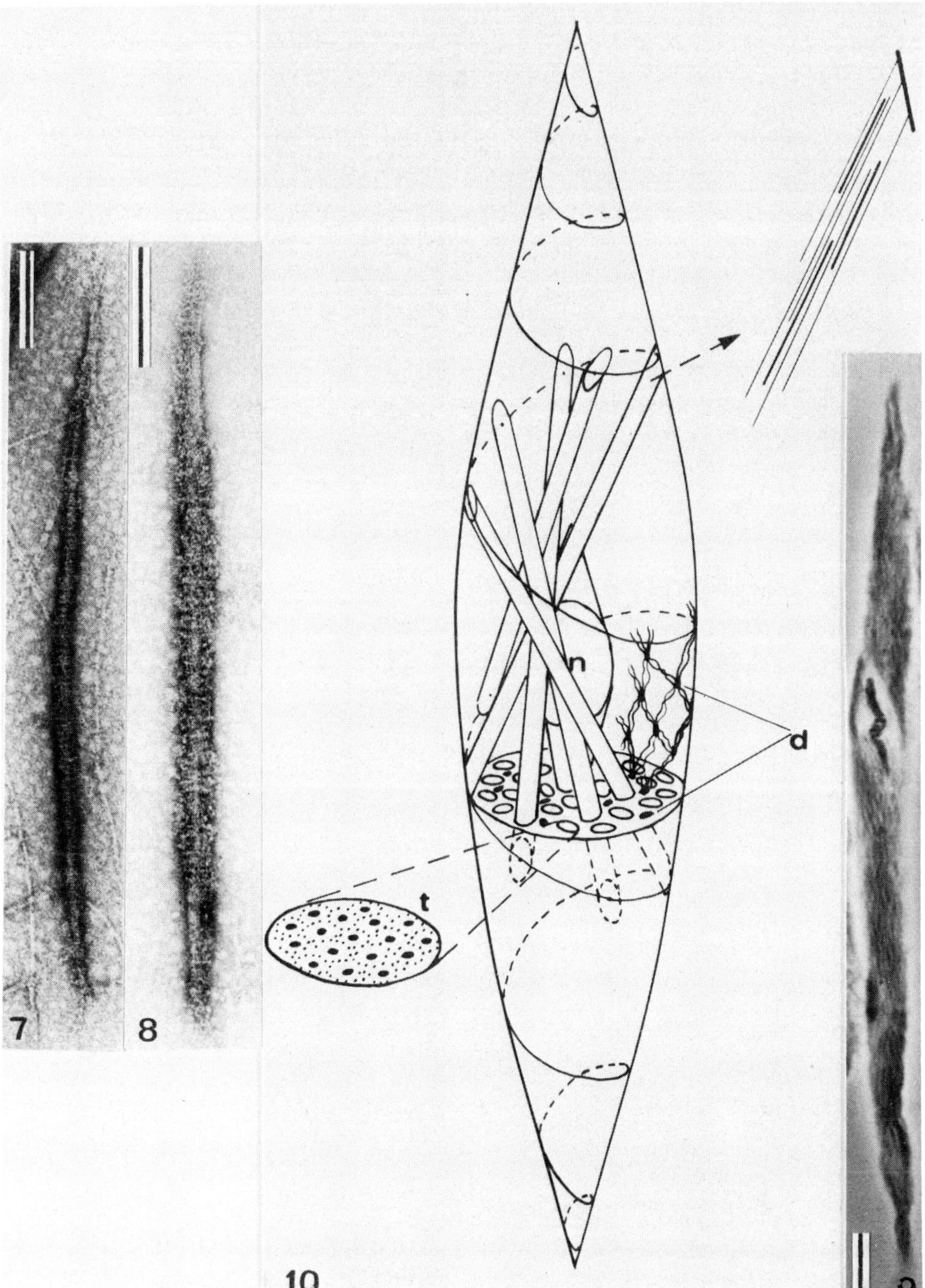
7
8
n
d
t
10
9

be arranged. This could be readily achieved with "contractile units" of the type shown in Fig. 10. In the shortened cell these units would be obliquely arranged while in the extended cell they would lie approximately parallel to the cell axis. That units of this kind might well exist in smooth muscle has been indicated from light microscope observations of cells in rigor (Small, 1974; 1977a and Fig. 9), but as yet seperate units have not been isolated intact from the cells. The presence of units of this kind is also consistent with the blebbing demonstrated to occur on the surface of isolated toad smooth muscle cells in response to electrical stimulation (Fay & Delise, 1973).

The complex network of 100Å filaments would appear to be arranged between the groups of contractile elements in such a way as not to impair their movement. The dense bodies, in this respect, may be necessary nodes in the network which serve to constrain the 100Å filaments such that there are well defined channels within which the contractile units may operate.

Further progress in our understanding of the structural basis of contraction in smooth muscle clearly requires an elucidation, in the first instance, of the structural polarity of the thick filaments. This should then give some indications of how the thin and thick filaments may be coupled to produce a functional contractile unit.

I am grateful to Mrs. M.R. Small for help in preparation of the manuscript and to the Volkswagen Foundation and Muscular Dystrophy Association for generous financial support.

REFERENCES

Ashton, F.T., Somlyo, A.V. & Somlyo. A.P. (1975). J. Mol. Biol. 98, 17-29.

Cooke, P. (1976). J. Cell Biol. 68, 539-556.

Fay, F.S. & Delise, C.M. (1973). Proc. Natl. Acad. Sci. U.S.A. 70, 641-645.

Groschel-Stewart, U., Chamley, J.A., Mc Connell, J.D. & Burnstock, G. (1975). Histochemistry 43, 215-224.

Heumann, H.G. (1971). Cytobiologie 3, 259.

Lowy, J., Vibert, P. J. Haselgrove, J.C. & Poulsen, F.R. (1973). Proc. R. Soc. Lond. B. 265, 191-196.

Rice, R.V., Moses, J.A., Mc Manus, G.M., Brady, A.C. & Blasik, L.M. (1970). J. Cell Biol. 47, 183-196.

Small, J.V. (1974). Nature, Lond. 249, 324-327.

Small, J.V. (1977a). In "Biochemistry of Smooth Muscle" (ed.N.L. Stephens) pp. 379-411, University Park Press, Baltimore.

Small, J.V. (1977b). J. Cell Sci. 24, 327-349.

Sobieszek, A. (1972). J. Mol. Biol. 70, 741-744.

Sobieszek, A. (1977). In "Biochemistry of Smooth Muscle" (ed.N.L.Stephens). pp. 413-443, University Park Press, Baltimore.

Sobieszek, A. & Bremel, R.D. (1975). Eur. J. Biochem. 55, 49-60.

Somlyo, A.P., Devine, C.E., Somlyo, A.V. & Rice, R.C. (1973). Phil. Trans. R. Soc. Lond. B. 265, 223-229.

Excitation-Contraction Coupling in Smooth Muscle, R. Casteels et al. eds.

ULTRASTRUCTURE OF THE CONTRACTILE APPARATUS: CONTROVERSIES RESOLVED AND QUESTIONS REMAINING

Andrew P. Somlyo and Avril V. Somlyo
Pennsylvania Muscle Institute
Presbyterian-University of Pennsylvania Medical Center
51 North 39th Street
Philadelphia, Pennsylvania 19104 U.S.A.

There is now substantial agreement about the ultrastructural organization of the contractile proteins in vertebrate smooth muscle. Earlier controversies have been settled and the view that myosin in vertebrate smooth muscle is organized into approximately 15-18nm diameter filaments having a relatively regular lateral spacing [1-3] is now generally accepted [4-6]. Myosin filaments are present in both relaxed and in contracted smooth muscle [7], and earlier difficulties in demonstrating them have been related to their susceptibility to destruction by chemical fixatives, particularly in swollen smooth muscle [8, 9]. The large ribbon-like structures observed in hypertonically treated smooth muscles are now also accepted to be due to aggregation of thick filaments [1, 8, 10]. The number of thick filaments contributing to the production of a ribbon appears to be similar to the number forming the "mini-sarcomere" - like units visualized in serial sections [11]. The myosin filaments in portal-anterior mesenteric vein are approximately 2.2μm long and tapered [11]. The actin/myosin ratios determined through, respectively, electron microscopy or biochemical techniques are comparable [12 and for review see 13] and there are no ultrastructural grounds for expecting to find significant amounts of soluble myosin in differentiated smooth muscle. However, dedifferentiated smooth muscle, like skeletal myoblasts [14] and non-muscle cells [for review see 15], may contain a type of myosin that is not organized into thick filaments.

The major unanswered question remaining about the ultrastructure of the myosin filaments in vertebrate smooth muscle is their precise molecular assembly including the organization of the cross-bridges. A definite bare zone is not seen in conventionally fixed sections [11], although this may be due to technical difficulties. Tail-to-tail assembly of myosin molecules [16] can result in the formation of short (~0.6μm) bipolar segments of artificial smooth muscle myosin filaments [17, 18]. *In vitro* filament formation may also occur by packing mechanisms that yield longer filaments with considerable fine structure and no central bare zones [18-20]. Therefore, the observations on artificial myosin filaments do not provide sufficiently unequivocal evidence for determining the in vivo molecular assembly of myosin in vertebrate smooth muscle. The presence of a 14.3nm meridional x-ray reflection suggestive of cross-bridge repeat has been demonstrated, at least under some conditions, by two groups [10, 21, 22].

The possibility of using isolated cells to obtain more rapid fixation has led to hopes that this method will better preserve the fine structure of the myosin filaments [6, 23]. Unfortunately, this approach is subject to the potentially distorting effects of the proteolytic enzymes (used for the isolation of single cells) on myosin filament structure: trypsin and/or collagenase treated smooth muscle cells often contain tactoids, rather than filaments, of myosin [24, 25 and Somlyo and Somlyo, unpublished]. Thick filaments showing cross-bridges having some helical symmetry have been observed in fragments of isolated cells [23] and in apparently well preserved single cells such filaments may show considerable fine structure [6]. However, grossly dissimilar filament structures also occur in adjacent, isolated cells [Somlyo and Somlyo, unpublished]. Therefore, there is considerable uncertainty as to which of the fine structures observed in isolated cells represents the correct, in vivo organization of myosin filaments.

Cryoultramicrotomy represents an alternative approach for the preservation of fine structure, since it circumvents the use of the most destructive fixative, osmium. In negatively stained transverse cryo sections of rabbit portal-anterior mesenteric vein smooth muscle obtained after glutaraldehyde fixation but without cryoprotectants, rosettes consisting of a central thick and surrounding thin (actin) filaments are present (Fig. 1). Fine threads, suggestive of cross-bridges, emanate from the thick filaments that are relatively round in cross section. These preliminary observations are promising, but have not yet allowed a conclusion regarding the precise assembly of myosin. Intermediate (10nm) filaments and dense bodies are also demonstrable in cryo sections.

The thin (5-8nm diameter) actin filaments are present in a ratio of approximately 15:1 to the thick filaments in portal vein smooth muscle. The length of the filamentous actin has not yet been directly measured, due to technical difficulties. This information, of obvious relevance to smooth muscle mechanics, and the polarity of the actin filaments emerging from opposite poles of the dense bodies are among the ultrastructural questions remaining to be answered.

The role of the cytoplasmic dense bodies as attachment sites for the actin fila-

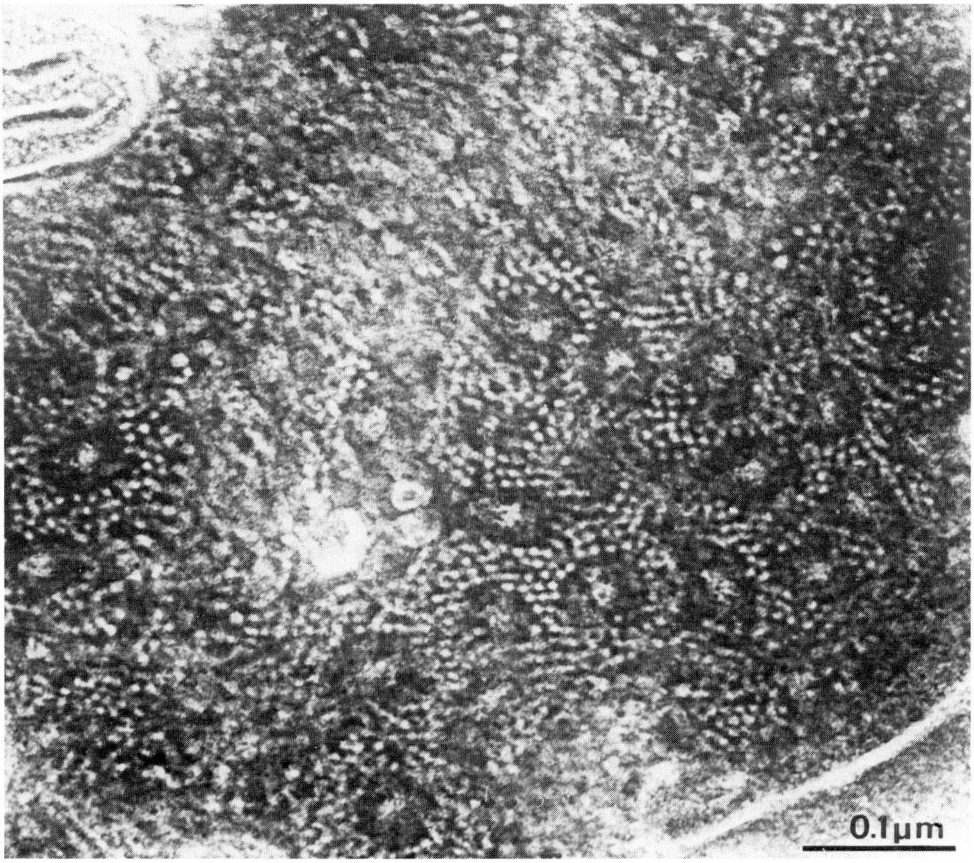

ments, while not universally agreed upon, is supported by intermediate high voltage stereo electron micrographs showing the emergence of actin filaments from dense bodies in situ [11], by the presence of actin filaments in isolated dense bodies [26, 27] and by the binding of anti-α-actinin to dense bodies [28]. Nevertheless, in view of the rather non-specific staining characteristics and shape of the dense bodies, they may not represent a biochemically or functionally homogeneous structure.

Intermediate (10nm) filaments frequently course in long bundles and often surround dense bodies (e.g., Plate XIV in 11). They have been compared [1, 29] to the cytoskeleton of obliquely striated muscles, but their function in smooth muscle has not been definitely established. The fact that they are not part of the myosin or actin filament [1, 11] is now generally accepted. Intermediate filaments have been isolated [29-31] and contain a relatively major protein component of approximately 55,000 dalton molecular weight. Similar filaments that are at least antigenically related are present in a variety of cell types [13, 32, 33] and can be induced in large numbers with drugs such as cytochalasin and colcimide [34, 35]. The irregular ("scrambled") appearance of such filaments and the variability of their numbers that in abnormal smooth muscle fibers may crowd out the thick and thin filament lattice (e.g., Fig. 4 in 2), is also reflected in the variable biochemical yield of the 55,000 dalton protein [31], and makes it difficult to draw valid conclusions about their function.

In conclusion, earlier descriptions of the myosin filament in vertebrate smooth muscle have been verified and extended to include a more precise determination of filament length. The mode of molecular assembly of myosin into filaments and the resultant cross-bridge arrangement that determines the detailed mechanism of force generation remain to be established.

Supported by HL-15835 to the Pennsylvania Muscle Institute.

REFERENCES

1. Somlyo, A.P., Somlyo, A.V., Devine, C.E. and Rice, R.V. (1971) Aggregation of thick filaments into ribbons in mammalian smooth muscle. Nature New Biol. 231:243-246.
2. Somlyo, A.P., Devine, C.E., Somlyo, A.V. and Rice, R.V. (1973) Filament organization in vertebrate smooth muscle. Phil. Trans. Roy. Soc. B. 265:223-229.
3. Rice, R.V., McManus, G.M., Devine, C.E. and Somlyo, A.P. (1971) A regular organization of thick filaments in mammalian smooth muscle. Nature New Biol. 231:242-243.
4. Squire, J.M. (1975) Muscle filament structure and muscle contraction. Ann. Rev. Biophys. Bioengr. 4:137-163.
5. Shoenberg, C.F. and Needham, D.M. (1976) A study of the mechanism of contraction in vertebrate smooth muscle. Biol. Rev. 51:53-104.
6. Small, J.V., this volume.
7. Devine, C.E. and Somlyo, A.P. (1971) Thick filaments in vascular smooth muscle. J. Cell Biol. 49:636-649.
8. Jones, A.W., Somlyo, A.P. and Somlyo, A.V. (1973) Potassium accumulation in smooth muscle and associated ultrastructural changes. J. Physiol. London 232:247-273.

9. Somlyo, A.P., Devine, C.E. and Somlyo, A.V. (1971) Thick filaments in unstretched mammalian smooth muscle. Nature New Biol. 233:218-219.

10. Lowy, J., this volume.

11. Ashton, F.T., Somlyo, A.V. and Somlyo, A.P. (1975) The contractile apparatus of vascular smooth muscle: intermediate high voltage stereo electron microscopy. J. Mol. Biol. 98:17-29.

12. Murphy, R.A., Herlihy, J.T. and Megerman, J. (1974) Force-generating capacity and contractile content of arterial smooth muscle. J. Gen. Physiol 64:691-705.

13. Somlyo, A.P., Somlyo, A.V., Ashton, F.T. and Vallieres, J. (1976) Vertebrate smooth muscle: ultrastructure and function; in Goldman, Pollard and Rosenbaum Cell Motility; 3rd ed., pp. 165-183 (Cold Spring Harbor Conference on Cell Proliferation, Cold Spring Harbor Laboratory, U.S.A.)

14. Chi, J., Fellini, S. and Holtzer, H. (1975) Differences among myosin synthesized in non-myogenic cells, presumptive myoblasts and myoblasts. Proc. Natl. Acad. Sci. 72:4999-5003.

15. Pollard, T.D. and Weihing, R.R. (1974) Actin and myosin and cell movement. Crit. Rev. Biochem. 2:1-67.

16. Huxley, H.E. (1963) Electron microscope studies on the structure of natural and synthetic protein filaments from striated muscle. J. Mol. Biol. 7:281-308.

17. Kaminer, B. (1969) Synthetic myosin filaments from vertebrate smooth muscle. J. Mol. Biol. 39:257-264.

18. Wachsberger, P.R. and Pepe, F.A. (1974) Purification of uterine myosin and synthetic filament formation. J. Mol. Biol. 88:385-391.

19. Sobieszek, A. (1972) Cross-bridges on self-assembled smooth muscle myosin filaments. J. Mol. Biol. 70:741-744.

20. Craig, R. and Megerman, J. (1977) Oblique filaments from vertebrate smooth muscle myosin. Biophys. J. 17:121a.

21. Lowy, J., Polsen, F.R. and Vibert, P.J. (1970) Myosin filaments in vertebrate smooth muscle. Nature 225:1053-1054.

22. Shoenberg, C.F. and Haselgrove, J.C. (1974) Filaments and ribbons in vertebrate smooth muscle. Nature 249:152-154.

23. Somlyo, A.V., Ashton, F.T., Lemanski, L., Vallieres, J. and Somlyo, A.P. (1977) Filament organization and dense bodies in vertebrate smooth muscle; in Stephens The biochemistry of smooth muscle; pp. 445-471 (University Park Press, Maryland).

24. Rosenbluth, J. (1971) Myosin-like aggregates in trypsin-treated smooth muscle cells. J. Cell Biol. 48:174-188.

25. Gabella, G. and Raeymaekers, L. (1976) Effect of collagenase on mechanical activity and fine structure of an intestinal smooth muscle. Cell Tiss. Res. 173:29-44.

26. Panner, B.J. and Honig, C.R. (1967) Filament ultrastructure and organization in vertebrate smooth muscle: contraction hypothesis based on localization of actin and myosin. J. Cell Biol. 35:303-321.

27. Nonomura, Y. and Ebashi, S. (1975) Isolation and identification in smooth muscle contractile proteins; in Daniel and Paton Methods in pharmacology; 3rd ed. pp. 141-162 (Plenum Press, New York).

28. Schollmeyer, J.E., Furcht, L.T., Goll, D.E., Robson, R.M. and Stromer, M.H. (1976) Localization of contractile proteins in smooth muscle cells and in normal and transformed fibroblasts; in Goldman, Pollard and Rosenbaum Cell Motility [A]; 3rd ed., pp. 361-388 (Cold Spring Harbor Conference on Cell Proliferation, Cold Spring Harbor Laboratory, U.S.A.).

29. Cooke, P. (1976) A filamentous cytoskeleton in vertebrate smooth muscle fibers. J. Cell Biol. 68:539-556.

30. Rice, R.V. and Brady, A.C. (1973) Biochemical and ultrastructural studies on vertebrate smooth muscle; in The mechanism of muscle contraction; 37th ed., pp. 429-438 (Cold Spring Harbor Symposia on Quantitative Biology, Cold Spring Harbor Laboratory, U.S.A.).

31. Small, J.V. and Sobieszek, A. (1977) Studies on the function and composition of the 10nm (100Å) filaments of vertebrate smooth muscle. J. Cell Sci. 23: 253-268.

32. Ishikawa, H., Bischoff, R. and Holtzer, H. (1968) Mitosis and intermediate-sized filaments in developing skeletal muscle. J. Cell Biol. 38:538-555.

33. Blose, S.H., Shelanski, M.L. and Chacko, S. (1977) Localization of bovine brain filament antibody on intermediate (100Å) filaments in guinea pig vascular endothelial cells and chick cardiac muscle cells. Proc. Natl. Acad. Sci. U.S.A. 74:662-665.

34. Holtzer, H., Croop, J., Gershon, M. and Somlyo, A.P. (1974) Effects of cytochalasin-B and colcimide on cells in muscle cultures. Am. J. Anat. 141:291-296.

35. Holtzer, H., Croop, J., Dienstman, S., Ishikawa, H. and Somlyo, A.P. (1975) The effects of cytochalasin-B and colcimide on myogenic cultures. Proc. Nat. Acad. Sci. 72:513-517.

Excitation-Contraction Coupling in Smooth Muscle, R. Casteels et al. eds.

SUMMARY OF DISCUSSION ON "STRUCTURAL ASPECTS"

H. E. Huxley
MRC Laboratory of Molecular Biology
University Postgraduate Medical School
Cambridge, CB2 2QH, U.K.

There now appears to be general agreement that myosin in vertebrate smooth muscle is present in the form of cylindrical filaments under normal physiological conditions. It appears that the earlier difficulties with the electron-microscopy of these muscles, first in seeing any myosin aggregates at all, and then in deciding whether they were cylindrical or ribbon-shaped, arose from changes in the tissue during fixation. It is not yet clear whether the lability of smooth muscle myosin filaments is of physiological significance or not. Dr. Somlyo reported that he had been able to observe cylindrical filaments in freeze-fractured preparations, providing further evidence for their existence in normal tissue. Dr. Lowy reported that optical diffraction studies on myosin filaments in sections of smooth muscle indicated that they had helical symmetry. He also pointed out that under conditions where the myosin in smooth muscle aggregates into ribbons, in hypertonic Ringer, the muscle will still exert up to 50 % of its maximal tension. There was some discussion between Drs. Small and Somlyo as to whether "dense bodies" in vertebrate smooth muscle have actin filaments connected to them (Dr. Small says no, Dr. Somlyo, yes) but the matter was not resolved. Finally there was a discussion about the exact location of tropomyosin to the groove in the actin filaments of smooth muscle. It was pointed out that if a muscle was myosin regulated only, then the tropomyosin would be expected to lie nearer the centre of the groove even in a relaxed muscle, whereas in certain smooth muscles (for instance ABRM) this did not seem to be the case and a change in position appeared to occur upon contraction. This paradox was not resolved, but it was pointed out that the measurements of tropomyosin position by three-dimensional image reconstruction and optical filtering methods had always been done using vertebrate striated muscle actin and myosin, and that it would be desirable to repeat some of these studies using smooth muscle proteins.

Excitation-Contraction Coupling in Smooth Muscle, R. Casteels et al. eds.

REGULATORY PROTEINS OF SMOOTH MUSCLE

S. Ebashi, T. Mikawa, M. Hirata, T. Toyo-oka and Y. Nonomura

Department of Pharmacology, Faculty of Medicine,
University of Tokyo, Tokyo

INTRODUCTION

It was once considered that the regulatory system of smooth muscle would be essentially the same as that of skeletal muscle[1, 2], i.e., the smooth muscle contraction should also be regulated by the troponin-tropomyosin system. The discovery of a myosin-linked system in scallop muscle[3], however, has thrown a doubt upon the soundness of this conventional concept. Since then, the regulation in smooth muscle has been the subject of much controversy.

The following is our present view based on our recent studies[4 - 11]: The regulatory system of smooth muscles, viz., chicken gizzard, bovine stomach and bovine aorta, is composed of tropomyosin and a factor of a molecular weight about 80,000 dalton. The factor has an affinity for actin, not tropomyosin, and is fully effective at a molar ratio to tropomyosin of one tenth or less. Smooth muscle myosin and actin even with tropomyosin cannot exhibit a significant interaction in the presence of ATP. The activation of the interaction is observed only in the presence of both the 80,000 dalton factor and Ca ion; thus Ca ion and the regulatory system act as the real activator. This is in sharp contrast with vertebrate striated muscle, where troponin in collaboration with tropomyosin depresses the myosin-actin interaction in the absence of Ca ion; Ca ion acts as a de-repressor.

In the meantime, several research groups have proposed that the principal reaction underlying the regulation of smooth muscle is the phosphorylation and dephosphorylation of a myosin light chain by a Ca-dependent kinase[12 - 15].

One of the aims of this article is to find out the grounds for the discrepancy among investigators.

METHODS AND MATERIALS

The preparation methods and experimental procedure were essentially the same as described in previous papers[6, 8, 9]. As the source of muscle proteins, rabbit skeletal muscle, chicken breast muscle, chicken gizzard, bovine stomach and bovine aorta were used.

RESULTS AND DISCUSSIONS

A) Characteristics of smooth muscle contractile system

Studies on the regulatory system of smooth muscle have rather unexpectedly revealed many distinct properties of its contractile system from those of vertebrate skeletal muscle[4 - 8]. One of marked differences is the lack of the property responding to the increase in MgATP concentration in a 'biphasic' way; this property, so-called 'substrate inhibition' type response, has been considered as one of fundamental properties of vertebrate skeletal muscle actomyosin* (Fig. 1). The result with superprecipitation of gizzard myosin B was shown in a previous paper (cf. Fig. 3 in the ref. 7; in this case the response in the absence of Ca ion showed a faint tendency of biphasic response). This is more definitely demonstrated in Fig. 2 in which the actomyosin ATPase activity is taken as the index of contraction. It is interesting that a steady increase in ATPase is still observed even at 1 mM MgATP.

We have shown that the ATPase activity as well as superprecipitation of smooth muscle myosin B is also much less sensitive to the change in the ionic strength than skeletal muscle myosin B (cf. Fig. 4B in ref. 7). As is well known, the sensitivities of skeletal contractile systems increase in the order: glycerinated fibers $<$ myofibrils $<$ myosin B $<$ reconstituted actomyosin $<$ acto-H-meromyosin. This is also true of smooth muscle. The reaction of fresh gizzard myosin B preparation is quite insensitive to the change in ionic strength, but reconstituted gizzard actomyosin (tropomyosin and the 80,000 dalton factor are both required for this kind of experiment) is fairly sensitive to the ionic strength, though still less sensitive than skeletal myosin B.

Another important feature of the smooth muscle contractile system is the contraction requirement for free Mg ion, as first pointed out by Filo et al.[16] using vascular muscle. Mg ion has a

---

* In the presence of Ca, the biphasic response of myosin B of scallop striated muscle is not so marked as that of skeletal muscle; thus the scallop muscle apparently resembles smooth muscle. In the absence of Ca ion, however, scallop myosin B shows a sharp 'substrate inhibition', i.e., at a low MgATP concentration, say, 10^{-5} M, it shows nearly full activation, not so much less than the maximum response to Ca ion, but with a slight increase in MgATP concentration, it shifts suddenly to a depressed state. Thus scallop muscle seems more akin to vertebrate skeletal muscle (see Table I).

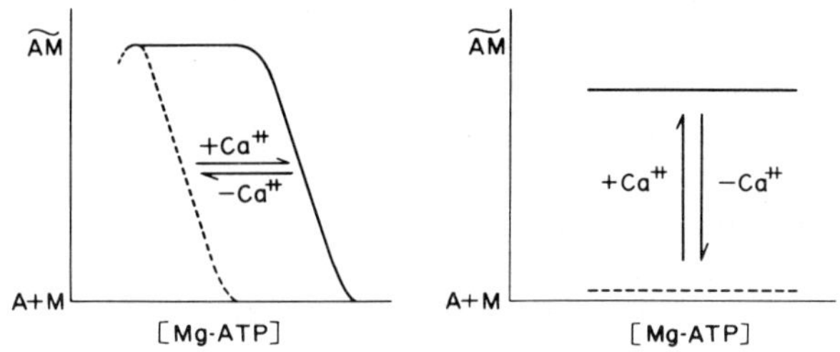

Fig. 1 Schematic illustration of the myosin-actin-ATP interaction as a function of the Mg-ATP concentration. AM = the state in which actin(A) and myosin(M) are actively interacting with each other. A + M = the state in which actin and myosin are dissociated.

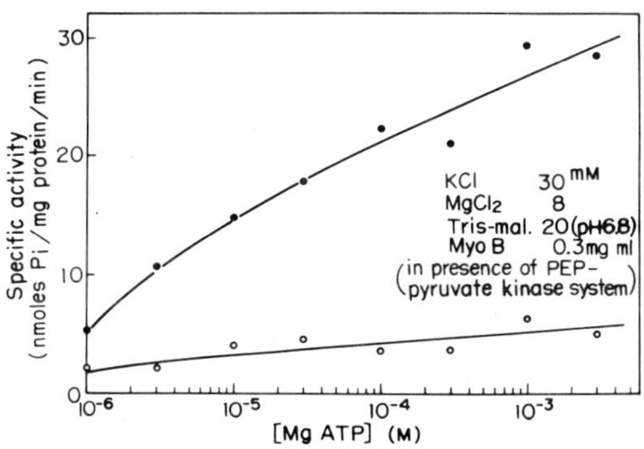

Fig. 2 The ATPase activity of gizzard myosin B versus the MgATP concentration.

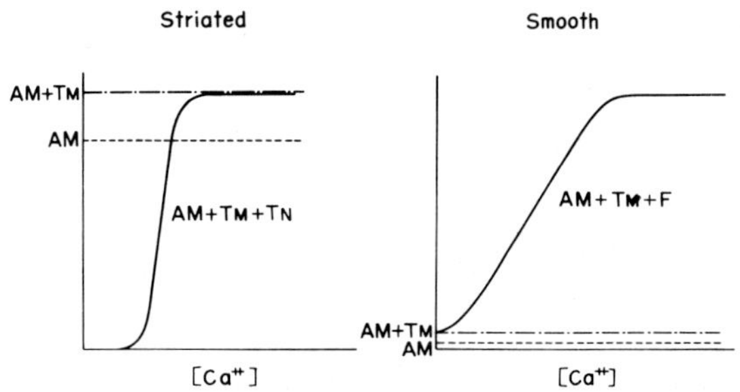

Fig. 3 Schematic illustration of the myosin-actin-ATP interaction as a function of the Ca2+ concentration. TM = tropomyosin. F = the 80,000 dalton component.

Table I

Interactions of myosin, actin and native tropomyosin derived from either rabbit skeletal muscle or chicken gizzard

Myosin		Actin		Native tropomyosin		ATPase nmoles/min.mg. myosin	
RS	G	RS	G	RS	G	$1x10^{-7}Ca^{++}$	$1.3x10^{-5}Ca^{++}$
x		x				373	363
x		x		x		78	392
x		x			x	324	438
x			x			251	228
x			x	x		71	139
x			x		x	290	301
	x	x				2.2	1.8
	x	x		x		1.2	1.1
	x	x			x	2.0	8.1
	x		x			4.1	4.2
	x		x	x		2.2	2.4
	x		x		x	5.5	44.2

RS: rabbit skeletal. G: chicken gizzard
For conditions see Fig. 5

Table II

Regulation Patterns

	Requirement for TM	Requirement for specific protein	Role of Ca	Biphasic response to MgATP	Sensitivity to ionic strength
Troponin-dependent	yes	yes (troponin)	de-repressor	yes	high
Myosin-linked	no	no	de-repressor	yes	moderate
Smooth muscle	yes	yes (80,000 factor)	activator	no	low

TM: tropomyosin

profound effect on the filament formation of gizzard myosin. This may partly explain the requirement for Mg ion (cf. plate 3 in ref. 7).

The most important characteristic of the smooth muscle contractile system is the virtual absence of the myosin-actin interaction in the presence of ATP[4 - 10], i.e., no significant superprecipitation and actin activated ATPase (Fig. 3 and Table I). Tropomyosin does not significantly alter the situation. The activation can be observed only in the presence of Ca ion and the factor responsible for Ca regulation. Thus the interaction of myosin, actin and ATP in smooth muscle is fundamentally different from that in vertebrate striated muscle and also from scallop muscle. This crucial nature is the real basis of smooth muscle regulation.

B) Regulatory Proteins

When we succeeded in separating 'native tropomyosin'[2] into tropomyosin and a protein fraction[4 - 8] by the same method as used for the skeletal muscle native tropomyosin[17], we believed that the protein fraction should have a similar quality to troponin.

The protein fraction was composed of a major component(s), having a molecular weight about 20,000 dalton, and a minor component of a molecular weight about 80,000 dalton[4, 5, 7]. We first thought that the major component might be the regulatory protein(s)[4], because it appeared something like a mixture of troponin I and troponin C. However, this was not true. The 80,000 dalton component, the minor component, was shown to be sole responsible factor[8] (Fig. 4). This conclusion is further confirmed by our recent experiment carried out for other purposes (Fig. 6).

As shown in Fig. 5, only a very small amount of the factor, less than one hundredth of actin in molar ratio (about one twentieth of tropomyosin) is required for full activation. In view of this fact together with other various findings, we are now of the opinion that the factor should not be called 'troponin'[8] (in our laboratory, we tentatively call it 'leiotonin'). In Table II the mode of action of three different kinds of regulatory system are compared.

C) Specificities of smooth muscle contractile proteins

All components involved in contraction are exchangeable among chicken gizzard, bovine stomach and bovine aorta, but the exchangeability between skeletal and smooth muscle proteins is rather complicated.

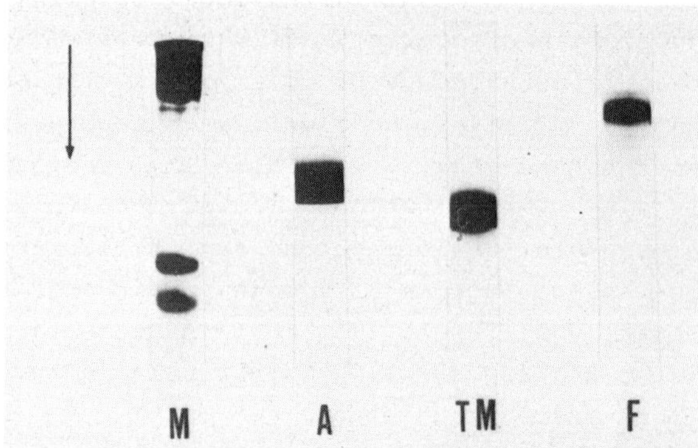

Fig. 4 Sodium dodecyl sulfate polyacrylamide gel electrophoresis of gizzard contractile proteins. For others see the legends to previous figures.

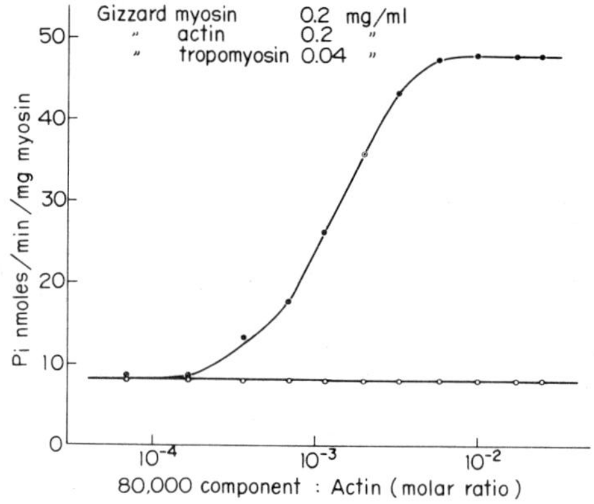

Fig. 5 The relationship between the amount of 80,000 component expressed as the molar ratio to actin and the ATPase activity of reconstituted gizzard actomyosin.

Actin had been considered to be common to all kinds of muscle. Quite unexpectedly, however, rabbit skeletal actin cannot replace gizzard actin in the contractile system of gizzard (Table I). This interesting finding, however, cannot be generalized. Chicken skeletal actin can be a weak substitute for gizzard actin and rabbit skeletal actin can replace aorta actin to a certain extent. Rabbit skeletal tropomyosin cannot replace gizzard tropomyosin in the gizzard system.

Smooth muscle myosin is very unique and even in the case where skeletal actin can replace smooth muscle actin, the resulting actomyosin behaves as a smooth muscle actomyosin. It is interesting, however, that smooth muscle myosin can be irreversibly converted to a skeletal muscle type by aging* or some deteriorating procedure. Such a converted smooth muscle myosin can interact with any skeletal actin in the presence of ATP without the aid of tropomyosin and the 80,000 dalton component. Troponin is effective on such an actomyosin system.

Skeletal myosin can interact with any smooth muscle actin rather markedly (Table I). Since the smooth muscle regulatory system is originally an activating system, it is quite reasonable that it has practically no regulatory effect on the interaction of skeletal myosin with actin in the presence of ATP(Table I).

Bremel[18] concluded from their result using 'dilution method' that regulation in gizzard is of myosin-linked nature. From the facts mentioned above, however, this cannot be an evidence against the actin-linked regulation, because added rabbit skeletal actin does not react with gizzard myosin, and added skeletal myosin interacts with gizzard actin irrespective of the presence of the gizzard regulatory system (Table I).

D) Light chain kinase

Hartshorne and his colleagues[12] have shown that our 'native tropomyosin' preparation[2, 6] contains a lot of Ca-dependent myosin light chain kinase, and the phosphorylation of myosin light chain by the kinase is the very mechanism underlying the activation by Ca ion. Adelstein and his colleagues[13], Sobieszek and Small[14] and Watanabe and his colleagues[15] have presented an idea along the same line.

We completely agree with Hartshorne et al. in that our preparation contains a lot of the kinase. However, the 80,000 dalton factor does not show any kinase activity. The kinase can certainly phosphorylate myosin, but it cannot activate the actin-myosin-ATP interaction in the absence of the 80,000 factor (Fig. 6). Thus the

---

* At an early stage of our experiments, reconstituted gizzard actomyosin behaved somewhat like skeletal actomyosin (cf. Fig. 9 in ref. 4). Now it is certain that the preparation at that time was deteriorated owing to immature technique. Smooth muscle tropomyosin has a promoting action on the superprecipitation and ATPase of such deteriorated preparations as it does on those of skeletal actomyosin.

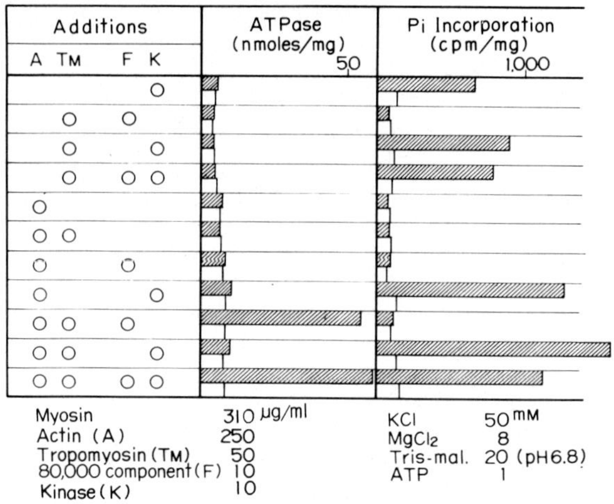

Fig. 6 The ATPase activity and the degree of phosphorylation of gizzard myosin. ▨ : 1.3 x 10^{-5} M Ca^{2+}. □ : 10^{-7} M Ca^{2+}.

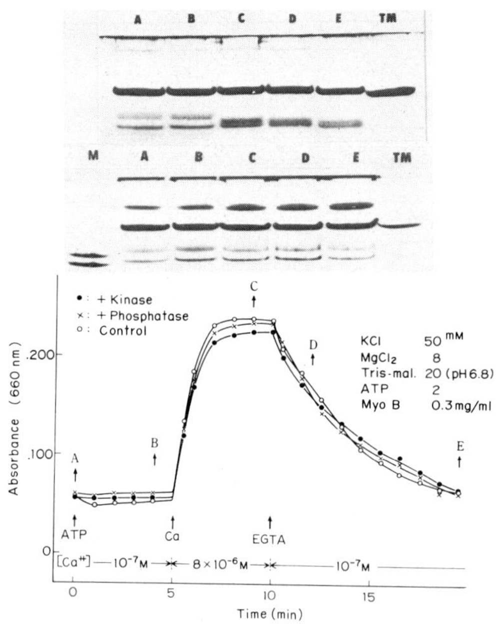

Fig. 7 superprecipitation of gizzard myosin B in the presence of myosin light chain kinase of phosphatase and in the absence of both enzymes. The inlet figures show the urea slab-gel electrophoretic patterns of myosin B without added enzyme (upper) and myosin B with phosphatase[19] (lower) sampled at the time indicated by the same letter.

activity of the kinase is not related to the contractile response.

This is more clearly demonstrated in Fig. 7. The superprecipitation of gizzard myosin B was carried out in the presence of plenty of the kinase, or phosphatase, or in the absence of both enzymes from without. Myosin B was incubated with ATP in the absence of Ca ion and then activated by Ca; after a while superprecipitated myosin B was dispersed, or relaxed, by removing Ca ion. Surprisingly, no difference in the course of superprecipitation was observed under the three different conditions. The degree of phosphorylation was then analyzed. The light chain of the control myosin B, i.e., without added enzymes, was not phosphorylated before the addition of Ca, but certainly phosphorylated upon addition of Ca ion, as if there were a parallelism between phosphorylation and contraction. Removal of Ca did not dephosphorylate the light chain, perhaps because of low intrinsic phosphatase activity, whereas superprecipitated myosin B was well dispersed or relaxed, showing a dissociation of contractile process from phosphorylation. The response in the presence of plenty of kinase was rather similar to the above, though the phosphorylation had started before addition of Ca ion. The most impressive was the result with phosphatase. As shown in Fig. 7, practically no phosphorylation took place during the whole course, but myosin B completed the contractile-relaxation cycle.

Thus the phosphorylation of myosin light chain has no direct relation with contractile processes.

CONCLUSION

Our view as regards the regulatory mechanism of smooth muscle described in 'INTRODUCTION' has been confirmed and extended, while myosin light chain kinase appears to play no significant role in the contractile processes.

The mechanism how the 80,000 dalton factor modifies the myosin-actin interaction in collaboration with tropomyosin is the urgent problem to be studied.

ACKNOWLEDGEMENT

Our thanks are due to Miss H. Nakasone for her technical assistance. This work was supported in part by research grants from the Muscular Dystrophy Associations of America, Inc., the Ministry of Science, Culture and Education, Japan, the Ministry of Health and Welfare, Japan, and the Iatrochemical Foundation, and the Toray Sicence Foundation.

REFERENCES

1. Ebashi, S. and Endo, M. (1968) Prog. Biophys. Molecul. Biol. 18, 123-183.
2. Ebashi, S., Iwakura, H., Nakajima, H., Nakamura, R. and Ooi, Y. (1966) Biochem. Z. 345, 201-211.
3. Kendrick-Jones, J. Lehman, W., and Szent-Györgyi, A.G. (1970) J. Mol. Biol. 54, 313-326.
4. Ebashi, S., Nonomura, Y., Kitazawa, T. and Toyo-oka, T. (1975) In Calcium Transport in Contraction and Secretion, eds. E. Carafoli et al., pp. 404-414, North-Holland: Amsterdam & New York.
5. Ebashi, S., Toyo-oka, T. and Nonomura, Y. (1977) Gizzard troponin, In The Biochemistry of Smooth Muscle, ed. N.L. Stephens, University Park Press, Baltimore, pp. 551-556.
6. Ebashi, S., Toyo-oka, T. and Nonomura, Y. (1975) J. Biochem. 78, 859-861.
7. Ebashi, S., Nonomura, Y. Toyo-oka, T. and Katayama, E. (1976) In Calcium in Biological Systems, ed. Duncan, C.J., pp. 349-360, Cambridge Univ. Press, London.
8. Mikawa, T., Toyo-oka, T., Nonomura, Y. and Ebashi, S. (1977) J. Biochem. 81, 273-275.
9. Ebashi, S. (1976) J. Biochem. 79, 229-231.
10. Ebashi, S., Mikawa, T., Hirata, M. and Nonomura, Y. (1977) Blool Vessels, in press.
11. Ebashi, S. (1977) Jap. J. Pharmacol. 27, 3P.
12. Aksoy, M.D., Williams, D., Sharkey, E.M. and Hartshorne, D.J. (1976) Biochim. Biophys. Res. Comm. 69, 35-41.
13. Sobieszek, A. and Small, J.V. (1976) J. Mol. Biol. 101, 75-92.
14. Chacko, S., Conti, M.A. and Adelstein, R.S. (1977) Proc. Natl. Acad. Sci. USA 70, 129-133.
15. Ikebe, M. Onishi, H. and Watanabe, S. (1977) J. Biochem., in press.
16. Filo, R.S., Bohr, D.F. and Rüegg, J.C. (1965) Science 147, 1581-1583.
17. Ebashi, S. and Kodama, A. (1965) J. Biochem. 58, 107-108.
18. Bremel, R.D. (1975) Nature, 252, 405-407.
19. Morgan, M. Perry, S.V. and Ottway, J. (1976) Biochem. J. 157, 687-697.

Excitation-Contraction Coupling in Smooth Muscle, R. Casteels et al. eds.

THE TROPONIN-LIKE COMPONENTS OF SMOOTH MUSCLE

R.J.A. Grand, S.V. Perry & R.A. Weeks

Department of Biochemistry,
University of Birmingham,
Birmingham B15 2TT, U. K.

Several investigators (1,2) have claimed to have isolated a troponin-like complex, similar to that present in striated muscle, from vertebrate smooth muscle whilst others have concluded that troponin is absent from this tissue (3-5). On the other hand, some investigators have described troponin-like preparations that appear to be very different from the troponin of striated muscle (5,6). As an approach to resolving the important problem of deciding whether a functional troponin system is present in smooth muscle we have attempted to isolate components corresponding to troponin C and troponin I of skeletal muscle by making use of specific properties of these well-defined skeletal proteins.

When extracts of whole smooth muscle in 6 M urea, 25 mM-Tris-80 mM glycine buffer, pH 8.6, were electrophoresed in polyacrylamide gels in the presence of EGTA, a band of protein migrated well in front of the bulk of the proteins that moved into the gel (7). A similar effect was observed earlier with striated muscle (8,9) and was shown to be due to the dissociation of a complex of troponin C and troponin I that was stable in 6 M urea in the presence of Ca^{2+} and which split into its components when the Ca^{2+} concentration was reduced to a low level by the addition of EGTA. This observation, which was made with extracts of rabbit and bovine uterus, bovine stomach and carotid artery and chicken gizzard, therefore suggested that these smooth muscles might contain proteins similar to the troponin C and troponin I present in striated muscle.

Troponin C-like component

By application of an affinity chromatographic technique using troponin I from fast skeletal muscle linked to Sepharose (7), a troponin C-like protein has been isolated and characterized from uterus of the rabbit and cow, bovine carotid and stomach and chicken gizzard. Although all the preparations of troponin C obtained from these smooth muscles had similar properties the most extensive studies were made on troponin C isolated from rabbit uterus. The properties of smooth muscle troponin C are summarized in Table 1 in which they are compared with troponin C from rabbit fast skeletal muscle and the troponin C-like protein from brain that activates phosphodiesterase (10), sometimes called the brain modulator protein (11).

Table 1. Comparison of the properties of troponin C from muscle and the brain modulator protein.

	Fast skeletal muscle	Smooth muscle	Brain
Molecular weight	18,000	18,500	18,000
Effect of EGTA on electrophoretic mobility	Decreases	Increases slightly	Increases
Troponin I interaction			
Complex formation	+	+ (Less stable)	+ (Less stable)
Neutralisation of the ATPase inhibition	+	+	+
Inhibition of phosphorylation	+	+	
Phosphodiesterase	Slight activation (1/600)	Activates	Activates
Trimethyl lysine	−	+	+
Calcium sensitivity + troponin I + tropomyosin	−	−	+

The smooth muscle troponin C neutralised the inhibitory action of fast skeletal muscle troponin I on the Mg^{2+}-stimulated ATPase activity of desensitized actomyosin from skeletal muscle in a similar manner to troponin C from fast skeletal muscle (Fig.1). It was likewise similar in behaviour to the latter protein in its ability to inhibit completely the phosphorylation of fast skeletal muscle troponin I catalysed by 3':5'-cyclic AMP-dependent protein kinase and phosphorylase kinase. It differed from fast skeletal troponin C, but resembled the brain modulator protein in containing trimethyl lysine (approx. 1 mol/mol). In the effect of EGTA on its electrophoretic mobility in non-dissociating conditions at pH 8.6, the smooth muscle troponin C was intermediate in behaviour between the fast skeletal muscle troponin C and the brain modulator protein. On addition of Ca^{2+} to the calcium-free form, fast skeletal muscle troponin C undergoes a marked conformational change which leads to an increase in electrophoretic mobility to the anode at pH 8.6. This change in mobility is the opposite to that expected from the increase in charge resulting from the binding of Ca^{2+} (9). In contrast, brain modulator protein decreases in mobility on binding Ca^{2+} as would be expected from the charge effect (12). The behaviour of smooth muscle troponin C was intermediate between these two extremes in that its mobility was slightly decreased on binding Ca^{2+}.

The fact that smooth troponin C can form a Ca^{2+}-dependent complex with fast skeletal troponin I that is stable in urea is apparent from its mode of preparation by the affinity method. This property was more readily lost by smooth muscle troponin C than is the case with fast skeletal muscle troponin C. Often after preparation by the affinity method and other procedures, smooth muscle troponin C would form a complex with fast skeletal troponin I only in non-dissociating buffers. In this property it resembled the brain modulator protein.

Although the smooth muscle troponin C was similar to brain modulator protein in possessing phosphodiesterase activating ability (kindly assayed by Dr. T.C. Vanaman) it resembled the striated muscle forms of troponin C in that it did not restore Ca^{2+} sensitivity to the Mg^{2+}-stimulated ATPase of desensitized skeletal actomyosin in the presence of skeletal tropomyosin and skeletal troponin I (Fig.1). Brain modulator protein, however, restores Ca^{2+}

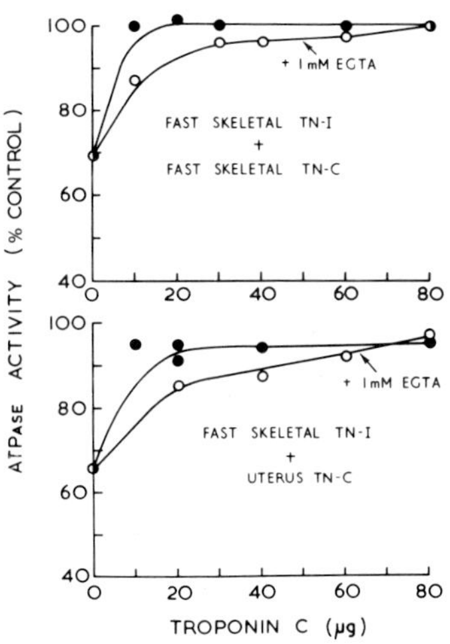

Fig.1. Effect of fast skeletal C and smooth troponin C on the Mg^{2+}-stimulated ATPase of skeletal desensitized actomyosin inhibited by fast skeletal troponin I. Assay conditions; 300 µg desensitized fast skeletal actomyosin, 15 µg fast skeletal troponin I, 2.5 mM-$MgCl_2$, 25 mM-Tris-ATP, 25 mM-Tris-HCl, pH 7.6, 1 mM EGTA where indicated. Incubated 5 min at 25°C.

sensitivity to muscle actomyosin systems in the presence of striated muscle troponin I and tropomyosin, i.e. in contrast to the homologous skeletal muscle system, troponin T is not required (12).

Thus smooth muscle troponin C appears to possess properties that are intermediate between those of the striated muscle and the brain forms of troponin C.

Troponin I-like component

The electrophoretic studies with extracts of whole smooth muscle indicated that the troponin C-like protein formed a complex with some other protein (or proteins) that was stable in 6 M urea in the presence of Ca^{2+}. By analogy with striated muscle this property suggested that in whole tissue extracts of smooth muscle the troponin C was complexed with a troponin I-like protein.

Polyacrylamide gels of extracts of whole smooth muscle were not well resolved in the region where, from experience with striated muscle, the band of troponin C/troponin I complex would be expected to migrate. The position of the complex was determined by adding a sample of smooth muscle troponin C that had been carbamoylmethylated by treatment with ^{14}C-labelled iodoacetamide. As no change in charge resulted from this modification the ^{14}C-labelled troponin C migrated with the same mobility as normal smooth muscle troponin C and was still able to form a complex with skeletal troponin I. By adding known amounts of the ^{14}C-labelled troponin C of known specific activity to a whole homogenate of the tissue and applying the isotope dilution technique described by Head et al., (7), a value of 0.43 ± 0.15% of total protein N was obtained for the troponin C content of whole uterus. For this estimation the electrophoresis of the whole muscle extract was carried out in the presence of 10 mM EGTA under which conditions all the radioactivity was shown to be located in the troponin C band (Fig.2). If electrophoresis was carried out in the

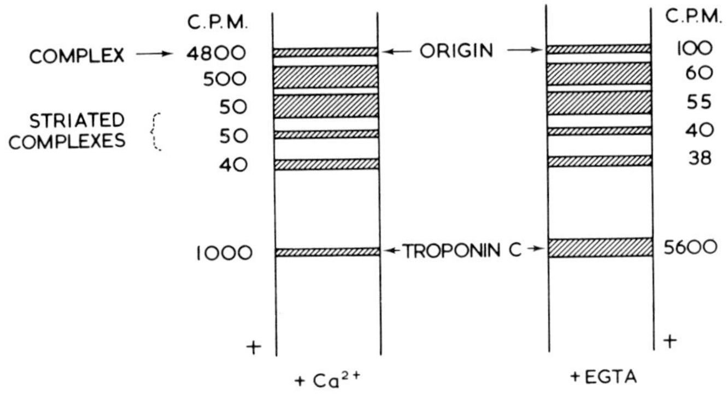

Fig.2. Diagrammatic representation of polyacrylamide gels of homogenates of whole rabbit uterus in 6 M urea, 25 mM-Tris-80 mM-glycine, pH 8.6, with added ^{14}C-labelled smooth muscle troponin C. ^{14}C carbamoylmethylated smooth muscle troponin C (50 µg) added to 5 ml of homogenate and dialysed overnight against 8 M urea, 50 mM-Tris-HCl, pH 8.0, 15 mM 2-mercaptoethanol with 1 mM $CaCl_2$ or 10 mM EGTA for $+Ca^{2+}$ and +EGTA samples respectively.

presence of Ca^{2+} most of the radioactivity was located at the origin implying that the troponin C had formed a Ca^{2+}-dependent complex which did not migrate at pH 8.6. In this respect its mobility differed from that of the complexes present in extracts of whole skeletal and cardiac muscle or that formed by smooth troponin C and skeletal troponin I. In the presence of Ca^{2+} the latter complexes had mobilities of 0.50 - 0.35 of that of the troponin C determined in the presence of EGTA. The results suggested that the smooth troponin C was complexed with a protein (or proteins) that was more basic than the troponin I from striated muscle. If several different proteins were involved they were rather similar in charge.

Direct application of the affinity chromatographic method using troponin C linked to Sepharose (13) to the whole extract of uterus gave a preparation of 'troponin I' which consisted principally of two proteins, one of 23,000 to 25,000 molecular weight and the second of 17,000 molecular weight. Smaller amounts of protein of 80,000 molecular weight were also present. When the muscle extract was passed through a DEAE cellulose column before affinity chromatography the preparation of 'troponin I' consisted of material of 25,000 and 17,000 molecular weight only, with the relative proportions of the components varying in different preparations. Preparations obtained by both methods inhibited the Mg^{2+} ATPase of desensitized actomyosin from skeletal muscle. The inhibition obtained was increased 3 times in the presence of tropomyosin. The best smooth 'troponin I' preparations were not as effective as fast skeletal troponin I when tested against the same sample of desensitized skeletal actomyosin and usually the maximum inhibition obtained was less than that obtained with skeletal troponin I. If the 'troponin I' preparations obtained by direct affinity chromatography of whole smooth muscle were incubated with bovine cardiac 3':5'-cyclic AMP-dependent protein kinase and [γ-^{32}P]ATP radioactivity was obtained in the 25,000 and 17,000 bands.

The method of isolation of the 'troponin I' by affinity chromatography implied that a Ca^{2+} sensitive complex had been formed by a protein fraction in smooth muscle with the Sepharose bound troponin C. This fact coupled with the increase of the inhibition of the Mg^{2+}-stimulated ATPase obtained with the fraction in the presence of tropomyosin and its phosphorylation by 3':5'-cyclic AMP-dependent protein kinase suggested that bovine and rabbit uterus contained a protein similar to the troponin I found in striated muscle, which also possesses all of these properties. The 'troponin I' from smooth muscle, like troponin C from the same tissue, appeared to be less stable than its striated counterpart. Often after isolation and storage the 'troponin I' of smooth muscle lost its inhibitory activity and appeared to be degraded readily to material of lower molecular weight during purification.

Conclusions

These limited studies suggest that uterus and probably all vertebrate smooth muscles contain components similar to troponin C and troponin I of striated muscle. The troponin C-like protein is present in significant amounts and if it is assumed that it is restricted to smooth cells that make up about 60% of the whole uterus, it is present at about the same concentration as in fast rabbit skeletal muscle. The properties of smooth troponin C are intermediate between those of bovine brain modulator protein and fast skeletal troponin C, perhaps slightly more like the former than the latter.

The troponin I-like protein is much less well defined and appears to be rather unstable and readily degraded. When active, however, it possesses similar properties to those associated with skeletal troponin I. Smooth muscle 'troponin I' exists in homogenates as a complex with troponin C, although it is possible that troponin C also complexes with other proteins in smooth muscle. Indeed the results with affinity chromatography suggest that this is the case.

At this stage it cannot be decided finally whether a troponin complex similar to that found in striated muscle exists in smooth muscle and if so whether it is associated with the actin filaments. Nevertheless troponin C and troponin I-like components are present in significant amounts. We have not characterized a component corresponding to troponin T, although bands of the appropriate molecular weight have been observed on gel electrophoresis of some crude preparations of smooth muscle troponin. All conclusions about the enzymic properties of the troponin components of smooth muscle have been made by studying their effects on desensitized actomyosin of skeletal muscle. Although final conclusions about their function must wait until they are tested on smooth muscle actomyosin systems it is likely that they will show similar properties under these conditions. Certainly skeletal troponin preparations can produce Ca^{2+} sensitivity in smooth muscle actomyosin (14). In our hands also crude preparations made from smooth muscle and which are probably comparable to skeletal troponin preparations will produce Ca^{2+} sensitivity in skeletal desensitized actomyosin.

It is tempting to conclude from these studies that troponin-like systems are widespread in all tissues and that the properties of the components of the system in smooth muscle resemble those of the corresponding proteins in non-muscle cells as is now well documented to be the case with myosin.

REFERENCES

1. Carsten, M.E. (1971) Arch.Biochem.Biophys. 147, 353-357
2. Sparrow, M.P. & van Bockxmeer, F.M. (1972) J. Biochem. 72, 1075-80
3. Bremel, R.D. (1974) Nature (London) 252, 405-407
4. Sobieszek, A. & Bremel, R.D. (1975) Eur.J.Biochem. 55, 49-60
5. Aksoy, M.O., Williams, D., Sharkey, E.M. & Hartshorne, D.J. (1976) Biochem. Biophys.Res.Comm., 69, 35-41
6. Ebashi, S., Toyooka, T. & Nonomura, Y. (1975) J.Biochem.(Tokyo) 78, 859-861
7. Head, J.F., Weeks, R.A. & Perry, S.V. (1977) Biochem.J. 161, 371-382
8. Perry, S.V., Cole, H.A., Head, J.F. & Wilson, F.J. (1972) Cold Spring Harbor Symposium Quant.Biol., 37, 251-262
9. Head, J.F. & Perry, S.V. (1974) Biochem.J. 137, 145-154
10. Cheung, W.Y. (1970) Biochem.Biophys.Res.Comm. 38, 533-538
11. Vanaman, T.C., Harrelson, A.G. & Watterson, D.M. (1975) Fed.Proc. 34, 307
12. Amphlett, G.W., Vanaman, T.C. & Perry, S.V. (1976) FEBS Letts. 72, 163-167
13. Syska, H., Perry, S.V. & Trayer, I.P. (1974) FEBS Letts. 40, 253-257
14. Sparrow, M.P. (Personal communication).

Excitation-Contraction Coupling in Smooth Muscle, R. Casteels et al. eds.
© 1977 Elsevier/North-Holland Biomedical Press

REGULATORY LIGHT CHAINS IN MYOSIN

John Kendrick-Jones and Ross Jakes
MRC Laboratory of Molecular Biology,
Hills Road, Cambridge CB2 2QH
England

Contraction in molluscan muscles is regulated by the direct interaction of calcium with the myosin[1]. In these muscles calcium regulation of contractile activity is mediated by specific regulatory light chains, associated with each myosin head[2]. The regulatory function of these light chains has been clearly demonstrated by the selective release of one of the two regulatory light chains from scallop myosin, which results in a complete loss of calcium regulation and a partial loss of calcium binding, i.e. the myosin is desensitized and no longer requires calcium for interaction with actin. The isolated light chain readily recombines with this 'desensitized' myosin and both calcium regulation and calcium binding are completely restored[3]. In vertebrate smooth muscle there is now compelling evidence that a myosin-linked calcium regulatory system is also present[4], although the precise nature of this system remains uncertain. It is generally agreed that actin-myosin interaction in smooth muscle requires phosphorylation of the myosin regulatory light chains by a specific calcium dependent light chain kinase[5-7]. One of the intriguing questions is whether the regulatory system in smooth muscle is analogous to that in molluscan muscles involving a direct calcium switch mediated by the light chains with phosphorylation as an additional modulatory step or whether calcium regulation is mediated solely by a calcium regulated phosphorylation-dephosphorylation of the light chain?

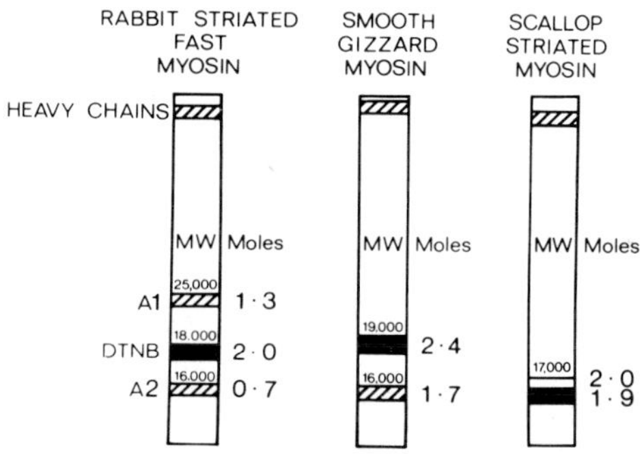

Fig. 1. Diagrammatic representation of the subunit structure of the myosins. The stoichiometry of the light chains (moles light chain/myosin) was established by gel densitometry and verified in the case of the rabbit myosin by radioactive labelling[9]. Each myosin contains about two moles of regulatory light chain (solid bars).

To attempt to answer this question, we have carried out comparative studies on the primary structure and functional capabilities of regulatory light chains isolated from vertebrate striated, vertebrate smooth and molluscan myosins[8].

The subunit structure of these myosins is similar with two moles of regulatory light chain per myosin molecule, although the size of the vertebrate light chains are slightly larger (fig. 1). The possible 'regulatory' function of these light chains has been demonstrated by the following experiment. Since scallop myosin is only calcium regulated myosin from which a regulatory light chain can be reversibly detached without denaturation, it has been used to test the calcium regulatory capabilities of these light chains (table 1)[9].

TABLE 1

EFFECT OF THE REGULATORY LIGHT CHAINS ON 'CALCIUM REGULATION' IN SCALLOP MYOFIBRILS

	Regulatory LC content	Calcium sensitivity ratio of ATPases (EGTA/Ca^{2+})	Calcium binding No. of sites
Myofibrils	1.9	0.03	1.9
Desensitized myofibrils	1.0	0.85	1.1
Desensitized myofibrils + scallop-LC	1.9	0.07	1.9
Desensitized myofibrils + gizzard-LC	2.0*	0.06	2.0
Desensitized myofibrils + rabbit-LC	2.0*	0.09	1.1

Desensitized scallop myofibrils were incubated with the various regulatory light chains (at a 1:1 light chain/myosin molar ratio, assuming the myosin content of the myofibrils is 65%) as described previously[9]. Actin-activated Mg^{2+} dependent ATPase activities in the presence of 0.1 mM EGTA and 0.1 mM Ca^{2+} and calcium binding measurements were performed by the procedures outlined by Kendrick-Jones et al.[9]. The ratio of the ATPases (EGTA/Ca^{2+}) indicates the calcium sensitivity of the preparations, i.e. low values indicate calcium regulation of actin-myosin interaction. The number of calcium binding sites, with affinity constants in the range 3 - 7 x 10^6, were obtained from Scatchard plots. The regulatory light chain content of the preparations was determined by gel densitometry[9].
* The regulatory light chain content in these hybrids is the sum of the remaining scallop light chain and added vertebrate light chain.

All three light chains bind tightly to 'desensitized' scallop myofibrils (replacing the released scallop light chain) and calcium regulation of actin-myosin interaction is completely restored (which directly implies that the function of all three light chains is 'regulatory'). However they differ in their ability to restore full calcium binding to desensitized scallop myofibrils and accordingly can be divided into two groups: (1) the light chains from vertebrate smooth and molluscan myosins, where some 'form' of myosin-linked regulation is present[1,4], restore both calcium regulation and calcium binding, i.e. they are completely functional, whereas (2) the light chain from vertebrate striated myosin where biochemical evidence indicates the absence of myosin-linked regulation, although it restores calcium regulation to desensitized scallop myosin when it is complexed with actin, it has no effect on calcium binding (table 1). The remarkable feature that these vertebrate light chains crossreact with scallop myosin implies that the light chains and the regions on the myosin head involved in binding are relatively conservative and have changed little during evolution. Therefore the differences in calcium binding affinity of these two groups of regulatory light chains must be due, in part, to differences in the light chains themselves, and may be apparent at the level of their primary structures.

Primary structure of the light chains

The three dimensional structure of carp parvalbumin[10] has provided a structural basis for comparing the striking sequence homologies that have been shown to exist between the 'regulatory' light chains[11] and the other muscle calcium regulatory proteins, troponin C[12], and parvalbumin[13]. Parvalbumin contains two basic structural units, each composed of a calcium binding site in a pocket surrounded by two helical regions on either side and the whole structure is stabilised by the packing of non polar residues within these helical areas into a buried hydrophobic core.

In fig. 2 the sequence of the scallop light chain[14] has been drawn out in diagramatic form to correspond to the main structural features of parvalbumin. Four regions composed of a pair of helices surrounding a potential calcium binding site are clearly recognisable and the asymmetric distribution of non-polar residues which form the core of parvalbumin has been retained in the scallop light chain sequence. Comparing the sequence of the scallop light chain with those of the rabbit and gizzard light chains (only the N terminal portion of the gizzard light chain sequence is yet available) (fig. 2) indicates considerable homology especially in the N terminal halves of the molecules. In this region 42% of the residues are identical in the scallop and rabbit light chains (marked in black in fig. 2) whereas in the rabbit and gizzard light chains 73% of the residues are identical and in many cases the substitutions are conservative. The important question is whether these sequence homologies imply 'structural' homology. The extensive sequence homology between these light chains and other muscle calcium binding proteins, i.e. parvalbumin and troponin C, suggests that they have evolved from a common primitive calcium binding protein by successive gene duplication and mutation[15,16]. However, since they have evolved to perform quite distinct functions within muscle cells and *in vitro* evidence indicates that they are not interchangeable[17], considerable modification of their three dimensional structures must have occurred. Hydrodynamic measurements indicate that isolated molluscan light chains are extremely asymmetric[18], with lengths ranging from 10 to 14 nm, about the same length as a single myosin head. It remains to be established

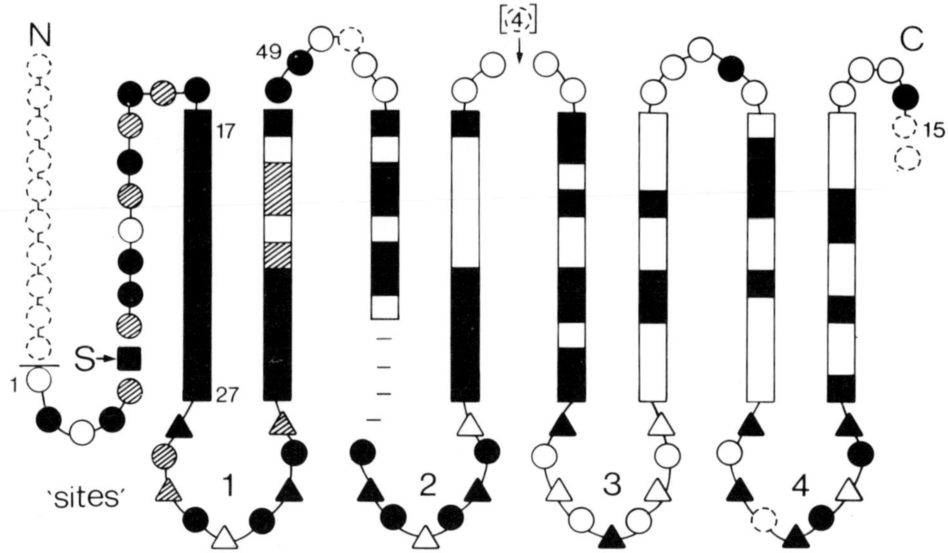

Fig. 2. The sequence of the scallop light chain[14], aligned with the sequence of the rabbit light chain[11] and troponin C[12], and drawn out in diagramatic form to correspond to the main structural features of carp parvalbumin[10] with four pairs of helices (vertical columns) surrounding four 'possible' calcium binding sites. The positions of the calcium ligating groups in each site are represented by the triangles. The residues in black are identical in the scallop, rabbit and gizzard light chain sequences (partial sequence of gizzard left chain residues 1-49) while those in stripes are further identical residues in the rabbit and gizzard sequences. The dotted circles in the scallop sequence represent the positions of the additional residues in the rabbit light chain sequence, most of which occur at the N terminal end of the molecule. S represents the position of the serine residue phosphorylated in the rabbit and gizzard light chains.

whether the structure of these light chains in their isolated state is identical to that when associated with the myosin head, and if other regulatory light chains have similar shapes. Obviously the precise structural relationship between these light chains will depend on the determination of the three dimensional structure of their parent myosins.

Calcium binding sites

The precise location of the calcium binding sites in scallop myosin still remains in doubt. Scallop myosin contains two high affinity calcium binding sites and selective removal of one of the regulatory light chains results in the loss of one calcium binding site[3], which would imply a direct relationship between the light chain content of the myosin and the number of calcium binding sites. The isolated light chain, however, does not bind calcium, but when recombined with the 'desensitized' myosin this lost calcium binding site is restored. These results suggest that the calcium binding site is located either on the light chain, whose conformation is altered by interaction with the myosin head or is composed of residues with the correct configuration donated by both light and heavy chains.

The three dimensional structure of the calcium binding unit of parvalbumin (referred to by Kretsinger as an 'EF hand') has been extensively used to predict calcium binding regions in many proteins. In parvalbumin, calcium is coordinated to oxygen atoms donated by six different amino acids which form the vertices of an octahedron (positions of the amino acids in capital letters in fig. 3 and triangles in fig. 2) and within these ligating positions four acidic residues are present. Using these criteria, only site 1 in both the rabbit light chain[11] and scallop light chain[14] sequences satisfy these requirements. Sites 2, 3 and 4 in fig. 2 can be ruled out as potential calcium binding sites since they contain less than four negatively charged groups and show deletions and substitutions, e.g. proline residues, which would distort these sites. Comparing the sequences within the potential calcium binding sites (site 1) in all three light chains with that of carp parvalbumin (fig. 3) reveals strong similarities; however there is one distinct difference which might explain the different calcium binding affinity of these light chains. Four acidic residues are present within the six coordina-

	site									helix			
	1	2	3	4	5	6	7	8	9	1	2	3	negative charges
Scallop LC	ASP	val	ASP	arg	ASP	gly	PHE	val	SER	lys	asp	ASP	4
Gizzard LC	ASP	gln	ASN	arg	ASP	gly	PHE	ile	ASP	lys	glu	ASP	4
Rabbit LC	ASP	gln	ASN	arg	ASN	gly	ILE	ile	ASP	lys	glu	ASP	3
Carp CBP	ASP	gln	ASP	lys	SER	gly	PHE	ile	GLU	glu	asp	GLU	4
Calcium ligating positions	X		Y		Z		-Y		-X			-Z	

Fig. 3. Comparison of the sequences of the 'potential' calcium binding regions (site 1 in fig. 2) of the regulatory light chains and carp calcium binding parvalbumin[19]. The residues in capitals, by analogy with the three dimensional structure of carp parvalbumin are the positions of the calcium ligating groups which lie at the vertices of an octahedron (X, Y, Z, etc.). The number of negative charges within these six coordinating positions in each site is indicated.

ting positions in the sites of parvalbumin, and the gizzard and scallop light chains and since both light chains restore the high affinity calcium binding site ($K \sim 10^6 M^{-1}$) when they combine with desensitized scallop myofibrils (table 1) it was assumed that this site is the high affinity calcium binding site responsible for the calcium regulatory switch[20]. The calcium binding site in the rabbit light chain contains only three negatively charged residues within the coordinating positions which could explain the lower calcium affinity of this site ($K \sim 10^5 M^{-1}$)[21] and the failure of this light chain to restore the high affinity calcium binding site when it binds to 'desensitized' scallop myofibrils (table 1). Using electron paramagnetic resonance (EPR) spectroscopy to characterise the divalent metal ion binding sites of rabbit myosin, Bagshaw[22] has tentatively identified site 1 in the rabbit light chain as a less specific divalent metal binding site which under physiological conditions would not be competent to act as a calcium switch. Preliminary results with the gizzard and other molluscan light chains indicate that they also contain this less specific divalent metal binding site[23], which raises the possibility that the region around the N terminus (containing site 1) which appears to be highly conserved in all three light chains may not be involved in the calcium switch but could be the region involved in permanent attachment to the myosin head. It has long been acknowledged that magnesium ions appear to play an essential role in stabilising the interaction between heavy chains and regulatory light chains (cf. EDTA treatment of scallop and rabbit myosins - see discussion Kendrick-Jones et al.[9]); could site 1 be the divalent cation binding site involved in this interaction? The precise location of the calcium binding site in molluscan and maybe vertebrate smooth myosins which operates the calcium ON/OFF switch, therefore remains unclear; it could be site 1 on the light chain or it could be composed of groups donated by both heavy and light chains or could be located in another region of the light chain whose three dimensional conformation is dependent on interaction with the heavy chain.

Phosphorylation

Phosphorylation as a form of covalent modification is a favourite mechanism for switching on or off metabolic pathways[24]. In vertebrate smooth myosins phosphorylation of the regulatory light chains by a calcium regulated light chain kinase is required for the myosin to be able to interact with actin[5-7]. Although there is no direct evidence that phosphorylation is required to mediate actomyosin interaction in rabbit striated myosin, the regulatory light chains are similarly phosphorylated by a calcium requiring light chain kinase[25]. Sequence analysis of the phosphorylated rabbit and gizzard light chains has located the single phosphoserine residue near the N terminus of these molecules (fig. 4).

Sca. AC-ala-asp-lys-ala-ala-ser-gly-val-leu-thr-lys-leu-pro-gln-lys-gln
(1) (16)

Giz. (pro-arg-glx)arg-ala-thr-ser-asn-val-phe-ala-met-phe-asp-gln-ser-gln
P C

Rab.* ala-ala-glu-gly-gly-ser-ser-asn-val-phe-ser-met-phe-asp-gln-thr-gln
(9) (25)

Fig. 4. Comparison of the sequences near the N-terminus of the regulatory light chains with the positions of the specific phosphoserine residues indicated ☐. The residues in brackets in the gizzard light chain (Giz.) are the tentative sequence and the arrows indicate the positions of papain (P) and chymotrypsin (C) cleavage of this light chain. Residues which are 'identical' in the three sequences have been underlined. * rabbit light chain (Rab.) sequence established by Collins[11], and confirmed by us, with the positive identification of the phosphoserine residue.

Although there is a serine residue in the equivalent position in the scallop light chain, it is not phosphorylated. A possible explanation is that the difference in size between the scallop and vertebrate light chains is due to extra residues located mainly in the N-terminal regions of the vertebrate light chains (fig. 2) which may serve as a recognition site for the light chain kinase. This speculation is supported by the observation that when the N-terminal region of the gizzard light chain is removed by cleavage at the arginine close to the essential serine residue (fig. 4) this modified light chain fragment cannot be phosphorylated. The sequences around the single phosphoserine residue in both vertebrate light chains are identical which would support the observed high degree of specificity of the kinases from rabbit striated and gizzard smooth muscles[25,26].

When gizzard actomyosin is digested with proteolytic enzymes the regulatory light chain is rapidly cleaved (fig. 5). Cleavage of the light chain occurs,

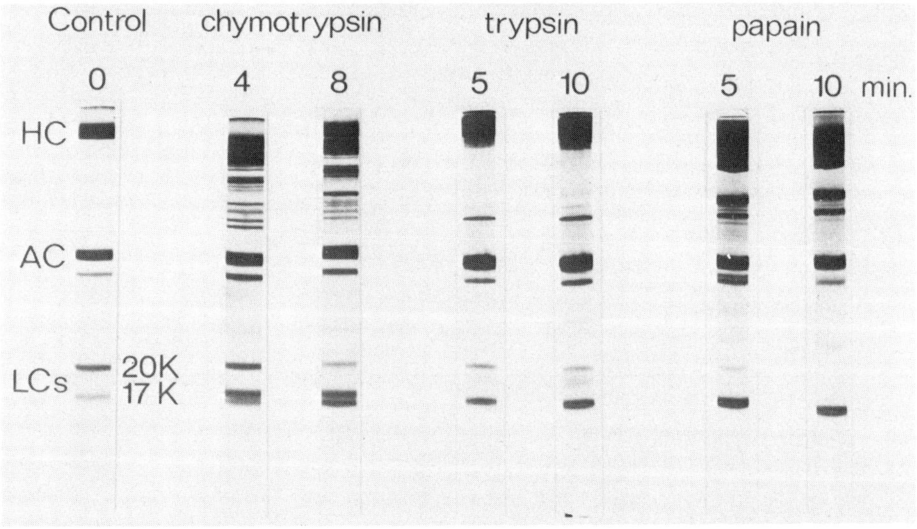

Fig. 5. Effect of proteolytic digestion on the regulatory light chains of gizzard actomyosin. Gizzard actomyosin (6-8 mg/ml) in 60 mM NaCl, 20 mM Imid. pH 7.0, 1 mM DTT in the presence of 2 mM EDTA or 2 mM divalent cation digested at 25°C with a) chymotrypsin (200:1 protein:enzyme ratio w/w), b) trypsin (1000:1 protein: enzyme), c) papain (2000:1 protein:enzyme ratio) under the conditions outlined previously[8]. Run on 10% SDS acrylamide gels with 0.1M Tris-bicine running buffer pH 8.1[8].

especially using papain or trypsin, before there is any significant fragmentation of the heavy chain. This feature should be borne in mind when using these enzymes to prepare 'native' heavy meromyosin or S-1 head fragments of this myosin. In scallop and rabbit striated myosins, divalent cations have been shown to protect the regulatory light chains against proteolysis[3,27], with the inference that magnesium ions stabilise the light-heavy chain interaction, however no such protection by divalent cations was observed with gizzard actomyosin. It is possible that phosphorylation could stabilise the association of the light chain with the gizzard myosin head in an analogous way or in conjunction with divalent cation binding, although we have no conclusive evidence yet to support such a speculation. Characterisation of the light chain fragments produced by brief chymotryptic or papain digestion indicates a loss of about 20-26 residues mainly from the N-terminus of the molecule. It is likely that this portion of the light chain is freely accessible and not tightly associated with the myosin head, a speculation supported by the observed rapid rate of phosphorylation of the serine in this region[7]. Although these fragments cannot be phosphorylated either when attached to the myosin or in the isolated state, they are fully functional when complexed with desensitized scallop myofibrils (table 2). Both fragments readily combine

TABLE 2

FUNCTIONAL CAPABILITY OF THE GIZZARD LIGHT CHAIN FRAGMENTS

	ATPase activity (Mmol/min per mg)		Ca^{2+} sensitivity ratio of ATPases	Calcium binding
	0.1 mM EGTA	0.1 mM Ca^{2+}	($EGTA/Ca^{2+}$)	number of sites
Scallop myofibrils	0.002	0.283	0.01	1.88
Desensitized myofibrils	0.180	0.224	0.80	1.16
Desensitized myofibrils + gizzard LC (papain fragment)	0.028	0.352	0.08	2.28
Desensitized myofibrils + gizzard LC (chymotrypsin fragment)	0.037	0.302	0.12	1.82

For procedures see the legend to Table 1.

with desensitized scallop myofibrils and restore both calcium regulation and calcium binding, indicating at least in the scallop myosin system the N terminal residues in the gizzard light chain are not functionally involved. The size of these fragments are about the same as that of the scallop regulatory light chain, which would suggest that they are the basic calcium regulatory subunit and during evolution, phosphorylation has been acquired, for some reason not yet understood, as an additional calcium dependent modulator of actin-myosin interaction in vertebrate smooth muscles.

Discussion

If the mechanism of calcium regulation in molluscan and vertebrate smooth muscles is basically similar, then there are a number of features of the molluscan regulatory system which should be emphasised and tested to see if they apply in vertebrate smooth muscle. Regulatory light chains in molluscan muscles act as inhibitors of actin-myosin interaction, and this inhibition is relieved by calcium. The presence of both regulatory light chains is required to maintain this 'OFF' state of the myosin in the absence of calcium. Calcium regulation therefore requires cooperation, between either the two regulatory light chains or the two S-1 heads. The evidence for this supposition is provided by the following observations: 1) although there are two identical regulatory light chains in scallop myosin, the loss of only one leads to a complete loss of calcium regulation[3]. 2) Molluscan heavy meromyosins (myosin subfragment composed of the two heads joined by a short tail) are calcium sensitive, whereas single 'headed' S-1 preparations containing intact regulatory light chains, are not[2]. A plausible explanation for the ability of rabbit regulatory light chains to restore calcium regulation to scallop myofibrils without calcium binding (table 1), is if regulation is restored as a result of cooperative interactions between this added vertebrate and the remaining 'operative' scallop light chain. A simple model of calcium regulation mediated by these regulatory light chains could be envisaged if the light chains contain two distinct regions; one involved in permanent attachment to the myosin head and the other in calcium induced interactions. The permanent heavy chain attachment site, probably stabilised by magnesium ions, is a highly conserved region which has been retained in all the regulatory light chains so far tested. The corresponding region on the heavy chain, which also must be highly conserved, is distinct from the actin binding site since it remains freely accessible for the release and recombination of light chains when the myosin is complexed with actin and the light chains remain attached during contraction. A different region of the light chain is involved in calcium modulated interactions either with the heavy chain or the other light chain which alters the relative position of these light chains or the orientation of the two heads and results in the myosin being switched either ON or OFF. The absolute requirement for 'cooperative interactions' between the light chains or the two myosin heads is a crucial feature of the molluscan regulatory system and it will be interesting to see whether such a 'cooperative' mechanism operates in vertebrate smooth muscle.

References

1. Kendrick-Jones, J., Lehman, W. and Szent-Györgyi, A. G. (1970) J. Mol. Biol. 54, 313-326.
2. Kendrick-Jones, J., Szentkiralyi, E. M. and Szent-Györgyi, A. G. (1972) Cold Spring Harbor Symp. Quant. Biol. 37, 47-54.
3. Szent-Györgyi, A. G., Szentkiralyi, E. M. and Kendrick-Jones, J. (1973) J. Mol. Biol. 56, 239-258.
4. Bremel, R. D. (1974) Nature (London) 252, 405-407.
5. Chacko, S., Conti, M. A. and Adelstein, R. S. (1977) Proc. Natl. Acad. Sci. 74, 129-133.
6. Aksoy, M. O., Williams, D., Sharkey, G. M. and Hartshorne, D. J. (1976) Biochem. Biophys. Res. Commun. 69, 35-42.
7. Sobieszek, A. (1977) Eur. J. Biochem. 73, 477-483.
8. Jakes, R., Northrop, F. and Kendrick-Jones, J. (1976) FEBS Lett. 70, 229-234.

9. Kendrick-Jones, J., Szentkiralyi, E. M. and Szent-Györgyi, A. G. (1976) J. Mol. Biol. 104, 747-775.
10. Kretsinger, R. H. and Nockolds, C. E. (1973) J. Biol. Chem. 248, 3313-3326.
11. Collins, J. H. (1976) Nature (London) 259, 699-700.
12. Collins, J. H. (1974) Biochem. Biophys. Res. Commun. 58, 301-308.
13. Kretsinger, R. H. and Barry, C. D. (1975) Biochim. Biophys. Acta 405, 40-52.
14. Kendrick-Jones, J. and Jakes, R. (1977) International Symp. Myocardial Failure (Tegernsee, Munich, June 1976) in press.
15. Weeds, A. G. and McLachlan, A. D. (1974) Nature (London) 252, 646-649.
16. Tufty, R. M. and Kretsinger, R. H. (1975) Science 187, 167-169.
17. Hitchcock, S. E. and Kendrick-Jones, J. (1975) in Calcium Transport in Contraction and Secretion (Carafoli, E. et al., eds) pp. 447-458, North-Holland Publishing Company, Amsterdam.
18. Stafford, III, W. F. and Szent-Györgyi, A. G. (1976) Biophys. J. 16, 70a.
19. Coffee, C. J. and Bradshaw, R. A. (1973) J. Biol. Chem. 248, 3305-3312.
20. Kendrick-Jones, J. and Jakes, R. (1976) Trends Biochem. Sci. 1, 281-284.
21. Morimoto, K. and Harrington, W. F. (1974) J. Mol. Biol. 83, 83-97.
22. Bagshaw, C. R. (1977) Biochemistry 16, 59-67.
23. Bagshaw, C. R. and Kendrick-Jones, J. (unpublished observations).
24. Cohen, P. (1976) Trends Biochem. Sci. 1, 38-40.
25. Perry, S. V., Cole, H. A., Frearson, N., Moir, A. J. G., Morgan, M. and Pires, E. (1975) 26th Mosbach Colloquium on 'Molecular Basis of Motility' eds. Heilmeyer et al., Springer-Verlag, pp. 107-121.
26. Frearson, N., Focant, B. W. W. and Perry, S. V. (1976) FEBS Lett. 63, 27-32.
27. Margossian, S. S., Lowey, S. and Barshop, B. (1975) Nature (London) 258, 163-168.

Excitation-Contraction Coupling in Smooth Muscle, R. Casteels et al. eds.
© 1977 Elsevier/North-Holland Biomedical Press

THE ROLE OF THE REGULATORY LIGHT CHAIN IN PIG CAROTID SMOOTH MUSCLE ATPASE

U. Mrwa and J. C. Rüegg
II. Physiologisches Institut der Universität
Im Neuenheimer Feld 326
D-69 Heidelberg
West Germany

INTRODUCTION

The Mg^{2+}-dependent ATPase activity of arterial smooth muscle actomyosin depends on calcium ion concentration in the range of 10^{-8} to $10^{-4}M$[1,2]. Addition of unregulated, i.e. troponin- and tropomyosin-free actin to such actomyosin preparations does not enhance ATPase activity at low Ca^{2+} concentrations ($10^{-8}M$) nor does it affect the ratio of ATP-splitting rates in presence and absence of trace Ca^{2+}. This implies the presence of a myosin linked regulatory system[3,4]. As described earlier[3,5], we have been able to pinpoint the Ca sensitivity to the subfragment 1 of arterial myosin. The possible involvment of the 20,000 Dalton light chain (L 20) of these muscles was also documented by the finding that the analogous light chain from chicken gizzard was able to restore calcium sensitivity to desensitized molluscan myosin[6]. A myosin light chain kinase has also been implicated in myosin linked Ca regulation[8-9]. However, we will show in the following that the L 20 light chain confers Ca sensitivity to actin activated subfragment 1, from which myosin light chain kinase has been removed. It will also be shown that a phosphorylated light chain is not required for actin-myosin interaction, since this occurs even with desensitized S1 devoid of the L 20 light chain.

RESULTS AND DISCUSSION

1. Ca sensitivity of the actin-activated, Mg-dependent S1 ATPase.

A simple system containing only S1 from pig carotid muscle and purified actin exhibits a Ca^{2+} sensitive, Mg^{2+}-dependent ATPase activity, which can be inhibited 50% by addition of EGTA (Fig. 1). The S1 used in this assay was obtained by mild proteolytic digestion of Ca sensitive actomyosin and subsequent purification on an affinity column[7]. It still contained the L 20 light chain. To exclude the possibility that the actin conferred Ca sensitivity to the system, it was assayed with skeletal S1: no Ca sensitivity could be observed in this case. To rule out the presence of a Ca-sensitive, ATP-splitting system, consisting of a Ca-dependent kinase transferring phosphate to the light chain and a Ca-independent phosphatase splitting off this phosphate again[8], smooth muscle S1 was assayed for Ca sensitivity in the absence of actin. As shown in Fig. 1 the ATPase activity of the S1 alone was the same in absence and presence of Ca^{2+}. This excludes the presence of a "pseudo" ATPase. S1 lacking the L 20 light chain is prepared from desensitized actomyosin. This actomyosin was washed and precipitated as described by Sparrow[2].

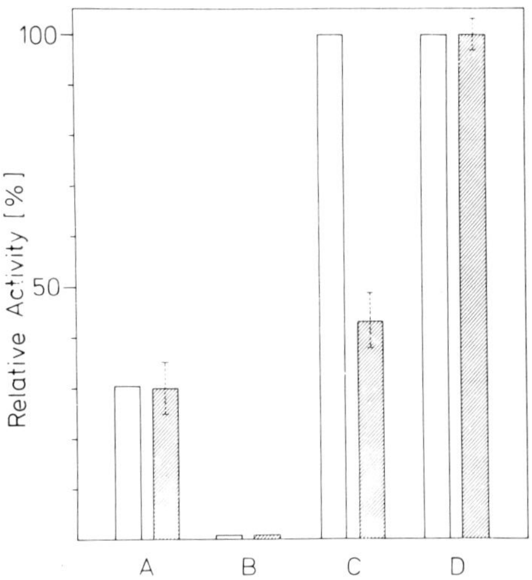

Fig. 1: ATP-ase activity in percent of activity of acto-S1 in presence of Ca^{2+} (10^{-4}M) ☐ or with 10^{-8}M Ca^{2+} ▨ . A: Subfragment-1 (S-1), B: actin, C: acto-S-1, D: Desensitized acto-S-1. Conditions: 50mM KCl, 10mM imidazole, 5mM $MgCl_2$, 2mM ATP, 0.1mM Ca^{2+} or 2mM EGTA, pH 7, 20°C.

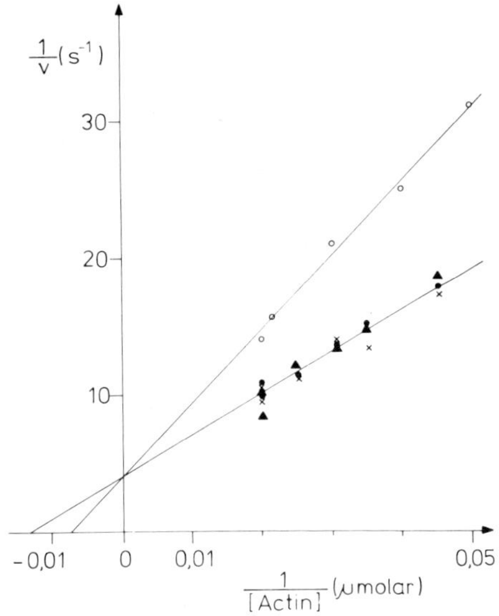

Fig. 2: Actin activation of natural and desensitized S1. Double reciprocal plot of activity versus actin concentration. Conditions: 50mM KCl, 10mM imidazole, 5mM $MgCl_2$, 2mM ATP, 0.6 μM S1, pH 7, 20°C. ● natural S1 (10^{-4}M Ca^{2+}); × desensitized S1 (10^{-4}M Ca^{2+}); ▲ desensitized S1 (10^{-8}M Ca^{2+}); ○ natural S1 (10^{-8}M Ca^{2+}).

Its ATPase is no longer Ca sensitive and it does not show superprecipitation, but it still has the L 20 light chain in contrast to the S1 preparations. When desensitized S1, lacking the L 20 light chain is incubated with actin, the resulting ATPase activity cannot be inhibited by addition of EGTA. The removal of the light chain thus results in loss of Ca sensitivity.

2. Kinetics of actin activation:

The Ca-regulated, Mg-dependent ATPase of S1 and that of desensitized S1 are actin-activated ATPases. When actin is added to desensitized S1 in the presence of EGTA, there is a marked increase of ATPase activity whereas no change appears upon subsequent addition of Ca^{2+} (Fig. 3, B). In the case of regulated S1 containing the L 20 light chain the actin activation in presence of EGTA is rather low; however, there is a large increase when Ca^{2+} is added. The actin activation kinetics of S1 which still contains the light chain greatly resemble these of the desensitized S1 lacking the light chain (Fig. 2). The double reciprocal plot suggests that the affinity of S1 to actin is not affected by the presence of the light chain. The apparent dissociation constant (K_{app})[9] for the acto-S1 complex calculated from the abscissa intercept is about 60 µM actin in both cases. In the desensitized system, the V_{max} extrapolated from Fig. 2 (0.2 s^{-1}) is the same in presence and absence of Ca^{2+} as the V_{max} in the regulated system under Ca-activating conditions. In presence of EGTA, actin activation is reduced. V_{max} seems to be unchanged, while the affinity for actin[9] is lowered (K_{app}: 130 µM actin). We therefore conclude that the L 20 light chain acts by inhibiting the actin activated ATPase in absence but not in presence of Ca^{2+}. The question arises as to whether Ca acts directly on subfragment 1 or whether it acts indirect via a light chain as suggested by Sobieszek[10].

3. Ca activation in relation to phosphorylation of the regulatory light chain.

Although the ATP splitting rate is increased after addition of Ca (10^{-4}M) (Fig. 3A) there is no detectable incorporation of labelled phosphate into S1 (Fig. 4). This shows clearly that Ca activation can occur without extra phosphorylation of the L 20 light chain. Control experiments with native actomyosin revealed Ca-dependent phosphorylation of myosin. A maximum of 0.7 mole phosphate incorporated per mole of myosin was obtained after 1 min[11]. Ca activation in smooth muscle acto-S1 is therefore evidently not mediated by myosin light chain kinase. It is due to a release of the inhibition caused by the (presumably phosphorylated) L 20 in the absence of Ca^{2+}. Furthermore we confirmed for arterial muscle the finding of Sobieszek[12] that the contaminating light chain kinase activity is lost on desensitation by repeated precipitation of smooth muscle actomyosin[2]. During incubation of desensitized actomyosin with γ-labelled ATP there is no incorporation of phosphate into myosin detectable (Fig. 4). Incorpo-

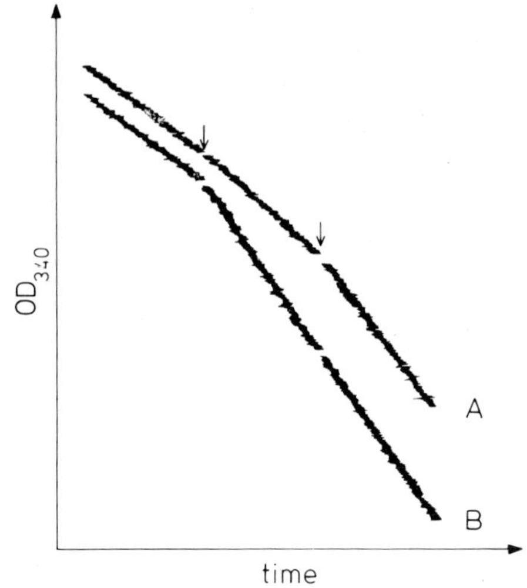

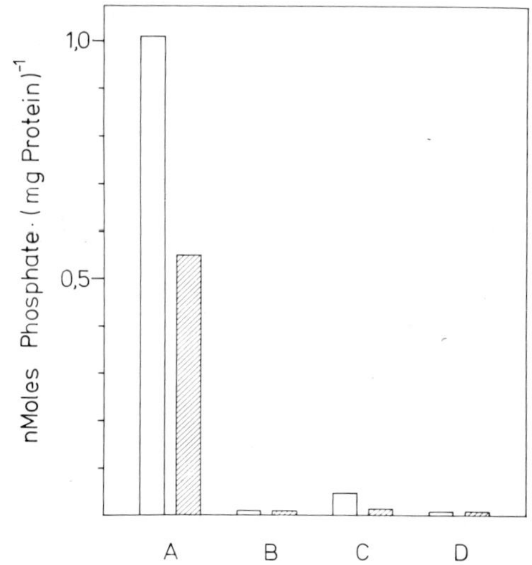

Fig. 3: Linked enzyme assay for ATPase activity determination. The assay starts with a medium containing 50mM KCl, 10mM imidazole, 5mM $MgCl_2$, 1mM ATP, 0.6 µM S1 and 2mM EGTA. At the first arrow actin is added (1.3 mg/ml); at the second arrow Ca^{2+} is added up to a concentration of 10^{-5}M. The pH was kept at 7, T = 20°C. A: NAM S1, B: desensitized S1.

Fig. 4: Phosphate incorporation into actomyosin and subfragment 1. A: natural actomyosin, B: Ca-insensitive actomyosin, C: natural S1, D: desensitized S1 (lacking the L 20 light chain). □ 10^{-4}M Ca^{2+}; ▨ 10^{-8}M Ca^{2+}. Conditions: actomyosin: 0.6M KCl, 50mM imidazole; S1: 50mM KCl, 10mM imidazole; 5mM $MgCl_2$, 2mM ATP, pH 7, 22°C, incubation time: 1 min.

ration does occur, however, after readdition of the supernatant which has been removed from actomyosin during the precipitation required for desensitation[12]. Desensitized actomyosin still contains the L 20 light chain, presumably in a dephosphorylated state[12,13] and this protein does not superprecipitate even with Ca^{2+}, suggesting that actin myosin interaction is inhibited. It would appear that, as in gizzard myosin, the dephosphorylated light chain acts as inhibitor of actin-myosin interaction (similar to troponinI) in the presence and absence of Ca^{2+}. If the (presumably dephosphorylated) light chain is removed by papain digestion and the S1 is purified by subsequent chromatography, the latter regains its capacity to interact with actin. However, this interaction is then no longer Ca-regulated.

ACKNOWLEDGEMENT

The financial support of the Deutsche Forschungsgemeinschaft (SFB 90) and the expert technical assistances of M.Troschka and C.Groß are greatfully acknowledged.

REFERENCES

1. Schirmer, R. H. 1965, Biochem. Zeitschr. 343, 269-282
2. Sparrow, M. P. and Bockxmeer, F. M. V. 1972, J. Biochem. 72, 243-257
3. Mrwa, U. and Rüegg, J. C. 1975, FEBS Letters 60, 81-83
4. Fredericksen, D. W. 1976, Proc. Natl. Acad. Sci. U 57 73, 2706-2710
5. Mrwa, U. and Rüegg, J. C. 1976, in: Ionic actions on vascular smooth muscle, ed. E.Betz, Springer Verlag, Berlin, Heidelberg, New York
6. Kendrick-Jones, J., Szentkiralgi E. M. and Szent-Györgyi A. G., J. Mol. Biol. 104, 747-775
7. Lamed, R. and Mrwa, U. 1976, Experientia 32, 1221
8. Morgan, M. S., Perry, S. V. and Ottaway, J. 1976, Biochem. J. 157, 687-697
9. Moos, C. 1972, Cold Spring Harbour Symp. Quant. Biol. 27, 137-144
10. Sobieszek, A. 1977, Eur. J. Biochem. 73, 477-483
11. Chako, S., Conti, M. A. and Adelstein B. S. 1977, Proc. Natl. Sci. USA 74, 129-133
12. Sobieszek, A. and Bremel, D. 1975, Eur. J. Biochem. 55, 49-60
13. Górecka, A., Aksoy, M. O. and Hartshorne, D. J., Biochim. Biophys. Res. Comm. 71, 325-331

Excitation-Contraction Coupling in Smooth Muscle, R. Casteels et al. eds.

Myosin Phosphorylation as a Regulatory Mechanism

Robert S. Adelstein, Mary Anne Conti, Stylianos P. Scordilis,
Samuel Chacko, Barbara Barylko and John A. Trotter
Section on Molecular Cardiology
National Heart, Lung, and Blood Institute,
Bethesda, Maryland USA 20014

SUMMARY

1) Smooth muscle myosin is similar in structure and functon to non-muscle myosin isolated from human blood platelets and rat proliferative myoblasts. Structurally, both myosins contain light chains of similar molecular weights. Functionally, the actin-activated ATPase activity of both types of myosin is related to the degree of myosin phosphorylation.

2) Differences in the manner in which phosphorylation regulates the actin-activated ATPase activity of non-muscle and vas deferens myosin include: a) the kinase catalyzing the phosphorylation of non-muscle myosin, unlike smooth muscle myosin kinase, is not dependent on Ca^{2+} for activity; b) the actin-activated ATPase activity of purified phosphorylated non-muscle myosin is not inhibited by the removal of Ca^{2+}, while that of phosphorylated vas deferens myosin is; c) tropomyosin increases the actin-activated myosin ATPase activity of phosphorylated vas deferens myosin, but does not increase this activity using phosphorylated platelet myosin.

INTRODUCTION

The contractile proteins actin and myosin are present in most, if not all, eukaryotic cells. In the vertebrate it has become common practice to group the contractile proteins into two broad categories: those found in muscle cells and those found in non-muscle cells. The muscle group can be further divided into subgroups of cardiac, smooth, and skeletal muscle contractile proteins. Less is known about non-muscle actin and myosin. Although the contractile proteins from different vertebrate cells are thought to be similar in structure and function, there is some evidence for differences in the primary structure of the myosin molecule.[1]

The purpose of this paper is to compare the properties of smooth muscle myosin with those of non-muscle myosin isolated from platelets and proliferative myoblasts. The major similarity between these two classes of myosin is that both must be phosphorylated in order for actin to activate the myosin ATPase activity. Phosphorylation of myosin is not a prerequisite for actin-activation in skeletal muscle.[2]

We shall first review the structural and functional properties of vertebrate non-muscle and smooth muscle myosin, emphasizing their differences from skeletal and cardiac myosin. We shall then explore the similarities and differences in the effect of phosphorylation of myosin from non-muscle and smooth muscle cells. Finally, a diagram summarizing our present ideas about phosphorylation in non-muscle and smooth muscle cells will be presented. This diagram presents a working model which extrapolates the *in vitro* experiments on actin-activation of myosin ATPase activity to actin-myosin interaction *in vivo*.

METHODS

Human blood platelets, Yaffe L-5 810 cloned rat proliferative myoblasts and guinea pig vas deferens were used as sources of myosin and the phosphorylation systems. The details of the preparation and purification of the proteins, as well as the techniques used for phosphorylating myosin and assaying myosin kinase, phosphatase and ATPase activities have previously been published: for platelets, see Adelstein and Conti;[3] for proliferative myoblasts, Scordilis and Adelstein;[4] and for vas deferens smooth muscle, Chacko, et al.[5] Partially-purified platelet phosphatase was prepared from an alcohol-ether powder[6] of platelets as described previously[7,8] and assayed as outlined in Chacko et al.[5]

Incorporation, release, and the exact location of ^{32}P, as well as protein purity, was monitored by 1% SDS - 7 1/2 or 10% polyacrylamide gel electrophoresis.[9,10]

RESULTS

Myosin (molecular weight 460,000) is composed of three pairs of polypeptide chains. Two are referred to as heavy chains (molecular weight 200,000) and four are light chains (15,000-27,000). Table 1 summarizes the molecular weights of the individual polypeptide chains for non-muscle and muscle myosins (as determined in SDS-polyacrylamide gels). Non-muscle and smooth muscle myosins differ from cardiac and skeletal muscle myosins in having light chains of similar molecular weights. Of note is that neither myosin contains a light chain with a molecular weight greater than 20,000 daltons (although data in Fig. 1 is presented for only two non-muscle myosins, similar light chain molecular weights have been established for numerous vertebrate non-muscle myosins).[1,11] Further evidence of a similarity between the light chains of platelet and smooth muscle myosins is their behavior in urea-polyacrylamide gels, in which they migrate more rapidly at pH 8.6 than either cardiac or skeletal muscle myosin light chains.

Table 1 also shows that the 18,500-20,000 dalton light chain of each myosin can be phosphorylated. Phosphorylation of non-muscle and smooth muscle myosin was carried out using an ammonium sulfate fraction prepared as outlined previously for human platelets,[3] rat proliferative myoblasts[4] and vas deferens smooth muscle.[5] This fraction contains actomyosin as well as myosin light chain kinase, myosin light chain phosphatase, and other proteins.

Actomyosin and purified myosin can be separated from the kinase and phosphatase by gel filtration on Sepharose 4B as shown in Fig. 1. This particular elution profile is for a 34-52% ammonium sulfate fraction of platelet actomyosin.[8] The first two peaks of ATPase activity (A_{720}), which contain ^{32}P radioactivity bound to the myosin light chain, are due to platelet actomyosin (45-60 ml) and

Table 1

Molecular Weights of Myosin Polypeptide Chains

	Heavy Chains	Light Chains
Platelet	200,000	20,000* 16,000
Proliferative Myoblast	200,000	20,000* 16,000
Smooth muscle	200,000	20,000* 16,000
Cardiac muscle	200,000	27,000 20,000*[19]
Skeletal muscle (fast)	200,000	25,000 18,500*[15] 15,000

*Can be phosphorylated

platelet myosin (65-85 ml) respectively. The third peak, which elutes at 125-135 ml contains both myosin light chain kinase and phosphatase activity.

Rechromatography of this platelet kinase-phosphatase peak on DEAE-cellulose shows the presence of a number of kinases and phosphatases (Fig. 2). The following peaks of enzymatic activity were eluted from this column: a) an initial peak (10-20 ml) which does not bind to the column and catalyzes the breakdown of p-nitrophenyl phosphate (p-NPP) and which does not catalyze the dephosphorylation of ^{32}P-labeled myosin; b) two peaks (43-68 ml) with cAMP-dependent kinase activity which catalyze the incorporation of ^{32}P from γ-AT^{32}P into an F1 histone fraction but which are not active when platelet myosin light chains are used as a substrate; c) a peak of phosphatase activity (labeled P) which dephosphorylates ^{32}P-labeled platelet myosin light chains as well as ^{32}P-labeled platelet myosin but has no activity when p-NPP is used as a substrate; d) a peak (75-95 ml) with platelet myosin light chain kinase activity which catalyzes the incorporation of ^{32}P from γ-labeled AT^{32}P into platelet myosin light chains but not into histone. This kinase, which has been purified and characterized[12] differs from the enzyme purified from skeletal[13] and smooth muscle[5,14] sources in not being dependent on trace Ca^{2+} (10^{-6} - 10^{-7}M) for its activity. Its activity is also unaffected by cAMP, as is that of all the myosin light chain kinases tested to date.[12,15]

The effect of phosphorylation on the actin-activated ATPase activity of non-muscle and smooth muscle myosin is summarized in Table 2. The myosin used in these experiments was purified from the ammonium sulfate fraction by gel filtration on Sepharose 4B (see peak 2 from Fig. 1). The values in Table 2 are for phosphorylated myosin, dephosphorylated myosin (i.e., myosin that was phosphorylated and subsequently dephosphorylated by endogenous or exogenous myosin light chain phosphatase), and unphosphorylated myosin (myosin isolated directly in the non-phosphorylated form).

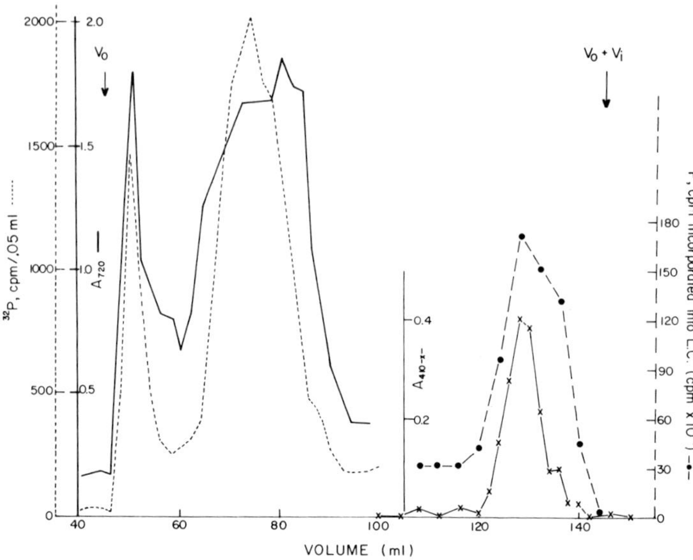

FIG. 1. Profile of Sepharose 4B gel filtration of the 34-52% ammonium sulfate fraction of platelet myosin.[8] Twelve mg of protein were applied in 2 ml to a 1.5 x 90 cm column of Sepharose 4B equilibrated and eluted with 0.5 M KCl, 15 mM Tris-HCl (pH 7.5), 1 mM EDTA and 2.5 mM DTT. Fractions were monitored for ATPase activity (A_{720}), ^{32}P radioactivity (dotted line), kinase activity (^{32}P incorporated into platelet myosin light chain) and phosphatase activity (A_{410}).

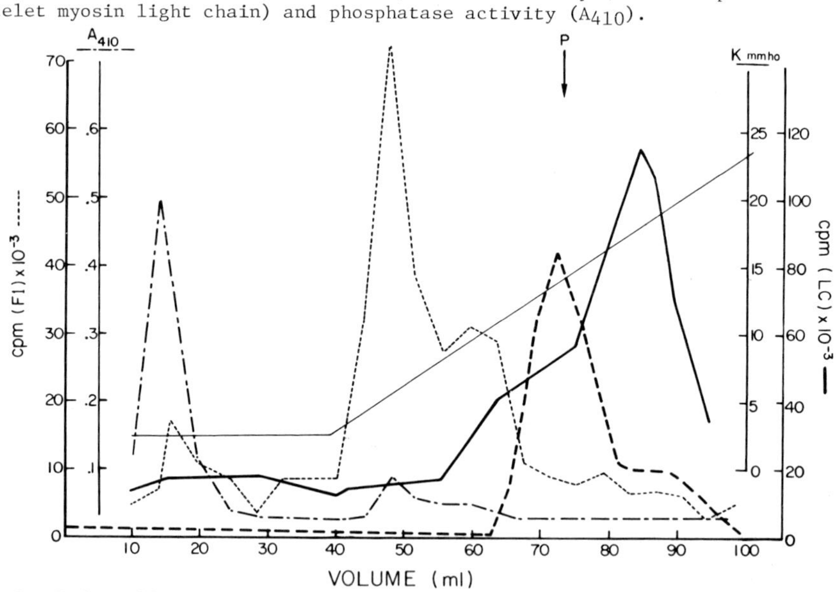

FIG. 2. DEAE cellulose chromatography of the pooled kinase-phosphatase fraction from Sepharose 4B. Ten mg of protein were applied to a 0.9 cm x 15 cm column of DEAE cellulose equilibrated with 20 mM KCl, 20 mM Tris-HCl (pH 7.5), 1 mM EDTA, 1 mM DTT. The limit buffer was the same except it was 0.4 M KCl. The column was eluted with the KCl gradient (indicated by K mmho and thin solid line) at a flow rate of 12 ml/hr. The enzyme activities eluted are described in the text.

Table 2

Myosin Phosphorylation and the Actin-Activated ATPase Activity

Myosin Source	Phosphorylated ATPase	Phosphorylated P_i	Dephosphorylated ATPase	Dephosphorylated P_i	Unphosphorylated ATPase
Platelets	.21	.8-.9	.05	.3	.03
Proliferative Myoblasts	.18	.8			.02
Vas Deferens	.10	.4-.6	.01	<.05	.01

ATPase: μMoles P_i/mg myosin/min at 37° (see legend for Table 3 for assay conditions).

P_i: Moles P_i bound/mole of 20,000 dalton light chain

There is a general correlation between the amount of myosin phosphorylation and the actin-activated ATPase activity. Unlike the dephosphorylation of vas deferens myosin, it has been difficult to dephosphorylate the intact platelet myosin molecule below 0.3 moles of phosphate/mole of light chain.

Recent experiments with smooth muscle myosin have resulted in myosin phosphorylation approaching 1 mole/mole of 20,000 dalton light chain, and this has resulted in increased ATPase activity suggesting that fully phosphorylated smooth muscle myosin, similar to non-muscle myosin, has an actin-activated ATPase activity of approximately 0.2 μmoles P_i/mg myosin/min at 37°.

Table 3 presents data showing two important differences between phosphorylated non-muscle and phosphorylated vas deferens smooth muscle myosins: a) the actin-activated ATPase activity of vas deferens myosin is inhibited by the removal of Ca^{2+} — this effect is not seen with platelet and myoblast myosin, b) rabbit skeletal muscle tropomyosin enhances the actin-activated ATPase activity of vas deferens muscle myosin but has little effect on the actin-activated ATPase activity of platelet and myoblast myosin (the latter is an unpublished observation of S.P.S.).

Table 3

The Effect of Ca^{2+} and Tropomyosin on the Actin-Activated ATPase Activity of Phosphorylated Myosin

Myosin Source	Actin-Activated ATPase Activity + EGTA	$+Ca^{2+}$	+ Tropomyosin
Vas Deferens-guinea pig^{+}	.025	.042	.12
	.028	.085	.16
	.034	.105	N.D.
Platelets - human	.18	.18	.19
Proliferative Myoblasts - rat	.18	.18	N.D.

$^{+}$individual determination N.D. = not determined

ATPase: μMoles P_i/mg myosin/min at 37°.

Assay conditions - 17-20 mM KCl, 10 mM Tris-HCl (pH 7.4) or 10 mM Imidazole-HCl (pH 7.2), 1 mM ATP, 1.4 - 1.6 mM $MgCl_2$ and either 1-2 mM EGTA or 0.1 mM $CaCl_2$. Rabbit skeletal muscle tropomyosin was assayed in the presence of 0.1 mM $CaCl_2$ (see references 3,4, and 5 for details).

DISCUSSION

Smooth muscle and vertebrate non-muscle myosins are markedly similar in structure and function. On the basis of these similarities smooth muscle myosin appears to be more closely related to vertebrate non-muscle myosin than to skeletal and cardiac muscle myosin. The marked structural similarities of these two myosins may well be the basis for the similar manner in which phosphorylation regulates the actin-activated ATPase activity.

There are, however, several important differences between non-muscle and smooth muscle contractile proteins in the manner in which phosphorylation regulates their interaction. These similarities and differences are summarized in Fig. 3, which diagrams how phosphorylation of myosin may act to regulate the actin-myosin interaction in these cells.

Part A summarizes the non-muscle and smooth muscle systems which have been found to have myosin phosphorylating systems. In non-muscle cells, myosin phosphorylation has been shown to regulate the actin-activated ATPase activity of platelet and proliferative myoblast myosin. In addition, astrocytic neuroglial cells,[16] as well as HeLa cells and rabbit macrophages, (unpublished observation, J.A.T.) have been shown to contain a myosin light chain kinase. Myosin light chain phosphatase has been identified in platelets,[7] myoblasts (unpublished observation, S.P.S.) and macrophages. Both myosin light chain kinase and phosphatase have been identified in vas deferens,[5] stomach[18] and gizzard[14,17] smooth muscle cells and this volume also records their presence in other types of smooth muscle.

In the smooth muscle we can postulate the following mechanism for control of the actin-activated ATPase activity: The Ca^{2+} concentration inside the smooth muscle cell is raised to 10^{-6}-10^{-7}M and the kinase is activated.[14] In the presence of MgATP this results in the phosphorylation of smooth muscle myosin producing P-Myosin and MgADP. Unlike unphosphorylated myosin, actin can activate the myosin ATPase activity of P-Myosin and smooth muscle contraction results. Relaxation would be initiated by lowering the Ca^{2+} concentration to less than 10^{-7}M, which would lead to inactivation of the kinase. Since the phosphatase is not dependent on Ca^{2+} for its activity, dephosphorylation would ensue, producing "Myosin" which cannot be activated by actin. This would correlate with smooth muscle relaxation. Although _in vitro_ experiments suggest[5,14] that the rate of myosin phosphorylation and dephosphorylation is rapid enough to account for smooth muscle contraction and relaxation _in vivo_, this important point will not be established until the proper physiological experiments are carried out.

A second mechanism by which Ca^{2+} appears to regulate actin-myosin interaction in vas deferens smooth muscle is listed in B. This particular mechanism, which may not be present in all smooth muscle cells, is manifested _in vitro_ by the decreased actin-activated ATPase activity of vas deferens myosin measured in the presence of EGTA[5] (see Table 3). In terms of the model, this would imply that relaxation of

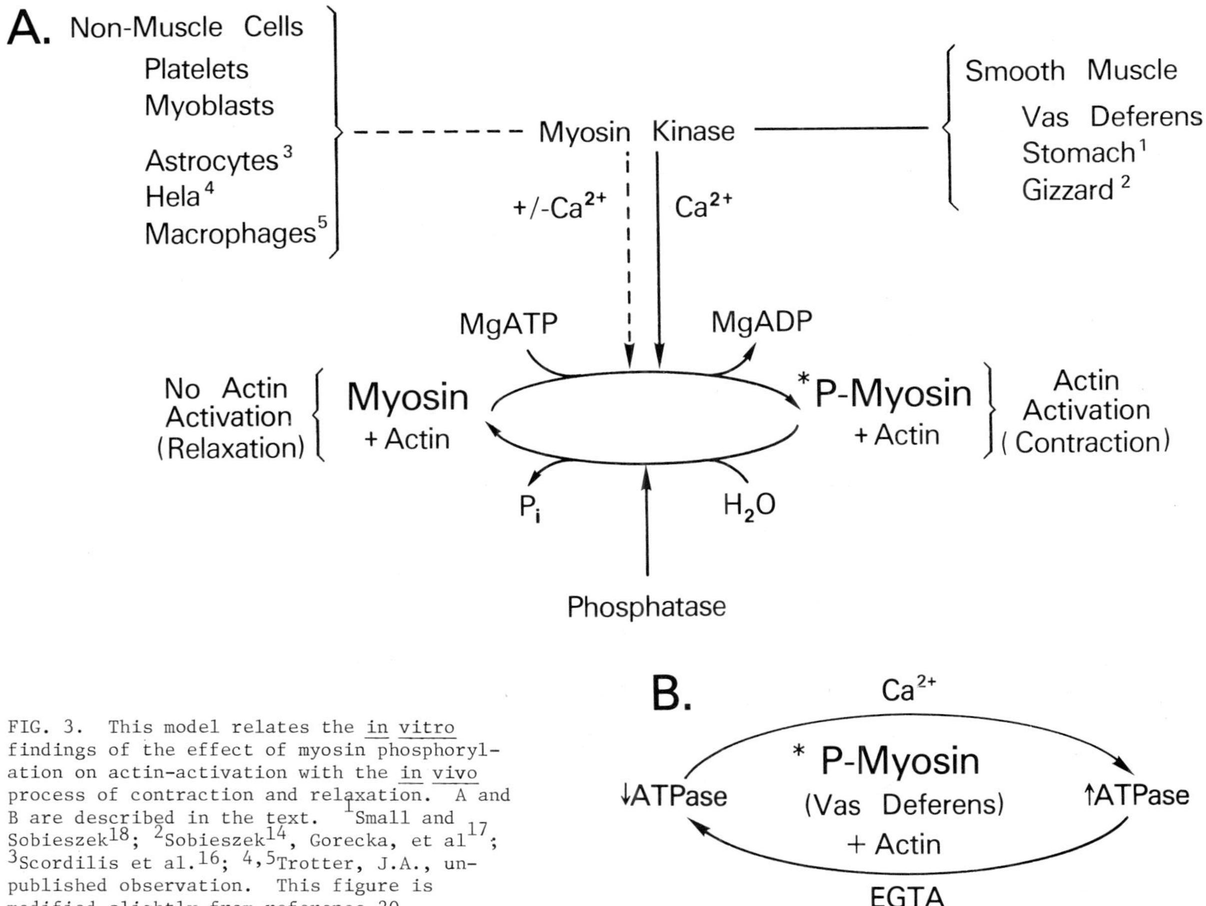

FIG. 3. This model relates the in vitro findings of the effect of myosin phosphorylation on actin-activation with the in vivo process of contraction and relaxation. A and B are described in the text. [1]Small and Sobieszek[18]; [2]Sobieszek[14], Gorecka, et al[17]; [3]Scordilis et al.[16]; [4,5]Trotter, J.A., unpublished observation. This figure is modified slightly from reference 20.

vas deferens muscle can also occur without prior dephosphorylation of myosin. The details of this mechanism have yet to be worked out.

The above model is more tenuous when applied to non-muscle cells. Here the terms "relaxation" and "contraction" do not at present relate to any well defined physiological functions. Moreover, important pieces in the puzzle such as how the kinase and phosphatase are controlled, remain to be filled in.

Myosin phosphorylation in non-muscle and smooth muscle cells is now emerging as an important new regulatory mechanism. Its presence and central role in regulating the actin-activated ATPase activity in non-muscle and smooth muscle cells indicate a marked similarity between the contractile proteins of these two different cell types. As we begin to discover the various factors controlling the kinases and phosphatases, we will increase our understanding of how the contractile proteins function in these cells, as well as what processes they control in non-muscle cells.

REFERENCES

1. Burridge, K. and Bray, D. (1975) J. Mol. Biol. 99, 1-14.
2. Morgan, M.W., Perry, S.V., Ottaway, J. (1976) Biochem. J. 157, 687-697.
3. Adelstein, R.S., and Conti, M.A. (1975) Nature 256, 597-598.
4. Scordilis, S.P. and Adelstein, R.S. (1977) Nature (in press).
5. Chacko, S., Conti, M.A., and Adelstein, R.S. (1977) Proc. Natl. Acad. Sci. USA 74, 129-133.
6. Cohen, I. and Cohen, C. (1972) J. Mol. Biol. 68, 383-387.
7. Barylko, B., Conti, M.A., and Adelstein, R.S. (1977) Biophysical J. 17, 270a.
8. Adelstein, R.S., Chacko, S., Barylko, B., Scordilis, S.P. and Conti, M.A. (1976) In: Contractile Systems in Non-Muscle Tissues. S.V. Perry, A. Margreth, and R.S. Adelstein (Eds), pp. 153-163, Elsevier North-Holland, New York.
9. Fairbanks, G.T., Steck, T.L. and Wallach, D.F.H. (1971) Biochemistry 10, 2606-2617.
10. Anker, H.S. (1970) FEBS Lett. 7, 293.
11. Adelstein, R.S. and Conti, M.A. (1976) In: Cell Molility, Cold Spring Harbor Conferences on Cell Proliferation, vol. 3. R. Goldman, T. Pollard, and J. Rosenbaum (Eds.) pp 725-738, Cold Spring Harbor Laboratory, Cold Spring Harbor, New York.
12. Daniel, J.L. and Adelstein, R.S. (1976) Biochemistry 15, 2370-2377.
13. Pires, E., Perry, S.V. and Thomas, M.A.W. (1974) FEBS Lett. 41, 292-296.
14. Sobieszek, A. (1977) Eur. J. Biochem. 73, 477-483.
15. Perry, S.V., Cole, H.A., Morgan, M., Moir, A.J.G., and Pires, E. (1975) In: Federation of European Biochemical Societies, Proceedings of the Ninth Meeting. E.N.A. Biro (Ed). pp. 163-176, Elsevier North-Holland, New York.
16. Scordilis, S.P., Anderson, J.L., Pollack, R., and Adelstein, R.S. (1977) J. Cell Biol. (in press).
17. Gorecka, A., Aksoy, M.D., and Hartshorne, D.J. (1976) Biochem. Biophys. Res. Commun. 71, 325-331.
18. Small, V. and Sobieszek, A. (1977) Eur. J. Biochem. (in press).
19. Frearson, N., Focant, B.W.W., and Perry, S.V. (1976) FEBS Lett. 63, 27-32.
20. Adelstein, R.S., Chacko, S., Scordilis, S.P., Barylko, B., Conti, M.A., and Trotter, J.A., 1977. In: Calcium Binding Proteins and Calcium Function. R.H. Wasserman, et al (Eds) (in press), Elsevier North-Holland, New York.

ACKNOWLEDGEMENTS

The authors wish to acknowledge the excellent technical assistance of William Anderson, Jr. and James M. Miles. We are very grateful to Exa Murray for typing the manuscript.

Excitation-Contraction Coupling in Smooth Muscle, R. Casteels et al. eds
© *1977 Elsevier/North-Holland Biomedical Press*

Phosphorylation and Ca^{2+} Regulation of Actin-activated ATPase Activity of Myosin Isolated from Cultured Aortic and Vas Deferens Smooth Muscle Cells

Samuel Chacko, Stephen H. Blose and Robert S. Adelstein
Department of Pathobiology, University of Pennsylvania,
Philadelphia, PA 19174
and
National Heart, Lung, and Blood Institute, Bethesda, MD 20014

SUMMARY

1) Cultured aortic smooth muscle cells are similar to non-muscle cells with respect to the absence of myosin filaments and Ca^{2+} independence of the phosphorylation system.

2) A myosin mediated Ca^{2+} regulation of the actin-activated ATPase activity exists in phosphorylated vas deferens myosin, in addition to the regulation via light chain kinase.

INTRODUCTION

It is now well established that phosphorylation of a light chain of myosin occurs in myosin isolated from both muscle and non-muscle sources.[1,2] In each case covalent phosphorylation occurs on a particular light chain of myosin with a molecular weight of 18-20,000 daltons (P-light chain) eg., white skeletal muscle 18,000; red skeletal muscle 20,000; cardiac muscle 20,000; smooth muscle 20,000; platelet 20,000.[1-3]

While the role of phosphorylation in striated muscle is poorly understood,[1] phosphorylation has been shown to increase the actin-activated ATPase activity of myosins isolated from platelet,[4] smooth muscle,[5-7] and cultured skeletal muscle myoblasts.[8]

Phosphorylation of myosin light chain is catalyzed by a specific enzyme, myosin light chain kinase, which differs in the case of non-muscle and muscle in that the former enzyme is independent of traces of Ca^{2+} for activity, whereas the latter enzyme requires Ca^{2+}.[1,3,8-12] Hence in smooth muscle Ca^{2+} regulation of actin-activated ATPase activity is mediated via kinase, since this myosin is activated by actin only when it is in the phosphorylated state.

Although the myosin in non-muscle cells and smooth muscle cells have several characteristics in common (e.g., molecular weights of light chains, lower ATPase activity than striated muscle myosin) they differ in the organization of the myosin molecule into filaments. Differentiated smooth muscle cells contain thick myosin (∿ 150 Å in diameter) and thin actin (∿ 60 Å in diameter) filaments, whereas, in non-muscle cells thick myosin filaments have not yet been visualized though, actin filaments are present. The absence of thick filaments serves as one of the distinguishing features between non-muscle and smooth muscle cells.

Aortic smooth muscle cells can be grown in culture for several generations.[13] These cells synthesize the same types and proportions of collagen[14] and glycosaminoglycans[15] as the aortic smooth muscle cells _in vivo_. However, unlike the cells in the intact aortic media the cultured aortic medial cells lack the thick myosin filaments.[16] Hence, they are more like the non-muscle cells because the myosin molecule does not organize into filaments. On the other hand, a large number of thin filaments are present in these cells.[16-18]

In this paper we report: (1) that the cultured aortic medial cells lack thick myosin filaments and they possess a phosphorylation system that is similar to non-muscle cells and (2) that a myosin mediated Ca^{2+} regulation of actin-activated ATPase activity exists in phosphorylated vas deferens myosin, in addition to the regulation via light chain kinase.

MATERIALS AND METHODS

Biochemical Methods: The biochemical techniques followed are in general similar to those used previously for our studies on vas deferens smooth muscle.[5] All procedures were carried out at 4°C unless specified otherwise and deionized water was used throughout. A 35-70% ammonium sulfate fraction of the cell extract containing myosin, kinase and phosphatase was used for phosphorylation, dephosphorylation and purification of myosin. The incorporation of ^{32}P into myosin samples at various intervals was determined as described by Daniel and Adelstein.[3] Phosphorylated and dephosphorylated myosin was purified by gel filtration on Sepharose 4B agarose columns as described before.[5] The ATPase, kinase and phosphatase assays were carried out as described by Chacko et. al.[5] The incorporation, release and exact location of ^{32}P was monitored by 1% SDS-7.5% polyacrylamide gel electrophoresis using gels prepared according to Fairbanks et. al.[18]

Rabbit skeletal muscle actin was prepared by the method outlined by Spudich and Watt.[20] Light chains isolated from gizzard actomyosin by dissociation in 8M guanidine-HCl as described by Perrie and Perry[19] was a gift from Dr.Apolinary Sobieszek. Protein concentrations were determined by the procedure of Lowry et. al.[21]

Cell Preparation and Electron Microscopy: Aortic medial cells were cultured from guinea pig as we described previously.[30] Aortas were fixed by the methods of Somlyo and co-workers.[22,23] Cells for electron microscopy were grown and processed as described before.[24] Cultured vascular smooth muscle cells were glycerinated _in situ_ and prepared for decoration with S-1 (heavy meromyosin subfragment-1) after the method of Ishikawa et. al.[25]

RESULTS

a. Morphologic Characteristics

The mammalian aortic media consists of a homogenous population of smooth muscle cells.[13,26-28] Figure 1 is an electron micrograph of an aortic smooth muscle cell

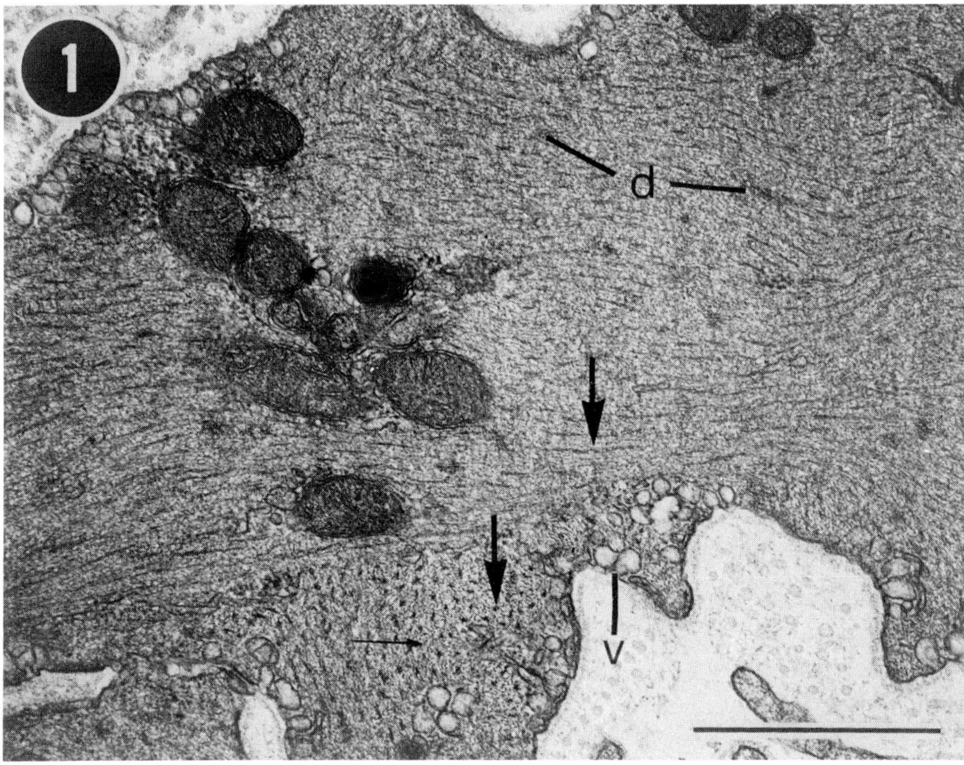

Figure 1: Electron micrograph of abdominal aorta fixed _in situ_, in cross section, The presence of thick (myosin, 150 Å diameter) filaments (large arrows), actin filaments (small arrows), dense bodies (d) and surface vesicles (v) are identified, Bar is 1 μm.

(fixed _in situ_) before culturing. The thick myosin and thin actin filaments, dense bodies and surface vesicles typical of smooth muscle cells are observed in this cell.[22,23]

Removal of the vascular media and growth _in vitro_ stimulates the proliferation and migration of smooth muscle cells from the explant.[13,28] These cells serve as a source of smooth muscle cells that synthesize the same type and proportion of glycosaminoglycans[15] and collagen[14] as the aortic smooth muscle cells _in vivo_.

These cells (Figs 2 and 3) contain bundles of longitudinally arranged actin filaments ($\sim$ 60 Å in diameter). No thick myosin ($\sim$ 150 Å in diameter) filaments are observed either within the bundle or in other regions of the cell. Figure 3a shows that the filaments within the bundles decorate with S-1. The S-1 decoration was removed by treating the cells with 10 mM Mg-ATP (Fig. 3b) unequivocally identifying these filaments as actin.

b. Biochemical Characteristics

1. Ca^{2+} requirement for phosphorylation: Figure 4 shows a time course for incorporation of ^{32}P when 35-70% ammonium sulfate fractions of extracts of vas deferens smooth muscle and cultured aortic medial cells, containing endogenous light chain kinase, phosphatase and myosin, were incubated with γ-$AT^{32}P$. The ^{32}P is incorporated into 20,000 dalton light chain of vas deferens myosin in the presence of Ca^{2+} but not in the presence of EGTA, indicating the Ca^{2+} dependency of the kinase. A similar result was obtained using an ammonium sulfate fraction of myosin from aortic smooth muscle prior to culture was used (data not shown). On the other hand, the endogenous kinase in the ammonium sulfate fraction of cultured aortic medial cells phosphorylated the myosin both in the presence and absence of Ca^{2+} (Fig. 1).

Having shown that the endogenous kinase present in the ammonium sulfate fraction phosphorylated the cultured cell myosin both in the presence and absence of Ca^{2+}, the Ca^{2+} requirement of the partially purified kinase was tested with exogenous light chain. Cultured cell kinase was separated from myosin by Sepharose 4B column chromatography as reported previously.[5] The ^{32}P incorporation into smooth muscle light chain, when incubated with isolated kinase, is the same both in the presence and absence of Ca^{2+}, confirming that the light chain kinase from cultured aortic cells is not dependent on Ca^{2+} for its activation. These experiments demonstrate that Ca^{2+} regulates the phosphorylation of myosin isolated from smooth muscle myosin but not of the myosin isolated from cultured aortic medial cells.

2. Evidence for myosin mediated Ca^{2+} regulation of actin-activated ATPase activity of phosphorylated vas deferens myosin: A number of reports[5,11,7] agree that the ATPase activity of smooth muscle myosin is activated by actin only when the myosin is phosphorylated. Since the kinase responsible for phosphorylation in smooth muscle is Ca^{2+} dependent, the actin-activated ATPase activity is regulated by Ca^{2+} indirectly.

It was of interest to determine whether Ca^{2+} regulates the actin-activated ATP-ase activity of myosin after the phosphorylation in addition to the regulation via kinase. For this purpose myosin was phosphorylated using endogenous kinase in the presence of Ca^{2+} and subsequently purified by gel filtration on Sepharose 4B agarose columns as previously published.[5] The myosin remains phosphorylated and

---

Figure 2: Electron micrograph of a cultured vascular smooth muscle in cross section. Actin filaments (60 Å diameter) arranged into bundles (singleheaded arrows) are found coursing along the upper (U) and lower (L) portions of the cell. Dense bodies are seen in cross section (doubleheaded arrow). Bar is 1 μm.

Figure 3 A and B: (A) Cultured aortic medial cells decorated with S-1. All of the filaments within the bundles are decorated with arrowhead complexes. (B) Control culture incubated with S-1 and 10 mM Mg-ATP. The ATP blocked decoration of the actin filaments. The arrow indicates 100 Å filaments running in parallel with the actin filament bundles. Bar is 0.5 μm.

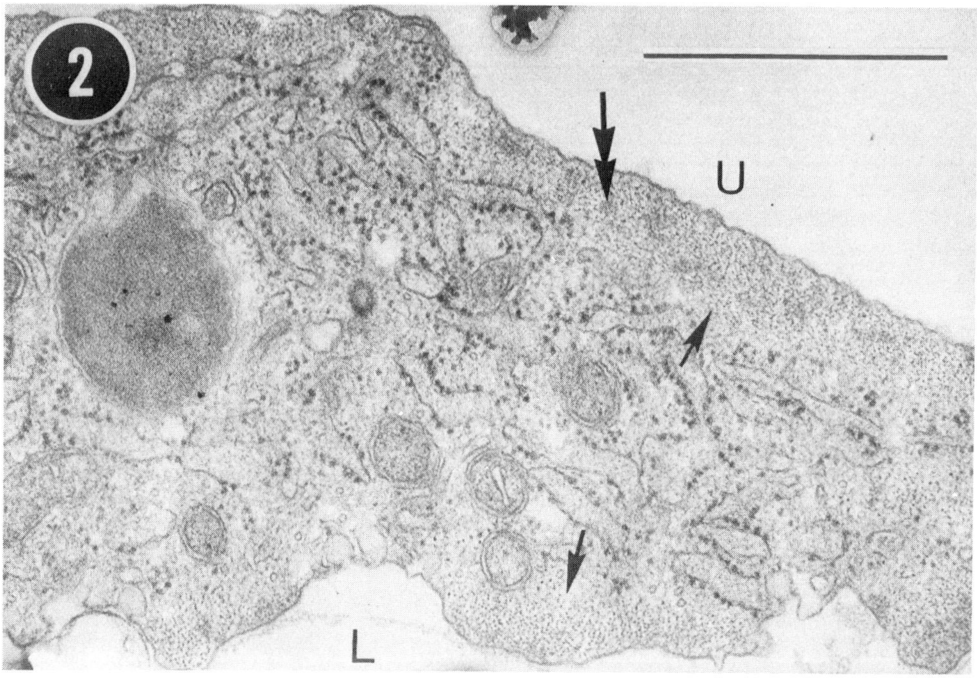
2
U
L

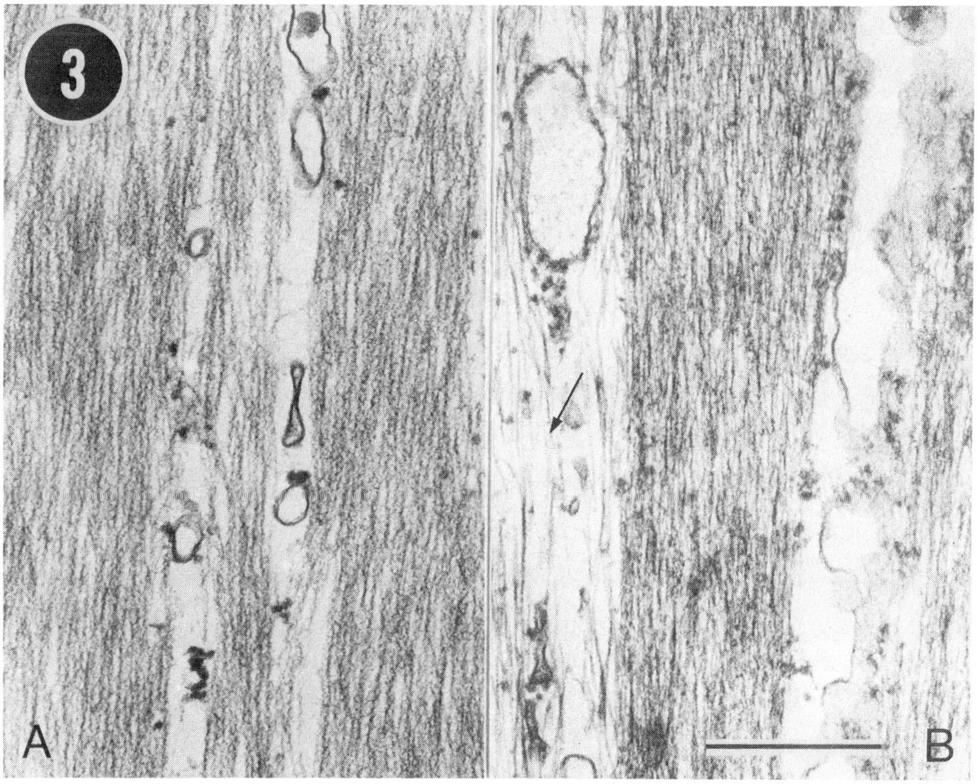
3
A
B

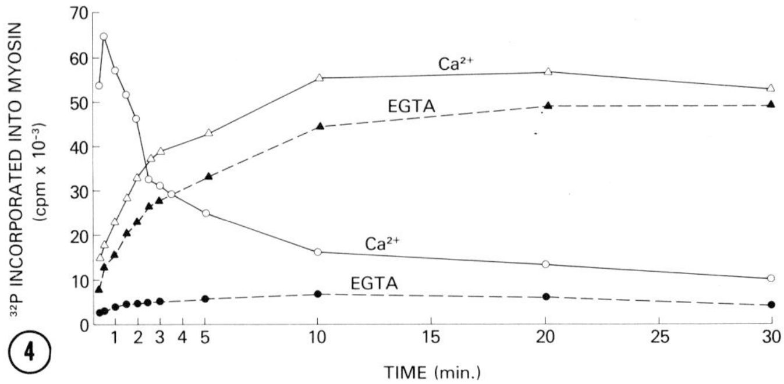

4

Figure 4: Time course for phosphorylation of vas deferens and cultured aortic medial cell myosin. The 35-70% ammonium sulfate fraction was incubated in the presence of 0.1 mM ATP and either 0.1 mM $CaCl_2$ or 2 mM EGTA. Aliquots were removed at indicated times and the incorporation of ^{32}P was determined as described by Chacko et. al.[9]

the kinase and phosphatase are separated from the myosin since they have lower molecular weights. Dephosphorylated myosin was prepared as previously described.[5] Using phosphorylated and dephosphorylated myosin, free of kinase and phosphatase, the effect of actin on the myosin ATPase activity was determined (Table I). The

Table I. The Effect of Phosphorylation and Ca^{2+} on Actin-activated ATPase* Activity

Prep	Myosin	+ Actin (0.02 M KCl) Ca^{2+}	+ Actin (0.02 M KCl) EGTA	Myosin Alone (0.5 M KCl) K^+-EDTA	Myosin Alone (0.5 M KCl) Ca^{2+}
1	Phosphorylated	.040	.017	1.64	0.89
1	Unphosphorylated	.012	.008	1.45	0.81
2	Phosphorylated	.085	.028	1.50	0.87
2	Dephosphorylated	.010	.009	1.36	0.96
2	Dephosphorylated	.007	.005	1.29	0.92
3	Phosphorylated	.105	.034	1.45	0.68

CONDITIONS:

Actin-activated ATPase: 20 mM KCl, 10 mM imidazole HCl (pH 7.2), 1 mM ATP, 1.6 mM $MgCl_2$ and either 0.1 mM $CaCl_2$ or 2 mM EGTA at 37°C, myosin 0.14 mg/ml, actin 0.5 mg/ml. K^+-EDTA and Ca^{2+} activated ATPase: 0.5 M KCl, 0.2 M Tris-HCl (pH 7.5), 1 mM ATP and either 2 mM EDTA or 10 mM $CaCl_2$ at 37°C, myosin 0.006 mg/ml.
*Activity is expressed as μ moles Pi released/mg myosin/min.

actin-activated ATPase activity of phosphorylated myosin is higher than that of dephosphorylated myosin. Furthermore the actin-activated ATPase activity of phosphorylated myosin is higher in the presence of Ca^{2+} than in the absence of Ca^{2+}, providing a direct evidence for myosin mediated Ca^{2+} regulation when the myosin is in the phosphorylated form. Note that the K^{+}-EDTA and the Ca^{2+} activated myosin ATPase activity measured in 0.5 M KCl is unchanged by phosphorylation.

The inhibition of actin-activated ATPase activity by removal of Ca^{2+} is not associated with dephosphorylation of light chain as shown in Table II. The specific radioactivity of the light chain is not changed either in the presence of Ca^{2+} or in the presence of EGTA during the ATPase assay.

Table II. ATPase Activity and ^{32}P Content of Myosin During the Assay

	ATPase activity (μmoles Pi/min/mg	^{32}P Content (μmoles Pi/moles LC 20)	
	+ Actin	0 time	20 minutes
Ca^{2+}	0.042	0.21	0.21
EGTA	0.025	0.21	0.21

DISCUSSION

Cultured aortic medial cells are similar to the cells in the intact aortic media with respect to the secretion of extracellular matrix.[14,15] However, these cells lack thick myosin filaments typical of differentiated smooth muscle cells making them similar to non-muscle cells. The finding that the myosin light chain kinase in the cultured cells is not dependent on Ca^{2+} for its activation suggests that a different mode of regulation exists in these cells.

The present data indicate that a change in the phenotype of these cells from smooth muscle to non-muscle with respect to the presence of myosin filaments may be associated with the "turning off" of the genes for smooth muscle specific kinase and the "turning on" of the genes for non-muscle kinase.

All published reports agree that a Ca^{2+} dependent kinase that phosphorylates the 20,000 dalton light chain of myosin is present in smooth muscle and is extracted with actomyosin preparation.[5-7] We have isolated this kinase in a partially purified form from vas deferens smooth muscle.[5] In all investigations, phosphorylation of the P-light chain increased the actin-activated ATPase activity. Removal of Ca^{2+} by EGTA from the myosin-kinase preparation inhibited the phosphorylation and lowered the actin-activated ATPase activity.[6,7] We showed that an endogenous phosphatase dephosphorylated the myosin in our preparation.[5] Dephosphorylated and phosphorylated myosin purified by gel filtration was free of kinase and phosphatase. The phosphorylated myosin was activated by actin, whereas, the dephosphorylated myosin was not.

Hence, in smooth muscle, Ca^{2+} regulates the actin-activated ATPase activity of myosin via phosphorylation by a Ca^{2+} dependent kinase. In addition, Ca^{2+} exerts a direct control on the actin-activated ATPase activity of phosphorylated vas deferens myosin.[5] The actin-activated ATPase activity of phosphorylated vas deferens myosin is inhibited by the removal of Ca^{2+} by EGTA. Therefore, in vas deferens smooth muscle, Ca^{2+} may regulate contraction by two mechanisms: (1) by controlling phosphorylation due to its effects on the activity of the light chain kinase; (2) by direct myosin mediated regulation of actin-activated ATPase activity of phosphorylated myosin. The second form of myosin mediated regulation by Ca^{2+} is similar to the molluscan muscle regulation described by Kendrick-Jones et. al.[29] However, there is no evidence to date that phosphorylation plays a role in molluscan system, while it is necessary for actin-activation in smooth muscle. The direct myosin mediated Ca^{2+} regulation of actin-activated ATPase activity observed by us for phosphorylated vas deferens myosin was not observed either by Gorecka et. al.[7] or Sobieszek and Small[6] working with chicken gizzard muscle. Further experiments will be directed to study the physiological significance of the two types of Ca^{2+} regulation in vas deferens smooth muscle.

References

1. Perrie, W.T., Smillie, L.B. and Perry, S.V., 1973, Biochem. J. 135, 151-164
2. Adelstein, R.S., Conti, M.A. and Anderson, W.J., 1973, Proc. Natl. Acad. Sci. 70, 3115-3119
3. Daniel, J.L. and Adelstein, R.S., 1976, Biochemistry 15, 2370-2377
4. Adelstein, R.S. and Conti, M.A., 1975, Nature 256, 597-598
5. Chacko, S., Conti, M.A. and Adelstein, R.S., 1977, Proc. Natl. Acad. Sci. 74, 129-133
6. Sobieszek, A. and Small, J.V., 1976, J. Mol. Biol. 102, 75-92
7. Gorecka, A., Aksoy, M.O. and Hartshorn, D.J., 1977, Biochem. Biophys. Res. Commun. 71, 325-331
8. Scordilis, S.P. and Adelstein, R.S., 1977, Nature (in press)
9. Chacko, S., Conti, M.A. and Adelstein, R.S., 1976, Fed. Proc. 35, 1581
10. Aksoy, M.O., Williams, D., Sharkey, E.M. and Hartshorn, D.J., 1976, Biochem. Biophys. Res. Commun. 69, 35-41
11. Sobieszek, A., 1977, Eur. J. Biochem. 73, 477-483
12. Frearson, N., Focant, B.W.W. and Perry, S.V., 1976, FEBS Lett. 63, 27-32
13. Ross, R., 1971, J. Cell Biol. 50, 172-186
14. Mayne, R., Vail, M.S., Miller, E.J., Blose, S.H. and Chacko, S., 1977, Arch. Biochem. Biophys. 181, 462-469
15. Wight, T.N. and Ross, R., 1975, J. Cell Biol. 67, 675-686
16. Blose, S.H., Lazarides, E. and Chacko, S., 1975, J. Cell Biol. 67, 34a
17. Jarmalych, J., Daoud, A.S., Landau, J., Fritz, K.E. and McElvene, E., 1968, Exp. Mol. Pathol. 9, 171-188
18. Fairbanks, G.T., Steck, T.L. and Wallach, D.F.H., 1971, Biochem. 10, 2606-2617

19. Perrie, W.T. and Perry, S.V., 1970, Biochem. J. 119, 3-38

20. Spudich, J.A. and Watt, S., 1971, J. Biol. Chem. 246, 4866-4871

21. Lowry, O.H., Rosebrough, N.J., Farr, A.L. and Randall, R.J., 1951, J. Biol. Chem. 193, 265-275

22. Devine, C.E. and Somlyo, A.P., 1971, J. Cell Biol. 49, 636-649

23. Ashton, F.T., Somlyo, A.V. and Somlyo, A.P., 1975, J. Mol. Biol. 98, 17-29

24. Blose, S.H. and Chacko, S., 1976, J. Cell Biol. 70, 459-466

25. Ishikawa, H., Bischoff, R. and Holtzer, H., 1969, J. Cell Biol. 43, 312-328

26. Ross, R. and Glomset, J., 1973, Science 180, 1332-1339

27. Wissler, R.W., 1968, J. Atheroscler. Res. 8, 201-213

28. Dzoga, K.F., Chen, R. and Wissler, R.W., 1974, in: Arterial Mesenchyme and Atherosclerosis. W.D. Wagner and T.B. Clarkson (Eds), pp. 299-310

29. Kendrick-Jones, J., Lehman, W. and Szent-Gyorgi, A.G., 1970, J. Mol. Biol. 54, 313-326

30. Garfield, R.A., Chacko, S. and Blose, S.H., 1975, Lab. Invest. 33, 418-427

Acknowledgements: The authors wish to acknowledge the excellent technical assistance of Caroline Han, William Anderson, Jr. and James M. Miles. S.C. is on scholarly leave to N.I.H.

Excitation-Contraction Coupling in Smooth Muscle, R. Casteels et al. eds.

ASPECTS OF THE REGULATORY MECHANISM IN SMOOTH MUSCLE

D. J. Hartshorne[1], A. Gorecka[2], M. O. Aksoy[1]
[1]Carnegie-Mellon University, Pittsburgh, Pennsylvania 15213
[2]Nencki Institute of Experimental Biology, Pasteur St., Warsaw, Poland

INTRODUCTION

It is well established that the contraction-relaxation cycle of all muscle types is controlled by the level of intracellular Ca^{2+}. In the case of skeletal and cardiac muscle the regulatory proteins are troponin and tropomyosin, which are located on the thin filaments. The regulatory mechanism in smooth muscle, however, is not characterized and there are several differing opinions. These are a) that a troponin-like mechanism exists (Head et al.[1]), b) that control is mediated via the phosphorylation of myosin, and c) that neither of the above are functional and an alternative mechanism operates (Mikawa et al.[2]). The majority of reports favor the phosphorylation theory and the following communication will deal only with aspects of this mechanism.

The original suggestion that phosphorylation was involved in the regulatory mechanism of smooth muscle came from Sobieszek[3] and Bremel et al.[4]. Based on their results and those of other investigators[5-7] it is possible to construct a tentative pathway to account for the regulation of smooth muscle activity. The critical aspects of this scheme are: 1) in the presence of Ca^{2+} ($\sim 1\times10^{-5}$M) a myosin light chain kinase is active and phosphorylates the 20,000 dalton light chains of myosin, 2) the Mg^{2+}-ATPase activity of the phosphorylated myosin is activated by actin and this leads to cross-bridge turnover and shortening or the development of tension, 3) in the absence of Ca^{2+} ($\sim 1\times10^{-7}$M) the kinase is not active and a phosphatase removes the phosphate groups, 4) the dephosphorylated myosin is not activated by actin, the actin-myosin complex dissociates and relaxation occurs. These features can be incorporated into a cyclic pathway as follows:

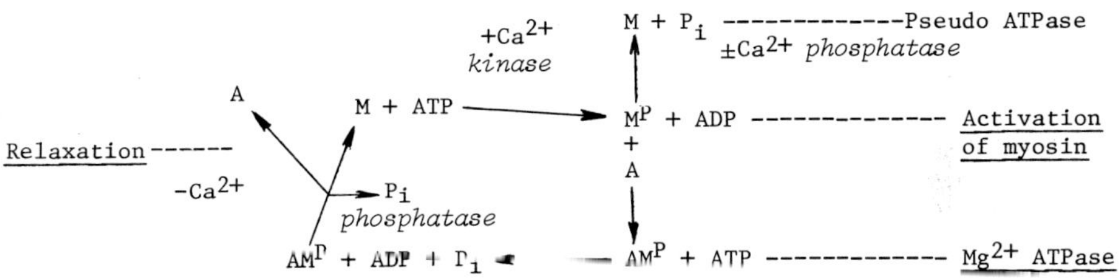

The three major phases of the cycle are activation of myosin, steady-state Mg^{2+}-ATPase activity and relaxation. The pseudo-ATPase activity is a reflection of the linked kinase-phosphatase activity, but the liberation of P_i by this mechanism is negligible compared to the steady-state ATPase activity (Gorecka, Aksoy and Hartshorne, unpublished).

In the following communication certain points of this pathway will be examined in more detail. The first is concerned with the influence of tropomyosin on the ATPase activity of actomyosin.[5] It has been reported that tropomyosin activates the Mg^{2+}-ATPase activity. This merits further attention as the regulatory mechanism is thought to be centered on the myosin molecule and tropomyosin is known to be a component of the thin filament. It was our intention to establish whether or not the activation by tropomyosin was accompanied by an increased level of myosin phosphorylation, and also to compare skeletal and gizzard tropomyosins.

The second part of our studies was focussed on the relaxation phase. The inhibition of Mg^{2+}-ATPase activity is a consequence of the removal of phosphate groups from myosin and this is accomplished when the Ca^{2+} level is reduced. It is known that both skeletal myosin [8,9] and gizzard myosin[10,11] bind Ca^{2+}, and therefore the possibility exists that the Ca^{2+}-myosin interaction plays an integral part in the regulatory mechanism. This view is supported by the studies of Chacko et al.[5]. The alternative and simpler hypothesis is that the dominant factor is whether or not the myosin is phosphorylated. If this is true then phosphorylated myosin would be active both in the presence and absence of Ca^{2+}. There are several ways to distinguish between the two possibilities and a full discussion will be published later. In this presentation only one approach will be considered, namely the limited proteolysis of the kinase. Our results indicate that at least with gizzard myosin the level of phosphorylation of myosin is the dominant factor in determining the extent of actin activation. These studies are pertinent also in that they confirm and extend the early observations by Ebashi et al.[12] who found that proteolysis of Ca^{2+}-sensitive actomyosin resulted in a loss of Ca^{2+}-sensitivity.

MATERIALS AND METHODS

Protein preparations: a) Ca^{2+}-sensitive actomyosin. Fresh or frozen chicken gizzards (Pel-Freez) are minced and homogenized with 20 vols. of 15 mM $MgCl_2$, 50 mM KCl, 10 mM tris-HCl, pH 7.6 (solution I) and centrifuged at $\sim$2,000 x g for 10 mins. The supernatant is discarded and the residue homogenized with 20 vols. solution I, followed by centrifugation. This step is repeated twice. The residue is then homogenized with 20 vols. of 0.1 M KCl, 10 mM tris-HCl, pH 7.6 and centrifuged as before. This step is repeated. Actomyosin is extracted for approx. 10 mins. in final solvent conditions of 5 mM ATP, 5 mM EDTA, 10 mM K phosphate, pH 7.6. The viscous solution is centrifuged at 25,000 x g for 20 mins. The supernatant volume is measured and the residue discarded. Up to this stage the preparation of actomyosin and myosin is identical. If actomyosin is required the supernatant is adjusted to 60% saturation with ammonium sulfate, the precipitate collected by centrifugation at 25,000 x g for 10 mins., suspended in 20 mM KCl, 10 mM tris-HCl, pH 7.6, 0.2 mM dithiothreitol, and dialyzed exhaustively against this solvent.

About 20 mg actomyosin per g wet weight gizzard is obtained.

b) Myosin. Our procedure is a modification of the method of Ebashi[13]. The supernatant from the actomyosin extraction (prior to the addition of $(NH_4)_2SO_4$) is adjusted to 0.1 M $MgCl_2$, 10 mM ATP, 15 mM K phosphate, pH 7.6, kept in ice for 20 mins., and centrifuged at 25,000 x g for 20 mins. The precipitate is desensitized actomyosin and is usually discarded. The supernatant is centrifuged at 55,000 x g for 12 hrs. to pellet F-actin. The resultant supernatant is diluted with 10 vols. of cold distilled water, to precipitate myosin, which is collected by centrifugation at $\sim$2,000 x g for 10 mins., dissolved and dialyzed against 0.5 M KCl, 10 mM tris-HCl, pH 7.6, 0.2 mM dithiothreitol. The myosin may be further purified following the washing procedures of Ebashi[13].

c) Actin. Skeletal muscle actin is used throughout, using the procedure described previously[14].

d) Kinase. The procedure is essentially that described by Ebashi et al.[12] for native tropomyosin from gizzards, as modified by Aksoy et al.[6]. Tropomyosin-free kinase was obtained as the supernatant following isoelectric precipitation at pH 4.8 in 0.2 M KCl, 0.2 mM dithiothreitol.

e) Tropomyosin. This was prepared from skeletal and gizzard muscle using the procedures described previously[14,15]. The tropomyosin from both muscle types was purified by chromatography on hydroxyapatite[16].

Proteolysis. Trypsin or α-chymotrypsin was used. Stock solutions at 10 mg/ml were prepared in 1 mM HCl. Tryptic hydrolysis was stopped by the addition of a 20 fold weight excess of soybean trypsin inhibitor. α-Chymotryptic hydrolysis was stopped by the addition of phenylmethylsulfonyl fluoride and diisopropylfluorophosphate, both at final concentrations of 0.5 mM.

Other procedures including: ATPase assays, ^{32}P incorporation and SDS polyacrylamide electrophoresis were as described previously[6,14].

RESULTS AND DISCUSSION

a) The effect of tropomyosin: The activation by tropomyosin of the Mg^{2+}-ATPase activity of gizzard myosin, gizzard kinase plus skeletal actin is shown in Fig. I. Skeletal and gizzard tropomyosins were used and both give essentially the same effect. The activation was maximum at a molar ratio of tropomyosin to actin of about 1:5. This is in the range of the expected stoiciometry of binding of tropomyosin to actin, i.e. 1:7, respectively. The extent of activation by tropomyosin was variable, between 1.5 and 3 fold activation of the control actin-myosin level.

These results suggest that the complex of actin and tropomyosin is a better activator of Mg^{2+}-ATPase activity than actin alone. An alternate hypothesis is that tropomyosin forms a complex with the kinase and that this is more active than the kinase alone. This possibility can be ruled out on consideration of the results shown in Fig. 2. At different concentrations of kinase the activation by tropomyosin was maximum at a fixed stoiciometry of actin to tropomyosin. This effect would not be expected if the interaction of the kinase with tropomyosin played a functional role. The phosphorylation data given in Table I, supports this conclusion, as it is shown that tropomyosin (gizzard and skeletal) does not significantly alter the extent of phosphorylation of gizzard myosin by the kinase.

It is apparent from the above results that the influence of tropomyosin is restricted to a modification of the co-factor, actin. An activation by tropomyosin was observed previously for mammalian skeletal proteins, although this occurred only at low ATP concentrations[17] or with chemically modified myosin[18]. It was suggested that the enhanced ATPase activity was as a result of rigor myosin-actin interactions, the effects of which were potentiated along the actin filament by the tropomyosin molecules[17]. Following this reasoning it was then proposed that the activation by tropomyosin of a Limulus system at higher ATP concentrations[19], was the reflection of a longer contact time of myosin with actin due to the lower rate of ATPase hydrolysis (i.e. a longer cross-bridge cycle). In this respect the actomyosin of smooth muscle and Limulus might be compared as both have relatively low rates of ATP hydrolysis. It was calculated that the cross-bridge cycle in arterial muscle at 21°C is about 0.75 secs.[20] which is about 30 times slower than the cross-bridge cycle of frog sartorius muscle[21,22]. Thus it is reasonable to assume that the absolute contact time of smooth muscle myosin with actin per mole of ATP hydrolyzed is longer than in the faster skeletal muscle systems. However, this by itself may not be the important factor, and we would like to suggest that it is the fraction of the total cross-bridge cycle

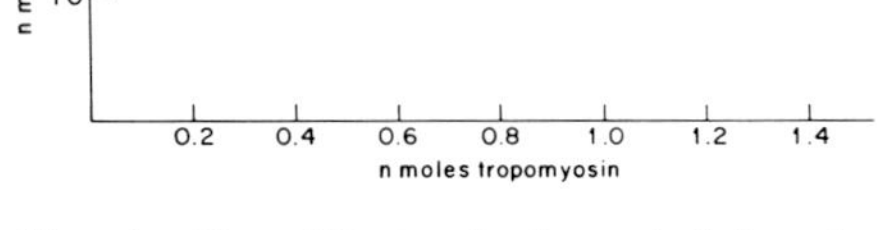

Fig. 1. The effect of gizzard (•) and skeletal muscle (o) tropomyosin on the Mg^{2+}-ATPase activity of gizzard myosin plus skeletal muscle actin. Assay conditions: 10mM $MgCl_2$, 50mM KCl, 10mM tris-HCl (pH7.6), 2.5mM ATP, 4.0nmoles actin, 2.1nmoles myosin. Approx. 0.03mg kinase isoelectric supernatant. Final volume 2ml.

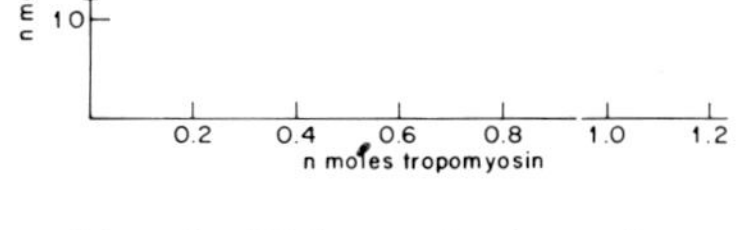

Fig. 2. Effect of gizzard tropomyosin on the Mg^{2+}-ATPase activity of gizzard myosin plus skeletal muscle actin at different levels of kinase. Assay conditions as in Fig. 1. Kinase isoelectric supernatant levels as follows: 0.015mg (▲); 0.03mg (•); 0.12mg (o).

that is spent in myosin-actin contact that is a more relevant consideration. This may be rephrased as follows: at any given time during a contraction the number of cross-bridge contacts will be higher for smooth muscle than for skeletal muscle. Of course this demands that the two kinetic pathways for the hydrolysis of ATP differ in some respects and the validity of our suggestion can be determined only when the details of the enzymic mechanism of smooth muscle actomyosin are established.

b) Proteolysis of the kinase: Tryptic digestion of Ca^{2+}-sensitive gizzard actomyosin destroyed the response to Ca^{2+}. This is shown in Fig. 3, and confirms the earlier results of Ebashi et al.[12]. A significant point that is evident from Fig. 3 is that Ca^{2+}-sensitivity was lost at an ATPase level close to the starting or control activity. If the kinase was destroyed completely then one could predict that the ATPase activity would also be lost, assuming that phosphorylation is a prerequisite for actin activation. Thus a possible explanation for the loss of Ca^{2+}-sensitivity is that limited proteolysis alters the regulatory site of the kinase, without destroying its ability to phosphorylate myosin. An alteration of phosphatase activity is not a reasonable alternative; a loss of phosphatase activity would not influence Ca^{2+}-sensitivity and an increase in activity would lower the Mg^{2+}-ATPase activity of the actomyosin.

The result of hydrolysis of the kinase with α-chymotrypsin is shown in Fig. 4. Limited proteolysis resulted in a loss of Ca^{2+}-sensitivity when the treated kinase was recombined with gizzard myosin and skeletal actin. These results are

Table 1

THE EFFECT OF TROPOMYOSIN (TM) ON THE PHOSPHORYLATION OF GIZZARD MYOSIN

	TM (nmoles)	TM:Actin (molar ratio)	^{32}P incorp. (moles ^{32}P/mole myosin)
Skeletal TM	0	--	1.76 1.79
	0.44	0.03	1.67
	0.88	0.06	1.59
	1.76	0.12	1.66
	3.68	0.24	1.81
	5.59	0.37	1.89
Gizzard TM	0	--	1.81 1.84
	0.74	0.03	1.91
	1.49	0.06	1.64
	2.97	0.12	1.64
	6.15	0.25	1.50
	11.07	0.40	1.64

Conditions of phosphorylations as in Fig. 1. 4 nmoles myosin per assay.

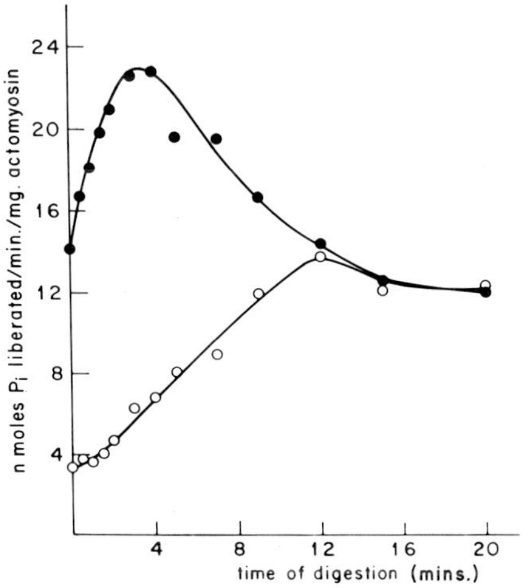

Fig. 3. Effect of tryptic hydrolysis on the Ca^{2+}-sensitivity of gizzard actomyosin. Proteolysis conditions: 1:1500, trypsin: actomyosin, 25°, pH 7.6. Assay conditions as in Fig. 1, Absence (●) and presence (o) of 1mM EGTA. Actomyosin 2.0mg per 2ml assay.

similar to those of Ebashi et al.[12] except that in the latter study skeletal actomyosin was used. Trypsin and α-chymotrypsin gave similar results. The extent of Ca^{2+}-sensitivity that remained with the actomyosin complex following proteolysis of the kinase was variable. In some instances Ca^{2+}-sensitivity was reduced but not completely destroyed, as in the case of the experiment shown in Fig. 4. In other experiments the regulation by Ca^{2+} was lost completely. These variations are the result of kinase contamination in the myosin preparations. It is very difficult to obtain myosin free of both the kinase and the phosphatase, and contamination by the kinase results in a Ca^{2+}-sensitive ATPase response.

The crucial experiment was to determine if the digested kinase still phosphorylated myosin. These results are shown in Fig. 5. With short periods of digestion it can be seen that the extent of phosphorylation in the absence of Ca^{2+} increased markedly, until the level of phosphorylation was the same in the absence and presence of Ca^{2+}. It is clear therefore that the kinase modified by proteolysis can still catalyze the transfer of phosphate groups from ATP to the myosin light chains, but that this kinase has lost the ability to be controlled by Ca^{2+}.

It was of interest to determine if myosin was altered under the conditions of proteolysis used to destroy the regulatory activity of the kinase. No change in the SDS-polyacrylamide patterns of myosin was observed, the heavy chains were not

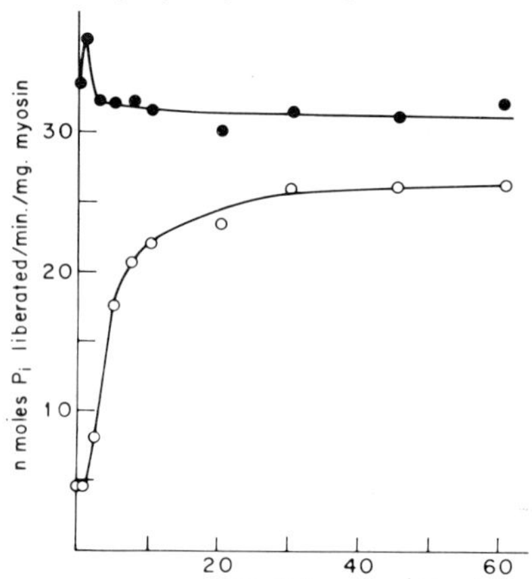

Fig. 4. Effect of α-chymotryptic hydrolysis on the crude kinase and the Ca^{2+}-sensitivity of the actin-activated ATPase activity of gizzard myosin. Proteolysis conditions: 1:3,000, α-chymotrypsin: kinase, 25°, pH7.6. ATPase assay conditions as in Fig. 1. Absence (o) and presence (o) of 1mM EGTA. Gizzard myosin 2.5nmoles, skeletal actin 8.0nmoles, crude kinase 0.27mg. Final volume 2ml.

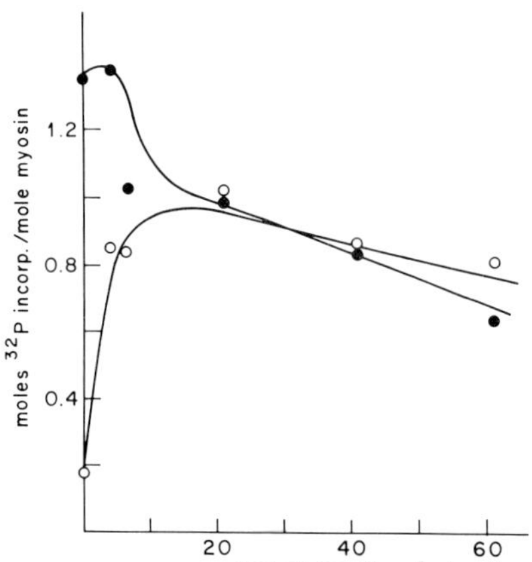

Fig. 5. Effect of α-chymotryptic hydrolysis on the crude kinase and the Ca^{2+}-sensitivity of phosphate incorporation into gizzard myosin. Proteolysis conditions as in Fig. 4. Phosphorylation condition as in Fig. 1. Absence (o) and presence (o) of 1mM EGTA. Gizzard myosin 5nmoles, crude kinase 0.54mg per 2ml assay.

degraded and no diminution of the 20,000 dalton light chain was apparent. The Ca^{2+}-ATPase activity of the digested myosin (the condition of proteolysis as given in Fig. 4) were not altered, and this is an indication that the myosin molecule was not appreciably degraded.[1]

The experiments described above show that the regulation by Ca^{2+} of the kinase activity is not a stable property and is lost following limited proteolysis. This is clearly a point of practical concern and during the isolation of kinase the protease activity of the tissue homogenate should be taken into consideration. The experiments also yield data which is relevant to the mechanism of relaxation. The partially degraded kinase will phosphorylate myosin in the absence of Ca^{2+}, as indicated in Fig. 5. In Fig. 4 it was shown that the ATPase activity in the absence and presence of Ca^{2+} becomes similar. Thus from a consideration of these two factors it can be concluded that the phosphorylated myosin was active in the absence of Ca^{2+}, i.e. the Mg^{2+}-ATPase activity was activated by actin. Since myosin was not subject to proteolysis the Ca^{2+}-myosin interaction could not be altered or disturbed in any way. It is obvious that phosphorylated myosin is active regardless of the Ca^{2+} concentration, and it is unlikely that Ca^{2+} binding to myosin plays a role in the regulation of Mg^{2+}-ATPase activity of gizzard actomyosin. The dominant controlling factor is phosphorylation.

SUMMARY

The activating effect of tropomyosin on the Mg^{2+}-ATPase activity of gizzard myosin plus skeletal actin is due to the tropomyosin-actin interaction. The complex of actin-tropomyosin is a better activator of ATPase activity than actin alone. Tropomyosin had no influence on the level of myosin phosphorylation. No difference was detected between gizzard and skeletal tropomyosins.

Limited proteolysis of gizzard actomyosin resulted in a loss of Ca^{2+}-sensitivity. The effect was traced to the proteolysis of the kinase which remained active but lost its ability to be regulated by Ca^{2+}. Since the ATPase activity of the phosphorylated myosin was not inhibited on the removal of Ca^{2+}, this suggested that Ca^{2+} binding to myosin does not play a dominant role in the regulatory mechanism, and thus the level of phosphorylation of myosin is a more crucial factor.

ACKNOWLEDGMENTS

This work was supported by the National Institutes of Health (Grant HL-09544)

---

[1]The Ca^{2+}-ATPase activity was measured at low and high ionic strength ($I \sim 0.03$ and ~ 0.6, respectively); with normal myosin the Ca^{2+}-ATPase activity is much higher at high ionic strength. This effect disappears when myosin is altered by proteolysis, as the low ionic strength ATPase level is increased.

REFERENCES

1. Head, J. F., Weeks, R. A. and Perry, S. V. (1977) Biochem. J. 161, 465-471.
2. Mikawa, T., Toyo-Oka, T., Nonomura, Y. and Ebashi, S. (1977) J. Biochem. 81, 273-275.
3. Sobieszek, A. (1977) Vertebrate Smooth Muscle Myosin. Enzymatic and Structural Properties. In: The Biochemistry of Smooth Muscle. N. L. Stephens (Ed.) pp. 413-443, University Park Press, Baltimore.
4. Bremel, R. D., Sobieszek, A. and Small, J. V. (1977) Regulation of Actin-Myosin Interaction in Vertebrate Smooth Muscle. In: The Biochemistry of Smooth Muscle. N. L. Stephens (Ed.), pp. 533-549, University Park Press, Baltimore.
5. Chacko, S., Conti, M. A. and Adelstein, R. S. (1977) Proc. Natl. Acad. Sci., U.S. 74, 129-133.
6. Aksoy, M. O., Williams, D., Sharkey, E. M. and Hartshorne, D. J. (1976) Biochem. Biophys. Res. Commun. 69, 35-41.
7. Gorecka, A., Aksoy, M. O. and Hartshorne, D. J. (1976) Biochem. Biophys. Res. Commun. 71, 325-331.
8. Morimoto, K. and Harrington, W. F. (1974) J. Mol. Biol. 88, 693-709.
9. Bremel, R. D. and Weber, A. (1975) Biochim. Biophys. Acta, 376, 366-374.
10. Hartshorne, D. J., Abrams, L., Aksoy, M., Dabrowska, R., Driska, S. and Sharkey, E. (1977) Molecular Basis for The Regulation of Smooth Muscle Actomyosin. In: The Biochemistry of Smooth Muscle. N. L. Stephens (Ed.), pp. 513-532, University Park Press, Baltimore.
11. Sobieszek, A. and Small, J. V. (1976) J. Mol. Biol. 102, 75-92.
12. Ebashi, S., Iwakura, H., Nakajima, H., Nakamura, R. and Ooi, Y. (1966) Biochem. Z. 345, 201-211.
13. Ebashi, S. (1976) J. Biochem. 79, 229-231.
14. Driska, S. and Hartshorne, D. J. (1975) Arch. Biochem. Biophys. 167, 203-212.
15. Hartshorne, D. J. and Mueller, H. (1969) Biochim. Biophys. Acta 175, 301-319.
16. Eisenberg, E. and Kielley, W. W. (1974) J. Biol. Chem. 249, 4742-4748.
17. Bremel, R. D., Murray, J. M. and Weber, A. (1972) Cold Spring Harbor Symp. Quant. Biol. 37, 267-275.
18. Shigekawa, M. and Tonomura, Y. (1972) J. Biochem. 71, 147-149.
19. Lehman, W. and Szent-Györgyi, A. G. (1972) J. Gen. Physiol. 59, 375-387.
20. Paul, R. J., Glück, E. and Rüegg, J. C. (1976) Pflügers Arch. 361, 297-299.
21. Curtin, N. A., Gilbert, C., Kretzschmar, K. M. and Wilkie, D. R. (1974) J. Physiol. 238, 455-472.
22. Paul, R. J. and Kushmerick, M. J. (1974) Biochim. Biophys. Acta 347, 483-490.

Excitation-Contraction Coupling in Smooth Muscle, R. Casteels et al. eds.
© 1977 Elsevier/North-Holland Biomedical Press

MYOSIN PHOSPHORYLATION AND CA-REGULATION IN VERTEBRATE SMOOTH MUSCLE

J.V. Small and Apolinary Sobieszek

Institute of Molecular Biology of the Austrian Academy of Sciences, 5020 Salzburg, Austria and Cardiology Branch, National Heart, Lung and Blood Institute, NIH Bethesda MD 20014, U.S.A.

INTRODUCTION

In common with certain invertebrate muscles[1,2] Ca-regulation in vertebrate smooth muscle is associated with the myosin molecule and in both cases this regulation is mediated via at least one of the myosin light chain components. It has also been shown that the 20,000 M.W. light chain from vertebrate smooth muscle can functionally replace the scallop light chain in scallop myofibrils[3]. But the results from recent studies also indicate that the details of the regulation process in these two systems may be fundamentally different. In this report we shall summarise data[4,5,6] obtained from two different vertebrate smooth muscles which provides strong evidence for Ca-regulation being mediated via a Ca-dependent phosphorylation and Ca-independent dephosphorylation of the myosin regulatory light chain. A similar phosphorylation-dephosphorylation process apparently does not take place in scallop but may well operate in non-muscle systems for which the myosins show a similar solubility and subunit composition to smooth muscle myosin [7,8].

The myosin light chain kinase

In purified chicken gizzard actomyosin the 20,000 M.W. light chain of myosin is the only component which shows a significant incorporation of inorganic phosphate on incubation with γ ^{32}-P ATP (Fig. 1). This phosphorylation, which is Ca^{2+}-dependent can be attributed to the presence of a specific myosin light chain kinase. Such a kinase, specific for the 20,000 M.W. light chain of myosin was first demonstrated in skeletal muscle[9] and was subsequently found to be present in cardiac[10] and in smooth muscle[4,11,12,13].

Using a light chain fraction[14] or the purified myosin 20,000 M.W. (L_{20}) light chain as substrate we found the light chain kinase to copurify in different amounts with gizzard actomyosin and the seperate contractile proteins[6]. Relative to myosin, gizzard actomyosin contained

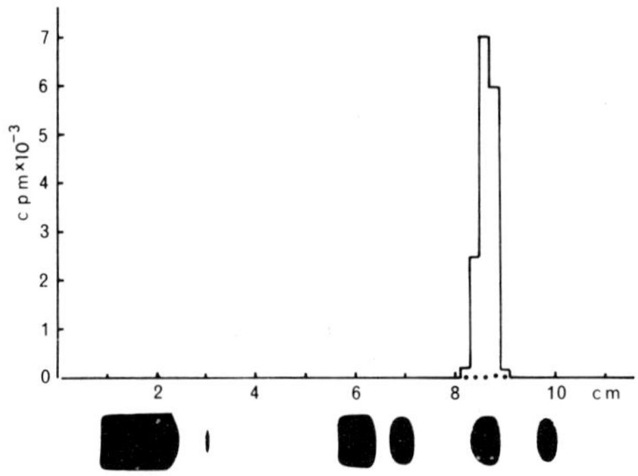

Fig. 1. Phosphate incorporation in gizzard actomyosin in the presence of Ca. The only component that becomes phosphorylated in the presence of (Mg) ATP and Ca is the 20,000 M.W. light chain of myosin. (From Sobieszek,1977).

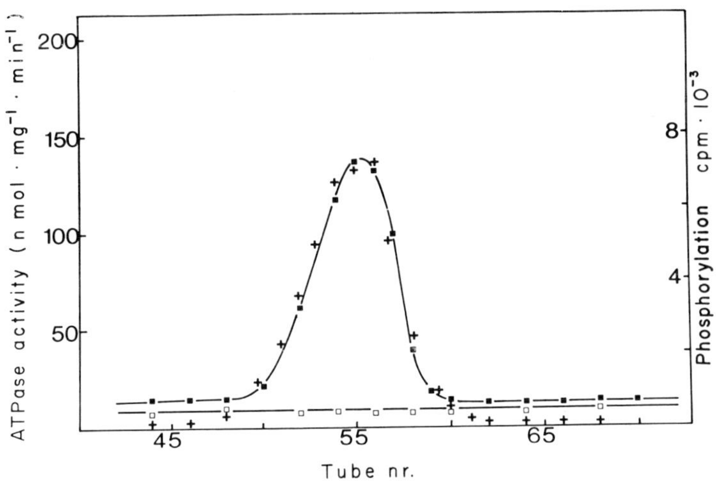

Fig. 2. Activation of the ATPase activity of dephosphorylated pig stomach actomyosin by the Sepharose 6B column fractions of the myofibril extract obtained from turkey gizzard (squares). Assays of the kinase activities of the same fractions using the isolated myosin light chain as substrate (crosses) showed this to closely parallel the ATPase activity. In this and the following figures closed and open symbols correspond to the activities in the presence and absence of Ca respectively. (From Sobieszek & Small, 1977)

6.5 times as much kinase and the thin filaments about half as much, per mg protein. Extracts showing very high kinase activity, 200-fold more active with respect to myosin, could be obtained by the extraction of homogenised and washed gizzard muscle (myofibrils) in the absence of ATP. By fractionation of such extracts on sepharose 6B the kinase was partially purified and could be stored at 4°C without an appreciable loss of activity for several days. Freezing or storage with antifreeze resulted in an irreversible loss in activity.

In contrast to the situation with gizzard the myosin light chain kinase from pig stomach was not co-purified with pig stomach actomyosin, a difference which proved useful in studies of the effect of the kinase on actin-activated ATPase activity.

Actin activated ATPase activity and myosin phosphorylation

The link between myosin phosphorylation and the actin activated ATPase activity of actomyosin could be shown directly for both chicken gizzard and pig stomach. This relationship was rather clearly illustrated using kinase-free pig stomach actomyosin purified by precipitation with high concentrations of $CaCl_2$ (20 mM). As purified, this actomyosin contained the myosin light chain in the dephosphorylated state and without added kinase showed negligible actin-activated ATPase activity (see[5]). Fig. 2 shows the results of an experiment in which the ATPase activity of this actomyosin was measured as a function of added column fractions obtained by gel filtration of the kinase-rich extract of myofibrils. Parallel measurements were also made of the kinase activity of the fractions using the isolated light chain. As shown in Fig. 2 the magnitude of the actin activated ATPase activity with Ca^{2+} present parallelled precisely the rate of phosphorylation of the myosin light chain.

The same type of effect could also be shown using reconstituted gizzard actomyosin from which both the kinase and tropomyosin were removed by ammonium sulphate fractionation[6]. Fig. 3 shows an experiment in which such kinase- and tropomyosin- free actomyosin was recombined with increasing amounts of the tropomyosin - kinase fraction obtained as a byproduct from the parent actomyosin. Again, in the presence of Ca^{2+}, the measured actin-activated ATPase activity parallelled the rate of phosphate incorporation into the myosin light chain. In this case the phosphorylation of the light chain was monitored directly using the recombination mixtures. As shown in Fig. 3 the phosphorylation rose more rapidly as a function of the added kinase than the ATPase activity. When tropomyosin alone was added in this type of experiment there was no phosphorylation or activation of ATPase activity.

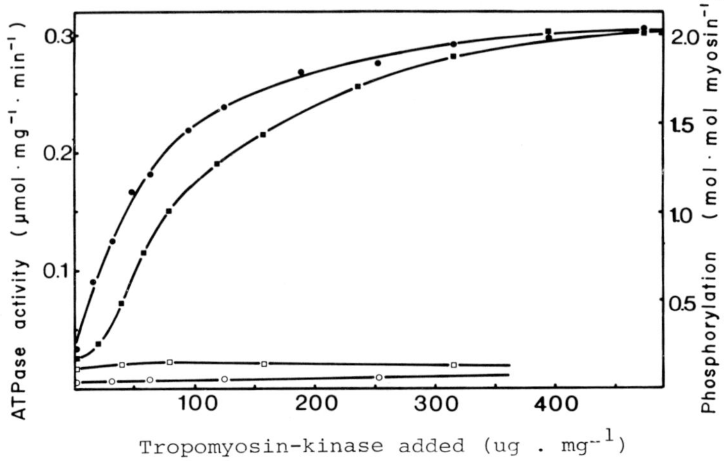

Fig. 3. Myosin phosphorylation (circles) and actin-activated ATPase activity (squares) for gizzard actomyosin. In this case increasing amounts of a tropomyosin-kinase fraction (see text) were added to a fixed amount of tropomyosin -and kinase- free actomyosin. (From Sobieszek, 1977 and Sobieszek & Small, 1977).

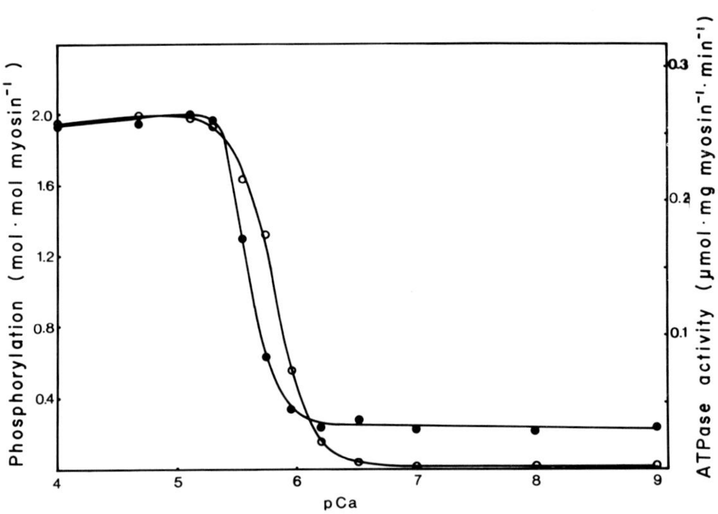

Fig. 4. Dependence of ATPase activity (●-●) and myosin light chain phosphorylation (o-o) on free Ca concentration for gizzard actomyosin. (From Sobieszek, 1977).

Using a Ca EGTA/EGTA buffer system the phosphorylation of the myosin light chain was further found to show the same dependence on free Ca^{2+} concentration as the ATPase activity of Ca-sensitive actomyosin (Fig. 4).

Measurement of the time course of phosphorylation of the myosin L_{20} light chain in gizzard actomyosin showed that the saturation of phosphate incorporation occurred within about 10 sec[4] with approximately 1 mol of P_i being bound per mol light chain. For the most active kinase preparations with the isolated L_{20} light chain as substrate phosphate incorporation was more rapid and reached saturation within 1-2 seconds.

Myosin dephosphorylation and relaxation

Using gizzard actomyosin we found in our initial experiments that while Ca^{2+} was required for the stimulation of actin-activated ATPase activity its subsequent removal did not affect the rate of splitting of ATP. This type of experiment is illustrated in Fig. 5. One series of points (●-●) shows the time course of P_i liberation with actomyosin in the presence of a CaEGTA/EGTA buffer (2×10^{-5}M free Ca^{2+}). The second series (triangles) shows data for a corresponding experiment in which excess EGTA (2mM) was added after 40 sec of incubation and the assay continued for different time periods thereafter, up to 80 sec. As seen in the diagram there was no measurable change in the rate of ATP splitting following Ca^{2+} removal. The actomyosin was therefore Ca-sensitive with regard to activation but was not deactivated by Ca-removal. This property could subsequently be explained by the retention of the light chain kinase during actomyosin purification and the corresponding loss during preparation of a light chain phosphatase. The presence of a light chain phosphatase in smooth muscle and the effect of the light chain dephosphorylation on actin-activated ATPase activity was shown in studies using pig stomach actomyosin.

While pig stomach actomyosin precipitated with high concentrations of $CaCl_2$ was obtained in a dephosphorylated state the use of micromolar concentrations of $CaCl_2$ for precipitation yielded an actomyosin in which the myosin light chain was more or less fully phosphorylated[5]. This actomyosin, which lacked any kinase activity, showed a high actin-activated ATPase activity but was essentially Ca-insensitive. Fig. 6 shows the result of an experiment of the type already illustrated in Fig. 2 using this phosphorylated actomyosin. For this experiment the phosphorylated actomyosin was combined with equal volumes of the column fractions from the myofibril extract and each mixture assayed for

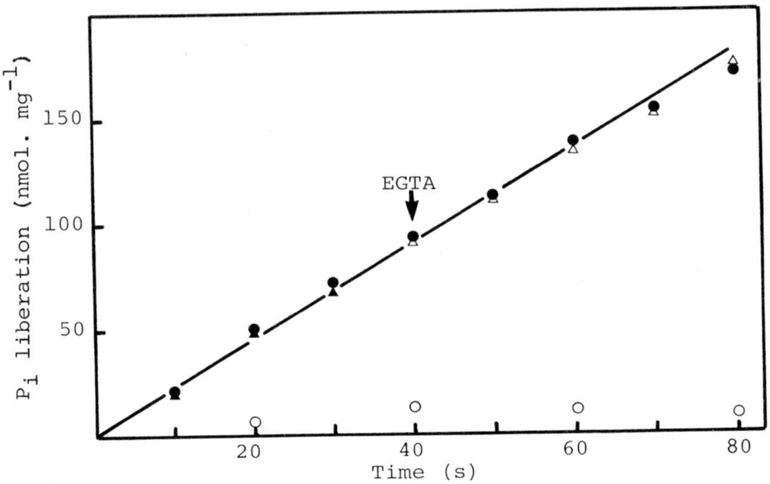

Fig. 5. Experiments showing effect of Ca removal on phosphorylated gizzard actomyosin. Closed circles show phosphate liberation in the presence of Ca ($2x10^{-5}M$) from 0 to 80 seconds. Triangles show the same experiment except that Ca was removed by the addition of EGTA (2mM) after 40 sec of incubation. Open circles show corresponding assays for which EGTA was added prior to initiation of the assay with ATP. As shown the removal of Ca at 40 sec did not affect the rate of ATP hydrolysis. Other measurements showed that the extent of myosin phosphorylation was also unchanged.

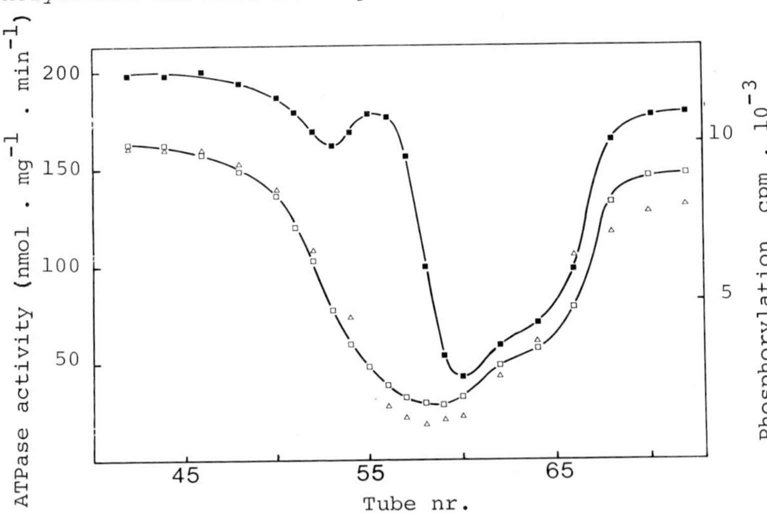

Fig. 6. Experiment similar to that in Fig. 2 showing the effect of the Sepharose 6B fractions of a gizzard myofibril extract on phosphorylated pig stomach actomyosin. Inhibition of the actomyosin ATPase activity in the absence of Ca coincided with the measured dephosphorylation of ^{32}P labelled myosin light chain by the same fractions (triangles). For details see Small & Sobieszek (1977).

ATPase activity in the presence of EGTA or Ca^{2+}.As shown, there was an inhibition of the ATPase activity with EGTA in a broad dip, with its centre shifted slighty to the right of the"Ca^{2+} peak"(i.e. the peak of kinase activity- Fig.2).In the presence of Ca^{2+} there was also an inhibition of the ATPase activity outside and to the right of the kinase peak. Using $\gamma\,^{32}$-P ATP-labelled phosphorylated light chain as substrate the fractions were also assayed for phosphatase activity. As shown in Fig.6 the measured phosphatase activity in the column fractions of the myofibril extract overlapped the dip in ATPase activity. This data thus provided evidence for a link between light chain dephosphorylation via a phosphatase and the deactivation or "relaxation" of the actin-myosin interaction.

DISCUSSION

From the present studies we have been able to demonstrate a direct correlation between the extent of phosphorylation of the myosin L_{20} light chain and the level of actin activated ATPase activity. This is shown both from the data on the phosphorylation of dephosphorylated actomyosin via the kinase and from that on the dephosphorylation of phosphorylated actomyosin via the phosphatase.The results thus point clearly to a kinase-phosphatase type regulatory mechanism in which the Ca^{2+}-sensitivity derives from the Ca^{2+}-dependence of the kinase.

A link between myosin phosphorylation and actin-activited ATPase activity in smooth muscle has also been indicated in the studies of Gorecka et al.[12] and Chacko et al.[13]. But the latter authors also claimed that their"phosphorylated myosin" from guinea pig vas deferens was Ca-sensitive with respect to its actin-activated ATPase activity, a result which is at variance with our own. As we have shown, phosphorylated actomyosin from pig stomach or chicken gizzard(Fig. 5) was Ca-insensitive.

While the phosphorylation of myosin is obligatory for the interaction of actin and myosin to take place it is also clear that tropomyosin plays an important role in potentiating the actin-activitated ATPase activity. As we have shown elsewhere[6] tropomyosin causes a 2-3 fold amplification of the activity exhibited by phosphorylated myosin and actin alone. A similar result was also obtained by Chacko et al.[13]. As with striated muscle [15,16] the F-actin-tropomyosin complex is a better substrate for myosin than actin alone.

The demonstration of a deactivation of the actin-myosin interaction by dephosphorylation of the myosin light chain suggests rather strongly that dephosphorylation via a phosphatase leads to the relaxation of smooth muscle in *vivo*. This conclusion must be reconciled, however, with the observed rates of phosphorylation and dephosphorylation of the myosin light chain. As we have noted the phosphorylation process is very rapid, with a time course which correlates well with the speed of contraction of smooth muscle. The dephosphorylation process, on the other hand is very slow and under the conditions we have used takes between 10 and 20 minutes at 25°C. Such a time course is observed even with extracts of whole muscle mince which we presume contain a large proportion of the endogenous phosphatase. Changes of pH from 6.0 to 8.0 did not dramatically alter the rate of dephosphorylation. If the light chain dephosphorylation is responsible for relaxation during phasic contraction of smooth muscle then the phosphatase must be considerably more active *in vivo*. How the activities of the kinase and phosphatase are modulated in vivo -in different smooth muscles and during tonic and phasic contraction- are clearly questions whose answers are fundamental to an understanding of the contractility of smooth muscle tissue.

We are grateful to Mrs. M.R. Small for help in preparation of the manuscript and to the Volkswagen Foundation and Muscular Dystrophy Association for generous financial support.

REFERENCES

1. Szent-Gyorgyi, A.G. (1975). Biophys. J. 15, 707-723.
2. Lehman, W. (1976). Internat. Rev. Cytol. 44, 55-92.
3. Kendrick-Jones, J., Szentkiralyi, E. and Szent-Györgyi, A.G.(1976). J. Mol. Biol. 104, 747-775.
4. Sobieszek, A. (1977). Eur. J. Biochem. 73, 477-483.
5. Small, J.V. and Sobieszek, A. (1977). Eur. J. Biochem. 76, 521-530.
6. Sobieszek, A. and Small, J.V. (1977). J. Mol. Biol. 112, 559-576.
7. Adelstein, R.S. and Conti, M.A. (1974) Exploratory Concepts in Muscular Dystrophy II (Milhorat, A.T., ed.)pp- 70-78, Excerpta Medica, Amsterdam.
8. Burridge, K. and Bray, D. (1975) J. Mol. Biol. 99, 1-14.
9. Perrie, W.T., Perry, S.T. and Thomas, M.A.W. (1974). FEBS, Letters 41, 292-296.
10. Frearson, N., Focant, B.W.W. and Perry, S.V. (1975) FEBS Letters 63, 27-32.
11. Frearson, N., Focant, B.W.W. and Perry, S.V. (1976). FEBS Letters 63, 27-32.
12. Gőrecka,A., Aksoy, M.O and Hartshorne, D.J. (1976). Biochem. Biophys. Res. Com. 71, 325-331.
13. Chacko, S., Conti, M.A. and Adelstein, R.S. (1977). Proc. Natl. Acad. Sci. USA. 74, 129-133.
14. Perrie, W. T., Smillie, L.B. and Perry, S.V. (1973).Biochem. J. 135, 151-164.
15. Bremel, R.D., Murray, J.M. and Weber, A. (1973). Cold Spring Harbor Symp. Quant. Biol. 37, 267-275.
16. Lehman, W. and Szent-Györgyi, A.G. (1972) J. Gen. Physiol. 59, 375-387.
17. Sobieszek, A. (1977)in "The Biochemistry of Smooth Muscle"(Stephens, N.L., ed) University Park Press, Baltimore.

Excitation-Contraction Coupling in Smooth Muscle, R. Casteels et al. eds.

GENERAL DISCUSSION

PHOSPHORYLATION AND THE REGULATION OF CONTRACTION IN SMOOTH MUSCLE

S.V. Perry
Department of Biochemistry
University of Birmingham
Birmingham B15 2TT, U.K.

The papers presented in the preceding session were almost entirely concerned with the role of phosphorylation of 20,000 dalton light chain of myosin, the P-light chain, in the regulation of contractile activity of the smooth muscle. It is important, however, to remember that other proteins of the contracto-regulatory system of muscle, namely troponin T, troponin I (1-4) and tropomyosin (5) have been shown, in striated muscle at least, to exist in phosphorylated forms.

Limited information is available about the enzymic mechanisms responsible for the phosphorylation of tropomyosin and troponin T but it is well established that the enzyme involved in the phosphorylation of the P-light chain of myosin, myosin light chain kinase, requires Ca^{2+} and hence its activation will be associated with excitation. On the other hand, the phosphorylation of troponin I in cardiac muscle in particular is catalysed by 3':5'-cyclic AMP-dependent protein kinase and is therefore under catecholamine control. In the case of cardiac muscle, phosphorylation of the troponin I leads to a decrease in the Ca^{2+} sensitivity of the Mg^{2+}-stimulated ATPase of natural actomyosin (6,7). If a similar system was present in smooth muscle, activation of protein kinase by an increase in the 3':5'-cyclic AMP concentration would be expected to bring about a fall in ATPase activity. Such a mechanism could explain the decrease in tension produced by the action of 3':5'-cyclic AMP on smooth muscle. Thus if the myosin and the troponin I phosphorylation systems are operative in intact smooth muscle they clearly will play important roles in the response of the tissue to changing concentrations of Ca^{2+} and 3':5'-cyclic AMP.

Although the evidence is now strong that contraction in smooth muscle results from the rise in intracellular Ca^{2+}, the actual step at which Ca^{2+} acts on the proteins of the contracto-regulatory system is by no means clear. The scheme illustrated in Fig.1 does not represent the processes occurring in any one muscle type but incorporates all the reactions involving Ca^{2+} in the regulation of contraction that have been described to date.

Two types of control are usually distinguishable. That involving the troponin system which is widely distributed in the thin filaments of vertebrate and invertebrate muscles, and the myosin-regulated system located in the thick filaments. The latter has best been characterized in molluscan adductor

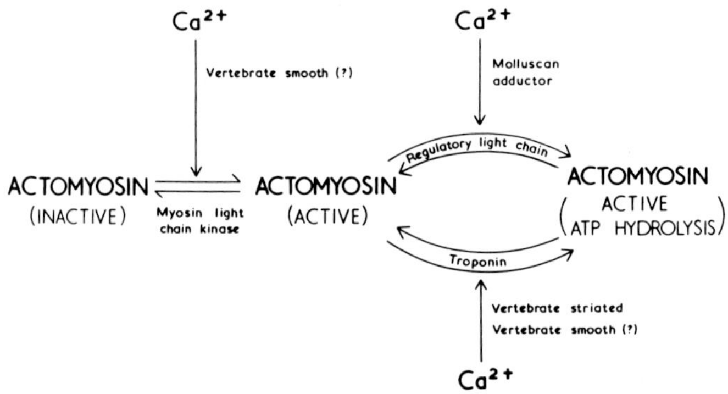

Fig.1 Scheme illustrating the mechanisms proposed for the Ca^{2+} regulation of ATPase hydrolysis and hence contraction in different muscle types. By 'active' actomyosin it is implied that the actin present is potentially able to activate the Mg^{2+}-stimulated ATPase of myosin whereas in 'inactive' actomyosin it cannot. In 'active' actomyosin of smooth muscle the myosin light chain of 20,000 molecular weight is phosphorylated, although this is still a matter of some controversy (see text).

muscle (8) although there is some evidence that myosin-linked regulation may also occur in rabbit skeletal muscle (9).

The demonstration that Ca^{2+} binding directly to myosin regulates the contraction of molluscan smooth adductor muscle coupled with the difficulties in identifying a troponin system have been taken to imply that vertebrate smooth muscle may also be myosin regulated, a view that has been supported by a number of experimental studies some of which have been described to-day. There is, however, no a priori reason for assuming that myosin regulation is the only type of control present in vertebrate smooth muscle. Indeed evidence is accumulating that both systems exist in vertebrate and invertebrate muscle systems. Further it would not be surprising if several regulatory systems were involved for it is general experience that more than one mechanism is required for the control of most biological systems.

An increase in the intracellular Ca^{2+} concentration will lead to the binding of the Ca^{2+} to troponin C in vertebrate striated muscles and to the EDTA light chain of molluscan myosin. The binding of Ca^{2+} to both these proteins leads to rapid conformational changes that enable actin to interact with myosin thus causing immediate hydrolysis of ATP. The effect of the binding of Ca^{2+} to the myosin light chain kinase is rather different in that it activates the enzyme which in turn phosphorylates the myosin light chain rather than directly switching on the ATPase activity. This type of response would be expected to be slower than that obtained through the troponin system and its speed will depend on the amount of enzyme present, which in smooth muscle is about one-tenth of that in skeletal muscle (10). If such a process is essential for actin activation of the Mg^{2+}-stimulated ATPase of myosin then it would override the troponin switch if the latter was present. If it is not, the myosin light chain kinase could act as an enzymic switch with somewhat different characteristics to the troponin system. The combined action of the two systems would provide a means by which an actomyosin ATPase system of variable V_{max} could be switched on or off by changes in Ca^{2+} concentration.

Evidence that phosphorylation of the P-light chain is essential for actin activation of the Mg^{2+}-stimulated ATPase of smooth muscle myosin has been reported by several speakers to-day. In this respect myosin from striated muscle appears to differ from that of smooth muscle as the Mg^{2+}-stimulated ATPases of myosins from skeletal and cardiac muscles are activated by actin when the myosin is completely dephosphorylated (11,12). There does not, however, appear to be complete agreement between the speakers about the effects of phosphorylation. On the one hand, Drs. Adelstein, Chacko, Hartshorne, Mrwa, Small and Sobieszek report that phosphorylation of the P-light chain is necessary for the actin activation of the Mg^{2+}-stimulated ATPase of smooth myosin whereas Dr. Ebashi reports that phosphorylation is not essential providing that an activator protein of 80,000 molecular weight is present.

Dr. Mrwa makes the interesting suggestion that the dephosphorylated P-light chain is an inhibitor of the Mg^{2+}-stimulated ATPase but that it is not essential for actin activation for both she and Drs. Marston and Taylor get actin activation of subfragment I from smooth muscle in which the P-light chain normally is missing. If this hypothesis is correct it follows that phosphorylation destroys the P-light chain's inhibitory activity. Irrespective of the mechanisms involved, all the contributors, with the exception of Dr. Ebashi, are in agreement as to the need for phosphorylation of the P-light chain to obtain actin activation of smooth muscle myosin in which the light chains are apparently intact. Differences exist, however, as to whether the actin-activated myosin ATPase is Ca^{2+} sensitive once phosphorylation has taken place. Dr. Adelstein and collaborators using actomyosin from vas deferens and platelets, and Dr. Mrwa using acto-subfragment I from artery in the presence of the P-light chain, find that their systems are Ca^{2+} sensitive after phosphorylation, observations which are not reported by workers on other smooth muscle systems.

Although Dr. Ebashi's results differ somewhat from the others presented at this meeting, if the 80,000 molecular weight protein in some way acted by neutralising the proposed inhibitory effect of the P-light chain, his findings would agree with those of Dr. Adelstein and collaborators and Dr. Mrwa. Indeed, the activator protein could be present in smooth muscle actomyosin systems and not readily detected because it is effective in relatively small amounts. The 80,000 dalton protein is not identical with myosin light chain kinase nor does the latter enzyme or myosin light chain phosphatase have to be present to produce the actin activation or Ca^{2+} sensitivity in Dr. Ebashi's system.

The lack of complete agreement between the results obtained by different workers clearly indicates that the processes involved in regulating the Mg^{2+}-stimulated ATPase of smooth muscle actomyosin have not been precisely defined. The need now is to characterize the systems used with more precision so that they will give similar results in different hands and provide a basis of accepted facts from which progress can be made towards understanding the mechanisms involved. It is difficult to come to a firm conclusion at this stage, but all the indications are that the system regulating the response of the contracto-regulatory system of smooth muscle to Ca^{2+} is complex and that several mechanisms probably operate, only one of which may involve the phosphorylation of myosin.

References

1. Bailey, C. & Villar-Palasi, C. (1971) Fed.Proc.Amer.Soc.Exp.Biol. 30, 1147
2. Stull, J.T., Brostrom, C.O. & Krebs, E.G. (1972) J.Biol.Chem. 247, 5272-5274
3. Pratje, E. & Heilmeyer, L.M.G. (1972) FEBS Letts. 27, 89-93
4. Perry, S.V. & Cole, H.A. (1973) Biochem.J. 131, 425-428
5. Ribolow, H. & Barany, M. (1977) Arch.Biochem.Biophys. 179, 718-720
6. Solaro, R.J., Moir, A.J.G. & Perry, S.V. (1976) Nature, 262, 615-617
7. Ray, K.P. & England, P.J. (1976) FEBS Letts. 70, 11-16
8. Kendrick-Jones, J., Szentkiralyi, E.M. & Szent-Gyorgyi, A.G. (1976) J.Mol. Biol. 104, 747-775
9. Lehman, W. (1977) Biochem.J. 163, 291-296
10. Frearson, N., Focant, B.W.W. & Perry, S.V. (1976) FEBS Letts. 63, 27-32
11. Morgan, M., Perry, S.V. & Ottaway, J. (1976) Biochem.J., 157, 687-697
12. Cole, H.A. & Perry, S.V. (Unpublished data).

Excitation-Contraction Coupling in Smooth Muscle, R. Casteels et al. eds.

STIMULATION OF ADENYLYL CYCLASE IN CORONARY SMOOTH MUSCLE BY ADENOSINE [+]

W.R. Kukovetz, A. Wurm, I. Rinner, S. Holzmann, G. Pöch
Institut für Pharmakodynamik und Toxikologie
der Universität A-8010 Graz / Austria

INTRODUCTION

Since previous results from our laboratory showed a close association between the relaxing effects and rises in cyclic AMP (cAMP)-levels, induced by adenosine in coronary arterial strips (hog, cattle), it was suggested that cAMP mediated this relaxation [1,2]. A dissociation found between relaxing (10^{-6}M) and cAMP-increasing concentrations of adenosine (10^{-4}M) in hog carotid arteries lead Herlihy et al. [3] to believe that cAMP was not involved in the action of adenosine. A similar conclusion was reached by McKenzie et al. [4], who did not find any stimulant effect of adenosine or isoprenaline on adenylyl cyclase (AC) from rabbit intestinal smooth muscle.

This paper provides further evidence in favor of a mediator function of cAMP in adenosine-induced relaxation of coronary arterial smooth muscle.

MATERIALS AND METHODS

a) <u>Coronary strips</u> (organbath): Circular strips of coronary arteries [5] from cattle and pig were suspended in a modified tyrode solution with 27 mM K^+. Relaxation was recorded by a mechano-electronic transducer and expressed as percent of maximum [6]. Cyclic AMP was determined in freeze-clamped strips by RIA [7].

b) <u>Adenylyl cyclase</u> (AC): Minced coronary arteries (cattle, pig) were homogenized with 10 mM Tris-HCl (pH = 7.5) for 30 sec in a microdismembrator. The washed 1000 x g particulate fraction, suspended in Tris-HCl, was used for the assay [8] (ATP regenerating system, 0.075 mM ATP, 8 mM papaverine). Cyclic AMP was determined by RIA [7].

+ This work was supported by the Austrian Research Council (Österreichischer Forschungsrat: Nr. 2857).

RESULTS

1. Correlation between relaxation and cAMP

Adenosine dose dependently induced relaxation of coronary strips from cattle and pig, which was accompanied by corresponding log increases in cAMP [8] ($r = 0.95$; $p < 0.05$; $N = 9$). When plotted against a probit ordinate the same and highly significant correlation was found between both parameters with adenosine and isoprenaline, and this was also seen when adenosine or isoprenaline were administered in the presence of papaverine or theophylline (fig. 1). With a linear ordinate (percent relaxation) a typical sigmoid dose response curve is obtained [8].

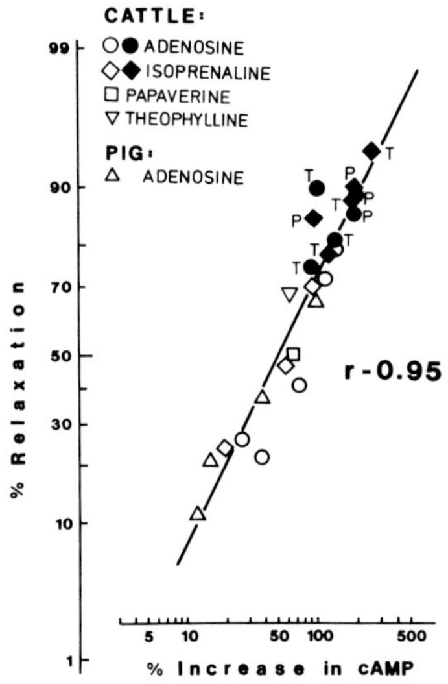

Fig. 1. Isolated circular strips from coronary arteries of cattle and pig heart: Significant correlation ($r = 0.95$; $p < 0.01$) between percent relaxation (ordinate: probit scale) and log increases in cAMP (abscissa), produced by 0.3 - 1500 µM adenosine, 4 - 40 nM isoprenaline, 11 µM papaverine, 1.1 mM theophylline, as well as isoprenaline and adenosine in the presence of papaverine (P) or theophylline (T). Linear regression line calculated from mean values (N = 6 - 66).

2. Influence of ß-adrenergic blockade on the effects of adenosine and isoprenaline

The adrenergic ß-blocking drug sotalol neither affected the relaxing nor the cAMP-increasing effect of 7.5×10^{-4}M adenosine at concentrations of 1.6×10^{-4}M, which completely abolished the actions of equieffective doses of 4×10^{-8}M isoprenaline on both parameters (fig. 2). These results clearly exclude any direct or indirect stimulatory effect of adenosine on adrenergic ß-receptors.

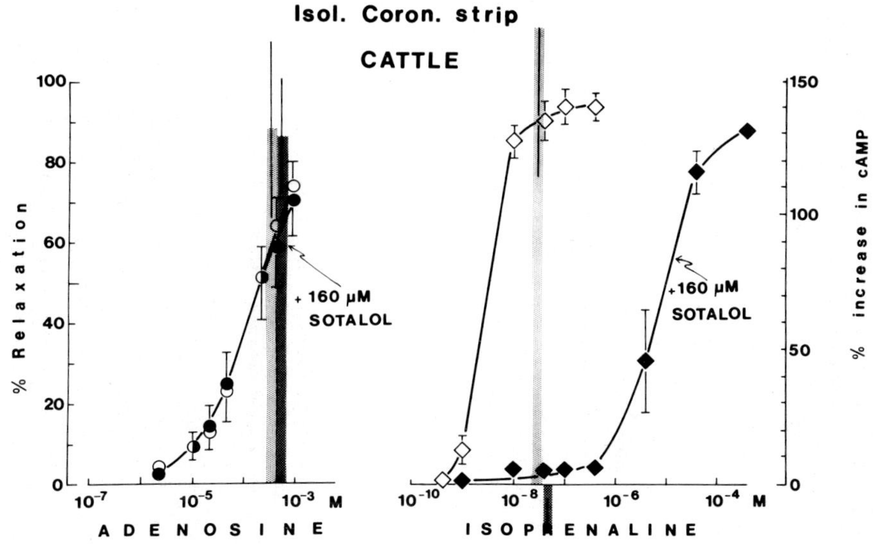

Fig. 2. Isolated coronary strips (cattle): Influence of ß-blockade. Relaxant and cAMP-increasing effects of adenosine (left) and isoprenaline (right) in the absence (open symbols, first columns) and in presence of 160 µM sotalol (filled symbols, second columns). Mean values ± S.E.M. of 3 - 8 experiments.

3. Influence of theophylline on adenosine and isoprenaline effects

It has been reported repeatedly that theophylline antagonizes adenosine induced relaxation, and Bünger et al.[9] showed that this effect was clearly competitive. Although such an action would be opposed and might be obscured by the inhibitory effect of theophylline on phosphodiesterase (PDE), it was investigated whether or not such an antagonism (at an adenosine receptor site) was demonstrable also by appropriate effects on cAMP.

Theophylline at 1.1×10^{-3}M concentration, which by itself decreased the tone of bovine coronary arteries by 61 ± 2.8 % (N = 12), attenuated the relaxant as well as the cAMP-increasing effect of adenosine, whereas similar actions of isoprenaline were enhanced (fig. 3). The latter effect of theophylline is due to its inhibitory effect on PDE (K_i = 0.1 mM [5]).

These results indicated that adenosine stimulates adenylyl cyclase by a specific receptor which can be blocked by theophylline. The interactions between adenosine and theophylline at the AC level were, therefore, studied more closely in vitro.

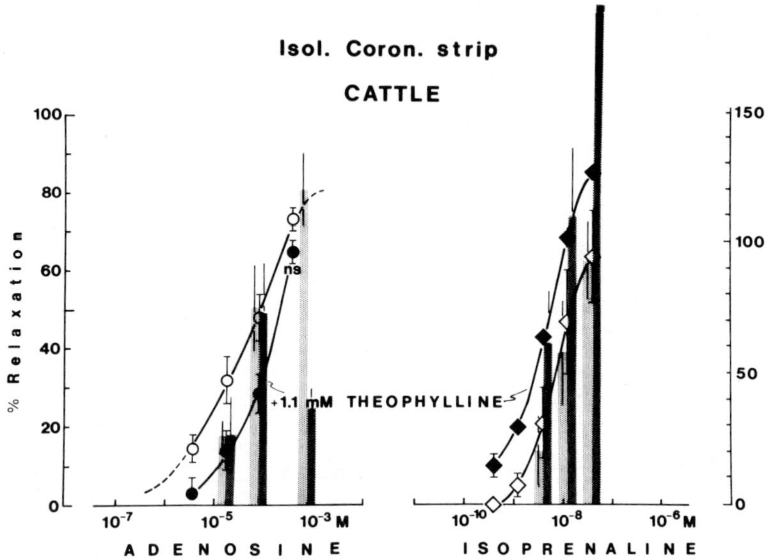

Fig. 3. Isolated coronary strips (cattle): Influence of theophylline. Effects of adenosine and isoprenaline as in fig. 2, in the absence and presence of 1.1 mM theophylline. Mean values ± S.E.M.; N = 5 - 26. Theophylline significantly lowered % net rises in cAMP induced by 750 μM adenosine but significantly enhanced those of 40 nM isoprenaline.

4. Stimulation of adenylyl cyclase by adenosine and isoprenaline.

Previous experiments had already shown, that adenosine at low concentrations of 10^{-8}M stimulated the activity of AC in particulate preparations from cattle and pig coronary arteries and that this action was blocked by theophylline [10]. In a more extensive study the actions of adenosine and isoprenaline on AC from cattle (Fig. 4) and pig coronary arterial tissue (fig. 5) were determined in the absence and presence of theophylline or sotalol. Both, adenosine and isoprenaline at low concentrations significantly increased AC-activity at 10^{-8} - 10^{-7}M. At concentrations above 10^{-6}M adenosine - in contrast to isoprenaline - began to depress AC. At 10^{-4}M adenosine AC-activity was significantly reduced to 52 ± 7.5 % of basal activity ($p < 0.01$; N = 8).

Theophylline (1×10^{-6}M) abolished the stimulatory effect of 10 - 50 nM adenosine on AC from both species but did not inhibit isoprenaline. Sotalol (100 μM), on the other hand, completely abolished the stimulatory effect of 0.01 - 1×10^{-7}M isoprenaline on AC from cattle and pig, but did not influence adenosine.

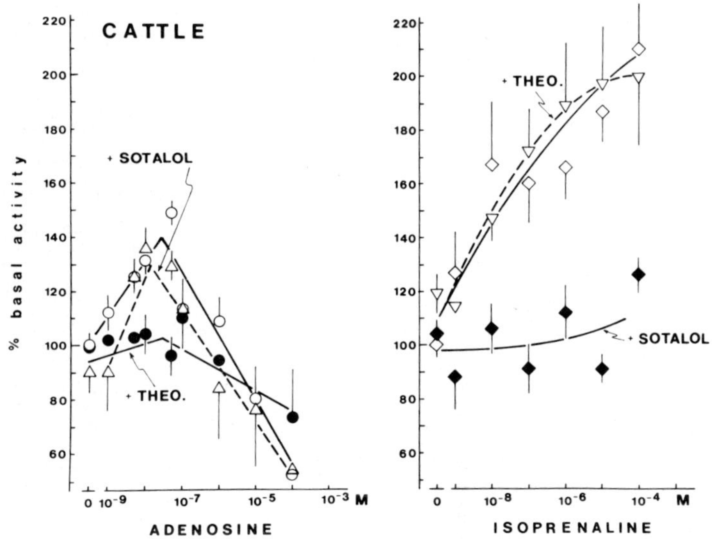

Fig. 4. Coronary artery cyclase (cattle): Formation of cAMP (% of basal activity) by 1 nM - 0.1 mM adenosine (left) and 1 nM - 0.1 mM isoprenaline (right) in the absence and presence of 1 µM theophylline and 100 µM sotalol, respectively: ○——○ adenosine (A), ●——● A plus theophylline, △- - -△ A + sotalol; ◇——◇ isoprenaline (I), ◆——◆ I + sotalol; ▽- - -▽ I + theophylline. Mean values ± S.E.M. of 3 - 35 experiments.

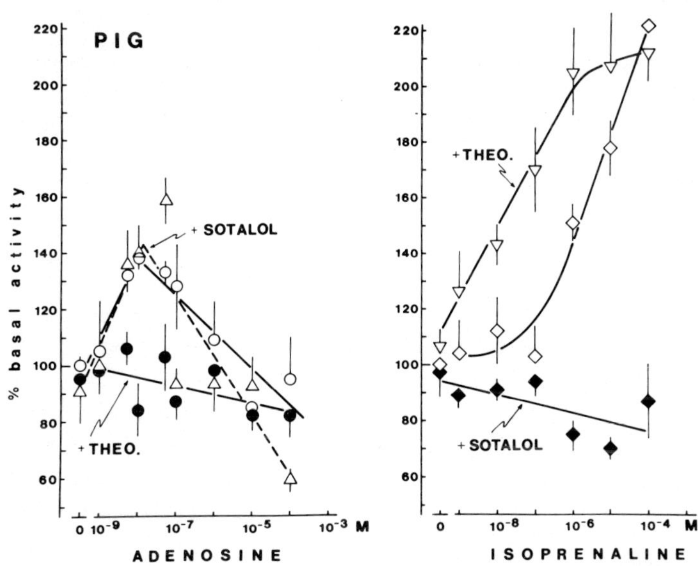

Fig. 5. Same as fig. 4 but with coronary cyclase from pig. Mean values ± S.E.M. of 3 - 29 experiments.

5. Effect of digitonin-solubilization and phosphatidylserine treatment on AC-response to adenosine, isoprenaline and fluoride.

Solubilization of AC by pretreatment with digitonin reversed the stimulant effect of adenosine into an inhibitory one and abolished the stimulation by isoprenaline but not that produced by sodium fluoride (fig. 6). Prolonged vigorous homogenization for 50 sec instead of 30 sec caused similar changes in the AC-response of particulate preparations to adenosine and isoprenaline [8]. The responsiveness to both drugs could be partically restored in the solubilized AC-preparation by addition of phosphatidylserine.

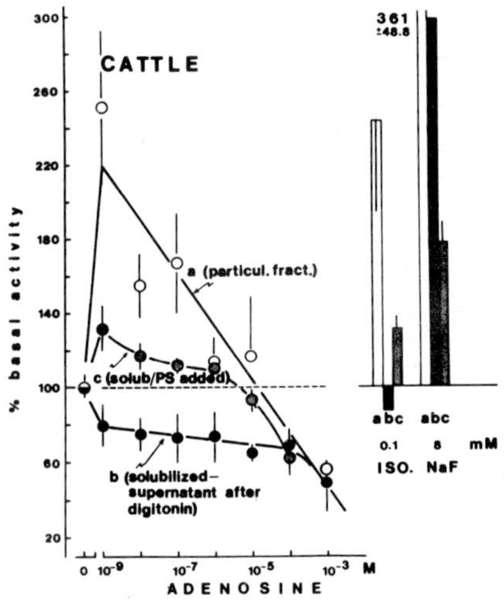

Fig. 6. AC from coronary arteries (cattle): response to adenosine, isoprenaline and NaF of

a) a particulate preparation of AC,

b) a supernatant fraction with solubulized AC obtained by tissue homogenization in the presence of 2 % digitonin which was then removed by column chromatography,

c) a supernatant fraction with solubilized AC after addition of phosphatidylserine [8].

AC-activity was determined in these experiments by measuring ^{32}P-cAMP formed from ^{32}P-ATP [8].

Mean values $\pm$ S.E.M. from 4 - 21 experiments.

DISCUSSION

The significant correlation between adenosine-induced rises in cAMP and relaxation was found to be identical with that produced by established AC-stimulating (isoprenaline, orciprenaline [1]) or PDE-inhibiting relaxants (theophylline, papaverine, eupaverine [1]). All of these drugs bear the characteristics for cAMP-mediated relaxant actions on various smooth muscles [1,11]. This finding supports the concept, obtained from dose-response experiments, time course and antagonist studies [1,2,8], according to which the relaxant effect of adenosine on coronary circular smooth muscle is mediated by cAMP.

Moreover, the observed stimulatory effect of adenosine in very low concentrations (10^{-9}M) on AC activity in vitro, which paralleled similar actions of 10^{-4}M isoprenaline, the abolition of these effects by mechanical- or by digitonin-induced [12] alteration of the enzyme and their partial restoration after addition of phosphatidylserine [13] to the digitonin-solubilized enzyme are evidence for the existence of an AC-attached adenosine receptor. Since AC-stimulation by adenosine was inhibited in the presence of theophylline whereas isoprenaline stimulation was inhibited by sotalol, but not vice versa, it is quite obvious that the adenosine receptor must be different from the adrenergic ß-receptor.

This adenosine receptor is obviously located at the surface of the smooth muscle cell in the coronary artery, as evidenced by experiments in open chest dogs in which oligosaccharide-linked adenosine (e. g. a molecule too large to penetrate the cell membrane) increased coronary blood flow, and similarly substituted theophylline prevented this effect [14].

The observation that receptor-dissociated AC, although still sensitive to fluoride stimulation, was not longer stimulated but only, dose dependently, inhibited by adenosine, indicates the existence of a second site of action of adenosine which is inhibitory in nature and is obviously located at the catalytic moiety of the enzyme. AC from blood platelets appears to react to adenosine in a similar manner [15].

REFERENCES

1. Kukovetz, W.R., Pöch, G., Wurm, A., Holzmann, S., Paietta, E. (1976) in: Ionic Actions on Vascular Smooth Muscle, 124 - 131. Ed. E. Betz. Springer, Berlin - Heidelberg - New York.
2. Wurm, A., Holzmann, S., Pöch, G., Kukovetz, W.R. (1976) Naunyn-Schmiedeberg's Arch. Pharmacol. Suppl. Vol. 294, R 13.
3. Herlihy, J.T., Bockman, E.L., Berne, R.M., Rubio, R. (1976) Am. J. Physiol. 230, 1239 - 1243.
4. Mc Kenzie, S.G., Frew, R., Bär, H.P. (1977) Eur. J. Pharmacol. 41, 193 - 203.
5. Kukovetz, W.R. and Pöch, G. (1970) Naunyn-Schmiedeberg's Arch. Pharmacol. 267, 189 - 194.
6. Kukovetz, W.R., Pöch, G., Holzmann, S., Paietta, E. (1976) Arzneim.-Forsch. (Drug Res.) 26, 1321 - 1330.

7. Steiner, A.L., Kipnis, D.M., Utinger, R., Parker, C.W. (1969) Proc. Nat. Acad. Sci. USA, 64, 367 - 373.

8. Kukovetz, W.R., Pöch, G., Holzmann, S., Wurm, A., Rinner, I. (1977) Adv. Cycl. Nucleot. Res. 9 (in press).

9. Bünger, R., Haddy, J.F., Gerlach, E. (1975) Pflüger's Arch. 358, 213 - 224.

10. Rinner, I., Wurm, A., Edlinger, R. (1977) Naunyn-Schmiedeberg's Arch. Pharmacol. Suppl. II, Vol. 297, R 12.

11. Pöch, G. and Umfahrer, W. (1976) Naunyn-Schmiedeberg's Arch. Pharmacol. 293, 257 - 268.

12. Caron, M. G. and Lefkowitz, R.J. (1976) Biochem. Biophys. Res. Commun. 68, 315 - 322.

13. Levey, G.S. (1971) Ann. N. Y. Acad. Sci. 185, 449- 457.

14. Olsson, R.A., Davis, C.J., Khouri, E.M., Patterson, R.E. (1976) Circul. Res. 39, 93 - 98.

15. Haslam, R.J. and Lynham, J.A. (1972) Life Sci. 11, II, 1143 - 1154.

Excitation-Contraction Coupling in Smooth Muscle, R. Casteels et al. eds.

CYCLIC-AMP AND RELAXATION MECHANISM IN MOLLUSCAN CATCH MUSCLE (ABRM)

Fernand Baguet
Université Catholique de Louvain
Unité de Physiologie Animale
Bâtiment Claude Bernard
Place de la Croix du Sud, 5
B-1348 Louvain-la-Neuve (Belgique)

SUMMARY

After the maximal tension of glycerol extracted and EDTA treated fibres has been developed in a high Ca^{2+} solution (10^{-6}M), the rate of relaxation induced by a low Ca^{2+} solution (10^{-9}M) depends on the Na^{+} concentration of the medium. The slowest rate is observed at a Na^{+} concentration of 1.5 10^{-4}M; in both types of preparation catch-state appears few minutes after the beginning of relaxation. In EDTA treated fibres, catch-state can be reversibly abolished by altering the Na^{+} concentration of the solution or by addition of 5HT or cyclic AMP. The presence of Na^{+}in 5-HT and cyclic AMP solutions might partly explain the relaxing effects observed. These relaxing effects are not observed on glycerol extracted fibres in catch-state.

INTRODUCTION

The anterior byssus retractor muscle (ABRM) of the mussel Mytilus, responds to Acetylcholine (ACh) or direct current (D.C.) by a tonic contraction : during the slow relaxation period, which may last for several hours, no active shortening is possible and the muscle, which exhibits the inert behaviour of a damped elastic body (6), is said to be in "catch". The application of 5-hydroxytryptamine (5-HT), the physiological relaxing mediator (7), induces a rapid relaxation and abolishes the catch-state. Similar relaxation can be obtained on tonically contracted ABRM after transfer from sea water to isosmotic sodium-free sea water (4); moreover, ABRM maintained in sodium-free sea water and stimulated by direct current, contracts normally but relaxes rapidly after the end of the stimulation.

Experimental evidences suggest that the relaxing effect of 5-HT is mediated by adenosine 3', 5'-monophosphate (cyclic AMP), the formation of which is triggered by 5-HT in the living muscle (1,2).

The aim of the present work is to investigate whether sodium (Na^+) ions can control the relaxation rate of ABRM by a direct action on the contractile proteins, independently of cyclic AMP.

In a first section, we have studied the ability of Na^+ ions to control the Ca^{2+} regulated activity of the contractile proteins of glycerol extracted fibres (3) and EDTA treated fibres (5). These preparations develop tension in presence of a high Ca^{2+} concentration (10^{-6}M) and relaxe when the free Ca^{2+} level is reduced (10^{-9}M). It is shown that the relaxation rate of the contracted fibres is strongly dependent on the Na^+ concentration of the relaxing solution.

In a second section, it is attempted to define the role of Na^+ ions in the relaxation mechanism; it is shown that, Na^+ ions at a very particular concentration can relax as well as cyclic AMP, the EDTA treated fibre in catch-state, but is ineffective, as well as cyclic AMP, on glycerol extracted fibres.

METHODS

A. The preparations

Mytilus edulis were collected from break waters, their natural habitat along the Belgian coast (Knokke), and kept in glass tanks of circulated and aerated sea water at 5°C. The ABRM dissected according the procedure described by Jewell (6) was used either for glycerol-extraction of for dissection of thin fresh bundles as described (3,5).

B. Force measurement

Bundles of about 100 microns diameter from glycerinated or fresh ABRM were transfered into a small chamber perfused with solutions. The preparation was tied to a strain gauge as previously described (3); isometric tension was recorded on a 20 cm strip-chart recorder.

C. Solutions

All experimental solutions contained ATP dipotassium salt (Sigma) 5 mM, $MgCl_2$ 5mM and Histidine buffer 10 mM; the pH was adjusted to 7.0 (20°C) with KOH instead of NaOH. The ionic strength of the solution was adjusted to 0.18 with 140 mM KCl.

The free Ca^{2+} level was controlled by varying the ratio (Ca-EGTA/EGTA) as described previously (3) using 4.6 10^6 as the value of the Ca-EGTA association constant at pH 7.0; stock solution of EGTA 200 mM was made by dissolving EGTA in KOH instead of NaOH in order to avoid any contamination by Na^+ ions. We essentially used a solution of low Ca^{2+} concentration (10^{-9}M) and a solution of high Ca^{2+} concentration (10^{-6}M).

The Na^+ content of the experimental solutions was checked by atomic absorptiometry; without external addition of NaCl, experimental solutions contain 5. 10^{-5}M Na^+. Four different NaCl concentrations were added to the experimental solutions giving the following Na^+ composition (M) : $1.5.10^{-4}$; 10^{-3}; 510^{-3} and 10^{-2}.

Fresh solutions of 5-HT and cyclic AMP were made before each experiment and diluted to the chosen concentration in the different experimental solutions.

D. EDTA treatment and standard procedure

The glycerol-extracted fibres washed free of glycerol were immersed in a solution of low Ca^{2+}(10^{-9}M) and low Na^+ concentration (5.10^{-5}M) at ionic strength 0.18 and pH 7.0. The fibre relaxed slowly in about 3 minutes and the final tension remaining was taken as a reference above which the tension developed in the contracting solution was measured.

To make the membrane of fresh fibres permeable to Ca^{2+} ions the preparation mounted in the chamber was washed free of glycerol with natural sea water and then immersed during 10 minutes in a solution whose composition is KCl 150 mM, EDTA 5 mM, $ATPK_2$ 5mM, Histidine 10 mM (pH 7.0 at 20°C). The residual tension is taken as a reference level above which the tension developed in the different experimental conditions is measured.

RESULTS

Glycerol extracted or EDTA treated fibres develop tension in presence of a high Ca^{2+} concentration (10^{-6}) and relaxe when the free Ca^{2+} level is reduced (10^{-9}M). The rate of force development and the maximal isometric force developed at ionic strength 0.18 are not affected by the Na^{+} concentration of the solution and are similar to the values reported previously (3,5). On the other hand the rate of relaxation induced by the low Ca^{2+} solution is strongly dependent on the Na^{+} concentration.

A. Sodium effect on the relaxation rate

I. EDTA treated fibres

In a first series of experiments, each preparation after EDTA treatment, is brought through 5 contraction-relaxation cycles each at a different Na^{+} concentration (I 0.18, pH 7.0) : after the steady force was developed, relaxation was brought about by the low Ca^{2+} solution (10^{-9}M) of corresponding Na^{+} concentration. The 5 sequences were applied on 10 preparations by the latin square design (3).

When the force values measured during the relaxation period are plotted on a semilog scale against time, all the points fall on a straight line. The effect of Na^{+} concentration can thus be quantitatively estimated by calculating the time constant of relaxation.

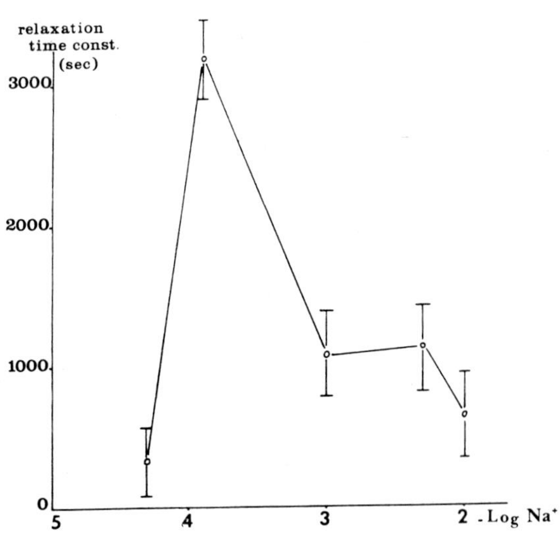

Fig. 1. Time constants of relaxation in seconds (mean values ± s.e.m.) (n = 10) of Ca^{2+} stimulated EDTA treated fibres as a function of the negative log. of the Na^{+} concentration in the relaxing medium.

In fig. 1, the mean values of the relaxation time constant are plotted as a function of the Na^+ concentration of the relaxing solution. At 5.10^{-5}M Na^+ concentration, the relaxation rate is rapid (time constant 370 ± 250 sec); for a 3 times increase in Na^+ content ($1.5.10^{-4}$M) the relaxation rate decreases dramatically by about 10 times (3185 ± 250 sec). For higher Na^+ concentration the relaxation rate increases and at 10^{-2}M, the relaxation rate (637 ± 250 sec) is not significantly different from that measured at 5.10^{-5}M.

II. Glycerol extracted fibres

The experiments on glycerol extracted fibres were performed in exactly the same way as on EDTA treated fibres.

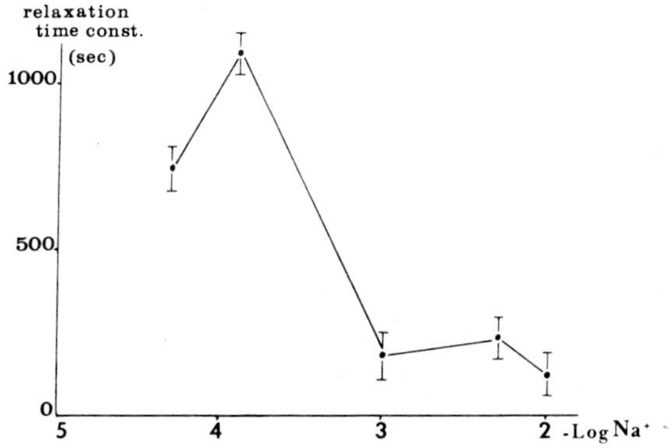

Fig. 2. Time constants of relaxation in seconds (mean values ± s.e.m.)(n = 10) of Ca^{2+} stimulated glycerol extracted fibres as a function of the negative log. of the Na^+ concentration in the relaxing medium.

In figure 2, the mean values of the different relaxation time constants are plotted as a function of the Na^+ concentration of the relaxing medium. Though there is also an obvious Na^+ concentration effect on the relaxation rate, there are mainly two differences when the time constants are compared with those measured on EDTA treated fibres :

(i) for a same Na^+ concentration, the relaxation rate is always higher in glycerol extracted fibres than in EDTA treated fibres.

(ii) the time constant value measured at 5.10^{-5}M Na^+ is very close to that measured at $1.5.10^{-4}$M Na^+.

It is concluded that in both kind of preparations, the rate of relaxation induced by a low Ca^{2+} solution (10^{-9}M) can be accelerated or delayed according to the Na^+ concentration in the medium. In glycerol extracted fibres as in EDTA treated fibres, the slowest rate occurs in presence of $1.5.10^{-4}$M Na^+.

3. Induction of the catch-state

In a living ABRM, during the maintenance of tonic tension, there is a progressive reduction of the active state (6) : as a matter of fact about five minutes after the development of tension, after a quick release, the drop of tension is not followed by a recovery of tension after the movement has stopped. This inability to redevelop a large fraction of the tension characterizes the presence of the "catch-state". The experiment reported in Fig. 3 shows that the

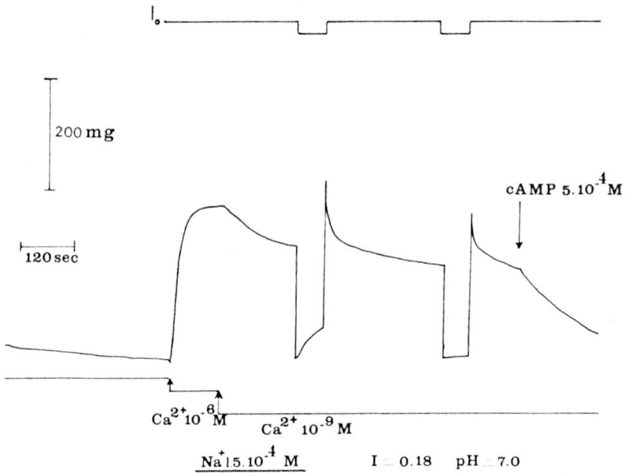

Fig. 3. Catch-state development in an EDTA treated fibre and relaxing effect of cyclic AMP. After the maximal isometric tension has been developed in the contracting solution (Ca^{2+} 10^{-6}M with Na^+ $1.5\ 10^{-4}$M), the corresponding relaxing solution (Ca^{2+} 10^{-9}M) is substituted and slow relaxation occurs. After the second released applied 7 minutes after the beginning relaxation, the preparation is in catch-state. The addition of cyclic AMP induces a rapid relaxation.

same sequence of events can be observed in EDTA treated fibre when the Na^+ concentration of the medium surrounding the contractile proteins is 1.5 10^{-4}M : the preparation is stimulated by $Ca^{2+}10^{-6}$M in a solution with 1.5 10^{-4}M Na^{+} concentration (I = 0.18 and pH 7.0). After the maximal isometric force has been developed, the Ca^{2+} level is decreased from 10^{-6}M to 10^{-9}M; a quick release (3 % of l_{o}) applied 3 minutes after the beginning of the slow relaxation is followed by a redevelopment of tension, suggesting the persistence of an active state. On the other hand when the same release is applied 4 minutes later, no tension redeveloped. However, after being stretched back, the preparation holds the same tension level as before. The same sequence of events can be observed in glycerol extracted fibres.

C. Relaxation of the catch-state

EDTA treated fibre in catch-state relaxes rapidly few seconds after adjusting the Na^{+} concentration of the medium either to 5.10^{-5}M, 10^{-3}M, 5.10^{-3}M or 10^{-2}M. Addition of cyclic AMP or 5-HT 5.10^{-4}M to the low Ca^{2+} solution containing 1.5 10^{-4}M Na^{+} (see fig. 3) also induces relaxation but at a slower rate.

We nerver succeeded to induce a similar rapid relaxation on a glycerol extracted fibre in catch-state nor by changing the Na^{+}concentration of the medium, nor by addition of cyclic AMP or 5-HT 5.10^{-4}M.

DISCUSSION

The present results give direct evidences that Na^{+} ions can regulate the relaxation rate of contracted glycerol extracted and EDTA treated fibres isolated from ABRM. In our experimental conditions the relaxation process is assumed to correspond to the dissociation of Ca^{2+} ions from the contractile proteins and its fixation to the EGTA present in the relaxing medium (Ca^{2+} 10^{-9}M). The relaxing medium containing Na^{+} ions at the particular concentration of $1.5.10^{-4}$M (EDTA treated fibres) or 5.10^{-5} and $1.5.10^{-4}$M (glycerol extracted fibres) applied on the contracted preparations induced a slow relaxation. At this concentration Na^{+} should prevent the dissociation of Ca^{2+} from the contractile proteins and afterwards reduce progressively the activity of the contractile mechanism leading to the development of the "catch-state". In EDTA treated preparations, "catch-state" can be reversibly abolished either by raising the Na^{+} concentration of the solution to 10^{-3}M or more, either by addition of 5-HT or cyclic AMP 5.10^{-4}M. However, the Na^+ content of these solutions measured by atomic absorptometry, is respectively about 5.10^{-5}M and 1.5 10^{-4}M. It is suggested that most of the relaxing effect of 5-HT and cyclic AMP is due to the presence of Na^+ ions in the solutions. Since we have not been able to relax glycerol extracted fibres in catch-state, it might be expected that 5-HT, cyclic AMP or higt Na^+ concentration do not act directly on those contractile proteins remaining after glycerol extraction but on some factor discarded by the extraction procedure.

REFERENCES

1. Achazi, R.K. (1973) Arzneimittel-Forsch (Drug Res.) 23, 624.
2. Achazi, R.K., Dölling, B. and Haakshorst, R. (1974) Pflügers Arch. 349, 19-27.
3. Baguet, F. (1973) Pflügers Arch. 340, 19-34.
4. Baguet, F. and Roegiers, A. (1973) J. Physiol. (Paris) 67, 238A.
5. Baguet, F. and Marchand-Dumont, G. (1975) Pflügers Arch. 354, 75-85.
6. Jewell, B.R. (1959) J. Physiol. (Lond.) 149, 154-177.
7. York, B. and Twarog, B.M. (1973) Comp. Biochem. Physiol. 44A, 423-430.

Excitation-Contraction Coupling in Smooth Muscle, R. Casteels et al. eds.

SUMMARY OF DISCUSSION ON "ROLE OF CYCLIC NUCLEOTIDES IN SMOOTH MUSCLE CONTRACTION AND RELAXATION"

V. A. W. Kreye

Universität Heidelberg
II. Physiologisches Institut
Im Neuenheimer Feld 326
D-6900 Heidelberg, F.R.G.

In his opening remarks, A. KATZ, New York, summarized the known effects of catecholamines on heart muscle which are probably mediated by cAMP. These are: 1. increased tension, 2. accelerated pace-maker activity, 3. accelerated relaxation, 4. shortened refractory period, and 5. increased energy production. The points of control for these various effects may be the cell membrane, the sarcoplasmic reticulum, the regulatory and contractile proteins, and other cell constituents such as soluble proteins or mitochondria. At present, the factors responsible for accelerated relaxation have been best studied. These are: 1. activation of the sarcoplasmic calcium pump resulting from cAMP-mediated phosphorylation of "phospholamban" (i. e. a protein component of the Ca^{++} transport ATPase), and 2. decreased calcium affinity of troponin as a result of a cAMP-dependent phosphorylation of troponin I. The cAMP-mediated effects of catecholamines are complex in nature, since in heart muscle the control points are manifold and some of the mechanisms mentioned above seem to interfere with each other. A similar complexity of the cyclic nucleotide system must be anticipated for smooth muscle.

Supplementing the presentation of J. DIAMOND, Chicago, on the role of cAMP and cGMP in smooth muscle contraction and relaxation, V.A.W. KREYE and K.H. SCHULTZ, Heidelberg, presented evidence that the lipophilic 8-bromo derivate of cGMP can relax vascular smooth muscle in a dose-dependent fashion. The vasodilator drug sodium nitroprusside strongly increases cGMP levels in vascular and vas deferens smooth muscle. These findings suggest a possible role of cGMP in relaxation. Such a relation, however, cannot be a simple one since nitroprusside relaxes vascular smooth muscle, but not rat vas deferens.

In experiments with rat myometrium, M. WORCEL, Paris, could not establish any clear-cut correlation between the levels of cyclic nucleotides and the state of contraction or relaxation. He suggested that the variations of cyclic nucleotides in this tissue may be secondary to variations of sodium or calcium concentrations.

Assuming a role of cAMP in the intracellular regulation of calcium levels, J. C. ALLEN, Houston, investigated calcium binding of aortic vesicles in relation to the state of phosphorylation of this preparation. On gel separation, a peak was found between 68,000 and 100,000 dalton probably representing a protein kinase. In presence of cAMP, the vesicles could be effectively phosphorylated. However, calcium binding or uptake proved to be independent of this cAMP-dependent phosphorylation of SR vesicles. At present, this does not rule out a possible correlation between calcium sequestration and phosphorylation of sequestering sites in the <u>intact</u> smooth muscle since factors responsible for a "link" might have been destroyed by the technical procedures of isolation.

In his oral presentation, F. BAGUET, Louvain, had suggested that the relaxation of "glycerinated" anterior byssus retractor muscle from Mytilus (ABRM) in "catch" state by cAMP may be artificial and result from the sodium ions present in Na_2-cAMP. However, R. K. ACHAZI, Münster, presented evidence that paramyosin, after cAMP-enhanced phosphorylation, was able to resolve "catch" of ABRM. This is in favour of a specific effect of cAMP in relaxation of this tissue.

Finally, G. KÖHLER, Heidelberg, showed that 5-HT induces relaxation of ABRM, and that it increases cAMP in this tissue. However, considerably higher 5-HT concentrations are necessary to increase cAMP levels than for relaxation, and the time course of relaxation is shorter than for increasing cAMP levels. In his opinion, this excludes a causal role of cAMP in relaxation of ABRM, but it might play a role as a "modulator" of contractility in this tissue.

From this session on cyclic nucleotides and the preceding ones on phosphorylation of regulatory proteins in smooth muscle the following conclusions can be drawn:

1. Several relaxant drugs can increase the intracellular concentrations of cyclic nucleotides.
2. cAMP can mediate or at least enhance phosphorylation of cellular constituents such as sarcoplasmic reticulum membranes, troponin, light chains, and paramyosin.
3. The causal links between the contractile state of smooth muscle and cellular cyclic nucleotides levels have yet to be established.

Excitation-Contraction Coupling in Smooth Muscle, R. Casteels et al. eds.

VARIATIONS IN ACTIN TO MYOSIN RATIOS AND CELLULAR FORCE GENERATION IN VERTEBRATE SMOOTH MUSCLES

R. A. Murphy, S. P. Driska, and D. M. Cohen
Department of Physiology
University of Virginia School of Medicine
Charlottesville, Virginia 22901, U.S.A.

SUMMARY

Maximum force generating capacities of smooth muscle tissues are variable, but often equal or exceed that of striated muscle. Two types of experiments indicate that measured tissue force normalized to cell cross-sectional area is a direct measure of cellular force generation. Several types of smooth muscle from the pig had the same estimated cellular myosin contents which were only a fourth of that in striated muscle. However, the thin filament protein contents were higher in arterial cells than in the other smooth muscles tested. There appears to be a relationship between high actin:myosin weight ratios and high force generating capacity in vertebrate smooth muscles.

INTRODUCTION

The force developed by fully activated striated muscles is proportional to the overlap between thick and thin filaments, i.e., to the number of cross-bridges acting in parallel. Since mammalian striated muscles have comparable contractile protein contents and filament lattice geometry, maximal force generated (F_o) at the optimum length (L_o) is reasonably uniform.[1] In contrast, there is considerable variability in estimates of F_o among different smooth muscle tissues.[2] This symposium documents the importance of intracellular [Ca^{++}] regulation by cellular membrane systems as a major determinant of contractile system output. However, the ability of some arterial tissues to develop high values of F_o with a low myosin (and potential cross-bridge) content suggests that the smooth muscle contractile system must be specialized for force generation.[3] The surprisingly high and variable F_o in smooth muscle tissues might reflect two features not present in striated muscles: (i) coupling of some cells in parallel to provide a mechanical advantage, or (ii) differences in the amounts, organization, or molecular properties of the contractile proteins. The studies described below support the latter possibility, and further suggest that some of the physiological diversity in smooth muscles reflects differences in the relative amounts of thin and thick filament proteins.

MATERIALS AND METHODS

Estimates of tissue F_o: Reported values from tissues meeting the following criteria[2] were accepted as valid estimates of tissue properties: (i) preparations had parallel cells in the axis of force measurement, (ii) passive elastic forces were obtained when the contractile system was functionally inactivated, and (iii) active force was estimated at the optimum tissue length with the most effective stimulus. Values were normalized to the cross-sectional area of cells to account for variations in the fraction of the tissue not occupied by smooth muscle cells.

Estimates of cellular F_o: There is presently no method for obtaining viable single cell preparations from mammalian smooth muscle tissues. However, the relationship between cellular and tissue properties can be estimated indirectly. Two approaches were taken.

In the first, intima-medial strips having a constant cross-section were obtained from pig carotid arteries.[4] They were mounted *in vitro* in a physiological salt solution and adjusted to L_o after the length-force relation to K^+-stimulation was obtained.[4] If cells anatomically arranged in series were coupled force-transmitting structures, force/cross-sectional area of cells in the tissue would be the same as that for individual cells and independent of the number of cells coupled in series.[2] That is, F_o would be independent of the length of the tissue segment from which force was measured. On the other hand, a dependence of F_o on the length of the tissue segments would reveal a functional parallel coupling of cells which were anatomically in series. With this arrangement, tissue F_o would be greater than that of the individual cells. A simple apparatus was designed to measure force from tissue segments of different lengths using additional clips to isolate smaller segments of tissue.[5]

The second approach was to determine the relationship between cell length and tissue length in contracting tissues at different points on the length-force curve. There would be a direct proportionality between cell and tissue lengths if smooth muscle cells anatomically arranged in series were coupled force-transmitting structures. Parallel coupling to give the tissue a mechanical advantage would lead to greater relative changes in cell length than in tissue length with stretch.[2] Tissues were adjusted to lengths of 0.4, 0.6, 0.8, 1.0, or 1.1 L_o, and fixed with glutaraldehyde during a maximal K^+-stimulated contraction. Individual cells were obtained by acid maceration and the lengths of 50 cells from each tissue were measured with a filar micrometer at 400x magnification.[5]

Estimates of cellular contractile protein contents: Few methods are suitable for quantitative comparative studies of diverse smooth muscles, as they must be applicable to small samples of tissues of varying composition. One feasible approach is to determine the actin, tropomyosin, and myosin contents of different tissues by quantitative densitometry of stained acrylamide gels after electrophoresis of sodium dodecyl sulfate (SDS) treated tissue homogenates.[3] The rationale for this method is that the amounts of a few proteins of differing mobilities which comprise the major fraction of the total protein entering the gel can be estimated with relatively small errors, even in very complex mixtures. The methodology and testing of this method on the pig carotid artery have been described.[3]

Tissues were obtained from pigs at slaughter, transported to the laboratory in cold physiological salt solution and carefully dissected to obtain the muscular layers. Arterial mediae and the muscular layers of the visceral tissues (esophagus and jejunum) were separated by careful dissection. Isolation of the trachealis[6] gave a pure airway smooth muscle. Myometrial muscle was obtained from a horn of the uterus.

Myosin,[7] actin (modified from Bárány and Bárány[8]), and tropomyosin[9,10] were prepared from pig carotid and intestinal smooth muscle. The isolated proteins were co-electrophoresed on 5.6% acrylamide gels with smooth muscle homogenates in order to identify the peaks containing these proteins. The six isolated proteins exhibited a linear relationship between protein applied and peak area of the stained gels[11] and myosin (heavy chains), actin, and tropomyosin all exhibited equivalent staining. The peak containing tropomyosin in gels from smooth muscle homogenates also contained a significant amount of another protein(s). Consequently, tropomyosin contents were estimated using a urea-SDS gel system in which tropomyosin has a lower mobility. This shifts its peak, and permits an accurate assessment of content.[3]

Carefully randomized electron photomicrographs were used to obtain extracellular space measurements by weighing cutouts of cell cross-sections. These values were used to calculate cellular protein contents from measurements of tissue contents.

RESULTS

Estimation of cellular F_o from tissue measurements: If a tissue of constant cross-section is stimulated at its optimum length (L_o), one would predict: (i) no change in F_o using progressively shorter segments of the original tissue if the cells anatomically arranged in series are coupled force-transmitting structures (horizontal line, Fig. 1); and (ii) a dependence of F_o on segment length if there is some parallel coupling. This dependence can take many forms, depending on the anatomy, but values would fall within the shaded area in Figure 1. The experimen-

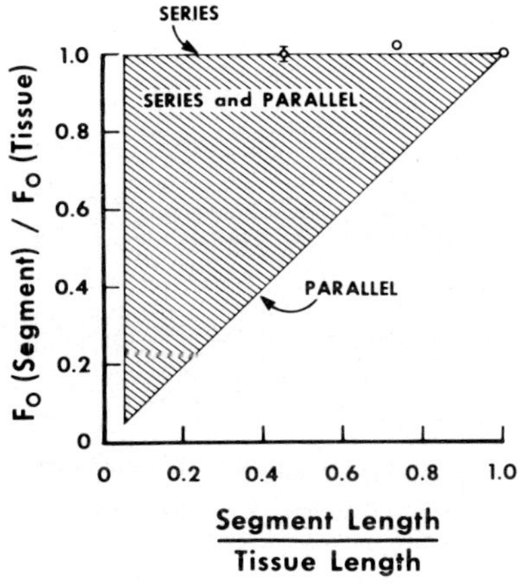

Figure 1. Maximum isometric force developed in response to K^+ by pig carotid media strips of constant cross-section held at L_o. F_o in the intact tissue (segment F_o/tissue F_o = 1) was 2.29 ± 0.18 (SEM) × 10^5 N/m^2 (N = 9). L_o in the intact tissue (segment length/tissue length = 1) averaged 10.94 mm. F_o was unchanged when segment length was reduced by 3 and 6 mm (open circles, bars show 1 SEM if greater than symbol size). The minimum segment length was 3 mm. See text for further details.

tal results (open circles, Fig. 1) follow the theoretical line for a series coupled tissue structure and suggest that active force/cross-sectional area of the tissue corrected for extracellular space is a valid index of cellular F_o. However, at very short segment lengths cell injury would produce a decline in force. Thus, this method would not detect parallel coupling restricted to short regions of the tissue (< 3 mm in our experiments).[5]

A second approach not subject to this qualification is to vary the length of entire tissues and determine the effect on cell length. The prediction is for a directly proportional change in cell and tissue lengths if cells are coupled in series (Fig. 2). Various types and degrees of parallel coupling would alter this relationship, but always in a direction such that cell lengths would increase more rapidly than tissue lengths with stretch (shaded area, Fig. 2). Our results (open circles, Fig. 2) indicate that cell length is directly proportional to tissue length. These results also suggest that tissue forces normalized to cell cross-sectional area, and tissue lengths and velocities normalized to L_o, provide reasonable estimates of cellular properties.

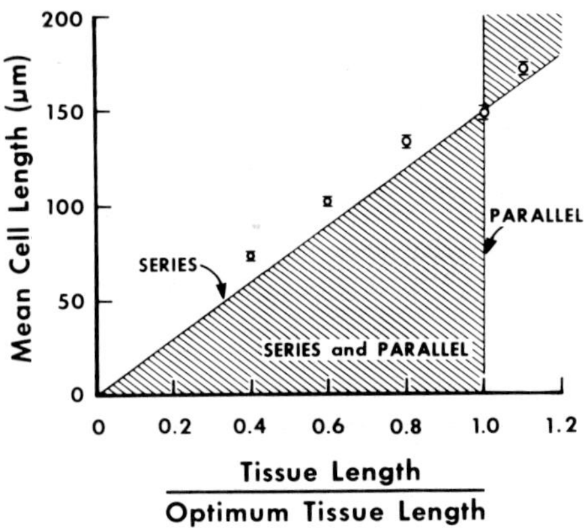

Figure 2. Mean cell lengths (± 1 SEM for 100-150 cells obtained from 2-3 tissues) measured after acid maceration of carotid media strips fixed at various lengths during a K^+ contracture. There was considerable variability in cell lengths in each sample. The theoretical lines (see text) were drawn to intersect the mean cell length at L_o. This value fell below the regression line for all the data in the experiments. Therefore, the other points are slightly above the theoretical line.

Estimates of cellular contractile protein contents: Eight smooth muscles representative of major organ systems in the pig were examined (Table 1). Each tissue had a similar myosin content, which was only a fourth of that in striated muscle. Actin and tropomyosin contents were higher in smooth than in striated muscle, and variable. However, they increased proportionally so that the actin:tropomyosin ratios were similar in all smooth and striated muscles. This implies that thin filament structures are comparable. Variations in thin filament contents are reflected in the ratios of actin:myosin. The averages of these estimates were 3 in the arterial tissues and 1.8 in the remaining smooth muscles (data from the uterus was quite variable and may reflect the hormonal status, which was unknown). The values for myosin should be multiplied by 1.19 to correct for the light chain weights which could not be estimated accurately in our system.

TABLE 1. ACTIN, TROPOMYOSIN, AND MYOSIN (HEAVY CHAIN) CONTENTS OF THE CELLULAR FRACTION OF VARIOUS PORCINE SMOOTH MUSCLES†

Tissue	Cellular Protein Contents (mg/g cell wt)			Weight Ratios	
	Myosin	Actin	Tropomyosin	Actin / Myosin	Actin / Tropomyosin
Arterial					
carotid	16.9 ± 2.8	48.5 ± 6.1	13.5 ± 1.8	3.1 ± 0.4	3.5 ± 0.2
aorta (thoracic)	15.4 ± 1.0	40.9 ± 3.6	12.5 ± 1.9	2.7 ± 0.2	3.7 ± 0.6
coronary (left ant. descending)	17.3 ± 2.9	60.6 ±11.9	16.0 ± 2.6	3.4 ± 0.2	3.4 ± 0.3
Non-arterial					
esophagus	18.5 ± 4.1	30.4 ± 5.4	7.7 ± 1.4	1.8 ± 0.1	4.2 ± 0.3
trachea	14.1 ± 2.1	25.2 ± 4.4	7.2 ± 1.2	1.8 ± 0.1	3.5 ± 0.2
intestine -circular	15.4 ± 1.4	29.0 ± 4.4	7.8 ± 1.2	1.9 ± 0.2	3.8 ± 0.3
-longitudinal	14.7 ± 1.9	29.2 ± 2.8	7.6 ± 1.0	2.1 ± 0.2	4.0 ± 0.2
uterus	18.8 ± 2.3	23.9 ± 2.9	8.1 ± 2.0	1.3 ± 0.1	3.5 ± 0.5
Means:					
arterial	16.5 ± 1.3	50.0 ± 4.7	14.0 ± 1.2	3.0 ± 0.2	3.5 ± 0.1
non-arterial	16.3 ± 1.1	27.5 ± 1.8*	7.7 ± 0.6*	1.8 ± 0.1*	3.8 ± 0.1
skeletal (rabbit)††	62	22	5	0.35	4.4

†Values are means ± SEM for duplicate determinations in six samples of each smooth muscle.
††Summarized from the literature[3]
*Significantly different ($p < 0.05$) from mean of arterial tissues

DISCUSSION

F_o in striated muscle provides useful estimates of cellular and contractile system function.[1] Our results indicate that F_o in the pig carotid media is an index of cellular F_o because tissue measurements sensitive to the mechanical linkage of cells are consistent with a coupling of cells anatomically arranged in series. Force developed by the rat portal vein is also independent of segment length,[12] and our results are further corroborated by measurements of cell length as a function of tissue length on the guinea pig taenia coli[13] and on the rabbit urinary bladder.[14] A similar conclusion was reached by Mulvany and Halpern[15] based on *in situ* measurements of cells in small rat mesenteric arteries. Thus, tissue force measurements in a diverse group of smooth muscles appear to reflect cellular properties.

Based on anatomical evidence for attachment of myofilaments to the lateral surfaces of the smooth muscle cells, cell-to-cell attachments, and connections to the collagen network, Gabella[16] concluded that parallel coupling of cells explains the high force development in the guinea pig taenia coli. However, such parallel coupling of adjacent cells may effectively put the contractile material of one cell in series with that of an adjacent cell. If this is true, the anatomical data on the taenia is not in conflict with our observation that the carotid media behaves as if the cells were series coupled. We conclude, therefore, that the cells of the pig carotid media and probably other smooth muscles have a high F_o which is due to some specialization in the organization or properties of the contractile proteins.

Comparative studies of contractile protein contents (Table 1) show that the aorta and coronary arteries have compositions similar to the carotid.[3] However, consistently lower contents of thin filament proteins were found for a variety of other smooth muscles. The technique employed has some variability, and systematic errors can arise in estimates of extracellular spaces or from the presence of proteins with mobilities on SDS gels corresponding to actin and myosin. However, we believe that the consistency of the results and the following factors argue for their general validity. (i) Two independent methods were used to estimate myosin in the pig carotid which gave results corroborating the electrophoretic technique. (ii) Because of its altered mobility on urea gels, tropomyosin can be estimated accurately from the different staining patterns in SDS and urea-SDS gels. (iii) The constancy in the actin:tropomyosin ratio in all muscle types, combined with confidence in the tropomyosin values, suggests that the actin estimates are reasonable if all vertebrate muscles have similar thin filament structures. X-ray diffraction evidence supports the latter assumption.[17] We conclude that there is strong evidence for heterogeneity in the ratios of thin:thick filament proteins in smooth muscles which has no counterpart in mammalian striated muscles. Further work is needed to determine whether this heterogeneity is limited to two discrete groupings, as suggested by our sample from pig tissues, or whether there may be other ratios or a continuum of values.

The carotid is the only porcine tissue which has been characterized in mechanical terms. This prevents any direct comparisons between physiological function and protein ratios. However, a certain correspondence between high force generation and high actin:myosin ratios is evident in data drawn from the literature on other smooth muscles (Table 2). Arterial tissues generally develop over 3×10^5 N/m^2 of cell cross-sectional area and the tested pig arteries have actin:myosin ratios of 3. Several non-arterial tissues generate about 2×10^5 N/m^2 and the average value for the actin:myosin ratio in non-arterial pig tissues was 1.8 (Table 1). The guinea pig taenia coli develops very high forces (Table 2). Tregear and Squire[18] obtained an actin:myosin ratio of 3.85 for this tissue using methods comparable to ours. Thus, data from the literature on the taenia coli are consistent with a relationship between high actin:myosin ratios and high forces, but implies that there may be no simple distinction between arterial and other tissues.

TABLE 2. MAXIMUM FORCE DEVELOPED BY SMOOTH MUSCLE TISSUES (37 C)

Tissue	Tissue F_o (10^5 N/m^2)	Cell x-sec (% of strip)	Cell F_o (10^5 N/m^2)	Ref.
Arterial				
carotid media (pig)	2.23	60	3.7	4
carotid artery (cow)	1.6	0.51 x 0.85 axial factor	3.7	19
carotid artery (dog)	1.1	31-45	2.4-3.5	20
mesenteric artery (rat)	0.72	20	3.6	15
Non-arterial				
tracheal (dog)	1.15	60	1.9	6
urinary bladder (rabbit)	1.25	67	1.9	14
superior mesenteric vein (calf)	0.73	37	2.0	21
taenia coli (guinea pig)	2.6	65	4.0	22
taenia coli (guinea pig)	4.16	68	6.1	16

In summary, our data lead to several conclusions and invite some speculations. First, it appears that in the pig carotid, and perhaps other smooth muscles, the high force generating capacity of the tissue is a property of the smooth muscle cell. Second, we find that a very diverse group of smooth muscles from the pig have fairly constant cellular myosin contents, but variable amounts of thin filament proteins. It seems likely that variation in the contractile system in mammalian smooth muscles is yet another factor involved in physiological differences in this heterogeneous group of tissues which has no counterpart in mammalian striated muscles. There is an analogous situation in many invertebrate striated muscles which show high and variable thin:thick filament ratios (Table 1.3, p. 25 in Huddart[23]).

The significance of these differences is unknown and filament ratios in mammalian smooth muscle may vary for quite different reasons than those in invertebrate striated muscles. Mammalian smooth muscles differ from vertebrate striated muscles in exhibiting variations in maximal force generating capacities, and high actin:myosin ratios appear to be associated with tissue types having high force generating capacities. However, biochemical and mechanical studies on the same tissues are required to validate this correlation.

ACKNOWLEDGMENTS

This research was supported by NIH grants HL14547 and HL19242. S. P. Driska was the recipient of a NIH Postdoctoral Fellowship (HL05180) and D. M. Cohen is a NIH Postdoctoral Trainee (HL05815). We greatly appreciate the many technical contributions of Mildred H. Smythers, Holly King, and Betty K. Haigh, and the constructive comments of Dr. Michael Mulvany.

REFERENCES

1. Close, R.I. (1972) Physiol. Rev. 52, 129-197.
2. Murphy, R.A. (1976) Blood Vessels 13, 1-23.
3. Murphy, R.A., Herlihy, J.T., and Megerman, J. (1974) J. Gen. Physiol. 64, 691-705.
4. Herlihy, J.T. and Murphy, R.A. (1973) Circ. Res. 33, 275-283.
5. Driska, S.P. and Murphy, R.A. (In Press) Blood Vessels.
6. Stephens, N.L., Kroeger, E., and Mehta, J.A. (1969) J. Appl. Physiol. 26, 685-692.
7. Megerman, J. and Murphy, R.A. (1975) Biochim. Biophys. Acta 412, 241-255.
8. Bárány, M. and Bárány, K. (1959) Biochim. Biophys. Acta 35, 293-309.
9. Hartshorne, D.J. and Mueller, H. (1969) Biochim. Biophys. Acta 175, 301-319.
10. Driska, S. and Hartshorne, D. (1975) Arch. Biochem. Biophys. 167, 203-212.
11. Fairbanks, G., Steck, T.L., and Wallach, D.F.H. (1971) Biochemistry 10, 2606-2617.
12. Sutter, M.C. and Ljung, B. (1977) Acta Physiol. Scand. 99, 484-495.
13. Cooke, P.H. and Fay, F.S. (1972) J. Cell Biol. 52, 105-116.
14. Uvelius, B. (1976) Acta Physiol. Scand. 97, 1-12.
15. Mulvany, M.J. and Halpern, W. (1976) Nature 260, 617-619.
16. Gabella, G. (1976) J. Physiol. (London) 263, 199-213.
17. Vibert, P.J., Haselgrove, J.C., Lowy, J., and Poulsen, F.R. (1972) J. Mol. Biol. 71, 757-767.
18. Tregear, R.T. and Squire, J.M. (1973) J. Mol. Biol. 77, 279-290.
19. Laszt, L. (1960) Arch. Kreislaufforsch. 32, 220-244.
20. Dobrin, P.B. (1973) Am. J. Physiol. 225, 664-670.
21. Paul, R.J. and Peterson, J.W. (1975) Am. J. Physiol. 228, 915-922.
22. Lowy, J., and Mulvany, M.J. (1973) Acta Physiol. Scand. 88, 123-136.
23. Huddart, H. (1975) The Comparative Structure and Function of Muscle, Pergamon Press, Oxford.

Excitation-Contraction Coupling in Smooth Muscle, R. Casteels et al. eds.

FACTORS INFLUENCING EXTENT AND/OR VELOCITY OF VASCULAR SMOOTH MUSCLE CONTRACTION

Ulrich Peiper

Physiologisches Institut der Universität
Röntgenring 9, D - 8700 Würzburg ·
Federal Republic of Germany

SUMMARY

The elementary process of contraction of vascular smooth muscle was studied by means of force-velocity relations of the tetanized rat portal vein. In spontaneously hypertensive rats the behaviour of the preparation is very similar to that observed at changed pCa: there is a distinct increase in force generation at constant velocity of shortening (polytropic effect). In renal hypertensive rats, however, a diminished speed of contraction takes place at constant force generation (tachytropic effect), like in the effect of temperature changes. In hyperthyroidism both extent and velocity of shortening are distinctly increased. The observed effects of these pathophysiological conditions tested in vascular smooth muscle are discussed with respect to the corresponding effects produced in cardiac muscle.

INTRODUCTION

The dynamics of muscle contraction is decisively influenced by two processes: (1) Force generation depends on the number of activated interaction sites between the contractile proteins. 'Polytropic Effects' (Peiper and Laven[1]) influencing the intensity of contraction by recruiting these acting centres are: initial tension (Peiper and Laven[2]), extracellular calcium concentration (Hellstrand et al.[3]; Peiper et al.[4]), extracellular pH (above pH 6.8), and noradrenaline (Peiper et al.[4]). (2) The speed of contraction is highly influenced by the velocity of moving cross-bridges. This process is influenced by 'Tachytropic Effects' (Peiper and Laven[1]). In vascular smooth muscle changes in temperature produce tachytropic effects only, an influence on force generation can be neglected (Peiper et al.[5]).

This distinction between polytropic and tachytropic effects in vascular smooth muscle seems to be suitable for analyzing the dynamics of contraction. According to in situ conditions the concentration of intracellular calcium-ions and their distinct polytropic effect are of great consequence, but tachytropic effects (besides the effect of temperature) have been scarcely observed. Thus in the vascular smooth muscle the solid speed of contraction contrasts with the varied contractility in the myocardium (positive inotropic effect of noradrenaline).

Under the pathophysiological conditions of renal hypertension, spontaneously hypertension, or in hyperthyreosis, characteristic and to some extent marked tachytropic and polytropic effects were described for cardiac muscle (Buccino et al. [6]; Strauer and Scherpe [7]; Jacob et al. [8]). The subject of the following studies was to examine, whether these inotropic effects in the myocardium can be demontrated as analogous tachytropic effects in vascular smooth muscle.

MATERIALS AND METHODS

The studies were carried out on the isolated tetanized rat portal vein. From isotonic contractions with varied afterload the values of force-velocity relations were obtained. By means of a digital computer (Telefunken TR 440) a Hill's hyperbola was approximated to the respective maximum values of tension development and velocity of shortening. The intersection of this curve and the velocity axis is defined as the unloaded shortening velocity V_o while the intersection of the hyperbola and the force axis is the isometric peak force T_o.

(a) Experiments on renal hypertensive rats (RHR): Hypertension was induced by clamping the left renal artery. In one group of test animals (RHR I, n = 14) a hypertension of 181 ± 11 mm Hg (measured by the maximum of cuff pulses in the rat tail artery) lasted for 74.8 ± 4.4 days, in the second group (RHR II, n = 4) hypertension of 172 ± 13 mm Hg lasted for 137.5 ± 4.5 days.

(b) Experiments on spontaneously hypertensive rats (SHR): The force-velocity relations of the portal vein of 15 spontaneously hypertensive Okamoto rats (blood pressure 172 ± 4 mm Hg, body weight 221 ± 9 g) were compared with those of 15 rats without genetic hypertension (blood pressure 85 ± 3 mm Hg, body weight 220 ± 11 g).

(c) Experiments on hyperthyroid rats (HR): By intraperitoneal injection of thyroxine or trijodthyronine for two weeks hyperthyroidism was induced in 38 Wistar rats, and the results were compared with those obtained from 38 control rats.

RESULTS

Renal hypertensive rats (RHR): In this experimental series a pronounced decrease of shortening velocity was generally noticed. Especially in the group of prolonged renal hypertension (RHR II) the unloaded shortening velocity V_o decreased from 0.865 ± 0.056 ML/s (control animals, n = 4) to 0.606 ± 0.085 ML/s (RHR II, n = 4, $P < 0.005$, Fig. 1), which means a decrease to 70.1 % (Tab. 1).

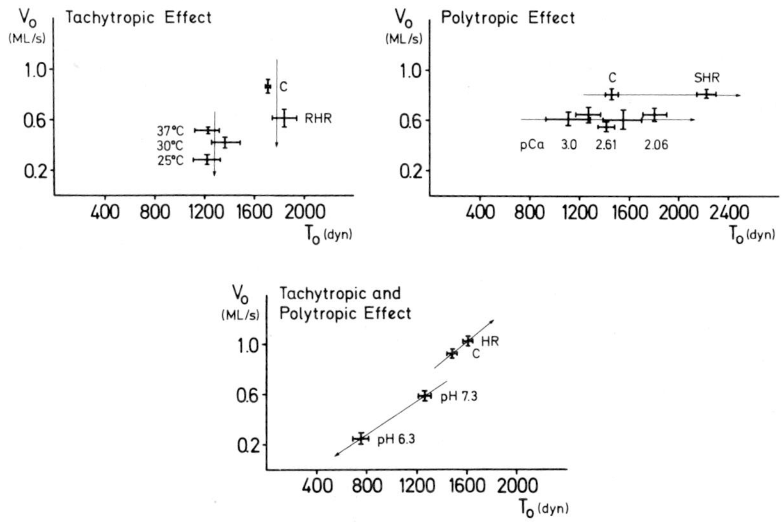

Fig. 1. Influence of some experimental conditions on the elementary process of contraction of the isolated rat portal vein. V_o = velocity of shortening at zero load. T_o = peak isometric tension. Both values are calculated by Hill's hyperbola. C = control rats. RHR = renal hypertensive rats. SHR = spontaneously hypertensive rats. HR = hyperthyroid rats. All values are $\bar{x} \pm s_{\bar{x}}$. Thin arrows indicate tendency of changes in V_o and T_o like tachytropic and/or polytropic effects.

After inducing renal hypertension for only 10 weeks the decrease of the speed of contraction to 87.3 % (from 0.852 ± 0.04 ML/s, n = 10, to 0.747 ± 0.079 ML/s, n = 14, $P < 0.005$) is distinctly less marked. Isometric force generation T_o was slightly augmented, but this increase could not be proved statistically. Altogether the maximum instantaneous mechanic power N_{opt} of the preparation resulted in a slight decrease from 1.44 ± 0.13 µW/mm ML to 1.38 ± 0.11 µW per mm ML (RHR I), or to 1.33 ± 0.10 μW/mm ML (RHR II). The corresponding Hill equations are: Control rats: V = 0.26 (1561-T)/(472+T); RHR I : V = 0.29 (1570-T)/(612+T); RHR II: V = 0.24(1845-T)/(732+T).

Spontaneously hypertensive rats (SHR): In contrast to experiments with RHR in this experimental series the force generation from the portal vein of hypertensive rats (T_o = 2236 ± 89 dyn, n = 15) was distinctly increased as compared with that of the controls (1468 ± 50 dyn, n = 15, P<0.0005, Fig. 1). The cross-

TABLE 1

Force velocity relations in the tetanized portal vein of renal hypertensive rats (RHR I = 10 weeks after clamping, RHR II = 20 weeks after clamping), spontaneously hypertensive rats (SHR), hyperthyroid rats (HR), or at changes of the bath solution in temperature (30°C resp. 25°C), or in pH (6.8 resp. 6.3), or in pCa (3.2 resp. 2.1). V_o = maximum velocity of shortening of the unloaded preparation. T_o = peak isometric tension. N_{opt} = maximum instantaneous power. The data are given in per cent of the control values.
(a) = P<0.05; (b)= P<0.005; (c) = P<0.0005;

	V_o (%)		T_o (%)		N_{opt} (%)	
a) Polytropic without tachytropic effects:						
pCa 3.2	103.7		60.5	(b)	68.9	(a)
pCa 2.1	117.1		120.0	(a)	148.6	(b)
SHR	98.1		152.3	(c)	127.3	(b)
pH 6.8	87.7		68.9	(c)	74.7	(a)
b) Tachytropic and polytropic effects:						
pH 6.3	44.9	(c)	56.7	(c)	49.5	(c)
HR	110.8	(a)	108.5	(b)	124.4	(c)
c) Tachytropic without polytropic effects:						
30°C	83.4		103.6		71.7	(a)
25°C	54.9	(c)	91.7		46.4	(c)
RHR I	87.3	(b)	100.6		93.8	
RHR II	70.1	(b)	107.5		97.0	

sectional area of the preparation was scarcely changed by hypertension: 0.43 ± 0.02 mm^2 (normotensive rats), 0.41 ± 0.03 mm^2 (SHR). The increase in force generation to 152.3 % of the control (Tab. 1) contrasts with a slight decrease of the unloaded shortening velocity from 0.81 ± 0.05 ML/s to 0.80 ± 0.03 ML/s, which means a diminution to 98.1 % (Table 1). Due to the increased force generation the maximum instantaneous power of the preparation, which was 1.43 ± 0.07 μW/mm ML in control rats, rose to 1.82 ± 0.10 μW/mm ML ($P < 0.05$). The force-velocity relations obtained in these experimental studies are: Control rats: V = 0.32 (1468-T)/(571+T); SHR: V = 0.22 (2236-T)/(622+T);

Hyperthyroid rats (HR): In this series force-velocity relations resulted in an increase in both extent and velocity of contraction of the portal vein. As compared with the control (0.93 ± 0.04 ML/s, n = 38), after inducing hyperthyroidism for 14 days the unloaded shortening velocity was distinctly ($P < 0.05$) increased to 110.8 % (1.03 ± 0.04 ML/s, n = 38, Fig. 1, Tab. 1). The isometric peak tension rose from 1480 ± 43 dyn to 1605 ± 41 dyn ($P < 0.005$), whereas the cross-sectional area of the preparation was slightly changed from 0.51 ± 0.02 mm^2 (euthyroid rats) to 0.50 ± 0.02 mm^2 (hyperthyroid rats). The increase in V_o and T_o produced an increase of the maximum instantaneous power from 1.35 ± 0.06 μW/mm ML to 1.68 ± 0.06 μW/mm ML, that is a rise to 124.4 % ($P < 0.0005$, Tab. 1). The corresponding functions of Hill's equation are: V = 0.24 (1480-T)/(388+T) for control animals, and V = 0.30 (1605-T)/(460+T) for hyperthyroid rats.

DISCUSSION

Previous studies of the isolated portal vein showed that under certain experimental conditions extent and velocity of contraction are influenced independently from each other. Polytropic effects (Peiper and Laven[1]) stand for influencing the force generation in particular, and can be produced by variations of the concentration of extracellular calcium-ions (Fig. 1) or by changes in extracellular pH not below pH 6.8 (Hellstrand et al.[3]; Peiper et al.[4]). Tachytropic effects are the result of changes in temperature for instance (Fig. 1), and they affect the velocity of contraction whereby the force generation is nearly unaffected (Peiper et al.[5]). At a low pH of 6.3 both force generation and velocity of contraction are reduced (Fig. 1) so that polytropic as well as tachytropic effects are detected (Peiper et al.[4]).

Polytropic effects are attributed to an influence on the number of activated interaction sites between the contractile proteins, i.e. a process of recruiting centres for force generation, whereas tachytropic effects are explained by an influence on the velocity of cross-bridge movement, that is the myosin ATPase activity.

Because of similar phenomenological indications the behaviour of the portal vein under pathophysiological conditions can be compared with observed effects of changes in pCa, pH, and temperature on the dynamics of vascular smooth muscle contraction (Fig. 1, Tab. 1):

(1) Spontaneously hypertensive rats showed no changes in the velocity of shortening at zero load (98.1 % of control), but an increase in force generation to 152.3 %, which cannot be explained by an increase of the cross-sectional area of the preparation. Sole polytropic effects also exist at varied pCa or pH not below pH 6.8 (Hellstrand et al. [3]; Peiper et al. [4]). A distinct increase in force generation of portal veins of SHR, independently of the kind of activation (noradrenaline, $BaCl_2$, KCl), was observed by Greenberg and Bohr [9], but contrasts with the findings of Hallbäck et al. [10] in venous and of Hansen and Bohr[11] in arterial smooth muscle. For explanation of this polytropic effect either an action localized to the membrane (improvement of the electromechanic coupling in SHR so that there is a higher concentration of intracellular calcium during the period of activation (Greenberg and Bohr [9])), or an effect localized intracellularly (augmentation of the sensitivity of myosin ATPase to calcium) has to be taken into account.

(2) A distinct negative tachytropic effect of renal hypertension is observed in cardiac muscle in principle. In myocardium a reduction of the specific actomyosin ATPase activity could be detected at renal hypertension (Jacob et al. [8]). A similarly diminished activity of the actomyosin ATPase in cardiac and vascular smooth muscle can be produced by temperature changes (Bárány [12]; Peiper et al. [5]).

(3) Certainly the tachytropic as well as the polytropic effect of hyperthyroidism is more pronounced in the cardiac muscle (Buccino et al. [6]; Skelton et al. [13]; Gunning et al. [14]; Strauer and Scherpe [7]) than in the portal vein. In myocardium the thyroxineeffect may be due to an influence on the sarcoplasmatic calcium exchange (Nayler et al. [15]), or an increase in the activity and/or a change in myosin structure (Strauer and Schulze [16]).

The main result of this report is the evidence of similar changes in the force-velocity relations by some pathophysiological conditions in myocardium and vascular smooth muscle. It must be stated, however, that only the same phenomena in both kinds of muscle have been described as yet, and it is an open question whether the effects are caused by influencing the same processes.

Extent and velocity of contraction can be varied independently from each other (polytropic and tachytropic effects). The intracellular calcium-ion concentration is a main factor in regulating force generation, whereas the mechanism responsible for the tachytropic effect in renal hypertensive as well as in hyperthyroid rats is still unknown.

REFERENCES

1. Peiper, U. and Laven, R., 1976: Noradrenaline and pH-effects at the membrane of vascular smooth muscle. In: Ionic Actions on Vascular Smooth Muscle. E. Betz (Ed.), pp. 56 - 60, Springer-Verlag Berlin-Heidelberg-New York.
2. Peiper, U. and Laven, R., 1976: Arzneimittelforsch. (Drug Res.) 26, 1235
3. Hellstrand, P., Johansson, B. and Ringberg, A., 1972: Acta physiol. scand. 84, 528-537
4. Peiper, U., Ehl, M., Johnson, U. and Laven, R., 1976: Pflügers Arch. 364, 135-141
5. Peiper, U., Laven, R. and Ehl, M., 1975: Pflügers Arch. 356, 33-45
6. Buccino, R. A., Spann, J. F. jr., Pool, P. E., Sonnenblick, E. H. and Braunwald, E., 1967: J. Clin. Invest. 46, 1669-1682
7. Strauer, B. E. and Scherpe, A., 1975: Basic Res. Cardiol. 70, 130-141
8. Jacob, R., Kämmereit, A., Medugorac, I. and Wendt-Gallitelli, M. F., 1976: Z. Kardiol. 65, 392-400
9. Greenberg, S. and Bohr, D. F., 1975: Circulation Res. 36, Suppl. 1, 208-215
10. Hallbäck, M., Lundgren, Y. and Weiss, L., 1971: Acta physiol. scand. 81, 176-181
11. Hansen, T. R. and Bohr, D. F., 1975: Circulation Res. 36, 590 - 598
12. Bárány, M., 1961: J. gen. Physiol. 50, 197 - 216
13. Skelton, C. L., Soleman, H. N., Wildenthal, K. and Braunwald, E., 1970: Circulat. Res. 27, 301 - 309
14. Gunning, J. F., Harrison, C. E. jr. and Colemann, H. N. III, 1974: Amer. J. Physiol. 226, 1166 - 1171
15. Nayler, W., Meerillees, N. C. D., Chipperfield, D. and Kurtz, J. B., 1971: Cardiovasc. Res. 5, 469
16. Strauer, B. E. and Schulze, W., 1976: Basic Res. Cardiol. 71, 624-644

Excitation-Contraction Coupling in Smooth Muscle, R. Casteels et al. eds

MECHANICS OF SINGLE ISOLATED SMOOTH MUSCLE CELLS

Fredric S. Fay
University of Massachusetts Medical School
Department of Physiology
55 Lake Ave., N.
Worcester, Massachusetts 01605

INTRODUCTION

Numerous studies have been performed over the past years into the mechanical properties of smooth muscle[1,2,3,4]. The evenual aim of such work is to utilize mechanical data to obtain insight into the process of force generation in the smooth muscle cell. All previous studies have been performed on segments of intact tissue and there are reasons to suspect that the properties of the tissue may not directly or simply reflect the properties of the individual cells. This might arise from the fact that individual smooth muscle cells within a tissue may well exist in a functionally heterogeneous state. In addition, even if homogeneity existed, the interconnection of cells by a complex network of intercellular connections may well obscure or blur fundamental cellular properties. For these reasons, I have chosen to develop methods for studying the mechanical properties of single isolated smooth muscle cells rather than whole tissue.

Single smooth muscle cells have now been isolated from a number of sources[5,6,7,8]. In my laboratory, work has concentrated on the properties of smooth muscle cells isolated by enzymatic digestion from the stomach muscularis of the toad, <u>Bufo marinus</u>. These cells are chosen because of their relatively large size (5-8 μm diameter, 200-400 μm length) which facilitates electrical and mechanical studies and because a considerable amount of information has now been accumulated which indicates that the cells are basically unaltered by the isolation procedure[9,10,11,12]. Furthermore, the increasing body of information regarding the structural[9,10], biochemical[12], and electrophysiological properties of these cells should prove extremely useful in analyzing the generation and regulation of force using these single cells as a model system.

MATERIALS AND METHODS

Single smooth muscle cells are isolated by enzymatic digestion[13] of thin slices of stomach muscularis from the toad, <u>Bufo marinus</u>. A single cell is then tied at either end to a glass probe of the sort shown in Figure 1. One probe is attached to an ultrasensitive force transducer[14] whereas the other is attached to two systems in series, one allowing for small rapid changes in cell length and a second allowing for larger but slower length changes. The cell is continuously observed with phase contrast optics through an inverted microscope and its image recorded on videotape. The cell is stimulated via a pair of silver plates that lie parallel to its length. Other details of the experimental arrangement have been reported elsewhere[15,16].

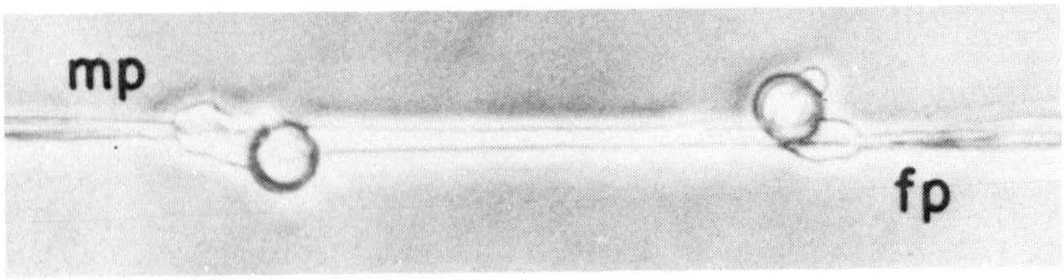

Fig. 1. Photomicrograph (phase contrast, X320) of single smooth muscle cell knotted at each end around a probe to facilitate isometric measurements of force. One probe (fp) was attached to a force transducer whereas the other probe (mp) was attached to a Burleigh Piezoelectric Translator (Model PZ-30) in turn mounted on a Kopf Hydraulic Microdrive. An eddy current sensor (Kaman Sciences, Model KD-2000-.5SU) was used to measure displacement amplitudes and timecourse. The microdrive was used for large amplitude stretches, and the piezoelectric device for small sinusoidal oscillations. Cell length was adjusted so that resting force was between 5 and 12 μg. Experiments were performed at room temperature (19-22°C). Most cells used were exposed to 10 μM isoproterenol to ensure that they remained relaxed during the knotting procedure. At this concentration isoproterenol has a potent relaxing effect on the isolated cells which can be overcome by electrical or other stimulation.

RESULTS AND DISCUSSION

A typical response of an isolated smooth muscle cell to a single 1 msec stimulus is shown in Figure 2. A peak force of 179 μg was achieved in 11.5 sec, whereas the relaxation occurred more slowly, having a $t_{1/2}$ of 7 seconds. In cells

Fig. 2. Isometric recording of contractile response of a single smooth muscle cell to a submaximal electrical stimulation. Stimulus was a 1-ms square wave of 14-mA amplitude and occurred at the arrow. Maximum force achieved was 179 μg, or 0.41 kg cm^{-2}. Cross-sectional area of individual fibers on which force measurements were made was calculated assuming the cells to have a circular profile. Cell diameter was measured midway along the length of the cell from video recordings through a microscope: the video image gave a X500 magnification of the cell. Cross sections of fixed single smooth muscle cells revealed a basically circular profile.

in which the length-force relationship has been investigated, peak forces of 3 kg/cm^2 are routinely obtained at the length for optimum active force development. This is comparable to peak forces measured from most intact tissue when the forces are normalized for the cellular cross-sectional area only[17,18]. Unlike intact tissue, however, the isolated smooth muscle cells exhibit low, almost constant, resting forces over a range of lengths extending to over twice that for optimal active force development[16,19].

The onset of active force development is considerably delayed following stimulation with a single 1 msec stimulus of supramaximal intensity. A typical delay between such stimulation and force development is illustrated by the record in Figure 3. In 60 cells subject to similar stimulation an average delay of 214 ± 13 msec was observed. This delay is more than 100X longer than that seen in skeletal muscle[20] and more than 10 X longer than that seen in cardiac muscle[21] as well. A series of experiments have been performed in order to pinpoint the cause for these extraordinary long delays in the isolated smooth muscle cells.

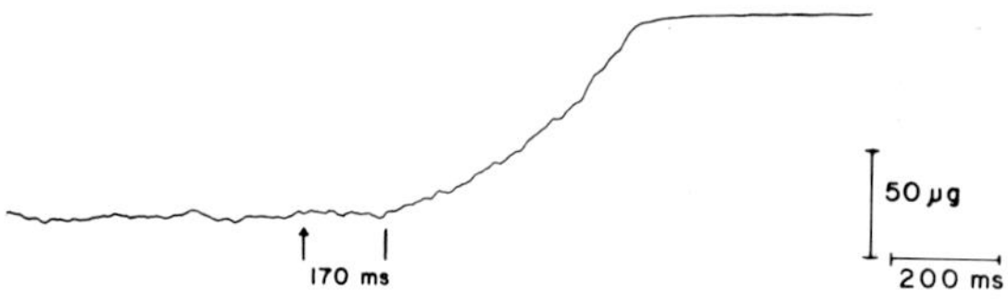

Fig. 3. Isometric recording of the contractile response of a single smooth muscle cell to supramaximal electrical stimulation. Stimulus (1-ms, 55-mA square wave) was initiated at the arrow on the record. 170 ms elapsed between stimulation and onset of measurable active force. Maximum force developed by this cell was 698μg (1.1 kg cm^{-2}). Due to the high gain used to facilitate early detection of active force development, amplifier saturates (flat right-hand portion of record) 700 ms after stimulation.

In order to decide if the delay resided in steps following or preceding cross-bridge attachment, experiments were performed to determine if cross-bridge attachement precedes the onset of active force development. Single cell stiffness was probed using either small (0.20-0.5% of cell length) sinusoidal (10-60 Hz) length perturbations or larger (5-10% of cell length) ramp changes in cell length. Both methods revealed that there is no change in stiffness prior to onset of active force development. The results of an experiment using the sinusoidal length perturbation technique are shown in Figure 4. Note that force and stiffness

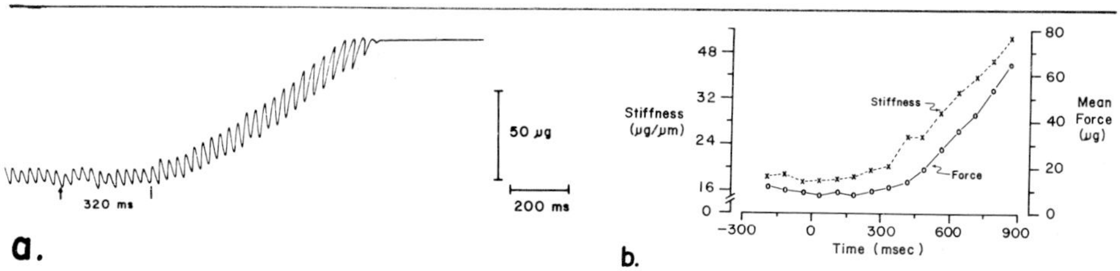

Fig. 4a. Isometric recording of the contractile response of a single smooth muscle cell to a supramaximal electrical stimulation while undergoing sinusoidal length oscillations. Stimulus (1-ms 62-mA square wave) was initiated at the arrow. Sinusoidal length oscillations (33 Hz) were 0.5 μm peak to peak. Cell length between knots was 186 μm. Flat portion of the record (right-hand portion of trace) indicates saturation of amplifier at high gain used. Maximum force achieved was 249 μg (0.56 kg cm^{-2}). 4b. Average force (0) and stiffness (X) as a function of time before and after electrical stimulation for the cell in Fig. 4a. Stimulation occurred at zero time. Mean force was taken as the midpoint of the force trace in Fig. 4a. The plotted values for stiffness represent the mean stiffness during successive 75-ms periods; the latency for the onset of active force development was 320 ms; stiffness did not increase until active force was measurable.

increase simultaneously. If cross-bridges had attached much prior to the onset of active force development, an increase in stiffness might be expected to preceed the increase in force development. This was not seen. In the experiments where larger ramp increases in length were executed, 40-100 msec after the stimulus, no change in the mean latency was observed (285 ± 20 msec) when compared to the control group (249 ± 33 msec). This makes it unlikely that the delay in force development reflects the time required for a rapidly activated contractile system to draw itself from a slack to a taut configuration. If the time for such a transition were responsible for the long delays, a stretch during the latent period should have shortened the observed delay. The delay thus appears to reside in steps prior to cross-bridge attachment.

One possible cause of the delay in this portion of the excitation-contraction sequence is that the delay in force production reflects the time required for Ca^{+2} levels at the myofibrillar sites to achieve same threshold required for activation. This might occur if Ca^{+2} release were a slow but continuous process following electrical stimulation. However, this explanation seems unlikely, since, as can be seen from Table 1, in a smooth muscle cell already producing

TABLE 1. EFFECT OF A PRECEDING SUBMAXIMAL ACTIVATION ON THE LATENCY BETWEEN SUPRAMAXIMAL ELECTRICAL STIMULATION AND THE ONSET OF ACTIVE FORCE.

Conditions	Latency (ms ± s.e.)
Supramaximal stimulus	201 ± 42 (n = 11)
Supramaximal stimulus, applied during submaximal contraction*	166 ± 26 (n = 8)

Latencies were determined after a supramaximal stimulus (1-ms 50-60-mA square wave) for the number of cells in parentheses. Only latency data from thos suspensions where the effect of a supramaximal stimulus alone and the effect of a supramaximal stimulus applied after a submaximal stimulus was tested have been included in the mean latencies.

* Submaximal contraction was initiated by a 10-15 mA stimulus, 1-ms square wave; supramaximal stimulus was applied during the plateau of the submaximal contraction.

some active force due to a preceding submaximal activation, the delay between a supramaximal stimulus and the onset of active force development is not significantly shortened relative to a control group given a supramaximal stimulus alone. In the first group of cells Ca^{+2} levels are presumably above the hypothetical threshold required for activation at the time of arrival of the supramaximal stimulus since some active force exists within the fiber at this time. The conclusion derived from these experiments is further strengthened by the observation that wide variations in extracellular Ca^{+2} concentration (10^{-7}-20 mM) have no effect on the delay in force development. On theoretical grounds[22], it seems unlikely that the time for diffusion of Ca^{+2} from its site of entry into the cytoplasm to its interaction with the myofibrils could be the source of the long delays. Even if Ca^{+2} had to diffuse from the surface membrane to a myofibril at the central axis of the cell (a 'worst case') diffusional equilibration would be 90% complete with 10 ms. Furthermore, the absence of an effect of changes in extracellular Ca^{+2} concentration on the latency also is consistent with the notion that the delay in the onset of force development can not be attributed to diffusional delays. Thus, one is left with only two possibilities. The first, that although Ca^{+2} arrives at the contractile machinery quite early in the latent period, considerable time is required to activate the contractile proteins. Recent evidence seems to indicate that the activation process in smooth muscle occurs via a phosphorylation reaction[23,24,25], triggered by Ca^{+2}. Thus, it might be that the kinetics of that reaction imposes a delay on the ability of the contractile machinery to produce force. As yet, no kinetic data appear to be available on the activation reactions, but the existence of a qualitatively different mechanism for activation of vertebrate smooth and striated muscles may well account for the differences in the lag between stimulation and force development between these muscle types. A second possibility that is as yet not excluded is that the delay in force development results from a delay between membrane depolarization and the opening of a Ca^{+2} gate, thus leading to an actual delay in the release of Ca^{+2} into the cytoplasm. Both possible causes of the long delays will have to be investigated in further studies.

Although comparison of stiffness within the isolated smooth muscle cells before and after stimulation provides a useful measure of the time required for cross-bridge attachment, comparison of the absolute value for single smooth muscle cell stiffness with that of whole tissue allows for a determination of the extent to which whole tissue stiffness reflects the properties of the individual smooth muscle cells. Such a comparison is made in Table 2. The stiffness of the isolated smooth muscle cells is higher than that measured in intact tissue preparations, although not too much higher than that of the smallest intact tissue studied[18]. This would seem to indicate that aside from the smallest preparations, the stiffness values measured in intact tissue probably reflect the combined properties of both non-cellular elastic elements and the smooth muscle cells. The stiffness of the isolated smooth muscle cells is also slightly higher than that of skeletal muscle when measured under similar conditions[26]. The finding of a somewhat higher stiffness in smooth muscle than striated muscle using low frequency length perturbations is not altogether surprising. If one assumes that stiffness reflects cross-bridge properties in both the single smooth muscle cells and in striated muscle, then at the lower frequencies cross-bridge reorientation and possibly even their dissociation may affect the stiffness measurement. The more rapidly that such changes can take place, the lower the measured stiffness

TABLE 2. COMPARISON OF STIFFNESS IN ISOLATED SMOOTH MUSCLE CELLS WITH INTACT SMOOTH MUSCLE TISSUE AND SKELETAL MUSCLE

Single Smooth Muscle Cells	Y/P	Stiffness (P_o/L_o units*)	Method***
Y(Young's Modulus) = 26.3 kg/cm^2 (n = 5)	81.3 ± 4.1 (n = 5)	100 ± 20 (n = 5)	SLP 20 Hz
P_{max} = 0.32 kg/cm^2 (n = 5) (Maximum Force/cross sectional area)			
Intact Smooth Muscle			
Mesotubarium, Rabbit (Meiss, personal communication)	65.8		SLP 80 Hz
Small Mesenteric Artery, Rat[18]	——	67 - 81	QR (4 ms)
Carotid Artery, Hog[1]	——	58	QR
Skeletal Muscle			
Whole Sartorius, Frog[26]	60.1	——	PWN < 20 Hz
Single Fibers, Frog[27]	136.0	——	QR (≃1 msec)

* Stiffness at peak force has been normalized using the cell length and maximum active force at the point on the length tension curve for optimum active force development.

** mean ± S.E.

*** Abbrevations used: SLP = Sinusoidal length perturbations
QR = Quick Release
PWN = Pseudorandom white noise

will be. Both measurements of myosin ATPase[17,28] and V_{max}[1,17] of smooth muscle suggest that cross-bridge turnover is much slower than in striated muscle. One might predict, therefore, that when low frequency length changes are used to measure stiffness, smooth muscle might be somewhat stiffer than striated muscles, as was found. As noted in Table 2 the stiffness of skeletal muscle is quite dependent on the speed with which the length perturbations are performed. Although not yet investigated, it is quite probable that measurements of stiffness in single smooth muscle cells will also exhibit dependence on the frequency or speed with which the length is changed. Thus, even the similarity in stiffness of isolated smooth muscle cells measured at 20 cps, and the stiffness of the small mesenteric artery preparation measured using more rapid length changes[18], may not be simply interpreted as indicating that in this tissue, stiffness values reflect those of the smooth muscle cells alone. In order to decide if that is the case, it will be necessary to compare measurements of stiffness under identical conditions.

ACKNOWLEDGEMENTS

This work was supported by a grant from NIH (HL 14523). F.S. Fay is the recipient of a Research Career Development Award (K04 HL 00048).

REFERENCES

1. Herlihy, J.T., and Murphy, R.A. (1974) Circ. Res. 34:461-466.
2. Gordon, A.R. and Siegman, M.J (1971) Am. J. Physiol. 221:1243-1249.
3. Meiss, R.A. (1971) Am. J. Physiol. 220:2000-2007
4. Csapo, A. (1954) Am. J. Physiol. 177:348-354.
5. Bagby, R.M., Young, A.M., Dotson, R.S., Fisher, B.A., and McKinnon, K. (1971) Nature 234:351-352.
6. Bagby, R.M., Thornsen, B., and McGaffee, L. (1976) Physiologist 19:113.
7. Kominz, D.R., and Groschel-Stewart, U. (1972) J. Mechanochem. Cell Motility 2:191-191
8. Small, J.V. (1974) Nature 249:324.
9. Fay, F.S., and Delise, C.M. (1973) Proc. Natl. Acad. Sci., U.S.A. 70:641-645.
10. Fay, F.S., Cooke, P.J., and Canaday, P.G. (1976) Physiology of Smooth Muscle (edit. by Bülbring, E., and Shuba, M.R.) 249-264 (Raven Press, New York).
11. Fay, F.S., and Singer, J.J. (1977) Am. J. Physiol. C144-C154.
12. Honeyman, T.H., and Fay, F.S. (1975) Fedn. Proc. 34:361.
13. Singer, J.J., and Fay, F.S. (1977). Am. J. Physiol. 232:C138-C143.
14. Canaday, P.G., and Fay, F.S. (1976) J. Appl. Physiol. 40:243-246.
15. Fay, F.S. (1977) Nature 265:553-556.
16. Fay, F.S. (1975) INSERM 50327-342.
17. Murphy, R.A. (1976) Blood Vessels 13:1-23.
18. Mulvany, M.J., and Halpern, W. (1976) Nature 260:617-619.
19. Fay, F.S. (1976) In: Cell Motility (ed. R. Goldman, T. Pollard, and J. Rosenbaum) 185-201 (Cold Spring Harbor Symposium).
20. Sandow, A., and Preiser, H. (1964) Science 146:1470-1472.
21. Morad, M., and Goldman, Y. (1973) Prog. Biophys. Molec. Biol. 27:257-313.
22. Hill, A.V. (1948) Proc. R. Soc. B135:446-453.
23. Sobieszek, A., and Bremel, R.D. (1975) Eur. J. Biochem. 55:49-60.
24. Sobieszek, A., and Small, J.V. (1976) J. Molec. Biol. 102:75-92.
25. Aksoy, M.O., Williams, D., Sharkey, E.M., and Hartshorne, D.J. (1976) J. Biochem. biophys. Res. Commun. 69:35-41.
26. Halpern, W.J., and Moss, R.L. (1976) Am. J. Physiol. 230:205-210.
27. Huxley, A.F., and Simmons, R.M. (1971) Nature 233:533-538.
28. Mrwa, U., Paul, R.J., Kreye, V.A.W., and Ruegg, J.C. (1975) INSERM 50:319-326.

Excitation-Contraction Coupling in Smooth Muscle, R. Casteels et al. eds.

MECHANICAL PROPERTIES OF SMOOTH MUSCLE CELLS

Michael J. Mulvany
Aarhus University
Biophysics Institute
DK-8000 Aarhus C
Denmark

INTRODUCTION

The mechanics of smooth muscle have been the subject of extensive investigations[1], but the relation of the parameters thus determined to the mechanical properties of the individual smooth muscle cells has not been clear. Unlike skeletal muscle fibres, which extend between the muscle tendons, smooth muscle cells are relatively short and are arranged in a mesh within the muscle tissue. It has therefore been thought that the arrangement of the intercellular connections and the mechanical properties of these connections could result in the mechanics of the tissue being very different from the mechanics of the individual cells. For this reason we[2] developed a preparation in which the individual cells could be visualized in the light microscope. Using this preparation we showed that there is apparently a close correlation between the mechanics of the individual cells and the mechanics of the preparation, and further evidence for this will be presented here. The preparation has also proved suitable for performing dynamic mechanical tests with a time resolution greater than that previously reported with smooth muscle, so that it has been possible to investigate the transient responses of the preparation with a time resolution of about 4 ms.

METHODS

The preparation used was a 150 μm artery taken from the mesenteric bed of a Wistar rat. The method of dissection and mounting has been described in detail elsewhere[3]. Briefly, a segment about 1 mm long was threaded onto two 32 μm

Fig. 1. Sketch of myograph used in this investigation.

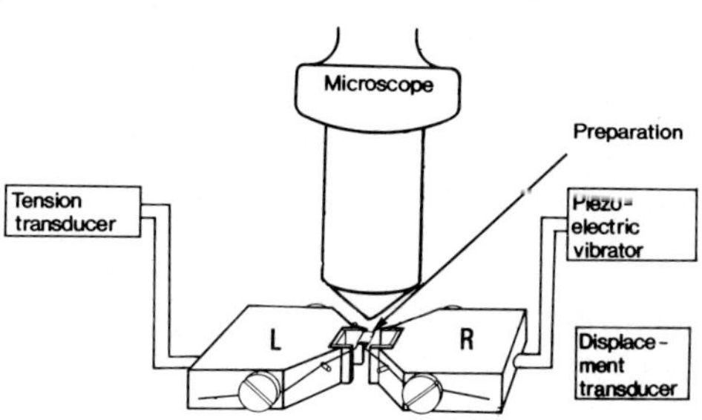

tungsten wires. These wires were attached (Fig. 1) to supports mounted, respectively, on a tension transducer (Kistler Morse, DSC6) and a displacement device (Burleigh, PZ-40), whose movement was monitored by an eddy current displacement transducer (Kaman 2300-. 5 SU). This myograph was mounted on the stage of a micrometer (Zeiss, photomicroscope II) equipped with Nomarski optics. The vessel image was photographed using high contrast copy film (Kodak). Vessels were held at an internal circumference $L_1 \approx 0.9\ L_0$, where L_0 is the internal circumference for which the maximum active wall tension was obtained. Vessels were maximally activated[3] by changing the immersion solution from normal Ringer's to a high K^+ Ringer's solution containing 5 mM Ca^{2+}. The 'quick release' and 'step' responses of the vessels were investigated by imposing 5 ms ramp changes in internal circumference (of up to 0.1 L_0) and monitoring the resulting changes in tension during and after the release on an oscilloscope (Tektronix, 5103 N/D11).

RESULTS

The small wall thickness of this preparation makes it possible to visualize the media in the light microscope without undue interference from the remainder of the preparation (Fig. 2a,b). The smooth muscle cells in the media are clearly visible and are seen to be oriented circumferentially. In many of the cells a number of intracellular structures can also be seen, and these have been used as markers to measure changes in cell length following imposed changes in internal circumference. Fig. 2a,b shows the media of a relaxed vessel before and after the internal circumference was decreased by 0.080 L_1. The measured fractional change in cell length was 0.077 ± 0.006 (n = 6). Similar results have been found in 7 other experiments using relaxed vessels and in 2 other experiments using activated vessels (Fig. 2c). In the latter experiments it has also been noted[4] that, following the quick release, there is no further change in cell length. Therefore,

---

Fig. 2. (a),(b). Photomicrographs using Nomarski interference contrast optics of the media of a resistance vessel (a) before, (b) after the internal circumference of the vessel was decreased. The vessel is threaded onto two 32 µm tungsten wires (W). The arrowheads point to a number of cells which appear in both pictures. The thin arrows point to 2 intracellular structures in one of these cells: such structures have been used as intracellular markers and the change in spacing between them from (a) to (b) is the basis for the measurements given in (c). Vessel held in Ringer solution to which 13% sucrose has been added to improve cell visibility.
(c). Measured changes in cell length plotted against imposed changes in internal circumference. Open symbols refer to measurements made with vessels relaxed in Ringer solution (squares) or Ringer solution plus sucrose (triangles) and were taken from micrographs similar to those shown in (a),(b). Closed symbols refer to measurements made with activated vessels of the change in cell length from immediately before to 10 msec after a quick release using video-recordings. Each point refers to one imposed change in internal circumference and shows the mean change in cell length of about 4 cells.

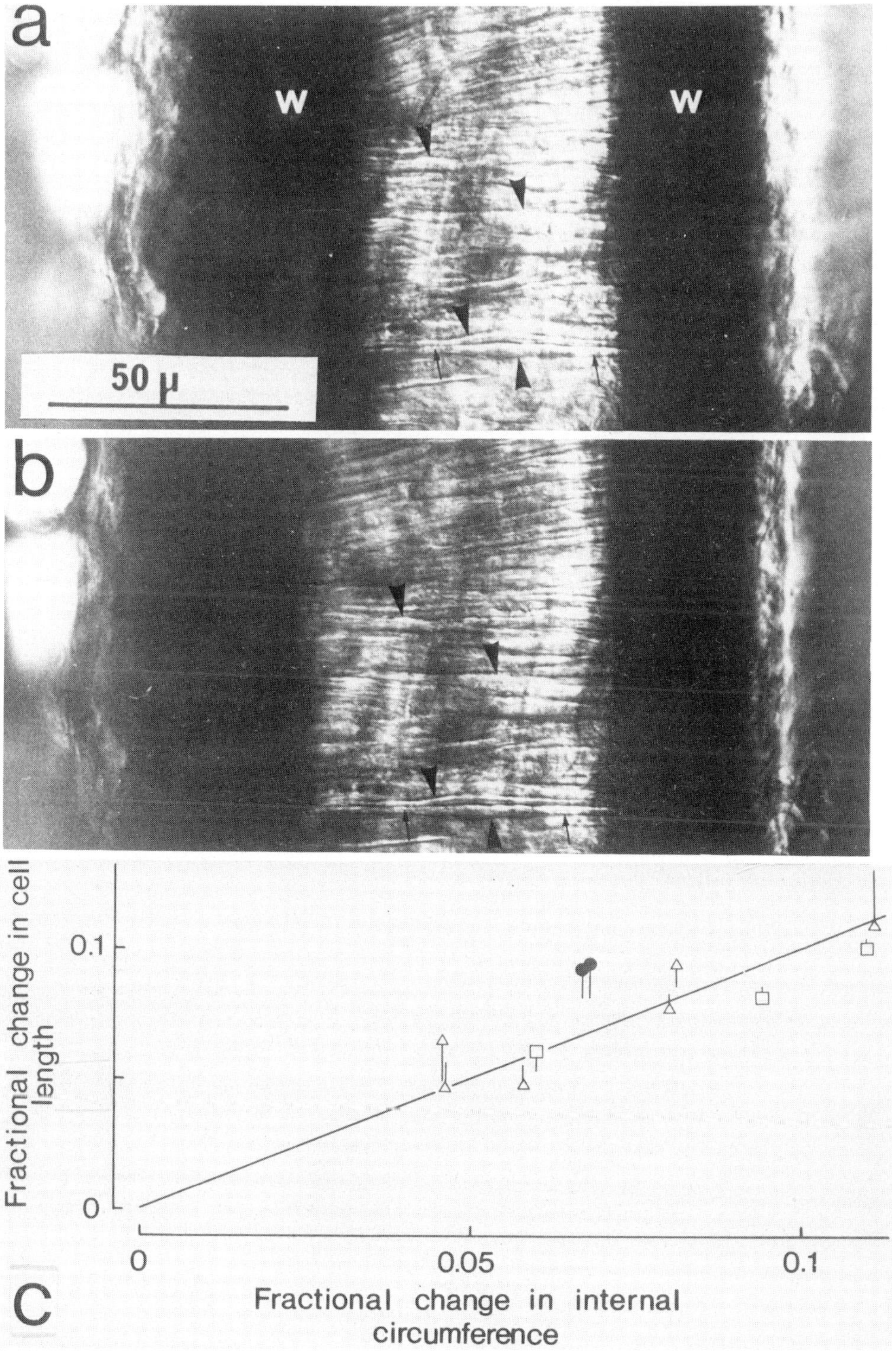

a
W
W
50 μ
b
Fractional change in cell length
0.1
0
0
0.05
0.1
c
Fractional change in internal circumference

in this preparation, both in the activated and the relaxed state, imposed changes in internal circumference produce the same fractional changes in cell length. From these experiments it is concluded (see Discussion) that the mechanical properties of the activated preparation are a true reflection of the mechanical properties of the activated smooth muscle cells it contains.

Two dynamic characteristics of the preparation have been investigated: The 'quick release' response (the relation between active wall tension and internal circumference when the latter is rapidly reduced) and the 'step' response (the time course of the recovery following a quick release). The quick release response of the preparation was found to be approximately exponential in shape (Fig. 3). The initial slope of this response gives the dynamic stiffness of the preparation and is equal to about 60 $\Delta T_1/L_1$ (ref. 3,5), where ΔT_1 is the active wall tension developed at L_1. The time course of the step response for smaller releases is seen (Fig. 4) to be approximately linearly related to the logarithm of the time following the initiation of the release. In previous experiments we[5] have shown that this is an almost exact relation for small (0.0016 L_1) releases. Here we see that the greater the release the more non-linear is this relation, the slope of the wall tension - log(t) curve increasing monotonically with time. As a measure of the rate of recovery the active half response time has been determined, that is the time taken for the wall tension to recover half the tension discharged during the release. This time increases with increasing release amplitudes; the time for small amplitudes is about 250 ms (ref. 3,5).

Previous experiments[5] have shown that when the active wall tension of the vessels was reduced by reducing the Ca^{2+} concentration of the activating solution, both the active dynamic stiffness and the active half response time were proportional to the active wall tension. For a given active wall tension level the dynamic stiffness was found to be temperature independent, but the active half response time increased with a Q_{10} of about 1.8.

DISCUSSION

It has been proposed recently[6] that the relatively high active stress of visceral smooth muscle may be due to a 'parallel' connection of cells, with cells which are situated in series in fact working in parallel. Such an arrangement would though also result in imposed changes in length producing proportionally greater changes in cell length. This was not the case in the vascular smooth muscle preparation described here: in relaxed vessels the fractional change in cell length was the same as the fractional change in internal circumference (Fig. 2). Thus in this preparation there does not seem to be any parallel connection of cells which are situated in series. Furthermore, because the same result

Fig. 3. Oscilloscope record of wall tension (vertical axis) and internal circumference (horizontal axis) during a 10 ms quick release for a preparation when activated (upper curve) and when it was relaxed (lower curve). L_l = 559 μm.

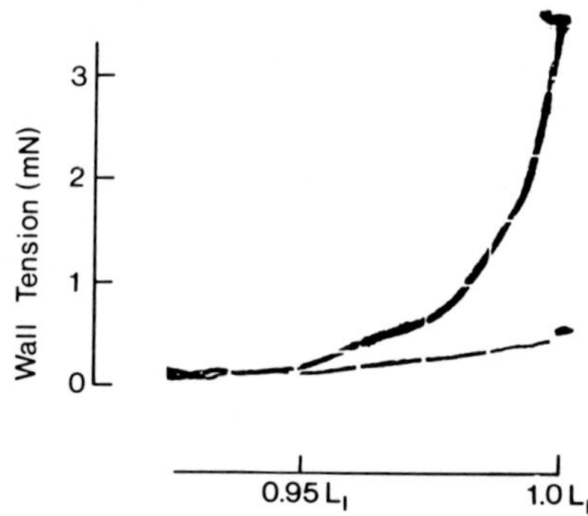

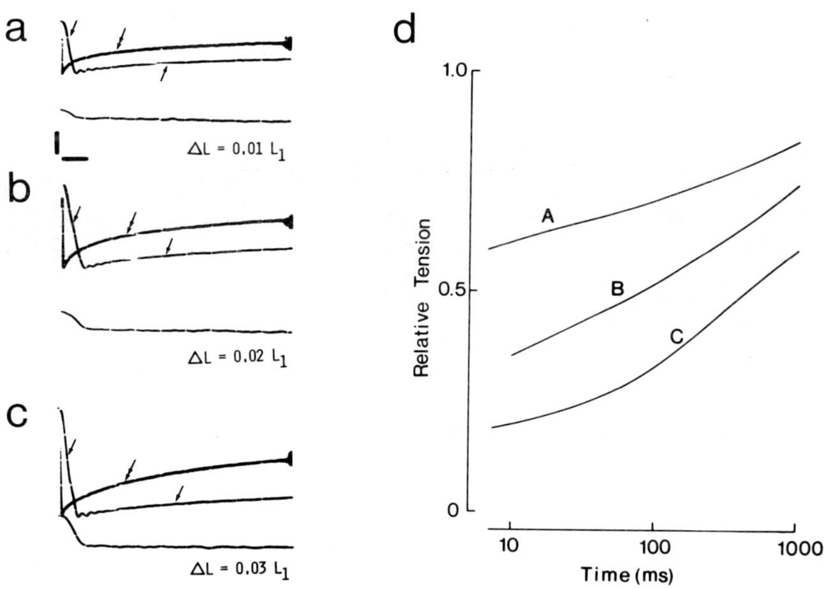

Fig. 4. (a),(b),(c). Oscilloscope records of the time course of quick releases and subsequent recovery for three different release amplitudes. Wall tension (upper two curves) recorded simultaneously with two different time bases: 10 ms/div (single arrowed curves) and 100 ms/div (double arrowed curves). The lower curve shows displacement recorded at 10 ms/div. All curves recorded at the sensitivities shown by the bars in (a), where vertical bar shows 0.5 mN (=0.16 ΔT_l) and 12 μm (=0.021 L_l) and horizontal bar shows 1 div.
(d). Curves show the tension recoveries in (a), (b) and (c), labelled A, B and C, respectively, plotted with time on a logarithmic scale.

has been obtained in activated vessels the overall compliance of the intercellular connections must be small compared with that of the activated cells. Therefore the mechanical responses of this preparation when activated are taken to be a close approximation to the mechanical responses of the active smooth muscle cells it contains.

On this basis the dynamic responses of smooth muscle cells appear to be very different to those of skeletal muscle fibres. The dynamic stiffness of skeletal muscle fibres[7] corresponds to about 200 $\Delta T_1/L_1$. The instantaneous elastic characteristic (IEC) is nearly linear[8] while the step response has at least two distinct phases[9]: a rapid exponentially shaped partial recovery with a time constant of about 2 ms (at 4°C) followed by a slower full recovery as the muscle adjusts to its new length. In the preparation described here the dynamic stiffness is about four times lower, the IEC is exponentially shaped (Fig. 3) and there are no readily discernible separate phases in the step response (Fig. 4a,b,c).

Ultrastructural studies of vertebrate smooth muscle have shown that both the actin[10] and myosin[11] it contains are organized in filamentous form. It is therefore concluded that, like skeletal muscle, vertebrate smooth muscle operates on a sliding filament principle. It might therefore be thought surprising that the dynamic responses of smooth and skeletal muscle are so different. However, although the basic contractile mechanism of smooth and skeletal muscle may be similar, the arrangement of the contractile filaments is apparently less organized in smooth muscle. In particular in smooth muscle the connections between the contractile filaments and the cell membrane give the appearance of being less rigid. It therefore seems likely that it is the compliance of these connections which gives rise to the relatively large compliance of smooth muscle and to the exponential form of the IEC, although it could also be due to the compliance of either connections between 'sarcomeres' (if these exist[12,13]) or of the filaments themselves. This suggests that a smooth muscle cell can be mechanically modelled by a contractile unit in series with an exponential spring. Fig. 5 shows how this concept can account for both the quick release and the step response of smooth muscle.

The mechanical characteristics of frog skeletal muscle fibres can be modelled fairly accurately by a contractile unit consisting of the series combination of a Voigt element[9] (to account for the initial response) and a force generator obeying Hill's equation[14] (to account for the subsequent recovery). The quick release and step responses of this model (Model I, Fig. 5a) are shown by the dashed curves in Fig. 5b,c and d, and are similar to those reported[7,9] for skeletal muscle fibres at 4°C. These responses are however substantially modified if an exponential spring is connected in series with the contractile unit (Model II, Fig. 5a), as shown by the full curves in Fig. 5b, c and d. The modified responses are seen to have the same form as the records shown in Fig. 4 as regards (i) the approximately exponential form of the quick release response, (ii) the linearity of the step response with respect to $\log(t)$ for small amplitude releases where t is the time following the start of the release, and (iii) the non-linearity of the step response with respect to $\log(t)$ for large amplitude releases. The recorded responses are about 10 times slower than those of the model, a factor

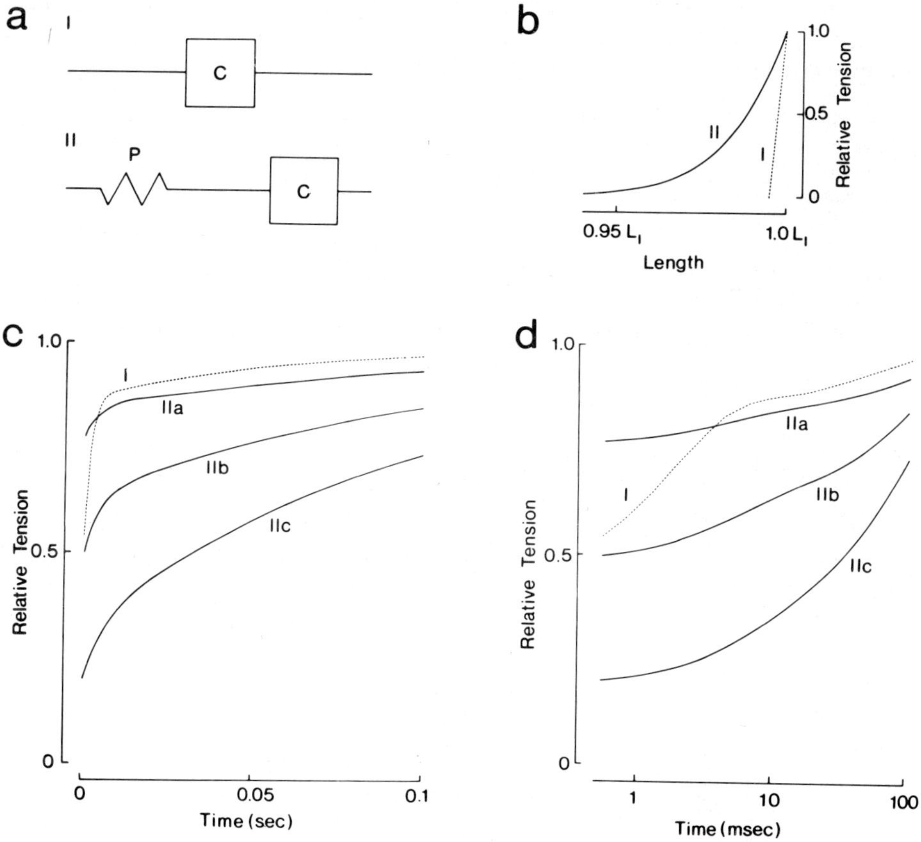

Fig. 5. Calculated responses of models described in the text showing effect of connecting an exponential spring in series with a muscle contractile unit.
(a) Model I shows model for muscle without passive compliance. The mathematical description of the contractile unit C is explained in the text. Model II shows model in which there is a passive element P (with an exponential characteristic) in series with C.
(b) Response of Models I (curve I) and II (curve II) to a 'quick release' plotted as in Fig. 3.
(c) Time course of recoveries from a quick release for Model I (curve I) and from quick releases of 3 different amplitudes for Model II (curve IIa, IIb and IIc).
(d) Curves shown in (c) replotted with time on a logarithmic scale.

which is consistent with the shortening velocity of the preparation being 10 times slower than that of frog skeletal muscle (at $4^{o}C$).

This model for the contractile apparatus of smooth muscle is clearly oversimplified, for no account is taken of the different stochiometry of the contractile proteins of smooth and skeletal muscles[15], or for possible differences in 'sarcomere length'. The purpose of presenting this simple model is, however, to demonstrate that the existence of a series passive compliance with an exponential characteristic would in itself be sufficient to account for the quick release and step responses of smooth muscle. A fuller description of the model will be presented elsewhere.

SUMMARY

The dynamic mechanical properties of the smooth muscle cells in a 150 μm artery have been shown by light microscopy to be similar to those of the whole preparation. Their quick release response is exponentially shaped. Their step response is linear with respect to log(time) for small releases, but non-linear for large releases. It is shown that both phenomena could be accounted for if there is an intracellular passive component with an exponential characteristic in series with the contractile units.

ACKNOWLEDGEMENTS

I would like to record my thanks to Ms. Kirsten Kandborg and Ms. Sanne Steinicke for technical assistance. This work was supported by the Danish Medical Research Council.

REFERENCES

1. Johansson, B., 1975, Experientia 31, 1377-1386
2. Mulvany, M.J. and Halpern, W., 1976, Nature 260, 617-619
3. Mulvany, M.J. and Halpern, W., 1977, Circ. Res., (in press)
4. Warshaw, D., 1977, private communication
5. Halpern, W. and Mulvany, M.J., 1977, J. Physiol. 265, 21-23 P
6. Gabella, G., 1976, J. Physiol. 263, 199-213
7. Julian, F.J. and Sollins, M.R., 1975, J. Gen. Physiol. 66, 287-302
8. Ford, L.E., Huxley, A.F. and Simmons, R.M., 1976, J.Physiol. 260, 28-29 P
9. Huxley, A.F. and Simmons, R.M., 1972, Cold Spring Habor Symp. Quant. Biol. 37, 669-680
10. Prosser, C.L., Burnstock, G. and Kahn, J., 1960, Am. J. Physiol. 199, 545-552
11. Lowy, J., Poulsen, F.R. and Vibert, P.J., 1970, Nature 225, 1053-1054
12. Ashton, F.T., Somlyo, A.V. and Somlyo, A.P., 1975, J. Mol. Biol. 98, 17-29
13. Small, J.V. and Sobieszek, A., 1977, J.Cell Sci. 23, 243-268
14. Hill, A.V., 1938, Proc. Roy. Soc. B 126, 136-195
15. Murphy, R.A., Herlihy, J.T. and Mergerman, J., 1974, J. Gen. Physiol. 64, 691-705

Excitation-Contraction Coupling in Smooth Muscle, R. Casteels et al. eds.
© 1977 Elsevier/North-Holland Biomedical Press

MECHANICAL AND ENERGETIC CORRELATES OF ISOMETRIC RELAXATION IN MAMMALIAN SMOOTH MUSCLE

M.J. Siegman, T.M. Butler, S.U. Mooers and R.E. Davies
Department of Physiology, Thomas Jefferson University, Philadelphia, PA 19107, and The School of Veterinary Medicine, University of Pennsylvania, Philadelphia, PA 19174, USA

INTRODUCTION

The decline of force during isometric relaxation is generally thought to be the result of a progressive termination of crossbridge cycling and actomyosin-ATPase activity as the intracellular free calcium concentration decreases. Mechanically, crossbridges presumably detach and filament overlap diminishes as elastic elements passively recoil. The similarity of the time course of isometric relaxation to that of stress relaxation in resting smooth muscles led Bozler to suggest that both are passive processes due to crossbridge detachment[1]. Recent studies from our laboratory[2,3] provided evidence for the existence of attached, but non-cycling crossbridges in a variety of resting, atonic smooth muscles. The present sutdy was intended to determine the energetics of relaxation and correlate this with mechanical events.

METHODS

Strips of taenia coli were isolated from rabbits (New Zealand white) as described previously[4]. The muscles were allowed to equilibrate for at least 2 hrs under 2 g-wt in a flowing Krebs-bicarbonate solution containing (mM): NaCl, 118; $NaHCO_3$, 25; KCl, 4.7; KH_2PO_4, 1.2; $MgSO_4 \cdot 7H_2O$, 1.2; $CaCl_2$, 1.9; glucose, 11, which was vigorously bubbled with 95% O_2-5% CO_2. The muscles were devoid of all spontaneous activity at 18-22°C. Muscle length was adjusted to either 90% L_o, where isometric force developed was 92% P_o, or to L_o, where maximum isometric force, P_o, was developed. Stimulation was with 60 Hz AC, 10 V rms pulses in a transverse field, which gave maximum mechanical activation.

For determination of the active state, muscle length was 3-4 mm at L_o, and servo-controlled quick-releases of 2% L_o at velocities exceeding 100 x Vmax[4] were imposed at precise intervals after cessation of 25 sec stimulation.

The maximum redeveloped force following quick-release was the active state intensity for each time interval. Control releases were made at rest and the values corrected accordingly. Releases of 1, 2 or 3% L_o did not change the active state decay when the time at which maximum redeveloped force occurred in each case was accounted for. The maximum redeveloped force during the peak of a tetanus was equal to P_o, hence represented 100% active state.

For energetics studies, the muscles were incubated anaerobically in 0.5 mM iodoacetate, 5 mM fluoroacetate, glucose-free Krebs solution for 30 min at 5^oC, in order to limit the resynthesis of ATP[5]. Following rewarming to 18^oC for 3.5 min the muscles were electrically stimulated according to the experimental designs shown in Fig. 1, then rapidly flattened and frozen with modified Wollenberger tongs precooled in liquid nitrogen. Control muscles were unstimulated. The muscles were subsequently extracted[6] and analyzed for ATP, ADP and AMP[3], phosphorylcreatine (PCr) and total creatine (Ct)[7].

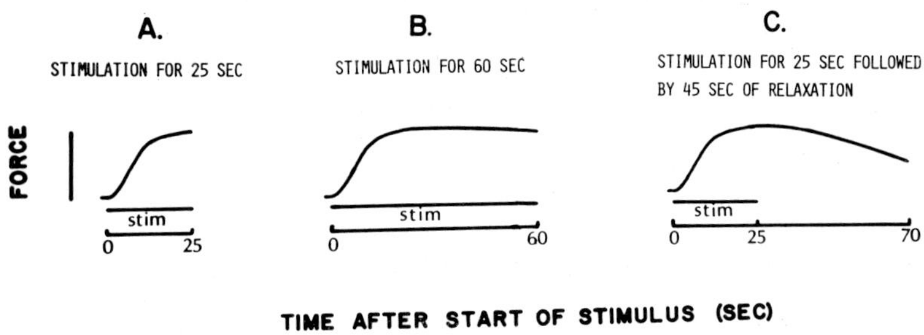

Fig. 1. Experimental designs for energetics studies (see text, above)

RESULTS AND DISCUSSION

The relative energy requirements for maximum isometric force maintenance during stimulation and during relaxation were determined first by comparing paired muscles that were stimulated for 60 sec (design B) with those that were stimulated for 25 sec and allowed to relax for 45 sec (design C). The energy usage was greater for muscles maintaining maximum isometric force than for those allowed to relax (ΔPCr/Ct for C - B = 0.05±).022, N=19, P<0.05). However, measurements of $\int$Pdt for both muscle groups showed that these were the same; for design B, $\int$Pdt = 144±11, and for design C, $\int$Pdt = 141±13, N=19.

These results suggest that force is maintained with less energy usage during relaxation than when maximum force is maintained during stimulation.

We may query, is force maintained more economically during relaxation in smooth muscle? A better estimate of the relative energy usage during relaxation and that occurring during maximum isometric force maintenance can be obtained from comparisons of the phosphorylcreatine breakdown determined in the experimental designs A, B and C. $\frac{B-C}{B-A}$ is the ratio of the difference between the energy usage for maximum force maintenance and that during relaxation to the energy usage for maximum force maintenance. Using the measured $\Delta PCr/Ct$ for each treatment in this calculation, we obtain 0.51±0.21, N=27. Therefore, 51% less energy is used during relaxation than during maximum isometric force maintenance. When similar comparisons are made for $\int Pdt$, we obtain a value 0.092±0.165, N=19. These results show that despite a 51% lower energy usage during relaxation than during maximum isometric force maintenance, there is no difference in $\int Pdt$ for the two conditions. This represents a significant change in the relationship between energy usage and force maintenance. Putting this another way, more force is maintained per ATP hydrolyzed during relaxation than during maximum force maintenance.

Is all of the force maintained during relaxation attributable to crossbridge cycling? The ability of the muscle to redevelop force following quick-release (active state) was determined at the peak of a tetanus during stimulation and at intervals following cessation of 25 sec stimulation (Fig. 2). The peak redeveloped force following quick-release (open circles) was normalized to P_o (100% active state = P_o). Values are means±SEM. The active state decays more rapidly ($t_{½}$ = 10 sec) than the force maintained by the muscle during isometric relaxation ($t_{½}$ = 19 sec). This suggests that not all of the force maintained during relaxation from a tetanus can be attributed to crossbridge cycling.

The active state has been taken as a determinant of the energy usage during relaxation. We find that the force maintained during the active state is equivalent to 15.35 sec of maximum isometric force. From the measured rate of

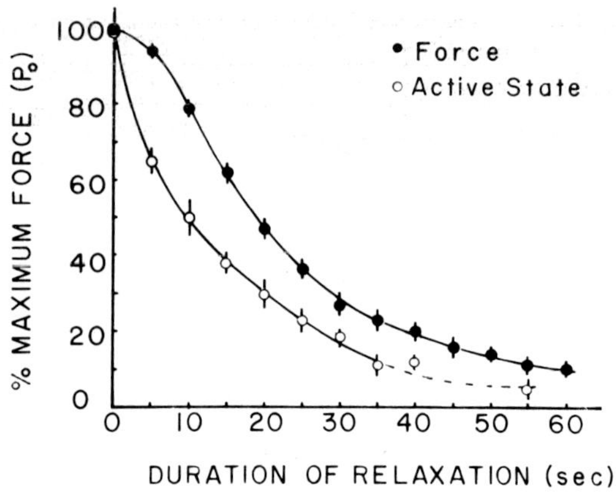

Fig. 2. Time course of isometric force and active state (redeveloped force following quick-release of 2% L_o at >100xVmax) during relaxation of the rabbit taenia coli at 18°C. 25 sec tetanic stimulation ceased at 0 time.

energy usage during maximum isometric force maintenance ($\Delta PCr/Ct \cdot sec^{-1} = 0.0028 \pm 0.0009$)[9], the expected energy usage during 45 sec relaxation is 0.0429±0.0147 mole PCr/mole Ct. The measured energy usage during 45 sec of relaxation is 0.06±0.04 mole PCr/mole Ct, N=10, and not significantly different. Therefore, the small energy usage correlates well with the extent to which force maintenance due to crossbridge cycling occurs during relaxation. The "extra" force maintained during relaxation may be due to the occurrence of a state in which force is maintained by attached, but non-cycling crossbridges[2].

We have also found that when contractile activity is prolonged by treatment with 1 mM caffeine and 20 mM nitrate, the active state is also prolonged, and this occurs in direct proportion to the increment in force maintenance. The rate of decay of the active state remains faster than that of isometric force. These results suggest that the time course of calcium removal is the primary determinant of the rate of relaxation.

During relaxation, the following scheme may operate: calcium removal from the vicinity of the contractile filaments results in gradual cessation of crossbridge cycling and associated energy input. As this occurs, the crossbridges revert to the attached state, in which force is transiently maintained[2]. In mammalian smooth muscle, the force maintained during relaxation is determined primarily by the time course of calcium removal and secondarily by stress relaxation of attached, but non-cycling crossbridges.

This study was supported by NIH grant HL 15835 to the Pennsylvania Muscle Institute and HD 06074 to MJS.

REFERENCES

1. Bozler, E., 1974/1976, Mechanical Properties of Contractile Elements of Smooth Muscle. In: Physiology of Smooth Muscle. E. Bulbring a. M.F. Shuba (Eds), pp. 217-221, Raven Press, New York.

2. Siegman, M.J., Butler, T.M., Mooers, S.U. and Davies, R.E., 1976, Am. J. Physiol. 231, 1501-1508

3. Butler, T.M., Siegman, M.J. and Davies, R.E., 1976, Am. J. Physiol. 231, 1509-1514

4. Gordon, A.R. and Siegman, M.J., 1971, Am. J. Physiol. 221, 1243-1249

5. Siegman, M.J., Butler, T.M., Mooers, S.U. and Davies, R.E., 1976, Physiologist 19, 366

6. Kushmerick, M.J., Larson, R.E. and Davies, R.E., 1969, Proc. Roy. Soc. Lond. B 174, 293-313

7. Kushmerick, M.J. and Davies, R.E., 1969, Proc. Roy. Soc. Lond. B 174, 315-353

8. Butler, T.M., Siegman, M.J., Mooers, S.U. and Davies, R.E., 1977, Fed. Proc. 36, 497

Excitation-Contraction Coupling in Smooth Muscle, R. Casteels et al. eds.

SMOOTH MUSCLE ENERGETICS

Richard J. Paul
University College London
Dept. of Physiology
London WC1 6BT
England

John W. Peterson
Universität Heidelberg
II Physiologisches Institut
6900 Heidelberg 1
Germany

Energetics, the study of biological energy conversions, in particular those between metabolically provided chemical free energy and such energy dependent processes as contractility and transport, is as yet in a relatively primitive state of development for smooth muscle. This in fact may be a good thing, for most of the recent progress in skeletal muscle energetics has been to undermine what once were considered to be solid facts (1). The experimental data that we would like to have for smooth muscle is the rate of ATP hydrolysis attributable to actomyosin interaction under the varying mechanical conditions faced by the muscle. One would then be in a position to relate the kinetics of the intact tissue to that of the isolated actomyosin ATPase (2), as well as to establish the thermodynamic bounds on smooth muscle performance. Another facet about which relatively little is known is the coupling between metabolism and the activation or excitation-contraction coupling processes: for example, how much of the total cellular ATP production is coupled to Ca^{++} or other ionic fluxes. Our work towards answering some of these questions has recently been reviewed (3,4).

One of the characteristics of smooth muscle metabolism is the low level of phosphogen stores compared with the rates of phosphagen utilization. Basal metabolism alone would exhaust the available phosphagen pools in only a few minutes, if ongoing metabolic synthesis were blocked. This can be used to advantage in smooth muscle studies for one can use non-tissue destructive techniques to measure intermediary metabolism, which in the steady state can be readily interpreted in terms of cellular ATP utilization. It is particularly useful for the very slow tissues from swine and cattle in which the development of mechanical and metabolic steady states have similar time courses, permitting rates of intermediary metabolism to be directly correlated with mechanical events.
The rate of oxygen consumption in bovine mesenteric vein was found to be linearly related to the level of maintained isometric force (5), under a variety of mechanical and pharmacological states.

This particular pattern has now also been observed for the hog carotid artery (6), by Hellstrand (7) in the rat portal vein, and is consistent with other smooth muscle studies. An obvious question then is how to partition these changes in total cellular metabolism between actomyosin ATPase, excitation-contraction coupling, and other cellular energy-dependent processes. The fact that smooth muscle is a muscle offers some help. As the dependence of isometric force on tissue length is now nearly unanimously accepted as being due to some arrangement of sliding filaments, changes in metabolism with tissue length also likely reflect the changes in the number of available actomyosin interaction sites. Figure 1a shows the relations between suprabasal oxygen consumption rate, isometric force, and tissue length in a single segment of mesenteric vein. In Figure 1b, the suprabasal rates of oxygen consumption are plotted against isometric force, illustrating the linearity described.

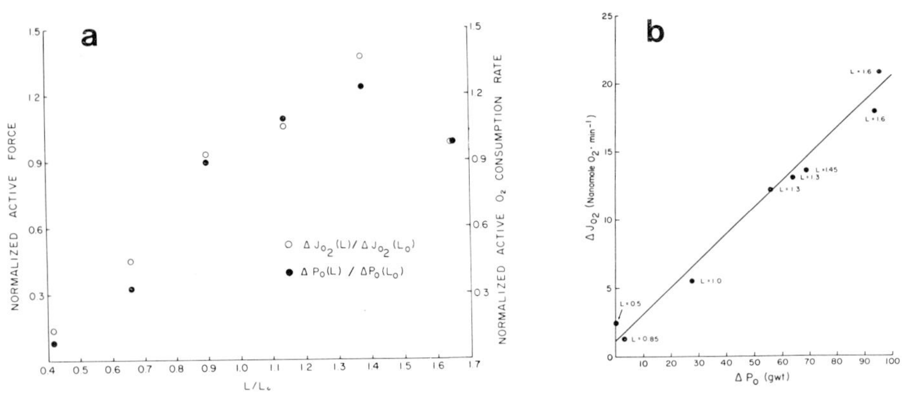

Figure 1 (a) Plot of suprabasal oxygen consumption (ΔJ_{O2}) and active isometric force (ΔP_o) against tissue length expressed relative to the zero passive force length (L_o). ΔJ_{O2} and ΔP_o are normalized to respective values at L_o. (b) Data taken from 1 (a); absolute ΔJ_{O2} plotted against active isometric force at indicated lengths.

The average (non-linear regression) relations found in 55 veins are shown in Figure 2. Though isometric force and oxygen consumption rate show a similar dependence on muscle length, the suprabasal oxygen consumption rate does not go to zero at zero active force levels in the stimulated muscle, as is seen at both long and short lengths. No statistical significance was found for the higher average value at the long-length ΔJ_{O2} zero force point compared to the shorter.

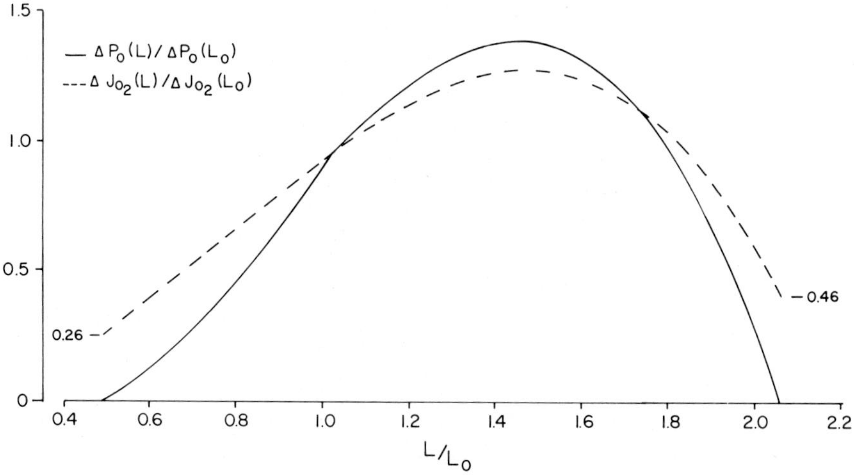

Figure 2 All data from oxygen consumption rate and isometric tension measurements with varying muscle length at maximal pharmacological activation were normalized to respective values at L_o and computer fit by least-squares non-linear regression. Results are given above.

This becomes clearer when metabolism is plotted directly against force as shown in Figure 1b and for the average of 55 veins in Figure 3. The upper curve shows conditions of maximum stimulation with the force being altered by changing muscle length; whereas, in the lower line, isometric force at a fixed length was altered by varying the level of pharmacological stimulation. The difference is metabolic energy input at the same level of maintained isometric force, which we'd like to attribute solely to differences in activation. This is particularly clear at L_{min}, where no isometric tension is developed, yet there is a finite increase in metabolism above basal levels upon stimulation. At maximum isometric force, about 40% of the total cellular ATP production is related to actomyosin ATPase and about 10% of the total or 20% of suprabasal metabolism is related to tension-independent or activation processes. In the tonic vessels of large animals, the tension-independent ATP utilization is about 0.15 μmole ATP/(min·g). Recent work following similar experimental lines (7) on rat portal vein, a preparation showing phasic spontaneous activity with relatively high shortening velocities gives a 5-10 fold higher figure for activation energetics; about 1 μmole ATP/(min·g).

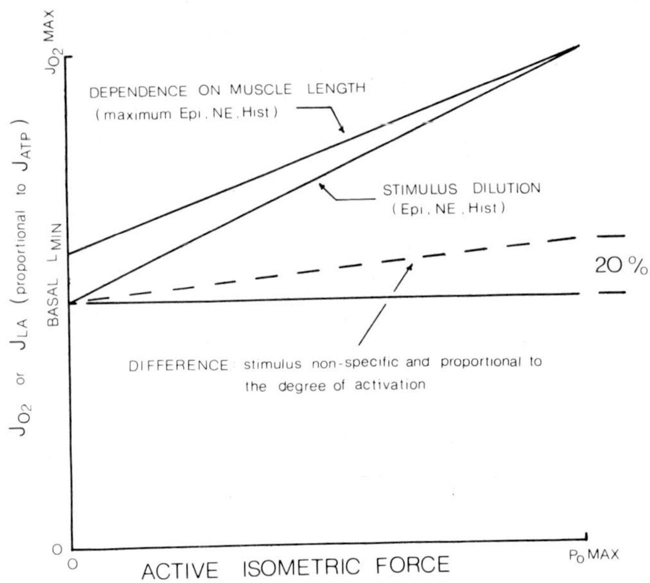

Figure 3 Data from 55 veins were averaged and results are shown above. The dashed line (difference of the two lines with and without varying degrees of activation) is taken to represent a linear increase in energy utilization for activation processes with increasing stimulation. In bovine mesenteric vein, the results were the same for both oxidative metabolism and aerobic glycolysis.

Before attempting more theoretical questions or interpretations of data from other mechanical conditions, our recent work has been directed in two areas which are critical to the interpretation of the evidence just summarized.

The first deals with an interpretation of the force-length data in terms of actomyosin interaction. Because of the large passive forces, smooth muscle data at very long muscle lengths are not very precise and metabolic measurements can often be masked by tissue damage. Thus, most of the evidence on the length-dependent metabolism is based on the ascending limb of the force-length relation; that portion in which the classical sliding filament models are ambivalent as to the mechanism accounting for the fall in tension. Though data at long lengths are consistent, it is not possible to achieve non-overlap lengths with certainty. One of the ways of testing the hypothesis that the drop in tension at short lengths is due to the removal of actomyosin interaction sites is to measure muscle stiffness. In these experiments, one imposes small (< 2%) and rapid (< 2 msec) mechanical length steps and measures concomitant changes in muscle tension. Though again the interpretation is not as straightforward as with single skeletal muscle fibres, it is probably a reasonable assumption that stiffness (changes in force for unit length change) is a measure of the instantaneous number of attached actomyosin cross-bridges (8).

In the hog carotid artery, stiffness has been measured at various levels of isometric force, altered by varying the external Calcium concentration during high-K^+ activation with added histamine. The resulting stiffness plots (Figure 4) are linear in the region of stretch, but show a non-linear tailing off in releases (much as has been observed in skeletal muscle studies).

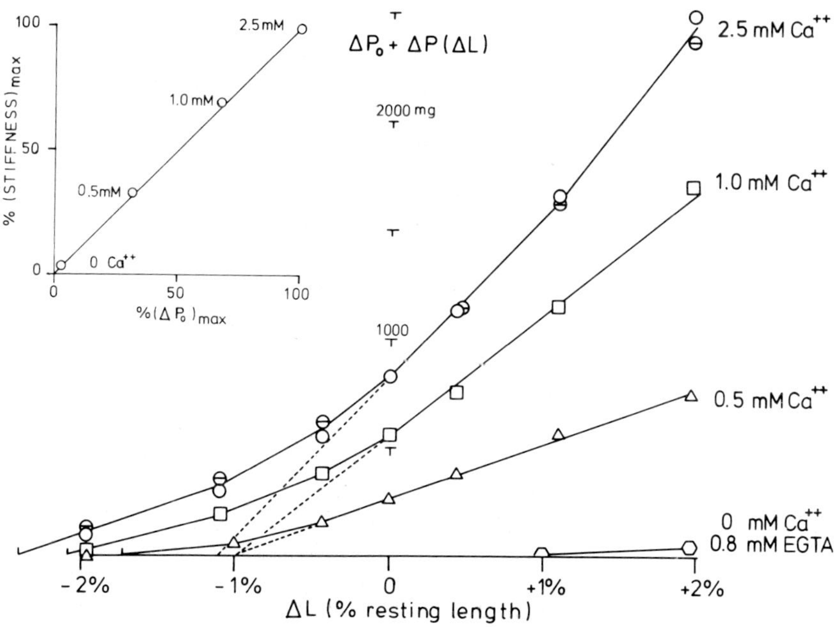

Figure 4 Stable isometric tension with varying external Ca^{++}, plus or minus the rapid peak changes in force upon stretch or release, are plotted against the imposed length change. The linear slope in the region of stretch is taken as a measure of stiffness. Broken lines show the extrapolation of the linear slopes. In the inset, relative stiffness at varying Ca^{++} dependent isometric tensions is plotted against the relative isometric tension. Data above is from a single arterial segment. Changes in force with varying Ca^{++} are reversible.

Most interestingly, the curves with varying stable isometric tensions all extrapolate to very nearly the same place (-1%). This indicates, again as has been found in skeletal muscle (9), that the rapid shortening necessary to discharge active force is independent of the absolute force generated. This, plus the fact that stiffness increases in direct proportion to increasing isometric force (inset) suggests rather strongly that increased isometric tension is simply due to increased numbers of active actomyosin cross-bridges formed. In addition, we have preliminary data indicating that when maintaining maximum activation but decreasing isometric force by shortening the tissue, stiffness again varies proportionally with the actively developed isometric force.

These experiments then give some confidence in the described partitioning of cellular ATP utilization, since active force appears to reflect simply the number of active cross-bridges. Furthermore, the metabolically determined values for actomyosin ATPase agree quite well with those estimated from isolated arterial actomyosin ATPase. The tension-cost (ATPase per unit force) of vascular smooth muscle is 200-300 times lower than in skeletal muscle counterparts. It is also interesting to note that tonic vascular tissues, which do not exhibit the mechanical characteristics of the so-called catch state, are nearly as efficient at maintaining tension as ABRM in the tonic state; and even more so at the cross-bridge level, if allowances are made for the longer sarcomere length of ABRM (10). One can use these numbers to estimate the average cross-bridge cycle time by dividing the length-dependent ATPase by the myosin content. One finds that the cycle time is about 30 times slower than in skeletal muscle, indicating that the 300-fold difference in tension-cost is not a simple scaling down in time. This is another view of the fact that smooth muscles maintain more force per unit myosin than skeletal muscle (11), which may be related to either structural factors or perhaps to the cross-bridge cycle itself in smooth muscle.

Another area of our attention is concerned with the aerobic glycolysis observed in all smooth muscles. Under fully oxygenated conditions, smooth muscle shows an appreciable lactate production rate with some 70-90% of the glucose uptake leaving the tissue as lactate. Though usually greater than or equal to the oxygen comsumption on a molar basis, in terms of ATP production, at most 30% can be attributed to aerobic glycolysis. Under pharmacological stimulants such as epinephrine, norepinephrine, and histamine, the lactate production shows a similar dependence on isometric force as the oxygen consumption (12), as evidenced in Figure 5. In fact, when the computed ATP productions of oxygen consumption and lactate production are summed, the resultant total ATP production is less variable than either parameter taken alone.

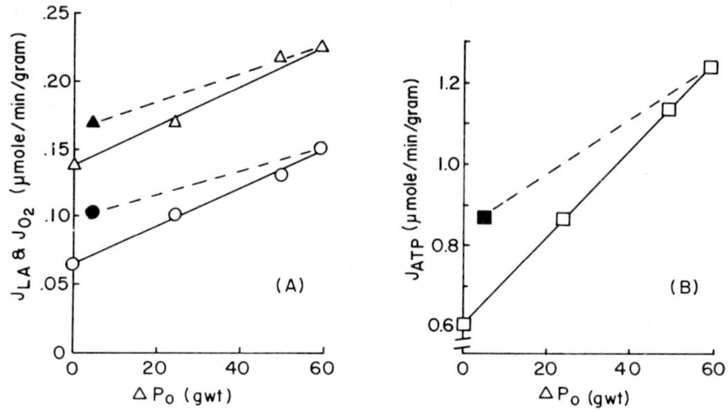

Figure 5 (A) Oxygen consumption rates (circles) and lactate production rates (triangles) are plotted against maintained active isometric force, varied by altering the level of stimulation (open symbols) or by allowing the tissue to shorten to L_{min} (closed symbols). (B) Net computed ATP production versus isometric force from data in A. Lactate production contributed 25.1%. All data is from a single bovine mesenteric vein segment

After many experiments documenting this parallel behaviour of lactate production and oxygen consumption, it was rather surprising to find that stimulation with K^+-substituted Na^+ medium gave no similar relation. Lactate production was not simply related to isometric force and in many cases was actually less than the basal rate (6).

As K^+-depolarization and pharmacological agonists may effect a contraction via different calcium sources, it is tempting to speculate that the aerobic glycolysis observed in these tissues is related to the energetics of activation processes. There is also some supporting evidence from Casteels and Wuytack's new work on taenia coli, showing a correlated increase in aerobic glycolysis when the sodium pump is restarted by adding potassium to K^+-depleted cells. Alternatively, it may be related to glucose transport, which in skeletal muscle, is inhibited by a K^+-substituted medium. An obvious experiment is to see if pharmacological agents can remove the inhibition of lactate production in a tissue already in K^+-depolarizing solution. Preliminary data have been somewhat conflicting and perhaps indicates that something even more complex is going on. While we hope to resolve this issue in the near future, there remains at present the intriguing possibility that activation processes may derive their energy from glycolysis, and that oxidative metabolism is more closely coupled to actomyosin energy requirements.

Acknowledgements

The authors extend their thanks to E. Glück, G. Pfitzer and J.-C. Rüegg for their invaluable contributions to and discussions of this work. Supported by the Deutsches Forschungsgemeinschaft and an American Heart Association Grant to RJP.

References

1. Curtin, N. and Woledge, R. (1977) Physiol. Rev. in press.
2. Marston, S. (1977) This volume.
3. Paul, R. and Peterson, J. (1977) The Biochemistry of Smooth Muscle, University Park Press, Maryland.
4. Paul, R., Glück, E. and Rüegg, J-C. (1976) Pflügers Arch 361, 297
5. Paul, R., Peterson, J. and Caplan, S. (1974) J. Mechanochem. Cell Motil. 3, 19
6. Glück, E. and Paul, R. (1977) Pflügers Arch. 370, 9
7. Hellstrand, P. (1977) Acta Physiol. Scand. 100, 91
8. Peterson, J. (1977) Proceedings of the Symposium on Vasodilatation Antwerpen, 1977 in press.
9. Huxley, A. and Simmons, R. (1972) Cold Spring Harb. Symp. Quant. Biol. 37, 669
10 Rüegg, J.-C. (1971) Physiol. Rev. 51, 201
11 Murphy, R., Driska, S. and Cohen, D. (1977) This volume
12 Peterson, J. and Paul, R. (1974) Biochim. Biophys. Acta 357, 167

Excitation-Contraction Coupling in Smooth Muscle, R. Casteels et al. eds.

HIGH ENERGY PHOSPHATE UTILIZATION DURING FORCE DEVELOPMENT AND FORCE MAINTENANCE IN MAMMALIAN SMOOTH MUSCLE

T.M. Butler, M.J. Siegman, S.U. Mooers and R.E. Davies
Department of Physiology, Thomas Jefferson University, Philadelphia, PA 19107, and The School of Veterinary Medicine, University of Pennsylvania, Philadelphia, PA 19174, USA

INTRODUCTION

The study of the chemical energetics of mammalian smooth muscle has as its goal a correlation of the mechanical responses of the muscle to the total quantity of high energy phosphate utilization. Three major conditions must be satisfied for such a correlation to be made: 1) the mechanical state of the muscle must be characterized and controlled, 2) the number of recovery reactions that regenerate ATP must be kept at a minimum or at least measured to permit a valid estimate of the total quantity of ATP used, and 3) the procedures used to limit the recovery reactions should not change the relationship between mechanical function and associated chemical energy input.

This study describes a preparation that has been used for determination of the chemical energetics of mammalian smooth muscle during force development and maintenance under isometric conditions.

METHODS

Segments of taenia coli were isolated from New Zealand White rabbits as described previously[1]. This muscle exhibits no spontaneous activity or tone at 18-22°C and can be fully activated tetanically utilizing 10 v rms 60 Hz ac field stimulation. After dissection, the muscles were incubated at 20°C for at least 2 hr under 2 g-wt in a modified Krebs-bicarbonate solution bubbled with 95% O_2, 5% CO_2 and which had the following composition (mM): NaCl, 118; KCl, 4.7; $MgSO_4 \cdot 7H_2O$, 1.18; KH_2PO_4, 1.2; $CaCl_2$, 1.9; $NaHCO_3$, 25; glucose, 11. The muscles were quick-frozen using modified Wollenberger tongs precooled with liquid nitrogen. Extraction[2] and analysis for ATP, ADP and AMP[3], phosphorylcreatine (PCr) and total creatine (Ct)[4] and lactate[5] were performed.

Muscle length was adjusted relative to the slack length (l_s) which was taken as the length at which passive tension was just zero. The experiments were done at either 1.3 times l_s or 1.45 times l_s where the active force was 82 ± 2% P_o or 90 ± 2% P_o, respectively. A 15 min equilibration period was allowed after each change in muscle length before proceeding with the experimental design.

RESULTS AND DISCUSSION

The chemical contents of the resting taenia coli are shown in Table 1.

TABLE 1

CHEMICAL CONTENTS OF THE RESTING TAENIA COLI AT 20°C. (N = 18)

A. Referred to wet weight (μmole/g) ± SEM

ATP	ADP	AMP	PCr	Ct
1.52 ± 0.05	0.67 ± 0.02	0.016 ± 0.001	1.83 ± 0.10	2.71 ± 0.14

B. Referred to total creatine (mole/mole Ct) ± SEM

ATP	ADP	AMP	PCr
0.57 ± 0.02	0.25 ± 0.01	0.0066 ± 0.0003	0.672 ± 0.006

There was no systematic variation in chemical contents along the length of the taenia coli. The high phosphorylcreatine to total creatine ratio and the low AMP content of the tissue indicated that the muscles were in good condition energetically. Since in the studies to be described the muscles are small and thin and difficult to weigh while frozen, chemical contents are referred to the total creatine content of the tissue[6].

In order to stop oxidative phosphorylation and glycolysis without dramatically altering the chemical contents or mechanical response of the tissue, the muscle was purged of oxygen and glycolysis stopped at 5°C; at which temperature the resting ATP usage was greatly reduced. This was accomplished by incubation in an anaerobic, glucose-free Krebs solution containing 0.5 mM iodoacetate, 5.0 mM fluoroacetate at 5°C. After 30 min, the muscles were rewarmed to 18°C, and Fig. 1 shows PCr/Ct as a function of time after rewarming. All of these muscles were unstimulated. PCr/Ct remained near the value of 0.67 found also for untreated muscles until about 4.5 min. Since a rapid loss of phosphorylcreatine occurred at

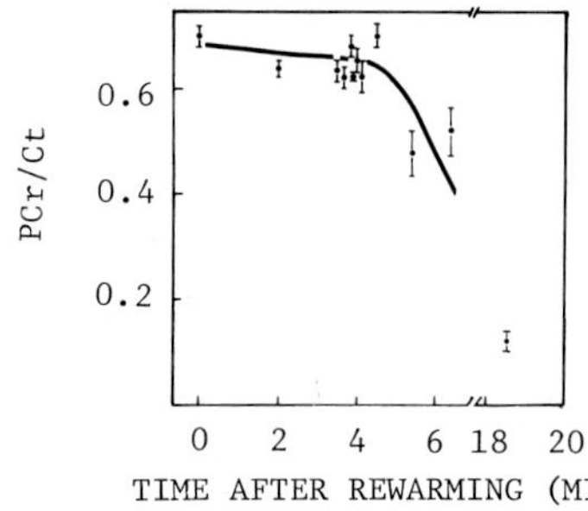

Fig. 1. Phosphorylcreatine content of the rabbit taenia coli as a function of time after rewarming to 18°C following treatment with metabolic inhibitors at 5°C.

longer durations, no experimental maneuver was allowed to exceed 4.5 min. The mechanical response of the muscle after 3.5 min of rewarming was determined. The ratio of the maximum force developed during a 35 sec tetanus after cooling in the metabolic inhibitors followed by 3.5 min of rewarming to that developed under normal untreated conditions was 1.1 ± 0.1, N = 7. The treatment also had no effect on the rate of force development or rate of relaxation.

Measurements of lactate production during an additional 15 min period of incubation (in the same solution after the 3.5 min of rewarming) showed that there was no detectable lactate in the muscle (0.009 ± 0.008 mole/mole Ct, N = 4) or in the bathing medium (0.000 ± 0.008 mole/mole Ct in the muscle, N = 4). It is concluded that glycolysis and oxidative phosphorylation are indeed stopped by this treatment and that the chemical contents and mechanical responses of the muscle are not significantly changed during these procedures.

Using this preparation, the chemical changes during isometric tetani were determined at 18°C at a length where the muscle developed 82% P_o active force. The muscles were quick-frozen immediately after the end of the stimulus and compared to unstimulated paired controls. The mean PCr/Ct of all of the control muscles was 0.65 ± 0.01, N = 33. The changes in PCr and ATP are shown in Fig. 2 along with a typical force response.

There is an initial breakdown of ATP which is significant ($P<0.02$) at 10 sec with a subsequent resynthesis ($P<0.05$) between 10 and 20-25 sec. No significant change in PCr content was observed at 10 sec or before, but an increase in the rate of PCr usage occurred between 10 and 25 sec ($P<0.05$) followed by a decrease in rate between 25 and 60 sec. These results show that the first sign of high energy phosphate usage in smooth muscle is the breakdown of ATP with subsequent resynthesis at a time when the rate of PCr breakdown is high. Once the ATP is totally resynthesized, the rate of PCr breakdown slows to a rate which equals that of the total energy usage in the muscle.

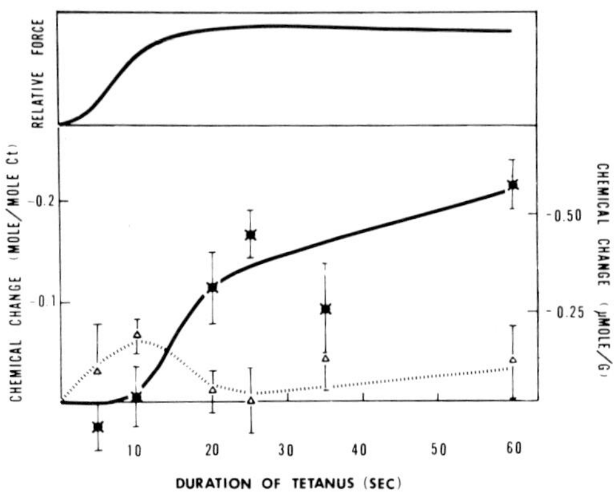

Fig. 2. ATP and phosphorylcreatine changes during an isometric tetanus in the rabbit taenia coli at 18°C. ■ ΔPCr; △ ΔATP. Values are mean ± SEM for experimental minus unstimulated control.

Measurement of phosphorylcreatine and ATP changes in frog skeletal muscle are, for the most part, consistent with the creatine kinase reaction being near equilibrium at all times[7,8]. These data on mammalian smooth muscles show that this is not the case during the early part of an isometric tetanus, but the observed changes in PCr and ATP after 25 seconds of stimulation are in agreement with an equilibrium hypothesis. This suggests either that there may be an allosteric effector for creatine kinase in smooth muscle or that there is a delay in the time at which ATP hydrolysis takes place compared to that at which ADP is made available to the enzyme and phosphorylcreatine.

Fig. 3 shows the total high energy phosphate utilization as a function of tetanus duration. Even though there are significant changes in the rates of phosphorylcreatine and ATP breakdown, the total high energy phosphate utilization shows a relatively linear relationship with time. The average rate of chemical energy usage is about 0.01 μmole/g·sec at 18°C.

In order to better examine the relationship between the chemical energy usage during isometric force development and that during subsequent force maintenance at a nearly constant value, a different set of internally controlled experiments were done at a muscle length where the force generated was 90 ± 2% P_o. Each taenia coli was cut into three segments; one was stimulated for 25 sec, another was stimulated for 60 sec, and the third was an unstimulated control. The rate of energy usage during the first 25 sec of stimulation, while force was being developed, (0.0078 ± 0.0019 mole/mole Ct·sec) was significantly greater ($P<0.05$) than that during the next 35 sec (0.0028 ± 0.0009 mole/mole Ct·sec) when near-constant force was being maintained.

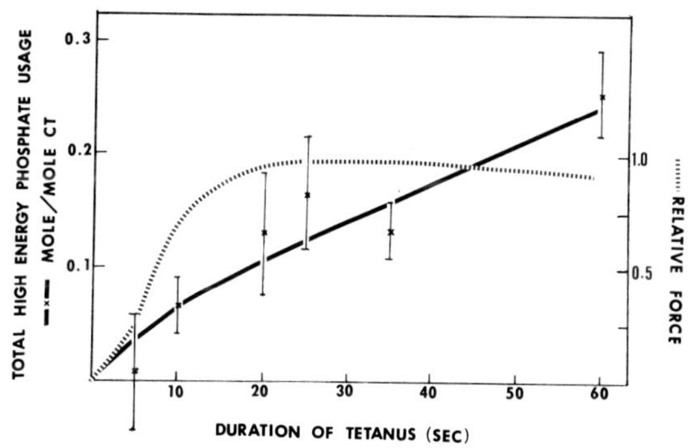

Fig. 3. Total high energy phosphate usage and relative force during an isometric tetanus in the rabbit taenia coli at 18°C. **X** - ΔPCr - ΔATP + ΔAMP, ± SEM. ''''''', relative force.

Comparison of the time course of force development and of chemical energy usage in the rabbit taenia coli at 18°C with that of the frog sartorius at 0°C[9] is shown in Fig. 4. At these different temperatures, the sartorius develops force about 45 times faster than the taenia coli and has about a 45 times higher rate of chemical energy usage. Corrected to 18°C with a Q_{10} of 2.5[10], the sartorius uses approximately 250 times more energy for force maintenance than the taenia coli. This agrees well with observations that smooth muscle contains about seven times less myosin[11] having a 30-40 times slower intrinsic rate of ATPase activity[12].

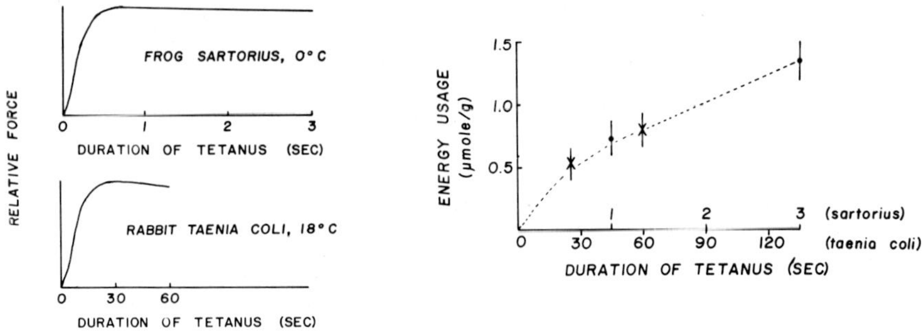

Fig. 4. Comparison of the relative rates of force development and associated energy usage in the frog sartorius at 0°C and rabbit taenia coli at 18°C. Note the differences in time scales. **X**, total high energy phosphate utilization for rabbit taenia coli at 18°C; • total high energy phosphate utilization for frog sartorius at 0°C. Mean ± SEM.

Both smooth and skeletal muscle used approximately 0.35 μmole/g more chemical energy for isometric force development than for constant force maintenance. The amount of work done against the series elastic component was approximately the same in both muscles. It is unlikely, however, that the work is the sole determinant of the extra energy usage, since the work would occur at a thermodynamic efficiency of only about 10%. Also, preliminary experiments on taenia coli suggest that under certain conditions force development can occur without the high rate of chemical change.

In summary, a smooth muscle preparation has been developed to study the chemical energetics of contraction under conditions in which the mechanical response and high energy phosphate content of the muscle are normal. During isometric tetani there is an initial hydrolysis of ATP with a subsequent resynthesis when the rate of phosphorylcreatine breakdown is high. The rate of chemical energy usage is higher during the development of force than during near-constant force maintenance in both the rabbit taenia coli and the frog sartorius. Even though there is a 250 times difference in the rate of energy utilization for force maintenance in these two muscles, both use approximately 0.35 μmole/g extra high energy phosphate during the initial period of force development.

This study was supported by NIH grant HL15835 to the Pennsylvania Muscle Institute and HD 06074 to MJS.

REFERENCES

1. Gordon, A.R. and Siegman, M.J., 1971, Am. J. Physiol. 221 1243-1249
2. Kushmerick, M.J., Larson, R.E., and Davies, R.E., 1969, Proc. Roy. Soc. Lond. B 174 293-313
3. Butler, T.M., Siegman, M.J. and Davies, R.E., 1976, Am. J. Physiol. 231 1509-1514
4. Kushmerick, M.J. and Davies, R.E., 1969, Proc. Roy. Soc. Lond. B 174 315-353
5. Hohorst, H.J., 1963, in Methods of Enzymatic Analysis. H.J. Bergmeyer (Ed) pp. 266-270
6. Carlson, F.D., Hardy, D. and Wilkie, D.R., 1967, J. Physiol. 189 209-235
7. Canfield, P., and Maréchal, G., 1973, J. Physiol. 232 453-466
8. McGilvery, R.W. and Murray, T.W., 1974, J. Biol. Chem. 249 5845-5850
9. Butler, T.M., 1974. Chemical and Mechanical Changes Upon Activation of Frog Skeletal Muscle and After Metabolic Depletion of Mammalian Smooth Muscle. (Ph.D. thesis) University of Pennsylvania, Philadelphia.
10. Kean, C. and Homsher, E., 1974, Fed. Proc. 33 1334
11. Murphy, R.A., Herlihy, J.T. and Megerman, J., 1974, J. Gen. Physiol. 64 691-705
12. Mrwa, U., Achtig, I. and Rüegg, J.C., 1974, Blood Vessels 11 277-286

Excitation-Contraction Coupling in Smooth Muscle, R. Casteels et al. eds.

HEAT AND FLUORESCENCE CHANGES IN RABBIT RECTOCOCCYGEAL MUSCLES

Colin Gibbs and Brian Chapman
Department of Physiology, Monash University, Clayton,
Victoria, Australia

SUMMARY

It is shown that the effect of a 10°C reduction in temperature produces mechanical and energetic effects in the rectococcygeal muscle which resemble those found in striated muscle. Force generation is practically unimpaired but shortening velocity, rate of isometric heat production, and resting heat rate are approximately halved. Lowering the temperature also decreases the rate of evolution of recovery heat, and it has been established by means of a fluorescence technique that recovery processes are activated early during a mechanical response such that, even at 17-19°C, the initial and recovery metabolisms are more closely coupled than in fast skeletal muscle. For a limited range of tetanic durations, at low temperatures, there is a linear relationship between the fluorescence-time integral and energy production. This relationship usually breaks down when the duration of tetanic stimulation exceeds about 3s and cannot be established in experiments at 27°C where the kinetics of the fluorescence transient become complex.

INTRODUCTION

In an earlier study (1) the mechanical performance and heat production of rabbit rectococcycus muscle was investigated. This preparation, which is composed of parallel longitudinal bundles, develops a force comparable with that developed by mammalian skeletal muscles. However, the estimated shortening velocity (muscle lengths/s) was about 50 times smaller than that predicted for frog sartorius muscle at the same temperature. Correspondingly, the heat rate : tension ratio was about 40 times less than that expected for frog sartorius muscle (2). In most respects the energetic data obtained in isometric and isotonic tetani resembled that obtained in skeletal muscle. Thus, isometric heat production was linearly related to force and to stimulus duration, and in isotonic studies the ratio of work/total energy production was maximally about 0.2 with loads close to 40% of the maximum isometric tension, P_o, developed at the optimal rest length, l_o. The ratio of total enthalpy to initial enthalpy was difficult to measure: different methods yielded different values (1).

In the present investigation we have examined the effects of temperature upon the mechanical and myothermic properties of rectococcygeal preparations. This study was prompted by the suggestion of Gordon & Siegman (3) that the force-velocity properties of smooth muscle might exhibit a large temperature dependence. We also hoped that a 10°C reduction in temperature would further inhibit spontaneous activity and allow more accurate determination of the time course of recovery heat. In order to obtain additional insight into recovery metabolism we have employed the fluorometric technique of Chance and colleagues (4, 5) to monitor the level of NADH as an indicator of the oxidation-reduction state of the NAD^+-NADH couple.

MATERIALS AND METHODS

The rectococcygeal muscles were obtained from female rabbits weighing between 1.75 and 2.5kg as reported previously (1). Tension and length measurements were made with a conventional magnesium-alloy lever (ratio 20:1) with strain-gauges bonded to the lever. Total system compliance was 0.4μm/mN.

The muscles were bathed in a modified Krebs solution, containing 11mM glucose as the normal substrate and aerated with 95% O_2:5% CO_2, and were stimulated transversely. A constant stimulus frequency, normally 5Hz was used throughout, with the total number of stimuli being varied. Usually experiments commenced with the temperature between 17 and 19°C. Later on the temperature was raised to 27°C. Stimulus heat was assessed as described previously (1) or as determined upon stimulation of a mechanically quiescent (calcium free) preparation. Velocity of shortening was always measured in after-loaded contractions and the experimental data were fitted by the method of least squares to the Hill equation (6).

Heat measurements were made with three thermopiles two of which have been described previously (1, 7). The third thermopile was another one of the integrating type with an output of 13.88 mV/°C. To produce fluorescence data the muscle preparations were irradiated with ultraviolet light (366nm) using a fluorometer similar to that described by Jöbsis, O'Connor, Vitale & Vreman (8). Both reflected (366nm) and fluorescent (>400nm) radiation were measured. The technique for making simultaneous heat and fluorescence measurements has been described previously (9).

The earlier investigation (1) showed that the stimulus procedure employed mainly activates the muscle indirectly via nerve stimulation. It was found in the present experiments that at temperatures near 17°C the nerve plexus becomes difficult to stimulate, possibly because the nerve sodium-potassium pump is inhibited in this temperature range. As direct stimulation of the muscle produced an unacceptably high heat artefact the bath temperature often had to be raised by one to two degrees to reactivate the nerve plexus. It was essential to maintain a stimulus frequency close to 5Hz even at the lower temperature otherwise there was a noticeable decline in the rate of force development and velocity of shortening.

RESULTS

Resting heat production: Table 1 shows the pronounced effect on the resting heat rate of altering the temperature by approximately 10°C (Q_{10} = 1.9).

TABLE 1

RESTING HEAT RATE (MEAN ± S.E., mW/g) AT DIFFERENT TEMPERATURES

Temperature	Normal Ringer	K-free Ringer
17 - 19°C	0.45 ± .04 (n = 13)	-
27°C	0.85 ± .06 (n = 12)	-
27°C	0.98 ± .10 (n = 7)	0.62 ± .10 (n = 7)

In experiments performed at 27°C attempts were made to inhibit the Na pump by reducing the extracellular K level to zero (10). One to two hours exposure to such a solution reduced the resting heat rate from 0.98 ± .10 to 0.62 ± .10 mW/g (mean ± SE.,n = 7). Re-exposure to normal solution returned the heat rate towards normal; the time course of this recovery was not followed.

Isometric experiments: Raising the temperature increased force generation slightly (see Table 2). This effect must be weighed against our doubt as to whether the preparations were maximally stimulated at the lower temperature (see Methods). In measurements made on the same muscles the isometric heat rate measured during a tetanic plateau was substantially increased by a 10°C temperature rise, changing from 2.56 ± .24 mW/g to 5.56 ± .44 mW/g (mean ± S.E.).

TABLE 2

FORCE VELOCITY PARAMETERS (MEAN ± S.E.) AT DIFFERENT TEMPERATURES

Temperature	$P_o l_o/M$*	V_{max}/l_o**	a/P_o	b/l_o**
17 - 19°	152 ± 6	0.116 ± .012	0.19 ± .02	0.020 ± 0.005(n=14)
27°	176 ± 8	0.298 ± .020	0.20 ± .02	0.059 ± 0.007(n=17)

Where P_o = tetanic tension, l_o = muscle length, V_{max} = maximum shortening velocity and a and b are Hill equation constants.

* Tension/cross sectional area (mN/mm^2)

** Data expressed in muscle lengths/s.

Isotonic responses: The after-loaded isotonic force-velocity data were fitted with the hyperbolic version of the Hill equation. The data are summarized in Table 2. Note that raising the temperature 10°C significantly increases V_{max} and the constant b : the relative Q_{10}s were 2.56 and 2.95 respectively. The ratio of a/P_o was not altered.

Recovery heat and fluorescence: Total heat production was measured in isometric tetani of various durations at both temperatures. As expected, recovery heat was produced more rapidly at 27°C. Hill (11) reports a Q_{10} of 2.4 for the recovery heat of amphibian skeletal muscle. Although the recovery heat phase appears to be exponential it is unlikely to be simply so since, as D.K. and A.V. Hill (11, 12) have emphasized, there are always some anaerobic processes occurring and these are more pronounced in the early stage of recovery heat evolution. Godfraind-De Becker (13) has shown that the recovery heat production of amphibian skeletal muscle can be more adequately fitted by a double exponential plot. There was evidence in some muscles that such a model would describe the data adequately but, for purposes of comparison, a single recovery heat time constant was determined at each temperature. In 4 muscles the mean values of the time constants were 0.54 and 1.18 min at the high and low temperatures respectively, so that the Q_{10} for recovery heat evolution was 2.2.

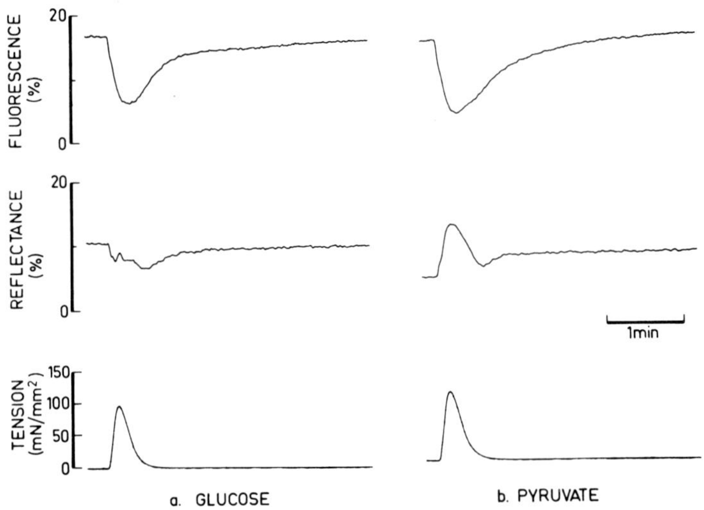

Fig. 1. Original records of fluorescence (upper), reflectance (middle), and isometric force (lower) in a rectococcygeus muscle stimulated 16 times at 5 Hz after equilibration in physiological saline containing different substrates. Muscle rest length, 25mm; mass, 63.1mg; temperature, 17°C.

In order to determine how soon recovery processes were contributing to heat production, combined heat and fluorescence experiments were undertaken. In spite of movement artifacts, it was clear that in this smooth muscle preparation, as in cardiac muscle, recovery metabolism, evidenced by a decrease in fluorescence, commences almost synchronously with mechanical activity and increased heat production. The experiments were always commenced by measuring the response to a single stimulus and then increasing the number of stimuli, usually at 5 Hz, until tetanic durations of 8 to 12s were reached. At low temperatures the fluorescence kinetics were usually simple, at least for tetanic stimulus durations up to 3s. It was often possible, within this range of durations, to obtain a reasonably linear relationship between the time-integral of the fluorescence change and either the tension-time integral or total energy production. Such a result offers experimental support for a suggestion put forward by Jöbsis & Duffield (14). This result is similar therefore to data we have collected at a similar temperature in cardiac muscle stimulated repetitively at low rates. In cardiac tissue the kinetics of the fluorescence transients are greatly influenced by the substrate provided (15), but no similar effect was observed for the rectococcygeus preparation when pyruvate was substituted for glucose (see Fig. 1)

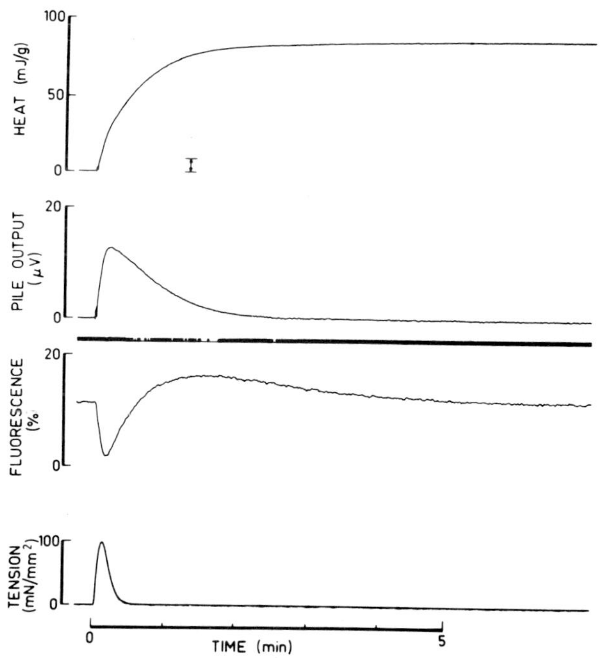

Fig. 2. Original records of, from top downwards, heat production; thermopile output; fluorescence and force development. Muscle stimulated 8 times at 5 Hz at 19°C. Notice that heat production plateaus before fluorescence signal returns to baseline. Magnitude of stimulus heat artefact shown on heat trace. Muscle rest length, 18.5mm; mass, 48.3mg.

In most muscles tetanic durations greater than 3s usually led to a curtailment of the period of fluorescence decrease (oxidation) and to the gradual appearance of fluorescence overshoots, i.e. an increase in fluorescence signals to levels above the original baseline. Frequently 4 to 8 min would elapse before the fluorescence trace returned to its original baseline; usually the heat trace had reached a plateau several minutes beforehand (see Fig. 2). As soon as fluorescence overshoots appeared, the linear relationship between fluorescence-time integral and energy production was lost (see Fig. 3), and there was a lack of correspondence between the evolution of recovery heat and the return of the fluorescence trace to baseline. At 27°C the problem was more pronounced : a decrease in fluorescence followed by a simple fluorescence transient was only seen when one or, at the most, two stimuli were employed, and even under these conditions the fluorescence trace appeared to return to baseline significantly before the evolution of recovery heat was complete.

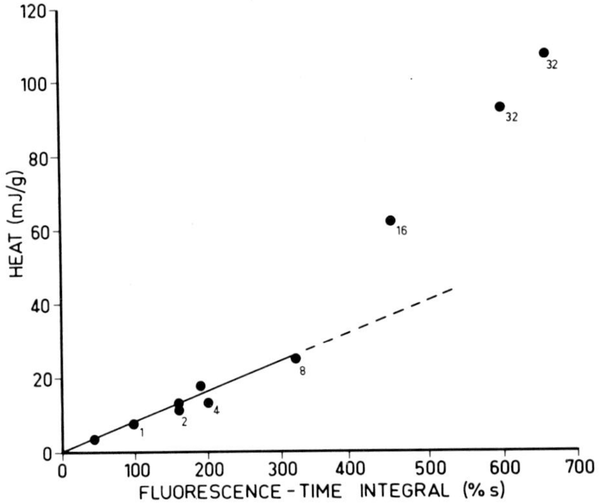

Fig. 3. Relationship between isometric heat production and fluorescence-time integral. Muscle stimulated at 5 Hz; number of stimuli indicated on graph. Notice linear relationship for up to 8 stimuli. Same muscle as shown in Fig. 1.

DISCUSSION

The present data reinforce the idea that smooth muscle shows energetic and mechanical behaviour similar to skeletal or cardiac muscle; a conclusion which accords with that reached in 1930 by Bozler (16). The mechanical results are broadly in agreement with a recent temperature experiment upon tetanized rat portal-anterior mesenteric vein (15), although the Q_{10} values for maximum shortening velocity and the Hill constant b are somewhat higher. It is likely that differences in stimulating procedures may account for the discrepancy.

The energetic data show the expected dependence upon temperature. The resting heat rate has a $Q_{10} = 1.9$, and it is interesting that procedures which should inhibit Na-K ATPase reduce the basal heat rate to almost half. These myothermic studies are thus in agreement with oxygen consumption studies on smooth muscle (10).

The rate of active heat production during a tetanus is markedly altered by a change in temperature, in agreement with the effects of temperature upon myosin-ATPase (18). However, mechanical efficiency, defined as external work/total energy production was not significantly changed in the few experiments where such measurements were made. The rate of evolution of recovery heat was markedly influenced by temperature change, the Q_{10} being close to 2. The time constant for recovery heat evolution appears to be somewhat higher than in amphibian sartorius muscles where other authors (11, 13) have reported time constants between the range 2.0 to 3.0 min at 20°C. The more rapid evolution of heat reported here may reflect the higher mitochondrial concentration of the rectococcygeus muscle - about double that found in amphibian muscle (1).

The fluorescence results are markedly similar to data we have obtained with cardiac muscle (19). Experiments by Chance and co-workers (5) have shown that when cardiac muscle is irradiated with ultraviolet light the intensity of the fluorescence emission is linearly related to the amount of NADH present. One of us has argued (15) that for cardiac muscle the technique specifically records fluctuations in the redox levels of mitochondrial NAD/NADH to the exclusion of any significant contribution from cytoplasmic NADH. In cardiac muscle at 20°C up to 20 contractions at 0.25 Hz can occur without distortion of the relationship between fluorescence integral and mechanical activity, i.e. $\int FL.dt/\int P.dt$ is constant. In isolated amphibian muscle such a correlation holds only for 3 to 5 twitches, or up to 7 twitches if glycolysis is inhibited (14). At high rates of energy expenditure (tetanic stimulation) this simple relationship breaks down in both amphibian and mammalian skeletal muscles (13, 20).

Rectococcygeal smooth muscle produces fluorescence responses (overshoots) which resemble those seen in cardiac muscle when the experimental conditions are such that energy flux is high. Although we are far less certain that the technique is primarily recording alterations in mitochondrial redox levels in smooth muscle, we suspect this to be the case and suggest that high rates of energy expenditure stimulate glycolytic mechanisms which modifies the fluorescence signal either by an indirect reductive pull via redox shuttles or by the production of large amounts of ATP that reduce the drive to oxidative phosphorylation. Such an explanation would be consistent with the data of Peterson and Paul (21) who found that aerobic glycolysis underwrites a substantial fraction of metabolism in smooth muscle. There is, however, the possibility that the fluorescence overshoots may be cytoplasmic in origin (13) and we have considered other less likely alternatives elsewhere (19).

REFERENCES

1. Davey, D.F., Gibbs, C.L. and McKirdy, H.C. (1975) J. Physiol. (London) 248, 207-230.
2. Hill, A.V. and Woledge, R.C. (1962) J. Physiol. (London) 162, 311-333.
3. Gordon, A.R. and Siegman, M.J. (1971) Am.J.Physiol. 221, 1243-1249.
4. Chance, B. and Jöbsis, F.F. (1959) Nature, London, 184, 195-196.
5. Chance, B., Williamson, J.R., Jamieson, D. and Schoener, B. (1965) Biochem. Z. 341, 357-377.
6. Hill, A.V. (1938) Proc.Roy.Soc.B. 126, 136-195.
7. Gibbs, C.L. and Gibson, W.R. (1972) Am.J.Physiol. 223, 864-871.
8. Jöbsis, F.F., O'Connor, M., Vitale, A. and Vreman, H. (1971) J. Neurophysiol. 34, 735-749.
9. Chapman, J.B., Gibbs, C.L. and Vogelsanger, H. (1975) Experentia 31, 445-447.
10. Casteels, R. and Wuytack, F. (1975) J. Physiol. (London) 250, 203-220.
11. Hill, A.V. (1965) Trails and Trials in Physiology, Arnold, London.
12. Hill, D.K. (1940) J. Physiol. (London) 98, 207-227.
13. Godfraind-De Becker, A. (1972) J. Physiol. (London) 223, 719-734.
14. Jöbsis, F.F. and Duffield, J.C. (1967) J.Gen.Physiol. 50, 1009-1047.
15. Chapman, J.B.(1972) J.Gen.Physiol. 59, 135-154.
16. Bozler, E. (1930) J. Physiol. (London) 69, 442-462.
17. Peiper, U., Lavern, R. and Ehl, M. (1975) Pflügers Arch. ges. Physiol. 356, 33-45.
18. Bárány, M. (1967) J.Gen.Physiol. 50, 197-216.
19. Chapman, J.B., Gibbs, C.L. and Loiselle, D.S. (1977) J.Molec.Cell Cardiol. 9, (in press).
20. Wendt, I.R. and Chapman, J.B. (1976) Am.J.Physiol. 230, 1644-1649.
21. Peterson, J.W. and Paul, R.J. (1974) Biochem.Biophys.Acta 357, 167-176.

Excitation-Contraction Coupling in Smooth Muscle, R. Casteels et al. eds.

NADH-LEVEL OSCILLATION AND OSCILLATORY ACTIVITY OF RAT PORTAL VEIN

Alfons Linke and Eberhard Betz

Institute of Physiology (I), University of Tübingen, Tübingen, West-Germany.

INTRODUCTION

Fluorometric measurements of intra-mitochondrial NADH have been performed in various types of tissues and isolated mitochondria including excised skeletal muscle tissue (1, 2). It is generally accepted that in working skeletal muscles the lactate concentration increases. After contractile work the rate of O_2-uptake and the NADH-oxidation level increase too (3, 4). It is difficult to record in vivo the rate of glycolysis, the variation of respiration and of the tricarboxcyclic acid cycle in vascular smooth muscle. The significance of the single components for the energy metabolism of the vessel walls during variations of vascular smooth muscle tension is still unclear, and the question which source of energy is mainly used for rephosphorylation of ADP - glucose or fatty acids - is also not yet clear. In order to elucidate some of the components involved in energetic processes we have measured the oxidation-reduction level of NAD/NADH in relation to the oscillatory mechanical activity under various metabolic states in intact vessel cylinders of rat portal veins. The question whether there exists a metabolic oscillatory NADH activity which corresponds to spontaneous contractile activity in the rat portal vein similar to oscillations in glycolysis found in yeast cells and cell extracts of rabbit skeletal muscles will be discussed.

MATERIAL AND METHODS.

Sections of veins, 7 to 10 mm long, of female Wistar rats weighing 400 g were prepared in a special preparation holder under microscopical magnification and then mounted in a holder with measuring devices. The holder was placed in a bath continuously perfused with Tyrode-solution and equilibrated with Carbogen-gas at constant temperature, constant flow rate and pH. Circular contractions of the vessels were recorded with a Statham transducer cell(force range $\pm$ 30 g, non linearity 0,15 %, displacement range $\pm$ 0, 06 mm, bridge resistance 350 Ohms, calibr. weight 25 g). Prior to the recording of NAD/NADH in relation to the development of mechanical tension, length-tension diagrams of the vessels were drawn up. The measuring devices permitted the recording of the tension-development of the vessel walls during gradual radial distension of the vessel cylinders. During the measurement of NAD/NADH the vessels were distended with a force of 0, 014 to 0, 017 N. This means they were distended to about

5o % above the circumference of the non-perfused vessel. By employing this manoeuvre gross movements of the structures in the field, seen by the microscope during contraction of the vessel wall, could be reduced to such an extent that the movements of the vessel walls did not disturb the recording. In our experiments the vessel wall contractions were isometric. The fluorometric measurements on the surface of the horizontally mounted vascular cylinders were performed with a microfluorometer, which we developed together with Zeiss(Oberkochen). Exciting light of a high pressure Mercury-Arc-Lamp and emitted fluorescence light from the vessel walls were limited by interference filters. The emitted signal was received by a photomultiplier, amplified and continuously recorded (fig. 1)

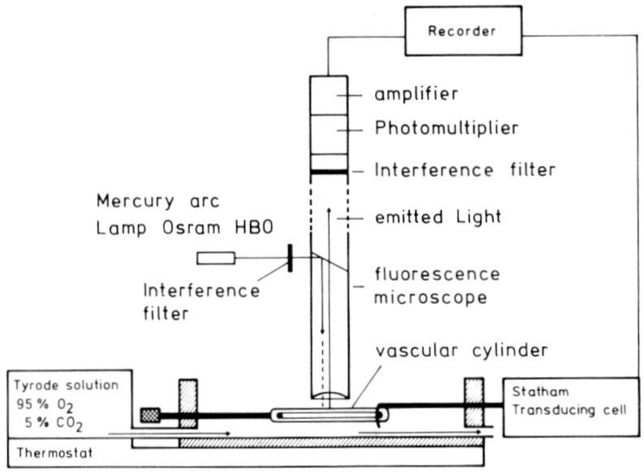

Fig. 1
Block scheme of the microfluorometer and the apparatus used for simultaneous measurement of the tension of the vessel walls.

RESULTS

Changes in the redox-level of NADH are dependent on the metabolic and the contractile state of the tissue. Fig. 2 illustrates a length tension diagram (of a portal vein cylinder) during stretch and release at 37 ^{o}C and 17 ^{o}C. Both curves show an exponential slope. At 17 ^{o}C the slope is not as steep as at 37 ^{o}C. The amplitudes of spontaneous contractions show a ratio near 2 : 1 between 37 ^{o}C and 17 ^{o}C. In fig. 3 the spontaneous amplitudes are shown during increasing tension caused by gradually increasing stretch. Fig. 4 shows the relation between contraction of the portal vein and the oxidation-reduction state of the measured nucleotides at different temperatures. During decreasing temperature the basic tension and the amplitudes of contractions also decrease (from 450 dyn to about 25 dyn). The NADH-fluorescence decreases by about 11 %. In normal preparations with spontaneous mechanical smooth muscle activity, the NADH oscillates concomitantly with the mechanical oscillations

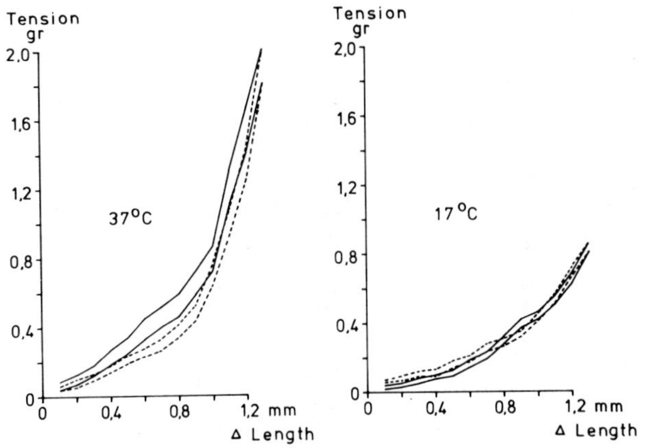

Fig. 2

Length tension diagram of a portal vein cylinder during gradual stretch ——— and release ------- .

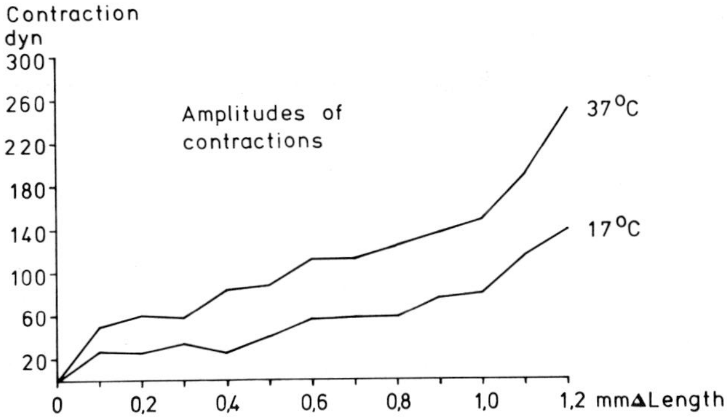

Fig. 3

Amplitudes of contractions of portal vein cylinders during increasing diameter of the vessel.

Δ length designates the relative change of the diameter.

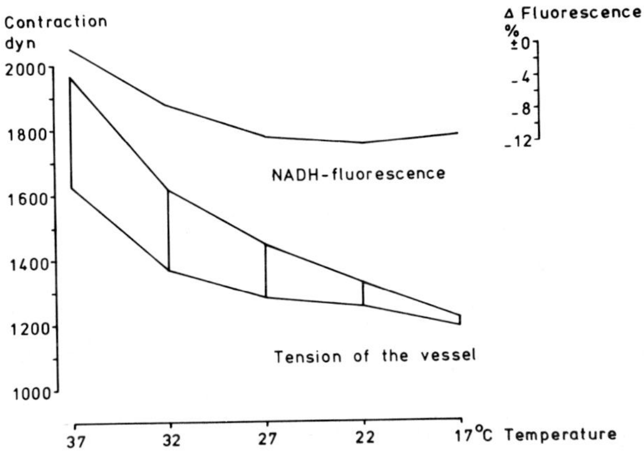

Fig. 4

Relation between contraction, oxidation-reduction state of NAD/NADH and temperature of the rat portal vein. Upper part of the graph: fluorescence ; lower part: vertical lines show the amplitude of contraction so that the base line represents the change of basic tension.

(fig. 5). In order to produce an extreme metabolic state we perfused the system

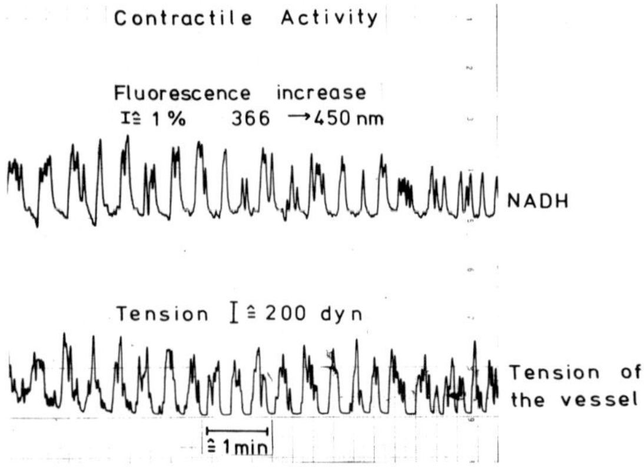

Fig. 5

NADH oscillations with simultaneous mechanical oscillations. Original record.

with Tyrode-solution free of glucose and pyruvate for 7o - 100 minutes and recorded the oxidation-reduction levels of NADH and flavoproteins. During glucose and pyruvate free perfusions the NADH was oxidized. If 5 mmol/l glucose and 2 mmol/l pyruvate were added to the perfusate, a reduction of NADH was observed, connected with an increase in the tension of the vessel walls for 8 minutes. Fig. 6 gives an example.

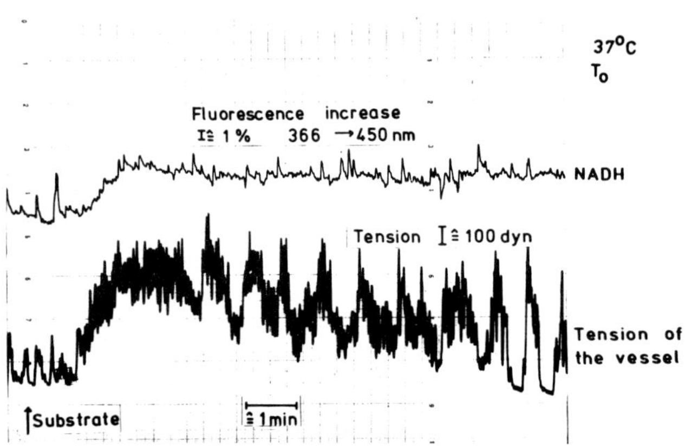

Fig. 6

Effects of substrates on substrate free perfused vessels. Original record.

The amplitudes of contractions in substrate free media are usually lower than after addition of substrates. The NADH-level in the vessel walls increases immediately when substrates are added. If one compares the contractions and the NADH-level during an application of ACh into the bathing solution in vessels perfused with or without substrates, one finds the following reactions: in both cases NADH-fluorescence increases to about the same extent, but the contraction force of the vascular muscles differs significantly. Portal veins without substrate supply develop a maximal tension of 1 100 dyn ; vessels supplied with substrate reach a maximal value of 1 750 dyn. The blocking of glycolysis with iodoacetic acid (JAA) causes a decrease of basic tension of the vessel and a stop of spontaneous contractions after some minutes. During a defined period of time after addition of JAA, ACh still elicits tetanic contraction, but after some time no contractile actions can be produced. NADH fluorescence decreases ; but in spite of that, a reduction of PN(pyridin nucleotides)could be seen during contractile activity. (ACh = acetylcholin).

DISCUSSION

An important methodical problem in experiments, in which contractile tissues are studied with spectrophotometry, is the movement of the tissue in the measuring-field as well as other sources of instability. Some techniques have been developed to avoid mechanical artifacts in such measurements. Photometric techniques like the "double-beam"-photometer of CHANCE et al. (5) have been developed in order to eliminate artifacts. Mechanical stability of the vessel walls was obtained in our experiments by means of a special fixation of the vessel in the holder. Fading effects, caused by high energetic UV-light, could be neglected.

Our observations show that there exist typical relations between the contractile state (basic tension as well as spontaneous contractions) of intact portal vein cylinders, the oxidation-reduction level of NADH and the temperature. NADH oxidized with decreasing temperature and during contraction despite high ATP-level and low activity of the respiratory chain. A possible explanation for the observed oxidation of NADH could be a relative tissue hypoxia in the beginning of the experiment combined with a pronounced reduction of NADH lowered by decreased metabolic activity. The NADH record shows that the reduction-oxidation level does not parallel the activity of contractions, but that a plateau is reached at 27 ^{o}C. Decreasing temperature per se causes an increase in the fluorescence-signal. Thus this temperature effect has to be added to the decrease of the signal in fig. 4. We assume that in vascular smooth muscle cells the same ratio between the cytosolic and the mitochondrial part of the NADH-fluorescence signal exists as in skeletal muscle (2). Here it is 1 : 10 to 1 : 20. Therefore glycolytic NADH will hardly contribute to the measured fluorescence-signal.

All vessels which were in the metabolic state 2 (6) after long lasting substrate-free perfusion showed a reduction of NAD during the addition of glucose and pyruvate. This finding leads to the conclusion that the mitochondria in the smooth muscle cells examined here do not possess substrates in excess and can not supply the demand during the increased activity of the respiratory chain following contractile activity. The tension of the vessel, as well as the amplitudes and frequency of the spontaneous contractions, increase for a short period of time simultaneous with the increased NADH-level upon addition of substrates.

In substrate free media changes in reduction-oxidation levels after activation of the smooth muscle by acetylcholine are almost identical with those obtained in vessels supplied with sufficient substrates. However, the mechanical force of the vessel walls with substrates exceed those without substrate-supply by about 3o %.

The coupling of NADH oscillations with spontaneous changes of mechanical activity, which we observed in some experiments, might be oscillations of the mitochondrial NADH caused by oscillations in cellular ADP that occur when the glycolytic rate oscillates. This, however, has to be proved.

REFERENCES

1. Chance, B. and Jöbsis, F. F. (1959) Nature, 184, 195 - 196
2. Jöbsis, F.F. and Duffield, J. C. (1967) J. Gen. Physiol., 5o, 1009 - 1047
3. Meyerhof, O. and Schulz, W. (1927) Pflüg. Arch. ges. Physiol., 217, 547-573
4. Hill, D.K. (194o) J. Physiol., 98, 207 - 227
5. Chance, B., Cohen, M., Jöbsis, F. F. and Schoener, B. (1962) Science, 137, 499 - 5o8
6. Chance, B. and Williams, G.R. (1955) J. Biol. Chem., 217, 383 - 427.

Excitation-Contraction Coupling in Smooth Muscle, R. Casteels et al. eds.
© 1977 Elsevier/North-Holland Biomedical Press

SUMMARY OF DISCUSSION ON "MECHANICS AND ENERGETICS"

J. Peterson

Universität Heidelberg
II. Physiologisches Institut
Im Neuenheimer Feld 326
D-6900 Heidelberg, F.R.G.

Several issues of importance to smooth muscle mechanics arose and were contested at various levels. It is perhaps most interesting to note that current work is progressing on three fronts related to defining the fundamental properties of the contractile apparatus in smooth muscle cells and the mechanical inter-relationship of the cells within tissues: 1) single cell studies; 2) very small preparations with visualizable cells and little connective tissue; 3) small segments of tonic blood vessels with more-or-less in vivo cell arrangements and connective tissue.

While single cell studies, as described by Fay, present a direct methodology for reducing mechanical studies to the most fundamental physiological unit, some controversy arose over possible major alterations in cell character due to the preparative procedures. The evidence that such cells are, however, intact viable structures was quite strong. Single cells appear to be stiffer than one would deduce from whole tissue studies, suggesting substantial inter-cellular compliances in the intact tissue; Mulvany's studies, on the other hand, would suggest that such compliances are quite small, the major portion of compliance residing in passive elements within cells. Peterson's studies, on the third hand, suggest that the bulk of the compliance is due to cross-bridge interactions themselves. Comparison of the data is perhaps not entirely fair however, allowing for possible large differences between frog stomach and mammalian blood vessels. The question remains then to what extent do stiffness measurements in smooth muscle reflect cross bridge properties (as in skeletal muscle) and to what extent do passive elements (either inter- or intra-cellular) play a role.

Much attention was devoted to the states of attachment of cross bridges in smooth muscle. Marston pointed out where the biochemical attachment/detachment cycle differs from skeletal muscle, indicating several slower steps, one involving a slow transition of the actomyosin-ADP complex within the attached cross bridge (as was suggested in earlier mechanical studies by Lowy and Millman for catch muscles); and which could play a role in the production of greater force per myosin (Murphy) and the lower tension cost (Paul) in smooth muscle.

Much debate focused on the role of attached, but non-cycling cross bridges in resting smooth muscle, as described by Siegman, and which play a role in the model for byssus retractor muscle, discussed by Twarog and originally proposed by Lowy and Millman. The discussion of mechanical studies focused on correlations with the biochemical actomyosin cycle and levels of intracellular calcium. In particular, while it is clear that a low to high Ca^{++} transition activates smooth muscle, there is growing evidence that only at extremely low values of Ca^{++} are bridges completely detached, as further indicated by Bose's report of Ca^{++}-free contractures in tracheal muscle. The role of Ca^{++} as activator has become complicated by the observation of obligatory phosphorylation steps in the activation processes, which are perhaps responsible for the latency observed between electrical stimulation and tension development in single smooth muscle cells, as observed by Fay. That the onset of increased stiffness shows the same latency as active tension development in single cells is somewhat opposed to Siegman's findings of attached cross bridges in resting muscle.

Johannson described initial contractile velocity transients upon quick release in the rat portal vein, perhaps related to an initial adjustment of bridges to the release in times shorter than the biochemical cross bridge cycle time; and basically similar observations were commented upon (Stephens, Mulvany, Peterson) in a variety of smooth muscles. In general, it seems hopeful that mechanical studies are coming increasingly closer to resolving the dynamics of cross bridge cycling in smooth muscle. A vital improvement in the smooth muscle situation is that both mechanical and biochemical studies have progressed sufficiently to allow a much closer correlation between the two to develop, so that the next several years might well witness dramatic increases in our understanding of smooth muscle contractility.

AUTHOR INDEX